Recycling the Resource
Ecological Engineering for Wastewater Treatment

School of Engineering Wädenswil ISW
Wädenswil, Switzerland

Centre for Applied Ecology Schattweid
Steinhuserberg, Switzerland

Stensund Ecological Center
Trosa, Sweden

Recycling the Resource

**Proceedings of the Second International Conference on
Ecological Engineering for Wastewater Treatment**
School of Engineering Wädenswil - Zürich
18. - 22. September 1995

in cooperation with

International Ecological Engineering Society (IEES)

Swiss Federal Institute of Technology (ETH), Zürich

DIANE Oeko-Bau

Editors:

J. Staudenmann, A. Schönborn, C. Etnier

TRANSTEC PUBLICATIONS
Switzerland • Germany • UK • USA

Cover design by Thomas Waldvogel, Wädenswil CH

Volumes 5-6 of
Environmental Research Forum
ISSN 1421-0274

Distributed in the Americas by

Transtec Publications Ltd
Whitman Distribution Center
10 Water Street, Room 310
Lebanon, NH 03766
USA

Tel. (603) 448 0317
Fax   (603) 448 2576

and worldwide by

Transtec Publications Ltd
Brandrain 6
CH-8707 Zuerich-Uetikon
Switzerland

Fax +41 (1) 922 10 33
e-mail: ttp@transtec.ch
http://www.transtec.ch

Printed on 100% recycled, acid-free paper
in the United Kingdom by
Hobbs the Printers Ltd,
Totton, Hampshire SO40 3YS

Foreword

The School of Engineering Wädenswil (ISW) is proud having been host for the *Second International Conference on Ecological Engineering for Wastewater Treatment*. It was a great pleasure to welcome ecologists, engineers, planners, government and university representatives and other specialists from Switzerland, as well as many other countries from Europe and all over the world.

These five inspiring days of knowledge exchange and educational highlights would not have taken place without the contribution in many different ways from our partner organizations. As representatives of the *School of Engineering Wädenswil* we therefore want to thank the *Centre for Applied Ecology Schattweid* (Steinhuserberg, Switzerland) and the *Stensund Ecological Center* (Trosa, Sweden) for the cooperation in this project. Especially we would like to express our estimation for the engagement and enthusiasm of the experts from all three organizations that gathered in the conference board: A. Schönborn, C. Etnier, B. Guterstam, J. Heeb, U. Ginsig and J. Staudenmann.

The School of Engineering Wädenswil (ISW) has its focal point in the fields of horticulture, viticulture, food science, and biotechnology. It has not been common for an educational organisation of this kind and size to host and coorganize such an event. The Swiss Schools of engineering are going through an evolutionary process, though. They will be upgraded in the near future and will have the same level as the Greman „Fachhochschulen". The ISW therefore wants to mark its position in this changing process and demonstrate the determination to thematically open itself towards new fields of research and development.

With this book, the work on the 1995 conference of ecological engineering for wastewater treatment comes to an end, although it is only an intermission for the developing process of ecological engineering meetings. The preparations for other sessions have already started again at other places. We are convinced that all delegates have returned home after the conference with many valuable friendships and new ideas for their work, and now, with this proceedings, even can hold a qualitatively high standing documentation in hands. At this point, we also would like to thank the editors C. Etnier, A. Schönborn, and J. Staudenmann for their engagement and hard work to bring together and revise the papers and make the book.

Dr. R. Grabherr
School Headmaster

J.-B. Bächtiger
Chief Department
Horticulture & Environment

School of Engineering Wädenswil, Switzerland

June 1996

Acknowledgments

The Second International Conference on Ecological Engineering for Wastewater Treatment and these proceedings were possible only with the help and sometimes enormous engagement of many people, institutions and private organizations.

First of all, we would like to thank the three organizing institutions, School of Engineering ISW Wädenswil, the Centre for Applied Ecology Schattweid and the Stensund Ecological Center for their support for this project and for the cooperation in the planning and performance of the conference. Only their generosity and contributions in many different forms made it possible. Special thank goes to the headmaster Rolf Grabherr for being the host of the meeting. Also, our thanks go to the Swiss Academy of Engineering Sciences (Zurich), the Foundation for Appropriate Technology and Social Ecology (Langenbruck) and the many private sponsors who generously helped to enable Schattweid's work, as well as to the non-profit foundation of Stensund Folk College and its director Per-Ola Jansson for Stensund's support.

We are also very thankful for cooperation of the International Ecological Engineering Society, the Swiss Federal Institute of Technology Zurich (ETHZ) with Stefan Schmuky, and DIANE Oeko-Bau who helped sponsor the conference. Many more colleagues and kindred organizations, especially Joan Davis, Martin Strauss, Tove Larsen, Willi Gujer and others from the Swiss Federal Institute for Environmental Science and Technology (EAWAG) as well as John Todd from Ocean Arks International, USA helped us with the preparations by having good ideas and criticizing goals and the program of the meeting. We highly value their contributions.

This conference would not have been possible without the generous financial support of several governmental institutions and private sponsors. Thanks to the generosity of the Swiss Agency for Development and Cooperation SDC and the Federation of Migros Cooperatives (Zurich) we were able to make the participation possible for about 40 delegates from low-income countries. We also wish to thank the Federal Office of Environment, Forests and Landscape (BUWAL), the ContinuingEducation in Technology CONTEC at the University of Berne, BASF Schweiz AG in Wädenswil, Dr.Lange AG in Volketswil, Sandoz AG in Basel, Metrohm Ltd. in Herisau, IBM in Zurich, the Lions Club in Wädenswil and Unilever in Zurich for their financial contributions. Our thank goes also to SWISSAIR in Zurich for special fare offers, to Waldvogel in Wädenswil for printing and the perfect graphical design of the conference logo on this book, to the Swiss Association of Professional Ecologists OeVS in Berne for the free shipment of announcements, and to the foundation Schweizer Jugend forscht in Winterthur for helping us with the poster panels/walls.

For the case studies we were dependent on the goodwill and collaboration of the hosts, and therefore want to thank Mr. Peter, Mr. Dubach, and Mr. Stampfli of EMMI Cheese Factory AG (Dagmersellen), Mr. Meier of FAT Tänikon, Mr. Bachmann and Mr. Aschwanden of the municipality of Wädenswil, Mr. Schmid of the FOCUS Housing Project, Mr. Dietrich of the Bischofszell Canning Factory, Mr. Lustenberger of the Lustenberger Butchery (Menznau), as well as Mr. Fismer and Mr. Wendeler of Prinzhöfte (Harpstedt, Germany) for their support and the time they took to introduce the groups to the problems. We hope that this „investment" has yielded good ideas and strategies as results at the end.

All the good ideas and financial resources alone would have been useless without working hands. We are therefore grateful to all the employees, students, interns and friends who engaged their skills and many hours of work - often during their free time - for helping us with the preparation of the conference and especially with the administration during the five "hot" days. For this we would like to thank by name Regula Keller-Nadler and Doris Huggler for managing the conference office and for their patience in every situation, Anuschka Heeb and Pernilla Rinsell who were often working late at night for keeping track of the "electronic extension", as well as Brigitte Eggmann, Martin Kunz, Regula Meyer, Patrik Müller, Karin Scheurer, Philip Stauffer, Regula Treichler, Urs von Rickenbach, Philippe Wyss and Brigitta Züst, and all the others in the background, who were always around when needed.

We also want to express our deep appreciation to all the delegates who contributed to the workshops and case studies as chairpersons and secretaries. And last but not least, thanks to the speakers and all participants from all over the world for their interest and participation at the conference, and especially for contributing to the warm and inspiring atmosphere that grew during the sessions. We are sure that the majority of them agree with us to one Indian delegate, who wrote in a letter after the conference: „Memories of my stay and participation in the Wädenswil Conference will remain "green" for many many more summers to come.

June 1996, the conference board:

Jean-Bernard Bächtiger[1], Carl Etnier[2], Ursin Ginsig[1], Björn Guterstam[3], Johannes Heeb[4], Andreas Schönborn[5], Jürg Staudenmann[1]

[1]School of Engineering ISW, Department Horticulture & Environment
P.O. Box 335, CH-8820 Waedenswil - Switzerland
100340.2504@compuserve.com

[2]Department of Economics and Social Sciences, Agricultural University of Norway
P.O. Box 5033, N-1432 Ås - Norway
carl.etnier@nlh10.nlh.no

[3]Stensund Ecological Center
S-619 91 Trosa - Sweden
stensund@nordnet.se

[4]Centre for Applied Ecology Schattweid
CH-6114 Steinhuserberg, Luzern - Switzerland
heebjohannes@ping.ch

[5]Centre for Applied Ecology Schattweid
CH-6114 Steinhuserberg, Luzern - Switzerland
ecoschattweid@access.ch

Preface

Många bäckar små gör en stor å.

"Many small streams make a large river," the Swedish proverb reads in translation. It has a rhyming lilt in Swedish, like the way Poor Richard (a.k.a. Benjamin Franklin) expressed the same sentiment: "Little strokes fell great oaks."

The small streams of ideas for new ways of treating wastewater are gathering into larger ones. They are not "mainstream" yet, but they are bubbling up from more and more springs, flowing through an increasing number of landscapes.

Like a raindrop that falls on Florida, flows to the Gulf Stream and evaporates to fall again on the plain in Spain, ideas for using wastewater as a resource travel freely over the globe. Fish production in wastewater has been used for centuries in China and for the past century in Germany. Now it is being imitated in greenhouse systems in Scandinavia and the U.S. Constructed wetlands and infiltration facilities have been tested extensively in countries like Germany, Denmark, and the U.S. and are being adopted and modified in many other countries, even north of the Arctic Circle. In Scandinavia, separate collection and treatment of blackwater and graywater, or even urine and faeces, is an idea bubbling out of the granite bedrock; water from this spring has begun to prime pumps in other parts of Europe.

Someday these new systems will form a broad river, flowing tranquilly through the coastal plain to the sea. "Wastewater" will be an antiquated expression, synonymous with "wasting water." Our present-day systems of throwing away the majority of the nitrogen, phosphorus, and other nutrients plus organic matter in wastewater will be regarded as primitive, like the early computer scientists "throwing away" so much space with their large, ungainly machines. It will be considered as natural to reuse wastewater's resources as it is now for farmers to plow their wheat stubble back into the ground.

That goal is still distant. The small streams are still in the mountains, splashing beautifully on the rocks, to be sure, but carrying little power. They can be easily held up by dams or disappear into sinkholes. Like the once mighty Colorado today, many will never reach the sea. But every so often there is a gathering of the streams of ideas into a sort of lake, or pond, perhaps, where one can see the potential they have. The conference proceedings you are holding is the record of one of those lakes.

This conference at the Ingenieurschule Waedenswil, overlooking Lake Zurich, is the second in a series on ecological engineering for wastewater treatment. The first was at Stensund Ecological Center in Sweden, in 1991. (A second edition of the proceedings is in press at CRC/Lewis Publishers, with the title Ecological Engineering for Wastewater Treatment.) The first conference brought together people who were mostly unfamiliar with each other's work and who were not quite sure of the boundaries of ecological engineering, the new field they thought united them. By the time many of these same people and others gathered in Waedenswil in September 1995, the field had become well established, with its own journal (Ecological Engineering, published by Elsevier) and society (International Ecological Engineering Society). The discussion had progressed from evaluating alternative approaches to wastewater treatment to figuring out optimal ways to recycle the resource in wastewater and, ultimately, eliminate waste.

The Waedenswil conference focussed on wastewater treatment that includes a recycling loop for water, nutrients, energy, and other resources in the wastewater. One workshop session was devoted especially to this topic. The target group was not only scientists but also engineers and public officials interested in new ways to deal with wastewater. The emphasis was on temperate climate facilities: design strategies that can be implemented in regions like central and eastern Europe, Scandinavia, and much of the USA. There were also special workshops on aquacultures for wastewater treatment, systems with floating plants, and the social and political sides of wastewater treatment.

A goal that influenced much of the planning was that the conference should also have an imaginative, open, and friendly atmosphere. An excursion by water to a local castle, with Swiss culture from the Alphorn to a yodler choir, helped provide much of this atmosphere. Case studies also provided a change of pace from the normal conference routine of papers and posters. A number of chapters in these proceedings grew out of case studies at the conference. Groups of participants spent one day studying places that required new approaches to industrial or domestic wastewater and suggested solutions. Though the participants had little or no previous knowledge of the particular cases, they were able to pool their experience to make concrete suggestions. The hosts found some of these ideas so interesting that the one-day exercises have turned into continuing dialogues with a number of the participants, and will perhaps lead to completed projects.

We invite you to turn the pages and plunge into the refreshing waters of "Lake Waedenswil." Follow water's path from myth to molecule, read about the little-known European history of wastewater-fed aquaculture, plan a hydrologically sustainable city in Germany, find out about the commercialization of duckweed products from wastewater ponds in Israel, and explore many of the other fascinating topics the participants bring up.

Flowing water can do unpredictable things, and it is equally hard to know where and when similarly intriguing ideas and actors might pool the next time. We know at least that the third conference in this series will be held in June 1998, at the Agricultural University of Norway.

By that time, we hope that economic costs and benefits of various systems of wastewater treatment will be clearer. The preliminary indications are that decentralized, recirculating systems of wastewater treatment can be significantly cheaper than building and maintaining some conventional systems. The network of pipes and centralized treatment plants that has taken root in industrialized countries dominates the wastewater landscape like a mighty oak on a hill. If research results continue to find ecologically engineered alternatives cheaper, the preface to the proceedings of the next conference may open with Poor Richard's proverb.

Carl Etnier
Ås, Norway

Jürg Staudenmann
Wädenswil, Switzerland

Andreas Schönborn
Steinhuserberg, Switzerland

June 1996

Table of Contents

WORKSHOP PRESENTATIONS

Workshop 1: Aquacultures for Wastewater Treatment

Workshop 2: Recycling Challenges and Solutions

Workshop 3: The Social Side of Wastewater Treatment: Politics and Markets

Workshop 4: Floating and Submersed Plants in Wastewater Treatment

POSTERS

CASE STUDIES

– Summary and Results by Chairperson & Secretary:

Environmental Research Forum Vols. 5-6 (1996) pp. 1-12
© *1996 Transtec Publications, Switzerland*

Key-note lecture:

Water through Time: Its Transition from Myth to Molecule

J.S. Davis

EAWAG (Swiss Federal Institute for Environmental Science and Technology),
CH-8600 Dübendorf, Switzerland

Introduction

Water's unique characteristics and its essential role for all life have inspired myths and influenced cultures, religions, art, and literature through all times.
In the past, respect, reverence and awe shaped man's dealings with water. But that is no longer so. These attributes are hardly part of today's cultural influences. And any remnants of values have long since been replaced by prices.
In today's society, water's essential value for life is overlooked. We see water only as a useful material. A cheap one at that. Its low price gives the signal of valuelessness: overuse and misuse are the results.

In the eyes of science, water does not fare much better: Here it is seen as nothing more than the molecule H_2O, a mere skeleton of its real self.
Together, science and society have stripped water of the protective cloak which myths and respect had laid around this element of life since the beginning of human history.

Water in myths and religions

Die Sonne entsteigt dem Meer

For all cultures water has been the most revered of the natural elements. For many, water is the beginning of all beginnings: Water is looked upon as having existed before any other parts of this planet were created.

Wolfgang Detel [1], in his discourse on Thales and water, quotes biblical references infering that water was not created by God, but had always been present. In this same vein of interpretation, water is seen in some cultures as having given birth to the sun.

For some of the oldest religious cultures we know - the sumerian-babylonian and egyptian - in the depths of water lay the source of everything.
In these rising cultures on the Euphrates and the Nile, the belief was held that in times primeval only water existed,.. in wild, chaotic state. The legend goes that these waters were then divided into male and female forces, representing the fresh and salt waters. Out of the merging of these two forces the world was created.

In other cultures, the waters were not only divided into fresh and salt waters, but there were many waters, often with names, stories and r™les within the culture.
A number of them are familiar to us: Okeanos, Poseidon and Neptune, his Roman counterpart. And then there are the Nereids, the playful nymphs of the seas. Of course the most alluring classical goddess associated with water is Aphrodite - the Venus, Greek goddess of love -, who was born in a seashell. But also Artemis, and her Roman counterpart Diana, has shaped the cultural and archetypical picture of water and waters. Artemis reigned over still waters and shady forests... embodying hidden aspects of the female psyche... an aspect represented often in paintings.

Even with such a brief and superficial glance at only a few beings associated with water, it becomes clear that any dealings with water in the past, were also dealings with a being, towards whom the culture showed utmost respect. This respect was reflected in dealings with water itself.

Water and its symbolism

The myths around water have given rise to symbols linked with water. The following selection [2] gives an indication of the wide range of them, which we associate with water.

The symbol of life

Since for many cultures water is seen as the ultimate source of all life, its use as the symbol of life itself was a common representation in ancient drawings and reliefs.
For the Egyptians, the association with water as a source of life was so strong, that they calmed their fears of death with the image that a life-giving flow of water went through the realm of death, thus connecting it with life.

The symbol of death

Water's strong ties with life has led to strong ties with death as well, in particular via floods and deluges. Sagas often tell of threats of floods and destruction... and of being dramatically rescued from them. Certainly the story of Noah's ark comes to mind here.

In some cultures - particularly in those where water was scarce - death by water was seen as something positive. Lytle Croutier [3] writes in her book 'Water, Elixir of Life' that in Egypt, floods were interpreted as the cleansing of the spirit. The highest tribute, that one could pay to a god, was to drown in the river, thereby uniting with the life-giving forces of water and the god. In Egyptian, the word 'drown' originally had the meaning 'to praise'. This conflicting association with life and death runs through many literary treatices on water.

Symbol of rebirth and regeneration

Water is strongly linked to symbols of rebirth and regeneration. This for several obvious reasons: It is associated both with the creation of life and the destruction of life and has as well the ability to 'rengerate' itself, in that it disappears and reappears, while changing its physical form: The liquid state as we most commonly know water, disappears into an evaporated form, not visible to us, and reappears as rain or snow.

Symbol of cleansing and cleanliness

The symbolism of rebirth is closely linked to the symbol of *cleansing*, both being associated with the hydrological cycle. And both are found in the role of water in baptism: Sins of the past are washed away, and life can - again - begin with a tabula rasa.

The link between cleansing and regeneration lies at the base of the cleansing rituals in essentially all religions. This symbolic aspect was also involved in the original use of baths, which in the meantime however are considered mainly for well-being and for social pleasantry.
While the baths in most cultures have lost 'only' their original symbolism, the hydrological cycle has lost that which *led* to the symbolism of cleansing: Today, water which evaporates and then precipitates out, is itself

not cleaned as it previously was, but instead cleans our polluted air, bringing back to earth acids, heavy metals and toxic chemicals.

Symbol - and element - of healing

The symbolism of healing is found in many forms in and around water. Historically, not only the direct physical role of water was considered important, but also the association, in the presence of water, with other symbols, spirits and myths surrounding water.
Coastal waters bring in an additional aspect: Here the air is continuously cleansed - and charged - by the water spray.
For any number of reasons, one indeed feels different near water.

Beyond the visual, associative and physical effect of water upon us, *sound* also plays a role. How calming, meditative is the effect of water splashing in a fountain, the sound of a babbling brook, or of peaceful waves lapping upon the shore.

Already in pre-historical times, the healing effect of water was recognized. Originally the process of healing was at the same time a religious process. In particular, springs were especially revered. Water coming from the depths of the earth was taken to be a gift of the earthly gods... and if the water had healing capacities, then doubly revered.
The reverence was often expressed in structures, found as far back as the ice age, as seen in the cult grottos near springs in the Pyrenees.

In addition, spirits or saints linked to the springs underscored the importance of water. This we see not only from Greek and Roman times, but as well in in other countries. This is particularly marked in India, China, and Japan.

In northern Europe there are many well-known springs associated with healing water, though Lourdes is the spring that most commonly comes to mind in this association. Around the world there are also a number of rivers accredited with healing powers: Among them, the Jordan, the Nile, the Ganges, Yangtsee and the Kaituna, the holy river of the Maoris.

Although water is used in many ways today to help with healing, it has lost its symbolism and significance in the original, deeper sense of the term. How can we expect a water, that we look at only as a chemical formula to have a healing effect? Water, which we no longer consider holy ... can it help us to again become whole... physically or spiritually?

Symbol of ephemerality

Although water's symbolism is associated with longevity, it is just as tightly linked to ephemerality, in the sense of *panta rhei*, everything flows, as phrased by Heraclitus. Looking at our world with this idea in mind, it is clear that not only water, but time, joys, sorrows, and life itself is included here. With water, the flowing is most visible... and thus it is the strongest symbol for the transitory life we all lead.

In today's all too stressed world, where no one has time to think about the ephemerality of our life, it seems particularly paradoxical, that we are polluting that element of ephemerality, with substances, whose pollution effect is anything but ephemeral... for example radioactive wastes stored on the ocean floor.
Just recently newspapers reported that the barrels used for storage of radioactive waste from European countries, placed 4500 m below the suface of the Atlantic are definitely leaking... and the Plutonium content of the water is increasing markedly in the area. How poorly we understand the significance of water. How much better it was protected by myths of the past than by us.

Symbol of the soul, of the unconscious

Another strong symbolic association with water is that of the soul - and/or the unconscious. This is dealt with in many myths and throughout literature.
For us, Sigmund Freud is probably the best known author of works and writings, which have given us insight into the symbolism of water in dreams, as a mirror of the unconscious.

Not only via dreams does water take on a tie with desires, with sensuousness, with love. The symbolism is also ubiquitous in art work of many cultures.

Today the 'life' of this life-giving element is threatened

The loss of respect for water has led to several threats of its 'life'. On the one hand, contamination with a myriad of chemicals and with radioactivity endanger all forms of life. And on the other hand, physical changes in river- and stream beds have lead to a loss of vitality of waters themselves around the world.

Neither water chemists nor engineers really see a problem here. The water chemists consider water to be in *good* shape, when the *bad* things are no longer in it: They understand too little of the difference between 'clean' and 'cleaned' water.
For them the attitude of the Maoris - and some other cultures - , who do not permit human waste to enter the rivers, is un-understandable, irrational and an unnecessary complication of water use. And neither engineers nor water-chemists understand what's wrong with altering stream- and riverbeds, channelling them, encasing them in cement pipes below the ground.

In spite of a lack of in-depth understanding of what is leading to water problems, and in spite of a lack of agreement as to what needs to be done, it is no longer an open question that *something* needs to be done: Water is endangered world wide, a status now recognized by all.

Rediscovering water

But how can we manage to move people to do something? The motivation up to now has been sadly lacking. Perhaps even understandably so. Water is nothing special to the average person. Politicians seldom see a need for protecting waters as such - rather only when water is so endangered as to be a critical issue. But even then, water is a 'thing', that can be 'managed'. The solutions are generally thought to lie in more money and more technology. Our water scientists are far from being able to see the pain on the faces of the spirits around a polluted river or lake. And far from hearing the moan of the suffering oceans, dumping place of the world, storage place for radioactive wastes. Nor are they disturbed by the recent use of the oceans as testing grounds for the effects of climate change: This employs continuous sonar beaming, which interferes with the delicate communication systems of whales, thereby endangering the very basis of survival of these gentle giants of deep waters.

Obviously we will not make progress in developing methods appropriate for dealing with the water problems, much less the *source* of the water problems, if we do not change our attitude towards water - and make progress in understanding it. As long as we see it only as a manageable physical substance we can hardly hope that even the most creative alternative technologies for both the use and treatment of water will be taken up and widely used, as is so desperately needed in all parts of the world.

Perhaps when we ourselves find a way to rediscover the essence of water as a life-giving element, we will be more successful in helping others to be aware of the gravity of the problem, of what needs to be changed, and how to contribute to the necessary changes in their own areas.
Surely we need to have the intensive cooperation not only of those people involved in cleaning up the problems, but those who are taking part in *causing* the problems in their every day life. And that means *all* of us.

So the question becomes, how can we be moved to *want* to make changes, to find ways to make them and to motivate others to do the same.

Perhaps a look to the past, to what moved our forefathers may give us a clue for our progress. Our ancestors were sufficiently awed by the physical characteristics of water, that they were moved to respect and protection. And perhaps, if we are still open to wonder and awe, becoming more aware of what water is, may help move us as well toward more commitment to the many issues now surrounding this element of life.

Characteristics of water

While the following characteristics of water are all familiar to us, a closer look at them helps us realize how exceptional this 'simple' element is.

Power

One of the most striking, most impressive associations we have with water, is the elemental feeling of force, of power, of creation, of destruction: tumultuous ocean waves or standing near a waterfall.
This power makes all the symbolism linked to life and death easily understandable.

Liquid form: between chaos and order

The fascination with the power of water is heightened by the recognition of the soft, yielding behaviour we observe in peacefully flowing or still waters.

The work of John Wilkes - the designer of the 'Flow Forms' - has pointed out very beautifully this aspect: Moving a narrow object slowly through a mixture of water and glycerine reveals the order, the coherence, the beauty of the movement in the liquid phase (Fig. 1).

Fig. 1 Moving a narrow object slowly through a water-glycerin mixture, covered with a fine layer of Lycopodium powder, creats a momentary structure between order and chaos.

Another view of water has been provided by the work of Hans Jenny [4]. Using oscillations, he has impressively shown the special structures and patterns which are formed as a result of the cohesion of the water molecules (Fig. 2).

Solid form: Many contributions to life

While we hardly pay any attention to the beauty in the behaviour of the liquid form, the structures of the solid form - as in snow flakes - is lookeded upon with fascination.
Not only are the forms of exceptional beauty, the diversity of shape and strucure is in itself amazing.

This point brings in several interesting aspects. Not only is this diversity an indication that water reacts to very subtle influences, which cause each flake to be different - though created under apparent identical conditions - it also points out the difficulty for scientific research on the subject. Science - as we know - insists upon reproducibility as a criterion for acceptance. Water often doesn't manage to do science the favor of providing reproducible results... not even in the simple case of snowflakes.

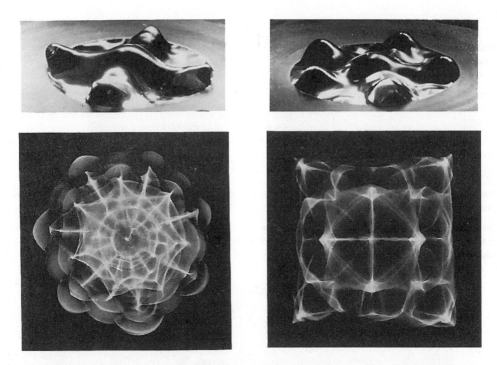

Fig. 2a&b Sound causes water to oscillate. Stroposcopic pictures reveal structure and dynamic.

In the eyes of science, water behaves indeed unscientifically... Not only are the results frequently non-reproducible, but also in a number of ways, water behaves abnormally. These abnormalities are however most critical: They have made life on this planet possible.

The most obvious example of an abnormality is the fact that water expands when it freezes, thus being less dense in the solid form than in the liquid.

If it weren't for this seemingly minor behavioural quirk, ice would sink to the bottom of a lake, and the freezing would progress up to the surface, thus eliminating any chance for the survival of fish and other aquatic forms of life during a cold period.

Physical characteristics

In fact, water doesn't show expected behaviour for quite a range of physical characteristics. Considering the same characteristics of compounds closely related to water, and calculating the values for water [5], the following has been observed:

- Water has a specific heat twice of what would be expected. This is significant for its role in the transport of heat... and the consequences for weather and climate.

- Water evaporates at 100°C instead of -100°C. It requires for this transformation almost 2½ times as much heat as would be expected. This also is of essential importance for weather and climate effects

- Water freezes at 0°C instead of at the calculated minus 120°C.

These few numbers let us see that if water would behave 'normally', there would be no 'normal' life on this planet.

No wonder that its behaviour has fascinated many throughout history. But why has it so lost its fascination for us?

The invisible presence of water

The fact that we no longer have water so visibly around us has decreased our awareness of it in general. Streams have long since been channeled underground in cities and towns, and in recent years, even in the country. In agricultural areas, brooks have been a victim of the consolidation of farming land: Streams which prevent a tractor from going straight across a field have been buried.

True, water remains in the air, but we are essentially unaware of that. The only visible reminders are when we see a heavy dew or experience the magic of a hoarfrost in winter.
Our ancestors were not only aware of the water in the atmosphere, but also knew of ways to collect it... a useful supplement during water shortage.

An interesting historical example is that of Masada, which had been cut off by the enemy from its water supply. Their rescue lay in the collection of water from the air: They had built walls with many niches on the inside. Upon this cool surface, the morning dew would condense, flow into a collecting channel, thus providing drinking water. For a long time, archeologists considered these strange walls simply to be pigeon holes. Only quite recently was this ingenius use of dew collection understood.

This same principle has also been used throughout southern France, where the usually round roofed structures are known as bories (Fig. 3a&b), and on the Greek Islands, where the elaborate structures are kown as Columbaria, again implying that they house pigeons. This they do... but this is only one of their functions.

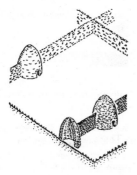   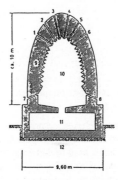

Fig. 3a: Bories with connecting walls   *Fig. 3b: Cross-section of borie (schematic drawings by H.Weber).*

1-8 air passage ways   9   wall
10   interior of borie   11 water containment
12   ground

In Lanzarote - as well as in the Wallis part of Switzerland, a wine producing area, essentially the same principle is used. However, in these places, the collecting surface is below ground, where soil has been removed and re-filled with crushed rocks. The dew collects on the cooled rocks and trickles down, contributint to the groundwater supply.

Influences of - and on - water

The previously mentioned 'abnormal' characteristics are of course recognized by science. It is however of interest to note that the basic scientific concept of water, i.e. existing as single molecules, does not allow for a satisfactory explaination of its marked deviation from expected behaviour: It is simply designated as 'anomalous'.
It is however another story, when looking at certain influences upon the behaviour of water, when the influences themselves are considered to be too weak to have the effects observed. As interesting as it would be to pursue the observations on this topic, science has done very little here: Not being able to imagine an effect (... due to the weakness of the influence), science doesn't look for one... and by not looking, they can't seen one... and by not seeing one they can't believe in one. A circulus vitiosus.

As scientists, we know that 'seeing is believing'. It is considered a sign of objectivity to be able to say: 'If I hadn't *seen* it I wouldn't have *believed* it.' However, a closer look at the selective way science deals with its data, gives rise to another phrasing: 'If I hadn't *believed* it, I wouldn't have *seen* it.' This rephrasing automatically raises the question as to whether science devotes its energies more to the confirmation of that which it has already accepted, rather than openly and constructively dealing with that which it doesn't - yet - understand.

Does all this have particular relevance to the topic of concern here? Indeed it does. It implies that science all too often refrains from looking at something until it has already accepted that it is possible. And since we - i.e. society in general - often tend to restrict our acceptance to that which science has accepted, we no longer act according to that which we feel is appropriate according common sense. This can mean, for example, that even though common sense tells us that it's not wise to allow dangerous chemicals and radioactivity to get into water, only problems already proved to be caused are really accepted by the scientific community or officials.

Not accepting what one hasn't understood (even though it exists...) has unfortunately hindered work on the effects of subtle energies on water. And further, what these effects can mean for living organisms, which are themselves mostly water: With this thought in mind, a closer look at the significance of subtle energies and water may help us toward a better understanding of, and deeper respect for nature in general.
Perhaps they help us as well to see new signs of water's specialness... of its vulnerability, and why we should be protecting it in all ways possible. A view that is hardly understandable, as long as we look at water only as an H_2O molecule.
Perhaps some of these not so well known effects may contribute to the fascination - and revival of respect - for water.

Moon

Changes in the behaviour of water, which fluctuate along with lunar periodicity have long been observed. Some of the more consistent reports on these changes deal with its varying ability to dissolve substances, and thus with its capacity to clean [6].

In this connection an interesting question arises as to whether there is also an effect on the water in living organisms. In any case, the effect of moon phases on the microstructure of living wood - and thus on its quality - has long been a factor used to determine the appropriate cutting times of trees to be used for specific purposes: A number of the qualities of wood - such as fire resistance, slipriness, pliability, etc. - vary during the moon phases. These effects are viewed more and more as stemming from the influences of the lunar phases upon the water in the wood. Though studies on such influences have been slow in developing, in recent years, research projects within the academic framework have provided insight into relevant biological associations, which have furthered interest in these phenomena.

Sunspots

A fascinating piece of work on the response of water to subtle influences is the work of Giorgio Piccardi [7]. Piccardi's research on sunspots stemmed from his observations that not even simple precipitation reactions - for example between bismuth and chloride - were always reproducable in their rates or in degree of completion. Reactions run in a Faraday cage showed less deviation than those run as control on the lab bench. Long term observations of the deviations showed a high correlation between the fluctuations and the strength of the sunspot activity (Fig. 4a). While his work initially brought about skepsis from the scientific community, in the mean time it has been confirmed from several institutes [8].

The extreme weakness of the fields associated with sunspots has hindered interes on research regarding their possible effects. But while scientists remain sceptical, radio amateurs know better: Radio reception is strongly disturbed when sunspot activity is high... a good indicator of influence.

Effects fluctuating with the sunspot periodicity have also been observed on living systems. One of the earlier reports on this comes from Harald Burr [9]. He noted that fluctuations in the electrical tension of tree trunks shows correlates with the sunspot activity (Fig. 4b). His work overlaps the same time period of Piccardi, which allows an interesting comparison. Also here the effect is suspected to be via the changes in water within the wood cells.

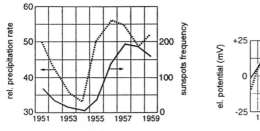 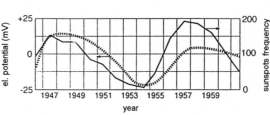

Fig. 4a&b:Two examples of fluctuations showing an eleven year periodicity: a chemical reaction (bismuth chloride) and the electrical potential measured in a tree trunk.

So much for the subtle *natural* influences.
Let's take a brief look at *applied* influences.

Magnets, electromanetic fields

The use of magnets has long been discussed in the treatment of water. More recently, water treatment with electromagnetic fields has found a wide range of applications in household and industry. The changes brought about in the chemical/physical behaviour of water have provided advantages for technical processes. For example, metals rinsed with treated water lend themselves much better to treating with plastic coatings or paint: The added materials adher much stronger. The same type of water treatment leads to a marked reduction in detergent needs for household washing machines and in commercial laundries.
Further studies, using electromagnetic fields in a number of wastewater treatment plants in Switzerland have demonstrated the prevention of the thick deposits within the pipes. These deposits can be a problem, since they strongly reduce the transport capacity for sludge, as well as the transfer of heat through the pipes, when it is warmed before entering the digestion tanks.

The use of electromagnetic fields in treating water for quite different purposes has been interesting to follow: While many industrial set-ups have been using these successfully for a number of years, many scientists maintain their skepticism regarding any effect that an electromagnetic field can have upon water... a skepticism, which seems suprising in view of the fact that water molecules are dipoles, i.e. have positive and negative charges.

The fact that there are measurable changes in physical characteristics of water after treatment with an electromagnetic field may open the door to more research in this field: Surface tension and electrical conductivity change significantly, giving evidence that major changes have taken place in the water. Changes have also been observed with NMR scanning.

Flow Forms

A physical - but non-technical - treatment of water, is the use of 'Flow Forms'[1] to improve water quality. While research is fairly new on this treatment, the results have been promising. Using these forms, a qualitatively higher biocenosis has been observed in the recieving pond, as compared to that recieving the water from a linear cascade. The differences were not due to oxygen content. Studies in Norway have shown another effect: The pond recieving the water treated by the Flow Forms freezes at a slightly lower temperature, as compared to the control.
The type of movement made by the water along the cascade itself seems to be the significant parameter in changing certain characteristics of water.

---

[1] A series of curved forms, connected in descent, through which water cascades, flowing with a lemniscate type of movement.

This last example ties in well to observations on the behaviour of some waters recognized as having a healing effect: These effects are not necessarily associated with any substances found in the water.

'Healing' waters

Springs that are known for their healing effects are visited by people who are often very ill. And although not everyone gets well, these people - in spite of their weakened condition - do not go away with new diseases. This is not of trivial significance. This observation has led to investigations as to what could prevent the spread of diseases via water. One of the explanations considered is that such healing waters can have a bacteriostatic effect. At least it has been observed that bacteria do not multiply in these waters at a rate that would be expected.
Interesting work [10] done along these lines on the water of healing springs in the Barage in France has not revealed any chemical substances that could explain the growth-inhibitory effect. Testing the water with nutrient media, normally supporting rapid growth of Streptococci, also did not yield the expected growth.

What exactly is behind such observations has not been intensively researched. But there are various indications that the type of structuring, or 'clustering' [11] as it is called, of the water molecules themselves play a significant role for the behaviour. One also speaks of the type of 'information' that these clusters can store.

There are a number of influences which can affect the state of clustering and information patterns in water. While many of these are quite subtle influences, less subtle, everyday influences are also of relevance. Among them is temperature. In lower temperature ranges, more molecules are grouped together: The number decreases upon heating. What type of heat source is used has an affect upon water structure and as a result, upon biological systems. Interesting research at the University of Fulda [12] has shown that water heated by wood, gas, electrical stove or microwave oven have significantly different effects upon the rate and type of growth of wheat seedlings. Both electrical heating forms produced effects judged as negative. For each type of heating there were differences in the ratio between leaf and root growth, though this was most significant between electrical or non-electrical heating. Seeing how strongly plants react to such influences should make us give more thought as to what might be the health effects for us, as a result of different treatment of water and/or food.

Summary of influences and energies

Such subtle influences and effects are of significance to our understanding of water. Also to our awareness of our role in influencing water. And with this awareness, how can we not feel more responsibility towards this element of life?

Conclusions

All of our activities have consequences for water... both qualitatively and quantitatively. We have long thought that the problems we thus create can be solved by simply developing and applying more technology. In recent times we have however been forced to recognize this will not suffice. In addition we need new ways of thinking and acting, which prevent the problems from developing in the first pace. This means deep-going changes in attitude... not only toward water itself, towards all that around us, but also towards ourselves: Our destructive treatment of nature is considered a reflection of how we treat ourselves.

The missing respect for nature, i.e. for life in general, has furthered a development that now threatens us all. How welcome would be the protective cloak that myths once gave to things of value. But today, not even understanding what there is to value, myths can hardly help us. Nor is it to be expected that the current approach of ÕmanagingÕ the environment will be successfull either: For this we understand too little about nature.
If our goals to protect the environment, to protect water are to be met, we must alter the emphasis of our work, of our endeavours from repair to prevention. This involves not only different priorities but also very different views than now previal. An essential key for a successful transition to another approach lies in rediscovering respect: For life, for water... the basis of all life.

References

[1]   Detel, W.: 'Das Prinzip des Wassers bei Thales', "Kulturgeschichte des Wassers", H. Böhme (Hrsg.); Suhrkamp Verlag, Frankfurt am Main, 1988.

[2]   Selbmann, S.: „Mythos Wasser"; Badenia Verlag, Karlsruhe, 1995.

[3]   Lytle Croutier, A.: "Wasser, Elixir des Lebens: Mythen und Bräuche, Quellen und Bäder"; Wilhelm Heyne Verlag GmbH & Co., München, 1992.

[4]   Jenny, H.: "Kymatik: Wellen und Schwingungen mit ihrer Struktur und Dynamik", Bd.2; Basilius Presse AG, Basel, 1972.

[5]   Mecke, R.: 'Kräfte in Flüssigkeiten'; Zeitschrift für Elektrochemie, 52:6, 269-282, 1948.

[6]   Paunnger, J. und Th. Poppe: "Vom richtigen Zeitpunkt"; Hugendubel, München, 1991.

[7]   Piccardi, G.: "The Clinical Basis of Medical Climatology", Springfield, Ill., 1962.

[8]   Eichmeier, J. und P. Büger: 'Über den Einfluss elektromagnetischer Strahlung auf die Wismutchlorid-Fällungsreaktion nach Piccardi'; Int. J. Biometeor, Bd.13, Nr.3/4:239-256, 1969.

[9]   Burr, H. S.: "Blueprint for Immortality: The Electric Patterns of Life"; C. W. Daniel Company Ltd, 1972.

[10] Tessier, F. et al: 'Au sujet du pouvoir antibacterien des eaux de Baréges'; Bulletin de la Société de Pharmacie de Bordeaux, 126:5-11, 1987.

[11] Dagani, R.: 'Water Cluster Cradles H_3O^+ Ion in Stable Cagelike Structure'; Chemical & Engineering News, April 8, 47-48, 1991.

[12] Fachhochschule Fulda, not yet published research results

Plenary Presentations

Environmental Research Forum Vols. 5-6 (1996) pp. 15-24
© *1996 Transtec Publications, Switzerland*

Ecological Engineering for Wastewater Treatment: Fundamentals and Examples

P.D. Jenssen

Department of Engineering/Centre for Soil and Environmental Research,
Agricultural University of Norway, N-1432 Åas, Norway

Keywords: Ecological Engineering, Wastewater, Gray Water, Organic (Household) Waste, Nutrient Removal, Recycling, Energy Use, Ecodesign, Economic/Social Aspects, Soil Infiltration, Constructed Wetlands, Pond/Hydroponic Systems, Decentralized Technology, Source Separation, Composting and Fermentation, Water Saving Technology, Urine-Separation

ABSTRACT

Constructing wastewater treatment systems based on ecological principles can minimize environmental problems and facilitate utilization of the resources in wastewater. Measures for optimizing nitrogen and phosphorus reuse often result also in decreased water consumption or waterreuse. Better nutrient management can also mean dramatic changes in the way the sewage pipe systems are built and used. This paper outlines some important principles of ecological engineering and gives some examples of possible solutions to the wastewater treatment problem in rural and urban areas.

1. INTRODUCTION

The consequences of not understanding ecology have been continuously demonstrated to human beings since their occurrence on earth. The Mediterranean region is an example of non-ecological human behavior, but it did not end with the Romans. Loss of fertile soil is threatening future food production and prosperity for mankind. Potable water sources are being depleted and for example parts of south-western USA are facing the same destiny as Northern Africa at the time of the Romans. But we are still flushing our toilets, using large amounts of potable water to transport and reallocate organic matter and plant nutrients, often with lake bottoms or the sea as the final destination.

This should not be necessary; we are terrific engineers. We control drilling of oil in the stormy North Sea and we can even send people to the moon. Why are we then facing such huge environmental problems and why does our technology often seem to collide with nature rather than work together with her?

Mitsch and Jørgensen [1] defined ecological engineering as "The design of human society with its natural environment for the benefit of both." This definition is very broad and envelopes all aspects of engineering. Think about it - for the benefit of both!

If a car runs out of gas, the result is predictable: it stops. The feedback mechanisms of nature are not predictable like a machine, they are often slow and the answer may be surprising. Do we want to risk these surprises or do we want to listen to nature and develop our society and technology for the benefit of both? I think this is necessary in order to avoid serious repercussions from Mother Nature.

How can then ecological thinking be applied to wastewater engineering, and what kind of systems will evolve?

One key word is holism. We have not always thought of the consequences or impact that a machine or treatment plant has on the larger system that it operates within. This holistic or overall system view is a basic concept of ecological engineering and one of the most interesting and challenging perspectives of future engineering.

Lack of holistic thinking is demonstrated by present practices of nitrogen removal. Increased N-removal from wastewater is mandated in many European countries due to the North Sea Treaty [2]. Conventional treatment plants are being rebuilt to add a denitrification step. This increases energy consumption in the removal process, and hence removal costs. If the electricity is produced from fossil fuels, the NOx emissions to the air from the power plant may exceed the N-removal from wastewater [3]. In addition, SOx is produced. N-removal may also produce significant quantities of N_2O, a very potent greenhouse gas [4]. By removing N from wastewater we have increased pollution of the air and increased the energy consumption. We have also lost a valuable plant nutrient to the atmosphere and increased the need for production of artificial fertilizer, which is also an energy-consuming process. The ecological soundness of this type of nitrogen removal can thus be questioned.

There have been few comprehensive assessments of the environmental effects of sewage treatment. However, Antonucci and Schaumburg [5] have documented the energy and chemical use at South Lake Tahoe tertiary treatment plant. They concluded that it was impossible to say whether the plant was an environmental benefit or liability.

Recycling is often a logical consequence of ecological thinking. But we need to recycle within spatially small loops and a time frame of one or a few growing seasons to obtain an ecologically sound solution. There are many possible ways to recycle and reclaim wastewater [6,7]. Recycling is facilitated by decentralized or on-site treatment [8]. Decentralized treatment offers better source control and sometimes shorter transport distances. Decentralized concepts can also be applied to large cities, as will be discussed below.

Wastewater treatment systems can be grouped according to degree of recycling and the energy use (Fig. 1). Extensive systems with high recirculation are the best ecological alternative, provided the purification efficiency is high enough. Several of the natural (soiland wetland) systems have, at the moment, little or no form of recycling, but recycling from these systems can be improved (see below). Figure 1 also indicates that today's conventional systems can produce a better ecological result by optimizing resource gains and minimizing resource use. Energy can be obtained from wastewater through using heat pumps and biogas generation. When optimized in this way, conventional systems are not necessarily ecologically inferior solutions, but the possibilities of optimizing these systems are not further considered in this paper.

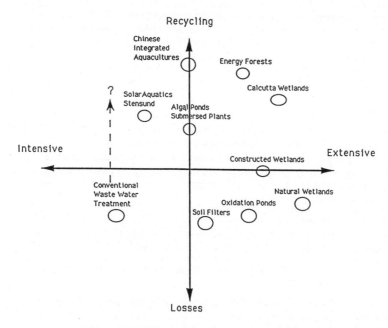

Figure 1: Wastewater treatment systems grouped according to energy use and degree of recycling [9].

It is unlikely that one single system can solve all future sewerage problems. Large investments have been made in conventional sewage systems which will be in operation for decades, but conventional systems will evolve as the principles of ecological engineering are communicated to the engineering society. Totally new systems, as well as hybrid or combination systems, will appear. With the present focus on environmental problems, the interest in ecological engineering is growing. Schools that teach ecological engineering and consultants and companies that implement ecological engineering will have advantages in the market because they can offer a broader range of solutions and solutions that more easily fulfill environmental requirements. A variety of systems are needed to meet the natural constraints of different regions, differing legislation, different sociological aspects, different budgets, personal needs, and preferences.

The speed of the progress toward the use of ecodesigned systems will depend on several factors, such as price, severity of the pollution problem, existing investments in sewage systems, and political decisions.But if ecologically designed systems cannot compete in today's market, we can only hope and wait until conditions force politicians to favor more ecological systems through legislation. If, however, the systems are competetive in the market, progress will be more rapid. Many systems are competetive today, especially in rural areas, but also in cities. Some examples will show this.

2. EXAMPLES OF SYSTEMS

In order to arrive at the most suitable wastewater treatment system under a given set of conditions, we need: 1) a proper assessment of the problem, 2) a number of proven systems, and 3) tools for selection of the most suitable solution based on local premises and environmental as well as actual system cost comparisons.

Such tools should involve system analysis, material flux analysis and economics. To date the various costs of different wastewater treatment methods have not been fully evaluated - neither the direct economic costs, nor environmental effects, and we are still far from an economic model which takes proper account of environmental and human habitat consequences of wastewater and waste disposal systems. This should be developed as means for selection of wastewater treatment systems.

In Norway there is a government goal to upgrade sewage systems in rural areas [10]. Often plans are made using conventional gravity sewers arriving at costs as high as 16 000 USD per home just for the piping system [11]. The piping is defended because local rivers are to be protected, especially from discharge of phosphorus which is the main cause of eutrophicaton. Along the coast, such pipelines often discharge sewage untreated into the sea. By the use of natural systems as soil infiltration or wetlands, costs can be cut and the local as well as the sea environment protected [11,12].

2.1. Soil infiltration systems

Soil infiltration systems [13] are low-cost, nature-based systems that generally operate with a loss of nutrients (Fig.1). There are many possibilties of optimizing the nutrient removal and overall purification performance of such systems [14,15]. The simplest way is to build the systems shallow. This increases the distance to the groundwater and allows percolation through the most oxidized and microbially active parts of the soil profile. In addition, the possibilities for nutrient uptake by adjacent plants increase. Freezing of shallow systems (i.e a soil cover of 40 cm) is not experienced as a problem in Norway.

An interesting design that allows for plant uptake is shallow infiltration in sloping terrain (Fig. 2). In many soils this allows for a significant horizontal flow component leading water through the root zone. A higher soil water content must then be acceptable downslope from the system. Experience from wetlands show that very efficient purification is acheived at saturated flow through the root zone [16]. With time the vegetation downslope from the system can be expected to change towards more moisture-tolerant plants and increased biomass production. The increased biomass can be harvested.

Ball [17] is essentially using the same principle in many of his installations, but he treats the wastewater in a sandfilter prior to shallow infiltration.

Figure 2: A shallow soil infiltration system in sloping terrain.

The treatment efficiency of properly designed and installed soil infiltration systems is high (Table 1).

	COD (%)	Total P (%)	Total N (%)	Fecal coli/100 ml
Setermoen rapid infiltration plant.*	90	99	73	$0-10^2$
Buried inf < 25 p.e.	95	100	30-70	very high
Sandfilter **	90	30-80	20-50	$0-10^2$

   *   The water quality is measured at the top of the water table 7m below and 15 m lateral distance from the infiltration basins.
   ** Sand type no 1 according to Norwegian regulations (MD 1992).

Table 1: Treatment efficiency of soil infiltration systems and sandfilters in Norway [15].

Setermoen rapid infiltration plant treats wastewater from 5000 people. The results from Setermoen are based on groundwater samples taken adjacent to the infiltration basins. The dilution in groundwater is accounted for by the use of chloride concentrations. The average depth to groundwater below the infiltration basins is 7 meters. Below each of the 3 basins the sediments in the unsaturated zone have an estimated phosphorus removal capacity equaling 12 years of constant loading. The nitrogen removal mechanisms are not completely revealed yet, but there are clues pointing towards biological nitrification/denitrification as the main processes [19].

The investment costs at the Bardu plant was less than 50% of a conventional mechanical plant and the operation costs are less than 10% of a conventional plant [20].

2.2. Constructed wetlands

If the plants were harvested, 10 - 15 % of the nutrient input to the constructed wetlands could be removed [21]. Plants are generally not harvested, hence removal and recycling of nutrients from these systems rely on other mechanisms. Constructed wetlands can be efficient denitrifying units [22], but nitrogen is lost to the atmosphere. The phosphorus removal in subsurface flow constructed wetlands is mainly dependent on adsorption and precipitation reactions [21]. If a porous media with a high P-adsorption capacity is used efficient, phosphorus removal is obtained (Table 2).

Parameter	System 1*	System 2**
BOD	93	85
Total - P	98	95
Total - N	48	59

* average over 2 years   **average over 3,5 years

Table 2: Removal efficiency (%) of the two constructed wetland wastewater treatment systems in Norway using natural sand rich in iron and Leca[1] as a P-adsorbent [23].

The removal efficiency has not been significantly different dependent on season. Theoretically these systems can adsorb phosphorus for more than two decades. When the P-removal drops below acceptable levels, replacement of the filter media will remedy the situation. The removed material constitutes a suitable soil conditioner, especially for heavy clay soils.The value as a P-fertiliser depends on the bioavailability of the associated phosphorus. Whether or not this will be a cost efficient P-removal method depends on the longevity of porous media as a P-adsorbent and its subsequent fertiliser value.

2.3. Pond systems

Various types of pond systems for wastewater treatment and utilization have been pioneered in China [24]. Aerated ponds are interesting low-energy treatment solutions in cold climate. Removal efficiencies of 30 and 65% for N and P respectively are obtainable in winter conditions [21] at a system treating wastewater from 100 persons. The P-removal is probably due to sedimentation. Recycling into plant production of P should be possible by excavation of the bottom sediments of the pond. In summer time, the effluent can be used for irrigation, giving a high nutrient utilization.

2.4. Decentralized- and hybrid systems and systems with source separation

Blackwater contains the majority of the resources in domestic wastewater. Of the blackwater components, it is the urine that contributes the greatest amount of nutrients (Table 3). Systems with source separation require the use of toilets that use little or no water (Table 3). These systems open a variety of new possibilities for wastewater and organic waste treatment. The systems briefly described above are less suited for utilization of the energy resources in wastewater and organic waste than the systems described below.

	Urine	Faeces	Greywater
Nitrogen	81	11	8
Phosphorus	48	24	28
Potassium	63	24	13
BOD7			58

Table 3: Distribution of resources in wastewater, Swedish standard figures [25].

Source separation requires a change of sewage system infrastructure. The system logistics depends on toilet type (Fig.3)

Using water-saving toilets requires a separate treatment line for toilet waste. The digestion of blackwater can be either aerobic (liquid composting) or anaerobic, yielding biogas [8,26].Water-saving toilets are very similar to the traditional water toilets and may be more easily accepted by the public than biological toilets.

The use of low-flush vacuum toilets has been chosen for a planned ecohousing project in Oslo (Fig. 3). The blackwater will be collected and trucked to a liquid composting facility located at a farm on the perimeter of the city. Shredded organic household waste will be added to the blackwater. The liquid composting facility will operate with 120 kWh of heat produced per cubic meter processed liquid and less than 1% nitrogen loss [27]. The graywater will be treated on site using biofilters [28], a hydroponic system, and a constructed

---

[1] Leca is a brand name for light expanded clay aggregates.

wetland. The final discharge of the graywater, which is projected to have swimming water quality, is to a local stream. The stream is currently running subsurface, but it is planned to reopen the stream aspart of the ecohousing project. More details on this project are given in Jenssen and Skjelhaugen [27] and by USBL [29].

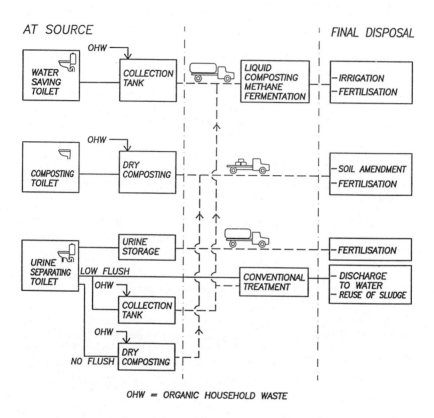

Figure 3: Logistics of blackwater and organic waste handling.

The system planned in Oslo requires trucking of the blackwater. Even when vacuum toilets using 0.7 liter pr.flush are installed, 2000 - 2500 liter/person/year must be transported to the treatment facility. The economic competitiveness of this system is therefore dependent on transportation cost.

The total annual investment, operation, and maintenance costs of the planned system in Oslo are calculated to be USD 400 per household [30]. The present annual fee in Oslo for sewage plus the calculated amount of the garbage fee that covers organic household waste is USD 230. The fees are based on a system of which parts are already completely depreciated. In a situation where both systems are being built from scratch, it is likely that the costs would be more in favor of the source separating system [30]. Adding the cost savings of produced energy would further favor the source separating system. A more complete system analysis is under way.

The Swedish car manufacturer Volvo is treating blackwater from low-flush (3-liter) toilets and organic household waste by fermentation with methane production. The nitrogen is fixed chemically after the anaerobic treatment. The nitrogen fixation uses phosphoric acid. The chemical fixation of nitrogen and the production of methane are both interesting possibilities, but this prototype system was very expensive. The use of phosphoric acid in nitrogen fixation also exacerbates the phosphorus problem in the supernatant, making a phosphorus precipitation step necessary afterwards.

The use of composting toilets results in the smallest amount of waste for transport from the house (Fig. 3). The composting toilet can also handle organic household waste (OHW) and addition of OHW to the toilet enhances the composting process because a better C/N ratio is obtained. Composting dewaters and reduces the volume to 4% of the original [31]. The end product is similar to organic soil and has much lower transportation costs than the end products from the other treatment lines. Composting toilets have up until now required a straight chute to the composting compartment, hence, they have not been suited for apartment buildings and urban areas. However, new systems are under way both for single-family and apartment houses. A prototype system that enables collection of toilet waste in an apartment house without using water has been made. It is proposed to treat the toilet waste in a cylindrical rotating dry-composting-reactor in the basement of the building. The composting reactor will also treat other organic household waste. Trial runs with components of this system have been performed in Norway and show promising results. The gas emissions from the reactor can be filtered through soil to retain escaping ammonia [32] and to decompose odorous substances.

With urine-separating toilets, urine and faeces need separate treatment lines. There are a variety of urine-separating toilets on the market. Most toilets use some water to flush the urine. The faeces can be flushed to a conventional collection system or to separate holding tanks, pending collection and trucking to a liquid composting or biogas plant. In apartment buildings, flushing of the faeces is the only practical solution today. On-site dry composting of the faeces is possible if a perpendicular chute from the toilet to the composting compartment can be installed. Urine-separating toilets where the composting of faeces takes place in a compartment in the toilet are under development.

Several urine-separating toilets are commercially available using dry and wet systems for treatment/handling of faeces. The dry system is suited for single-family dwellings. The wet system is more easily implemented in apartment buildings and retrofitted in existing houses. The faeces are then discharged to the existing sewer system, or they can be collected and treated separately, for instance by liquid composting. In Sweden several projects using urine-separating toilets are planned or underway [33,34].

If urine-separating toilets are used, the urine can be collected in tanks. These tanks can be equipped with a pump and a device controlling the urine discharge. The urine can then be released to the sewage system during hours when the flow is minimal (e.g. midnight). The relatively concentrated urine can be collected at the treatment plant for further treatment and reuse. This interesting system was proposed in Gujer [35].

The system offers the possibility of reclaiming 90% of the nitrogen since that is the fraction in the urine. The percentage of the phosphorus in the urine is lower. If high P-removal and recycling are to be obtained, a phosphorus removal step for the faeces and the graywater is necessary.

2.5. Graywater treatment

Compared to ordinary wastewater, graywater contains less organic matter and only small amounts of plant nutrients (Table 1; [36]). Hence, local discharge after simple treatment is possible [28]. With source separating systems, one can foresee decentralized treatment within cities and discharge to local, small streams and rivers. This reduces the need for large subterranean piping and collection systems. In present day sewage systems, the collection pipelines are the most expensive part of the system. It is estimated that 80% of the costs are associated with the transportation system [26]. In addition, the transportation systems are often in poor condition, and large investments are needed to bring the systems up to an appropriate standard. It can therefore be asked wether investing further in the present systems is wise. Alternatively, what are the costs of retrofitting new toilet systems and decentralized graywater systems and using the old leaky pipes for transport of graywater? If only graywater is to be transported, the service life of existing subterranean sewage systems increases. What are the environmental benefits of doing this? What are the social consequences? The last point may be perceived as the greatest obstacle, but people are adaptable and often motivated. In the field of solid waste, source separation has often been implemented with fewer problems and greater success than anticipated [36].

3. DISCUSSION AND CONCLUSION

The development of the present wastewater treatment infrastructure has taken a hundred years and gone through phases of different goals, with different technologies used to meet these goals. Pathogens, organic matter, phosphorus, and now nitrogen have successively been added to the list of things to be removed from wastewater. Now the concept of recycling some of the contents of the wastewater is once again remaking the goals and technology of wastewater treatment.

The Norwegian Agricultural University has initiated a four-year project to evaluate the technical and economical possibilities of many of these new systems in cold climates. After two years, the various components of the systems have been tested and we are now at the point of implementing full-scale systems in cities and rural areas.

4. REFERENCES

[1] W. J. Mitsch and S.E. Jørgensen, Ecological Engineering: An Introduction to Ecotechnology, New York: Wiley Interscience (1989).

[2] Danish EPA, Esbjerg Declaration: Fourth Int. Conf. on the Protection of the North Sea, Esbjerg, Denmark, 8-9 June 1995.

[3] J. Løgstrup, Danske renseanlegg på vej til at blive vore største forureningsindustrier. Unpublished.

[4] E.K. Jensen and G. Solheim, Biologiske nitrogenfjerningsprosesser - en kilde til lystgassutslipp? State Pollution Control Authority report 92:27. (1992).

[5] P. C. Antonucci and F. D. Schaumburg, Journ. of the Water Poll. Contr. Fed., 47(11), pp 2694-2701 (1975).

[6] USEPA, Municipal Wastewater Reuse, United States Environmental Protection Agency, Report no. EPA 430/09-91.022 (1991).

[7] C. Etnier and B. Guterstam (eds.), Proc. of the Int. Conf. in Ecol. Eng. for Wastewater Treatment, March 24-28, Stensund, Sweden (1991).

[8] P. D. Jenssen and A. Vatn, in H. Ødegaard (ed.), Proc. Int. Conf. Environment Northern Seas (ENS) Stavanger Norway. pp. 191-207 (1991b).

[9] B. Guterstam, in C. Etnier and B. Guterstam (eds.), Proc. of the Int. Conf. in Ecol. Eng. for Wastewater Treatment, March 24-28, Stensund, Sweden (1991).

[10] SFT. Opprydding av avløpsforholdene i spredt bebyggelse. Norwegian State Pollution Control Authority, TA-1182/95 (1995).

[11] H. Wekre and J.C. Køhler, Avløpsplan for bebyggelse langs deler av Rossfjordvassdraget, Centre for Soil and Environmetal Research Report no. 42/96. Ås, Norway (1995).

[12] J. C. Køhler, Avløp fra grendebebyggelser i Velledalen, Centre for soil and Environmetal Research Report no. 25/95, Ås, Norway (1995).

[13] P. D. Jenssen and R. L. Siegrist, Wat. Sci. Tech., 22 (3/4) pp. 83-92. (1990).

[14] P.D. Jenssen and R.L Siegrist. In Proc. Nordic Conf. Dec. 15-16, The Norwegian Institute of Technology, Tapir, Trondheim, Norway, pp. 114-128. (1988).

[15] P: D. Jenssen, Proc. Conf. Alternative Systems: Nutrient Removal and Pathogenic Microbes. Waterloo Centre for Groundwater Research. pp. 28-43. (1995).

[16] S.C. Reed, Subsurface Flow Constructed Wetlands for Wastewater Treatment. A technology assessment. US EPA, 832-R-93-001. (1993).

[17] H.L. Ball, Sand Filters and Shallow Drainfields. Orenco Systems Inc., Roseburg, OR. (1995).

[18] MD. Forskrift om utslipp fra separate avløpsanlegg. Ministry of Environment, Oslo, Norway. (1992).

[19] P. D. Jenssen, Oppfølging av Setermoen renseanlegg. Centre for Soil and Environmetal Research, Report no. 7.2400-05, Ås, Norway. (1992).

[20] Bardu Kommune, Troms, Norway

[21] P. D. Jenssen, T. Krogstad, and T. Maehlum, in C. Etnier and B. Guterstam (eds.) Proc. of the Int. Conf. in Ecol. Eng. for Wastewater Treatment, March 24-28, Stensund, Sweden (1991).

[22] R. M. Gersberg, B.V. Elkiens and C.R. Goldman, Wat. Res. 17/9, pp 1009-1014. (1983).

[23] P. D. Jenssen, T. Maelum and W. Warner, Proc. Seventh Int. Symp. on Indiv. and Small Community Sewage Systems, Atlanta, ASAE 18-94, pp. 137-145 (1994)

[24] J. Yan and S. Ma, in C. Etnier and B. Guterstam (eds.) Proc. of the Int. Conf. in Ecol. Eng. for Wastewater Treatment, March 24-28, Stensund, Sweden, pp. 80 - 95 (1991).

[25] Swedish EPA. Vad innehŒller avlopp från hushåll? Rapport 8825 (1995).

[26] P. D. Jenssen and A. Vatn. In: C. Etnier and B. Guterstam (eds.) Proc. of the Int. Conf. in Ecol. Eng. for Wastewater Treatment, March 24-28, Stensund, Sweden, pp. 148-161 (1991a).

[27] P. D. Jenssen and O.J. Skjelhaugen. In Eldridge Collins, ed.: On-Site Wastewater Treatment: Proc. of the Seventh Int. Symp. on Individual and Small Community Sewage Systems, Atlanta, ASAE, pp. 379-387. (1994)

[28] G. Rasmussen, P.D. Jenssen and L., Westlie. In: J. Staudenmann, et al. ed. Recycling the Resource: Proc. of the Second Int. Conf. on Ecol. Eng., Waedenswil - Zuerich, Switzerland, Sept. 18-22 1995. Trans Tech Publ. (1996).

[29] USBL, Økologiboliger på Klosterenga, programutredning, USBL/SiO (1995).

[30] A. Høyås and W.B. Løfsgaard. Økonomisk vurdering av avløpssytemer - tett tank og våtkomposteringsreaktor. Master Thesis, Department of Economics and Social Sciences, Norwegian Agricultural University. (1995).

[31] Ø. Engen, Personal communication. Engen has researched composting toilets at the Norwegian Agricultural University.

[32] O. J. Skjelhaugen, in Proc. Int. Seminar of the Technical Section of CIGR. on Environmental Challenges and Solutions in Agricultural Engineering, Agricultural Univ. of Norway, Ås, Norway, July 1-4, 1991. (1991).

[33] B. Guterstam and C. Etnier. In J. Staudenmann, et al. (ed.) Recycling the resource: Proc. of the Second Int. Conf. on Ecol. Eng., Waedenswil - Zuerich, Switzerland, Sept. 18-22 1995. Trans Tech Publ. (1996).

[34] L. Jarlöv, in J. Staudenmann, et al. (ed.) Recycling the Resource: Proc. of the Second Int. Conf. on Ecol. Eng. Waedenswil - Zuerich, Switzerland, Sept. 18-22 1995. Trans Tech Publ. (1996).

[35] W. Gujer, in J. Staudenmann, et al. (ed.) Recycling the Resource: Proc. of the Second Int. Conf. on Ecol. Eng. Waedenswil - Zuerich, Switzerland, Sept. 18-22 1995. Trans Tech Publ. (1996).

[36] R. Ready, "Economics of solid waste disposal". Paper presented at Norwegian Agricultural University, January 1996. (1996).

Environmental Research Forum Vols. 5-6 (1996) pp. 25-34
© *1996 Transtec Publications, Switzerland*

From Sewage Irrigation to Ecological Engineering Treatment for Wastewater in China

Z.Q. Ou and T.H. Sun

Institute of Applied Ecology, Academia Sinica, P.O. Box 417, Shenyang, 110015, P.R. China

Keywords: Ecological Engineering, Land Treatment, Water Pollution Control, Wastewater Treatment, Sewage Irrigation, Wastewater Irrigation, Water Resources

ABSTRACT

This paper reviews the development of ecological engineering treatment for wastewater from normal agricultural irrigation practice with sewage in China. The rapid development of wastewater irrigation began in the late 1950s and early 1960s as the increasing demand for irrigation water due to the rapid development of agriculture and the shortage of water resources. However, environmental problems associated with wastewater irrigation occurred because of the increasing amount of un-treated industrial wastewater in sewage and unawareness of the potential reversed effects of wastewater irrigation. Ecological engineering land treatment systems (LTSs) emerged consequently to simultaneously solve the problems of water pollution and water shortage. After fifteen years of studies and development in China, ecological engineering LTSs as a cost-effective and energy-saving technology for the treatment and utilization of wastewater and sewage have blossomed into various types based on different climatic conditions and purposes. Typical operation results of different types of LTSs in China are given in this paper. The differences between wastewater irrigation and ecological engineering LTSs symbolized with nine basic characteristics are summarized. Perspectives and strategy for future application of LTSs in China are also discussed.

INTRODUCTION

The practice of applying human wastes, animal manure and agricultural wastes on land to increase crop yields and soil fertility can be traced back to the history of ancient China more than two thousand years ago[1]. Irrigation with sewage or wastewater on farmland has also a rather long history in China. Although the modern ecological engineering treatment techniques as natural/ecological innovation/alternatives for the simultaneous treatment and utilization of wastewater has only 15 years of history in China, their advantages such as cost-effective, energy-saving and high efficiency have shown great promise and prospectives for future application to solve the problems of water pollution and water shortage, particularly in the northern China.

The purpose of this paper is to review the development of ecological engineering land treatment for wastewater from normal agricultural irrigation practice with wastewater in China. Why is it necessary and feasible and how will it be in the future?

WATER RESOURCES IN CHINA

China is a country that lacks for water resources with a total amount of fresh water resources of 2800 billion tons per year, which ranks the world sixth. However, it is only 1/4 of the world average per capita, which ranks eighty-eighth in the world[2]. The principal waterways of China are illustrated in Fig. 1.

Fig. 1. Fresh water system in China: 1 - Songhua River, 2 - Great Liaohe River, 3 - Haihe River, 4 - Yellow River, 5 - Huaihe River, 6 - Yangtze River, 7 - Pearl River, 8 - Dontin Lake, 9 - Boyan Lake, 10 - Chaohu Lake.

Water shortage has become an critical factor affecting the development of society, production and economy in China. A survey conducted in 1985 showed that, of the 324 cities investigated, 183 were short of water and 40 in seriously water shortage. The total amount of water shortage is 20 million tons per day on the basis of the whole nation[3]. The situation of water shortage is most serious in the northern China due to the non-uniform distribution of population and cultivated land in space and of fresh water in both space and time. Table 1 shows the distribution of fresh water resources in China. The part of China, north to the Yangtze River, with more than 45.3% of the national population and 64.1% of the total cultivated land shares only 19% of the total water resources. The basins of the Yellow, Huaihe, Haihe and

Table 1. Distribution of fresh water resources in China

Region	Area (km^2)	Annual total rainfall($\times 10^9$ m^3/year)	Water resource Amount ($\times 10^9$ m^3/year)	Water resource Percentage (%)
Heilongjiang Basin	903418	447.6	135.2	4.81
Liaohe Basin	345027	190.1	57.7	2.05
Haihe & Luanhe Basin	318161	178.1	42.1	1.50
Yellow River Basin	794712	369.1	74.4	2.65
Huaihe Basin	329211	283.0	96.1	3.42
Yangtze River Basin	1808500	1936.0	961.3	34.18
Pearl River Basin	580641	896.7	470.8	16.74
Rivers in Zhejian, Fujian & Taiwan Provinces	239803	421.6	259.2	9.22
Rivers in south-west China	851406	934.6	585.3	20.80
Rivers in inland regions	3374443	532.1	130.3	4.63
Total	9545322	6188.9	2812.4	100.00

Great Liaohe Rivers with 40% of the total cultivated area have only 9% of the total water resources. Water resource per capita in this region is only 1/9 of the national average. The precipitation decreases gradually from the south-east to the north-west. More than 45% of the total national area has annual rainfall of less than 400 mm[2]. More than 70% of the annual rainfall happen during the period of June to August, resulting in that irrigation is necessary in spring and autumn for the growth of most crops. 45% of cultivated land in China needs to be irrigated regularly, accounting for more than 80% of the total water consumption of the country each year. Agriculture becomes the largest consumer of water[4]. Therefore, municipal wastewater has become an important water resource to be reused for agricultural irrigation[5,6].

WASTEWATER AND WATER POLLUTION IN CHINA

The amount of wastewater, both municipal and industrial, is steadily increasing (Fig. 2) as the development of industry, agriculture and urban construction, and the rapid improvement of people's living standards[7]. The total amount of wastewater discharged annually on the basis of the whole nation has now reached 36.8 billion tons of which about 75% are industrial wastewater. The quantity of wastewater in the year of 2000 is estimated to be 77.6 billion tons which double the present level. However, wastewater treatment facilities have fallen far behind the increase of wastewater volume. At present, the percentage of treatment is only about 24% for industrial wastewater and less than 4% for domestic sewage[5,6]. Due to the absence of treatment facility (Table 2) and funds for operation of existing treatment plants, only 3.2% of municipal wastewater are treated in different degree before discharged into water bodies, and only 2.4% are subjected to secondary treatment.

Water pollution is getting seriously in China due to the direct discharge of more and more volumes of un-treated wastewater into aquatic environments and is threatening the health of people[5,6]. An investigation in 1989 shows that 93% of the 44 rivers flowing across 42 large cities and industrial areas were polluted and 79% heavily polluted[5]. A study conducted in 38 cities in 1989 revealed that 54 out of 85 drinking water sources were polluted accounting for 63.5%. The situation of ground water is also alarming. An investigation in 1980 showed that ground water of 41 cities out of 44 cities was polluted to different extent[5]. Some water resources of both surface and underground have lost their original functions such as potable, industrial, amenity or agricultural irrigation water supplies and aquatic production due to contamination by wastewater. Of the six principal fresh water systems in China, the water quality in the main courses of the Yangtze River is generally good but getting worse compared with that of the late 1980s; the water quality has slightly improved in the main courses of the Yellow River;

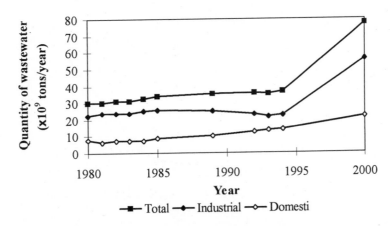

Fig. 2. Quantity of wastewater discharged in China

Table 2. Wastewater treatment facilities in China

Facility of 479 cities	Number	Percentage
Cities with drainage system	326	68
Cities with treatment plant	48	10
Treatment plant	87	
Secondary treatment plant	53	61

most reaches of the Pearl River remain good quality; the Huaihe River has deteriorated; the pollution in the Songhua River has become more serious; and the Great Liaohe River water system is heavily polluted. As a result, water pollution aggravate the situation of water shortage which has inflicted heavy economic losses upon the country[5,6].

DEVELOPMENT OF WASTEWATER IRRIGATION AND ITS ENVIRONMENTAL PROBLEMS IN CHINA

In China, irrigation with municipal wastewater has a rather long history. The rapid development in large scale and national wide, however, began in the late 1950s and early 1960s as organized by the government. Its potential reverse environmental effects were completely unaware of. In late 1960s and early 1970s, as environmental problems occurred, contamination of water, soil and crops due to wastewater irrigation was greatly concerned. But the area of wastewater irrigation has been actually still increasing, as illustrated in Fig. 3. Besides the increase of irrigation agriculture, the shortage of water resources, especially for agricultural irrigation, is the main cause for this rapid development of wastewater irrigation, particularly in the northern China. The total area of wastewater irrigation reaches 1.4 million hectares accounting for 1% of the total cultivated land of the nation[4].

Wastewater irrigation has brought great economic and environmental benefits to China, such as pollution control of aquatic environments, reutilization of wastewater as a water resource, and increasing of agricultural production. However, agricultural and environmental problems associated with wastewater irrigation emerged due to the following reasons:

(1) Unawareness of potential reverse environmental effects of un-proper wastewater irrigation,

(2) Un-adequate planning and unsuitable sites,

(3) Without strict control of irrigation water quality: The first quality standards for agricultural irrigation water of China was established as late as 1985[8]. Concentrations of most organic and inorganic pollutants were found much higher than the standard in some wastewater irrigation areas.

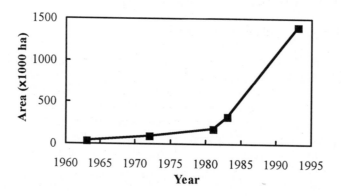

Fig. 3. Development of wastewater irrigation in China

(4) Unreasonable operation and management including cultivation methods and control of quantity of wastewater applied and leaching from wastewater storage ponds and conveyance canals.

Problems happened associated with wastewater irrigation include the followings:

(1) Contamination of soils as a result of accumulation of heavy metals (e.g. cadmium , mercury, lead, chromium and arsenic etc.) and persistent organic pollutants (e.g. mineral oil, policyclic aromatic hydrocarbons etc.): There are a total of 14000 ha of cultivated land contaminated by cadmium to different degree in China[9].

(2) Contamination of crops by heavy metals and organic pollutants,

(3) Contamination of ground water by both organic and inorganic pollutants,

(4) Deterioration of the quality of agricultural products, and

(5) Effects on people's health.

As the public awareness of those environmental problems of wastewater irrigation, some irrigation areas have been suspended for wastewater irrigation or even stopped for agricultural practice. For example, the Zhangshi wastewater irrigation area of 2800 ha, Shenyang, Liaoning province, which was the largest cadmium-polluted area was completely stopped for agricultural cultivation in 1993. Shenfu irrigation area (13300 ha), Liaoning province, the largest area irrigated with municipal wastewater dominated with petrochemical pollutants in China, was suspended for wastewater irrigation since two years ago.

DEVELOPMENT OF ECOLOGICAL ENGINEERING LAND TREATMENT SYSTEMS

For the treatment of municipal wastewater or sewage, modern treatment technologies, i.e. both secondary or advanced ones, can generally provide acceptable effluent. But as a developing country, China can not afford to build and operate many such expensive conventional treatment plants because of lack of funds, shortage of energy and insufficiency of skilled operating personnel and know-how. Therefore, in addition to paying more attention to speeding up the construction and operation of conventional treatment facilities as many as possible within the financial capacity, China has to put special emphasis on the research and development of natural/ecological innovation/alternatives or so-called appropriate technologies with such merits as low costs, energy saving, resources recovery, easy operation/maintenance and wide-spectrum and efficiency in removal of various pollutants. The most promising natural/ecological innovation/alternatives technologies that have been rapidly developing in China are ecological engineering land treatment and utilization systems, eco-ponds, and comprehensive ecological waste treatment and utilization systems[5]. Ecological engineering land treatment systems (LTSs) are not just for the disposal of sewage or wastewater, but also for the simultaneous use of nutrient elements and of water in sewage as reclaimed resources for plant growth[6,10].

Main types and principles of land treatment systems

Ecological engineering land treatment of wastewater is defined as the controlled application of wastewater onto land surface to achieve a designed degree of treatment through natural physical, chemical and biological processes within the plant-soil-water matrix[10,11]. There are five main types of LTSs, i.e. slow-rate infiltration (SR), rapid infiltration (RI), overland flow (OF), wetland (WL) and soil capillary (SC). Their basic characteristics and site requirements are summarised in Table 3. An LTS should be regarded as an ecological engineering which combines ecological principles and engineering techniques. The three main principles of recycling of materials, integral optimization and geographic diversity[12] should be applied to the planning, construction and operation of an LTS[1,6].

Recycling of materials: The rational underlying LTSs derives from the general and specific needs in utilization of soil and covering vegetation as an ecological engineering system for wastewater treatment.

Table 3. Basic characteristics and site requirements of land treatment systems[7]

Types of LTSs	Hydraulic loading rate (m/year)	Soil permeability (m/d)	Soil depth (m)	Depth of groundwater (m)	Surface slope (%)	Need for vegetation
SR	0.6-6	≥0.036	>0.6	0.6 - 3	<30/NL	Necessary
RI	6 - 150	≥1.20	>1.5	>1.0	<15	Unnecessary
OF	3 - 12	≤0.12	>0.3	NL	<15	Necessary
WL	3 - 30	≤0.12	>0.3	NL	<2	Necessary
SC	0.4 - 3	0.036 - 1.2	>0.6	>1.0	<15	Unnecessary

SR = slow-rate infiltration; RI = rapid infiltration; OF = overland flow; WL = wetland; SC = soil capillary.
* NL = no limit, <30 for agricultural systems, no limit for forest systems.

Utilization of the soil as a "living filter" is in fact the most ecologically appropriate treatment technique because recycling of materials is a very complex process in which a large number of interacting variables are involved. The fate of wastewater constituents in plant-soil system is determined by a great number of processes, including physical adsorption on solid surfaces, uptake by plants and microorganism, microbial mineralization, volatilization, leaching, chemical degradation and precipitation[13]. Theoretically, LTSs abide by the laws of ecological succession when undergoing development and material recycle.

Integral optimization: Each ecosystem has its own buffer ability and harmonious functions of its components. An ecological LTS is a systematic engineering including control of wastewater-discharging sources, transportation of wastewater, pretreatment, land application with strict control on influent quality, cultivation of selected vegetation, collection of reclaimed water and monitoring of environmental effects. Thus, a systematic design should be an integral optimization process of an ecosystem so that ecological purifying functions of the system can be fully realised and the purpose of treatment and utilization of wastewater can be met.

Geographic diversity: All sites available for land treatment have strict characteristics of geographic diversity. One differs from another on such aspects as land use, soil, climate, topography and geohydrology. LTSs of different types require different site conditions. Before any design of an LTS, extensive site investigation and adequate planning must be made to collect necessary information and to determine an appropriate type of LTSs and technological conditions.

Slow-rate infiltration land treatment systems (SR-LTSs) most closely related to sewage irrigation are the type of LTSs in which wastewater can be utilized as resources for crop/plant production to the greatest extent. Therefore, they are the main ecological engineering wastewater treatment technology applied to the reform of wastewater irrigation land in China. It should be pointed out that although deriving from traditional sewage irrigation, SR-LTSs have developed into an integral technology and differ from sewage farming at the following 9 main aspects which are summarised in Table 4[6]:

- **Purpose and utilization:** The main purpose of sewage farming is to utilize water and nutrients in wastewater as resources to increase crop yield while LTSs are designed for the treatment as well as utilization of wastewater for plant growth simultaneously.

- **Category:** Traditional sewage irrigation is an agricultural and hydraulic practice while LTSs are ecological engineering emphasising treatability of wastewater and paying great attention on systematic design for round year operation, hydraulic balances, application methods, hydraulic and pollutant loadings, selection of covering plants and harmonious function of subsystems to obtain optimum treatment results.

- **Operation:** Sewage farming operates only in the crop growth seasons without considering the continuous operation throughout the year. On contrary, acting as a wastewater treatment plant, an LTS must be designed and constructed to ensure continuous operation around the year.

- **Ecological structure:** LTSs with various vegetations are different from traditional sewage farming of single vegetation. With multiple vegetation, an LTS can meet the demand of balances and loading rates of water and pollutants at various seasons and conditions to ensure continuous and optimum operation of the system.

- **Control of pollutant loadings:** Wastewater is considered as water and fertilizer resources in sewage farming and thus emphasis is only focused on the amount of wastewater applied. Ecological treatment of wastewater places much emphasis on both pollutant loadings and environmental assimilative capacities, and loading rates of both wastewater and pollutants are determined based on the lowest limiting constituents (LLCs).

- **Site requirements:** Each type of LTSs has its own requirements on site conditions as mentioned above and illustrated in Table 3. On contrary, sewage farming emphasises only on soil fertility, climate and economic benefits.

- **Engineering design and management:** Sewage farming is usually managed to have highest crop yields. An LTS should be elaborately designed and managed to achieve planned goals of both treatment and utilization of wastewater.

- **Protection of receiving water bodies and re-utilization of reclaimed water:** The operation of an LTS must be properly managed and monitored with adequate adjustment of hydraulic and pollutant loadings to ensure that the quality of the effluent meets requirements and design and to protect receiving water bodies. The re-utilization of reclaimed water is also considered in LTS operation depending on the types of LTSs.

- **Water Movement :** Overflow usually occurring in traditional irrigation must be limited to minimum in LTSs except for OF.

Development of ecological engineering land treatment systems in China

Table 5 shows the history of LTS development in China which can be divided into three stages. The first stage started from 1981 and lasted for 5 year to 1985 for laboratory and small-scale studies. The first

Table 4. Differences between sewage farming and land treatment systems (LTSs)

Factors	Sewage farming	LTS
+ Purposes	- Increasing crop yields - Using water in sewage - Using nutrients in sewage	- Treatment of wastewater - Using water in sewage - Using nutrients in sewage
+ Category	- Irrigation/hydraulic engineering	- Ecological engineering
+ Operation	- Crop growth/irrigation seasons	- All year round
+ Ecological structure	- Single vegetation	- Multiple vegetation
+ Pollutant loading	- Not considered	- Considered as design and operation parameters
+ Site conditions	- Not strict	- Strict
+ Design and management	- Based on crop yields	- Crop yields - Effluent quality - Reutilization of reclaimed water - Environmental effects
+ Protection of receiving water body	- Not considered	- Considered
+ Water movement	- Overflow allowed	- Infiltration - Overflow not allowed

SR-LTS in China was built in 1982 in the southwestern suburb of Shenyang, Liaoning Province. Initial achievements were obtained by successful operation though it was a smell scale[1,13,14]. As s result of achievements in this period, the function of LTSs was affirmed by the Chinese authorities in 1986 in a document titled "*Regularities on Technological Policy for Water Pollution Control*". It suggested that local governments should consider LTSs as a priority based on their local conditions.

The second stage was in the period of 1986 to 1990 for joint pilot-scale studies. Several pilot-scale LTSs were set up over the country and sponsored by the Chinese National Environmental Protection Agency as joint state research projects. Four types of LTSs were intensively studied, i.e. an SR-LTS with a daily treatment capacity of 600 tons of municipal wastewater in Shenyang[15], another one with a daily treatment flow of 400 tons in Kunming, an RI-LTS of 500 tons/day in Beijing, an OF-LTS of 600 tons per day in Beijing, and a WL-LTS of 1000 tons/day in Tianjin. Their typical performance is shown in Table 6. Results indicate that LTSs had high removing efficiency, especially for organic compounds and nutrients. Studies also show that LTSs had rather high buffering capability through self-regulation and could maintain optimal performance in rather wide variation of hydraulic and pollutant loadings. Another important achievement in this period was the publication of the first Chinese technical guideline "*Design Manual for Land Treatment and Utilization of Municipal Wastewater*[1]".

The development of LTSs in China is now in its third stage starting from 1991 for production-scale application. Several production-scale systems of various types in different climate regions (Table 5) were built to treat and utilize various municipal wastewater, e.g. a large scale LTS with a treatment flow of 30000 tons/day in Xinjiang Province, a subsurface capillary seepage system for domestic sewage treatment in Shenyang, a forest SR-LTS of production scale with a treatment flow of 10000 tons/day in Huolingele city in Inner Mogolia, an SR-LTS of 15000 tons/day in Chuxiong city in Yunnan Province, and a WL-LTS of 10000 tons/day in Dagang harbour in Tianjin, etc. They operate now successfully and play an important role in pollution control for local communities. As a result of the 15 years' studies and achievements, ecological engineering LTSs for wastewater treatment and utilization have developed into a complete and true technology[6].

Table 5. Development of land treatment systems (LTSs) in China

1981-1985: Laboratory and small-scale studies
SR, RI, OF and WL

1986-1990: Pilot-scale

SR	600 tons/day	Shenyang, Liaoning
	400 tons/day	Kunming, Yunnan
RI	500 tons/day	Beijing
OF	600 tons/day	Beijing
NWL	1000 tons/day	Tianjin

1991-1995: Production scale

SR	10000 tons/day	Huolingele, Inner Mongolia
	15000 tons/day	Chuxiong, Yunnan
RI	30000 tons/day	Kuerle, Xinjiang
RI+SR	20000 tons/day	Atush, Xinjiang
NWL	10000 tons/day	Tianjin
NWL	100 tons/day	Panjin, Liaoning
CWL	1000 tons/day	Shenzhen, Guangdong
SC	50 tons/day	Shenyang, Liaoning

Note: NWL = natural wetland, CWL = constructed wetland.

Table 6. Typical operation results of LTSs in China

LTS		Shenyang SR-LTS	Kunming SR-LTS	Beijing RI-LTS	Beijing OF-LTS	Tianjin NWL-LTS	Tianjin CWL-LTS
pH	Infl.	6.6		7.7	7.6		
	Effl.	6.7		7.0	7.3		
F.coliform	Infl.	3.5×10^9		7.4×10^3	8.5×10^3		
(no./	Effl.	5.7×10^7		38.8	0.6		
100ml)	RR(%)	98.4		99.5	99.9		
BOD_5	Infl.	100.8	91.3	126.0	131.0	184.0	154.7
(mg/l)	Effl	3.5	4.8	5.9	19.8	18.5	9.5
	RR(%)	96.5	94.7	95.3	84.9	89.9	93.8
COD_{Cr}	Infl.	232.4	146.6	479.3	535.0		
(mg/l)	Effl.	29.8	4.8	38.8	107.0		
	RR(%)	87.2	96.7	91.9	80.0		
TOC	Infl.	66.8	69.7	78.6	87.7		
(mg/l)	Effl.	11.5	8.4	13.8	25.0		
	RR(%)	82.8	87.9	82.4	71.5		
Total	Infl	20.6	54.4	13.0	12.7	41.9	31.6
nitrogen	Effl..	3.4	12.4	2.7	4.9	21.6	9.2
(mg/l)	RR(%)	83.5	77.2	79.3	61.6	48.4	70.9
Total	Infl.	1.9	2.1	0.6			
phospho-	Effl.	0.2	0.1	0.5			
rous (mg/l)	RR(%)	89.5	95.2	16.7			
Suspended	Infl.	83	81	191	201	93	86
solids	Effl.	19	10	4	18	23	2
(mg/l)	RR(%)	77	88	98	91	75	98

Infl. = influent, Effl. = effluent, RR = removal rates.

PROSPECTIVES AND STRATEGY FOR FUTURE APPLICATION OF LAND TREATMENT SYSTEMS IN CHINA

Although researches on LTSs began quite late in China, important achievements have been obtained after 15 years of joint studies. LTSs of various types and different scale have been set up in different climatic regions in China. All these achievements have brought a wide prospect of the application and development of LTSs in future in China.   But further studies on design, operation and management of enlarged-scale systems, technologic and economic assessment indexes and public awareness are still needed. In addition, for a better and wider application of LTSs in China, the following feasible strategy is recommended according to results and experiences obtained so far.

(1) SR-LTSs are suitable for the treatment and reutilization of domestic sewage or municipal wastewater which contains no or acceptable low levels of heavy metals or non-readily degradable organic pollutants.

(2) LTSs can be built for medium and small cities or towns because their municipal wastewater is more simpler in characteristic and less in quantity.

(3) SR-LTSs are best developed in arid and semiarid areas with <500 mm of annual precipitation. In these areas in China, wastewater is always treated as a valuable water resource, economy is comparatively less developed, and wastewater is less in quantity and simpler in quality.

(4) Trees, wetland plants or dry land and long-stalked crops should be used for SR-LTSs. These plants or crops need less water than paddy rice, thus low pollutant loading can be implemented and an oxidation soil environment can be maintained which is necessary for the degradation of most organic pollutants. Further more, trees and wetland plants usually do not enter the food chain.

(5) SR-LTSs should be designed and operated as a utilization system other than treatment systems. The former has lower hydraulic and pollutant loadings than the later.

(6) Operation skill and management levels are needed to be improved: Comprehensive technical guideline and technological processes for LTSs should be established.

CONCLUSION

Ecological engineering land treatment systems have proved to be a cost-effective, low energy-consuming and high efficiency natural/ecological innovation/alternative to traditional wastewater treatment technique[1,5-7,10,14,15]. There is a great prospect for their widely application in China. A successful LTS can be guaranteed only when it applied based on local conditions including site, climate, wastewater, land use, economy and education level of local citizen. Rational operation and management are equally important for the continuous development of LTSs.

REFERENCES

[1] Z. M. Gao and X. F. Li (eds.), *Design Manual for Land Treatment and Utilization Systems for Municipal Wastewater*, Chinese Standards Press, Beijing, pp1-458(1991).

[2] J. S. Wang, *J. Appl. Ecol.*, **26**, 851-857(1989).

[3] T. Zhou, D. M. Liu, H. Y. Jiang, P. Ren and X. Z. Bao, *Water Pollution Control and Techniques for Utilization of Municipal Wastewater as Resources*, J. M. Jin (ed.), Science Press, Beijing, pp914-932(1993).

[4] C. R. Han, *J. Appl. Ecol.*, **26**, 803-812(1989).

[5] B. Z. Wang, *Proceedings of the International Post-conference Seminar on Low Costs and Energy Saving Wastewater Treatment Technology*, B. Z. Wang and Z. Y. Nie (eds.), Harbin, China, pp8-19(1990).

[6] T. H. Sun and Z. Q. Ou, *Proceedings of the 4th International Conference on Wetland Systems for Water Pollution Control*, Guangzhou, China, pp48-57(1994).

[7] UNESCO, Serial N°. FMR/ECO/93/229(FIT), FIT/509/CPP/40 Terminal Report, Paris, France, pp88-101(1993).

[8] NEPA of China, *Quality Standard for Agriculture Irrigation Water*, GB5084-85(1985).

[9] H. Y. Huang, D. M. Jiang, C. X. Zhang and Y. B. Zhang, *Studies on Pollution Ecology of Soil-plant Systems*, Z. M. Gao (ed.), China Science and Technology Press, Beijing, pp79-85(1986).

[10] Z. Q. Ou, Z. M. Gao and T. H. Sun, *Wat. Res.*, **26**(11), 1479-1486(1992).

[11] U.S.EPA, *Design Manual for Land Treatment of Municipal Wastewater*, USEPA 625/1-81-013(1981).

[12] S. J. Ma (ed.), *Perspectives in Modern Ecology*, Chinese Sciences Press, Beijing, pp5-10(1990).

[13] Z. M. Gao, A contribution paper to the Fourth International Congress of Ecology in Syracuse, New York, August 10-16, 1986.

[14] Z. M. Gao (ed.), *Studies on Pollution Ecology of Soil-plant Systems*, Chinese Science & Technology Press, Beijing, pp515(1986).

[15] Z. Q. Ou, S. J. Chang, Z. M. Gao, T. H. Sun, E. S. Qi, X. J. Ma, P. J. Li, L. S. Zhang, G. F. Yang, S. H. Han, Y. F. Song, H. R. Zhang, L. P. Ren, and X. R. Qu, *Wat. Res.*, **26**(11), 1487-1494(1992).

Environmental Research Forum Vols. 5-6 (1996) pp. 35-44
© *1996 Transtec Publications, Switzerland*

Waste Recycling in Freshwater Fish Culture, with Emphasis on Trace Metal Contamination

M.H. Wong

Department of Biology and Centre for Waste Recycling & Environmental Biotechnology,
Hong Kong Baptist University, Kowloon Tong, Hong Kong

Keywords: Waste Recycling, Polyculture of Fish, Animal Manure, Sewage Sludge, Trace Metals

ABSTRACT

Waste materials such as rice bran, distillers' by-products, animal manure, and even human excreta are used as pond fertilizers or supplementary feeds for freshwater fish farming in China. The advantages include a partial elimination of waste disposal problem, a supply of usable materials at low cost, the provision of cheap animal protein and good environmental practice.

Polyculture is the traditional fish farming method originally derived from China. It is stocked with different species of fish including common carp (*Cyprinus carpio*), silver carp (*Hypophthalmichthys molitrix*), grass carp (*Ctenopharyngdon idellus*), mud carp (*Cirrhinus molitorella*), bighead carp (*Aristichthys nobilis*), black carp (*Mylopharyngodon piceus*), and tilapia (*Oreochromis mossammbicus*), which adapt to feed on various organisms and materials within the fish pond to fully utilize all the resources derived from the waste materials.

This article describes the traditional fish farming method where waste materials are involved. The modes of action of wastes, especially animal manure and sewage sludge added into fishponds will be discussed. Emphasis will be made on the effects of incorporating animal manure and sewage sludge into fish feeds, and the possible adverse effects derived from trace metal contamination of the added wastes on the treated fish.

INTRODUCTION

Application of organic matter such as human excreta, animal manure, rice bran, distillers' by-products and various materials gathered in the vicinity of the ponds such as grain and snails is a general practice in Asian as well as East European countries [1].

Fish produce over a third of the animal protein consumed by human beings. Fish farming amounted to about 5% of the total yearly consumption of about 70 million tonnes of fish [2]. Traditional farming in Asian countries, particularly in China rely on polyculture (stocking ponds with different species of fish, with Chinese carp the main species), with manures the major nutrient input. This is an example of how substances contained in the waste materials can be effectively and beneficially promoted and circulated [3].

The total area of fish ponds is about 0.7×10^6 ha in China, compared with the total world area of 2.26×10^6 ha. However, Chinese fish ponds produce two-third of the world's yield of farmed fish (2.24×10^6 tonnes compared with the world total of 3.69×10^6 tonnes). The mean yield of 3.2 tonnes/ha/year in Chinese fish ponds is about 4 times higher than the world average [4].

Table 1 lists fish production in developing countries in four regions in 1990 and the predominance of Asia, especially China is apparent [5]. The major reasons for the high productivity of fish farms in China are due to rational use of manures, good management practices, suitable climate and water supply [2].

Table 1. Aquaculture production (t) of fish in developing countries of four regions. (Source: FAO data, [5]).

	Year	Asia (excl. China) (12 countries)	China	Sub Saharan Africa (29 countries)	West Asia/ North Africa (11 countries)	Latin America/ Caribbean (21 countries)	Totals
Carps and	1984	691,469	1,766,158	515	14,888	1,837	2,474,867
other	1985	545,916	2,838,786	389	40,293	1,147	2,971,531
cyprinids	1988	633,372	3,858,500	647	17,846	4,155	4,514,520
	1990	711,441	4,124,478	1,126	48,011	4,588	4,889,644
Tilapias	1984	71,768	18,100	4,141	3,328	26,311	123,648
and other	1985	105,325	77,120	7,006	26,745	17,976	234,172
cichlids	1988	136,091	39,000	7,726	1,469	31,736	216,022
	1990	154,716	160,369	8,799	30,555	26,622	381,061
Misc. spp.[a]	1984	459,259	9,359	4,430	26,589	7,422	507,059
	1985	410,380	118,683	3,185	2,117	19,605	553,970
	1988	508,938	32,671	2,843	60,098	7,500	612,050
	1990	661,049	204,434	4,251	15,691	51,864	937,289

[a] Includes all other finfish, such as catfish, milkfish, mullets and a wide range of freshwater, brackishwater and marine species.

It has been generally recognized that fish may be the cheapest animal protein when grown on wastes. Fish grow rapidly in the tropics and if wastes replaced the need for expensive supplementary feed, the production cost is minimized [6,7]. There is an urgent need to extend this practice to countries where malnutrition and hunger are prevalent. If handled and treated properly, recycling of wastes in this context will also ease the waste disposal pressure, and thereby mitigating the water pollution problem, as well as reducing insanitary disease.

The waste material is either applied to fish ponds as fertilizer to enrich the water or incorporated in fish feeds. Fish farming and animal husbandry are well integrated on a large scale in China and have gained rapid popularity in other places such as Taiwan, and more recently India, Indonesia, Malaysia, and the Phillipines [8,9]. There are considerable variations of these integrated systems, which included mulberry grove-fish, rice-fish, livestock/poultry-fish, vegetable-fish systems and combinations of these [14-17]. Trials have been conducted in culturing fish in the oxidation ponds of sewage treatment plants [10-12] and using sewage sludge as supplementary feeds on a smaller scale [13].

Most of the studies emphasized the nutritional values of the wastes, in particular animal manure [2] being used as pond fertilizer and their effects on the growth rates and yields of the treated fish. There is a comparatively lack of information related to the effects of incorporating wastes into fish feeds and their possible adverse effects of the added wastes, on the treated fish.

This article reviews the advances made in the recycling of wastes in the polyculture of fish, and the potential health hazards derived from the high concentrations of trace metals in sewage sludge when used as fish feeds will be discussed.

CHARACTERISTICS OF MANURE AND SEWAGE SLUDGE

Animal manure contains basic residual nutrients such as proteins, carbohydrates, lipids and minerals. Sewage sludge contains less nutrients, but a higher level of trace metals (Cu, Pb, Zn, Cd, etc.) and may contain other undesirable substances such as chlorinated hydrocarbons, hormones, enteric pathogens, etc. It has been claimed that the nutrient values and amino acid profiles of primary and secondary sludges are more than sufficient for exploration into their use in animal feed [18].

Table 2 compares the data for various trace metals, typical of purely domestic and domestic plus industrial sewage, with feed compounds [18]. Nevertheless, sludge produced by food processing industries, such as treatment of meatworks effluents might be a better choice as it contains a similar composition and amino acid profile as the activated sludge produced by the treatment of domestic sewage, but with much lower concentrations of trace metals such as Pb, Zn and Cd [19].

Table 2. Heavy metal analysis of two samples of sewage sludges compared with feed compounds [18].

Metal	Domestic sewage (μg/g)	Domestic with industrial sewage (μg/g)	Feed compounds (μg/g)
Cd	1 to 3	1 to 17	0.5 to 1.0
Cr	8 to 35	250 to 410	1.0 to 3.0
Cu	30 to 200	700 to 3550	10.0 to 15.0
Hg	Too low to measure accurately		
Ni	10 to 60	35 to 750	2.0 to 5.0
Pb	50 to 200	650 to 1100	2.0 to 10.0
Zn	40 to 650	850 to 4850	50.0 to 70.0

MODES OF ACTION OF WASTES ADDED INTO FISHPONDS

The predominant use of waste materials in fish farming in China and other countries is presumably associated with the unavailability of grain feeds. This constraint led to the polyculture of mixed species of fish (mainly carp) which adapt to feed on various organisms and materials within the fish pond to fully utilize all the resources derived from the waste materials.

When waste materials are added to fish ponds, there are three different routes:

1. The waste materials are consumed directly by fish.

2. Nutrients released from the waste materials accelerate autotrophic production and subsequently promote the growth of other organisms which are consumed by different fish with different feeding habits.

3. The waste materials can serve as a base for heterotrophic production where bacteria and protozoa are grown on the surface which are in turn eaten by fish.

Grass carp (*Ctenopharyngodon idellus*) and Silver carp (*Hypothalmichthys molitrix*) are herbivorous. The former consume aquatic plant as well as terrestrial plant residues such as vegetable top and grass clippings while the latter take in large quantities of phytoplankton, by means of their special filter apparatus. Bighead carp (*Aristichthys nobilis)* consume mainly zooplankton, black carp *(Mylopharyngodon piceus*) consume snails and other benthos, while common carp *(Cyprinus carpio)* consume benthic animals and detritus. Common carp stir up sediments during scavenging pond bottom. This creats turdity where organic particles are suspended and colonized by bacteria, which in turn served as food for silver carp. In addition to the danger of direct consumption of sewage sludge containing various undesirable substances as well as high concentrations of trace metals by fish, trace metals can be incorporated into phytoplankton efficiently and accumulated in different trophic levels along the food chain.

GROWING FISH IN SEWAGE EFFLUENT AND ADDING SLUDGE INTO FISH FEEDS

It has been noted that culturing fish in oxidation ponds could improve the water quality by reducing the growth of plankton and benthos, and at the same time increase the dissolved oxygen and the pH [10]. In fact, polyculture has been used in Arkansas, U.S.A. to upgrade sewage effluent by reducing biochemical oxygen demand, suspended solids, and fecal coliforms in order to meet the environmental protection standards [20].

Experiments carried out in the oxidation ponds of the Shek Wu Hui Pilot Sewage Treatment Plant, Hong Kong (rearing tilapia alone and a mixed stock containing silver carp, bighead, grass carp and common carp) indicated that no significant difference in trace metal concentrations (Cd, Cr, Cu, Ni, Pb, Zn) was detected in the fish muscle samples [11, 12]. It has been revealed that the concentrations of various trace metals (Cd, Cr, Cu, Pb, Zn) in the sewage effluent were similar to the 0.2% (v/w) sludge, followed by the sewage effluent while 0.4, 0.6 and 0.8% (v/w) sludge yielded poor growth rates which might be due to the higher concentrations of trace metals in these media (Table 3).

Adding sewage sludge into freshwater fish ponds could also raise plankton production and fish growth [22]. Besides plant nutrients, comparatively higher concentrations of trace metals from sewage sludge will also be released into the pond water and subsequently affecting the fish through accumulation along the food chain. Furthermore, a higher oral intake of trace metals will be envisaged as larger particles will be consumed directly by some species of fish.

Although the use of activated sludge as a feed for trout has been proved to be promising and no evidence of metal accumulation in the tissue was reported [23], other studies indicated that trace metals contained in sewage sludge might exert harmful effects on the treated fish [24-27]. This is especially true when sewage sludge is incorporated into fish feeds in a pellet form [28]. Table 4 (from [13]) indicates that trace metals differed from organ to organ of the treated fish.

Values were highest in those fish fed the highest proportions of sludge, especially in the viscera. Flesh and head contained lower concentrations of metals, in general. According to the trace metal standards in food products from fresh or marine waters (Table 5, from [29]), based on dry weight basis, it is apparent that all the metals tested (except Cr), based on dry weight basis, on the fish fed sludge-supplemented diets were within the limits. However, the rather high level of Cr even in the control fish needed further investigation.

Table 3. Metal contents in different media for fish rearing from [21].

Metal (mg/1)	0.2% sludge	0.4% sludge	0.6% sludge	0.8% sludge	Sewage effluent
Cd	0.04	0.05	0.05	0.06	0.04
Cr	n.d.	n.d.	n.d.	0.03	0.02
Cu	0.03	0.05	0.05	0.06	0.03
Fe	0.19	0.22	0.23	0.29	0.22
Pb	n.d.	n.d.	n.d.	n.d.	0.22
Mn	0.06	0.11	0.12	0.23	0.09
Zn	0.99	1.15	1.26	1.93	0.58
Ca	12.60	16.00	20.40	22.60	11.00
Mg	39.20	41.10	42.20	44.70	46.80
Hardness equivalent mg/1 as $CaCO_3$	194.80	211.20	226.80	242.70	222.50

Table 4. Trace metal concentrations (μg/g, dry weight basis) in different parts of tilapia fed various diets after 60 days (mean of 4 replicates) (from [13]).

Diet	Part	Cd	Cr	Cu	Ni	Pb	Zn
Commercial	Head	0.73	6.06	2.04	2.64	6.75	83.9
feed alone	Gill	2.23	16.9	4.65	1.53	n.d.	58.7
	Viscera	1.17	41.7	34.4	11.2	5.32	216
	Flesh	0.11	18.3	3.22	1.38	2.26	89.6
	Bone	1.27	19.6	4.92	5.99	5.97	119
Commercial	Head	0.97a	13a	19.4a	13.8a	15.2a	169a
feed 95% +	Gill	2.79	32.6a	36.9a	26.3a	8.05	269a
Sludge cake	Viscera	2.37a	40.6	304 a	25.9a	20.1a	265a
5%	Flesh	2.17a	22.6a	17.8a	16.9a	10.7a	161a
	Bone	2.21a	24.6a	15.5a	14.9a	16.6a	188a
Commercial	Head	1.71a	19.7a	56.7a	16.2a	18.3a	175a
feed 70% +	Gill	3.09a	31.6a	89.9a	36.8a	15.5a	280a
Sludge cake	Viscera	3.62a	87.1a	554 a	49.5a	32.6a	387a
30%	Flesh	2.10a	20.4	92.6a	26.7a	14.8a	188a
	Bone	2.98a	44.4a	54.2a	16.3a	15.9a	205a

Commercial feed consisted of 50% fish meal, 25% wheat flour, 20% peanut meal, 3% cod liver oil and 2% vitamins .
a indicates significant difference between the control (commercial feed alone) and treated fish with p<0.05.
n.d. = not tested

The same study indicated that sludge was contaminated with polychlorinated biphenols (PCBs) and a significant increase in the PCB concentrations was found in fish flesh fed the 30% sludge cake-supplemented diet. Total PCBs detected in the flesh was 61 ng/g (wet weight) and higher chlorinated PCB isomers were much more common than lower isomers [30].

Table 5. Median international standards and ranges in such standards (µg/g wet weight) for trace elements in freshwater and marine fish and shellfish (after [29]).

Element	Median standards Fish	Shellfish	Range		Number of countries with standards
Antimony	1.0	1.0	1.0 to	1.5	3
Arsenic	1.5	1.4	0.1 to	5.0	11
Cadmium	0.3	1.0	0.05 to	2.0	10
Chromium	1.0	1.0	1.0		1
Copper	20	20	10	to 100	8
Fluoride	150	-	150		1
Fluorine	17.5	-	10	to 25	2
Lead	2.0	2.0	0.5 to	10	19
Mercury	0.5	0.5	0.1 to	1.0	28
Selenium	2.0	0.3	0.3 to	2.0	3
Tin	150	190	50	to 250	8
Zinc	45	70	40	to 100	6

Limited studies on the histopathological assessments of the sludge-treated fish indicate that internal injury in gills, hepatopancreas and kidney as well as external damage, especially the particles observed around the gills were noted [26, 31-34]. Although it was suspected that the symptoms might be related to the high concentrations of trace metals and PCBs contained in the supplementary feeds, further studies are needed before a definite conclusion can be drawn.

Sewage sludge seems to be inferior for growing common carp [26, 27], but more promising for growing tilapia [34] which might be due to the fact that tilapia is a hardly species and has a wider spectrum of food intake [35]. It has been noted that tilapia had significantly lower coughing rate and ventilation rate than bighead when subjected to addition of sewage sludge in water [36]. The choice of fish is therefore important when sewage sludge is used as a supplementary food.

SOLUTIONS TO THE TRACE METAL PROBLEM

Pathogens contained in sewage sludge could be eliminated by composting or thermal treatment [37]. Aerobic composting system with good management is a viable, low cost sanitation technology [7]. However, trace metals contained in sewage sludge will remain a major problem when sludge is used as fish feed. Direct extraction of protein or other valuable products such as oils and fats is technologically feasible. Unfortunately, the extracts remain contaminated with trace metals. This factor should be either eliminated by restrictions or by employing chemical treatments for heavy metal removal.

Removal of trace metals from sewage sludge before it is used as fish feed is therefore essential. When treated with mineral acids, the metallic ions are solubilized and present in the aqueous phase which could be separated from the sludge solids [18]. However, the process involved is expensive and complicated.

Alternatively, sewage sludge could be used as a substrate for growing micro-organisms which in turn used as fish feed. Although it is commonly known that micro-algae absorb a considerable amount of trace metals when cultivated in water-extracts of sewage sludge [38, 39], the metals do not necessarily accumulate along the food chain, e.g. the results of a study involving two artificial food chains: 1. sludge extracts - *Chlorella pyrenoidosa* (green alga) - *Caridina sp.* (freshwater shrimp) - *Cyprinus carpio* (common carp) and 2. sludge extracts - *Chlorella pyrenoidosa* - *Cyprinus carpio* indicated that the concentration ratios of various trace metals (Mn, Fe, Cu and Zn) were high (355-22182) in *Chlorella pyrenoidosa* cultivated in sludge extracts, but low in shrimps and carp and in most cases the ratios were below 1. Carp and tilapia fed with sludge-grown algae had higher concentration ratios and absolute values of trace metals than those fed with algal-grown shrimps and cladoceras, indicating the elimination of trace metals when going up the food chain [40-42].

The moulted exoskeleton of the shrimps during the culture period also had a high concentration of trace metals which suggests that moulting may play an important role in bioelimination of the excessive toxic metals [43], in addition to other physiological mechanisms which will eliminate toxic metals. However, food processing wastes e.g. carrot waste [25] or sludge free of toxic metals (e.g. sludge from meatworks) would be the safest choice for rearing fish.

CONCLUSION

Polyculture of fish in wastewater-fed ponds has been practiced for centuries. Promotion of this practise to recycle nutrients from urban areas would alleviate eutrophication and produce animal protein [44]. Animal manure serves mainly as pond fertilizer once added into fish ponds, to enrich the pond water and accelerate the growth of different aquatic organisms. Incorporating manure and sewage sludge into commercial fish feeds normally results in poor yields which may due to the higher level of trace metals and other undesirable substances such as PCBs. Water quality is the most important factor in determining the quality of fish produced, and hence wastewater should be treated prior to introduction into fishponds in order to ensure the product is safe for human consumption [45]. There is an urgent need to study the public health issues associated with aquaculture operations which involved the use of waste materials.

ACKNOWLEDGEMENTS

This study is supported by the research grants from The Hong Kong Baptist University and Research Grant Council of the Universities Grants Committee of Hong Kong.

REFERENCES

[1] G. Wohlfarth. Proc. Conf. on Fish Farming and Wastes. University College, London, Janssen Services, 151 pp. (1978).
[2] G. Wohlfarth and G.L. Schroeder. Agric. wastes 1: 279 (1979).
[3] J. Yan and Y. Zhang. Ecol. Engin. 261 (1992).
[4] D.D. Tapiador, H.F. Henderson, M.N. Delmendo and H. Tsutsuiv. FAO Fish. Tech. Paper 168, 84 pp (1977).
[5] FAO. FAO Fish. Circ. 815 (Rev. 4), 260 pp (1992).
[6] P. Edwards. Aquaculture, 31: 261 (1984).
[7] P. Edwards. World Bank Technical Paper No. 36. The World Bank, Washington, D.C. 45pp (1985).
[8] T.P. Chen. Fishing News Books, Farnham, Surrey, U.K. 161 pp (1976).
[9] V. Jamandre. Joint FAO/UNDP/SCSP and SEAFDEC Regional Workshop on Aquaculture Engineering, CP10, 11 pp (1977).
[10] G.L. Schroeder. Water Res. 9:591 (1975).
[11] A.W. Sin and M.T.L. Chiu. Resoures and Conservation, 13: 217 (1987).
[12] A.W. Sin and M.T.L. Chiu. Resouces and Conservation 13: 231 (1987).
[13] M.H. Wong and S.T. Chiu. Environ. Technol. 14: 1155 (1993).
[14] E.T. Baconawa, D.O. Parawan, G.A. Bautista, H.B. Ovalo and D.P. Catbagan. Resources and Conservation 13: 265 (1987).
[15] S. Geeta, T.K. Mukherjee and S.M. Phang. In: Management and Utilisation of Agricultural and Industrial Wastes. Eds. Goh, S.H. et. al. Univ. of Malaya, Kuala Lumpur (1991).
[16] R.S.V. Pullin. In: Environment and Aquaculture in Developing Countries. Eds. Pullin, R.S.V. et al. International Center for Living Aquatic Resource Management, Manila (1993).
[17] K. Ruddle and G. Zhong. The Dike-Pond System of The Zhujiang Delta, Cambridge Univ. Press, Cambridge (1988).
[18] S.H. Smith and H. Rothman. Agric. Wastes, 3: 87 (1981).
[19] B.V. Kavanagh, L.S. Herbert and S.P Moodle. Agric. Wastes, 4: 305 (1982).
[20] National Research Council. National Academy Press, Washington, D.C., 154 pp. (1981).
[21] S.W. Yip and M.H. Wong. Environ. Pollut. 14: 127 (1977).
[22] M. Natarajan and T.J. Varghese. Agric. Wastes 2: 261 (1980).
[23] A.G. Tacon and P.N. Ferns. Nutrition Reports International, 13: 549 (1976).
[24] J.D. Lu and N.R. Kevern. Indian Farming, 28: 25 (1975).
[25] M.H. Wong and S.P. Cheung. Environ. Pollut. A23: 29 (1980).
[26] M.H. Wong, K.C. Luk, W.K. Liu and S.H. Kwan. Acta Anatomica, 103: 130 (1979).
[27] M.H. Wong, Y.H. Cheung and W.M. Lau. Toxicol. Letters, 12: 65 (1982).
[28] M.H. Wong and S.H. Kwan. Toxicol. Letters, 7: 367 (1981).
[29] C.C. Nauen. FAO Fish. Circ. 764, 102 pp (1983).
[30] M.S. Yang, W.F. Fong, T.C. Chan, V.W.D. Chui and M.H. Wong. Environ. Technol. 14: 1163 (1993).
[31] W.K. Liu and M.H. Wong. Environ. Res. 40: 164 (1986).
[32] W.K. Liu and M.H. Wong. Resources and Conservation 13: 255 (1987).
[33] W.K. Liu, M.H. Wong and Y.H. Cheung. Biomed. Environ. Sci. 3: 81 (1989).
[34] M.H. Wong, K.M. Chan and M.H. Wong. Conservation and Recycling, 7: 351 (1984).
[35] P. Edwards and K. Kaewpaitoon. Asian Inst. Tech. Bangkok, 44 pp (1984).

[36] M.S. Yang and M.H. Wong. Biomed. Environ. Sci. **7**: 383 (1994).

[37] C. Polprasert. Conservation and Recycling **7**: 199 (1984).

[38] Y.H. Cheung and M.H. Wong. Agric. Wastes, **3**: 109 (1981).

[39] M.H. Wong and K.M. Chan. Microbios, **43**: 261 (1985).

[40] M.H. Wong and F.Y. Tam. Arch. Hydrobiol. **100**: 423 (1984).

[41] K.M. Hung, S.T. Chiu and M.H. Wong. Environ. Manage. (in press) (1995).

[42] M.H. Wong, K.M. Hung and S.T. Chiu. Environ. Manage. (In press) (1995).

[43] M.H. Wong and Y.H. Cheung. Agric. Wastes. **13**: 1 (1985).

[44] P. Edwards. In: Environment and Aquaculture in Developing Countries. Eds. Pullin R.S.V., et al. International Center for Living Aquatic Resources Management, Manila. (1993).

[45] M.H. Wong, Y.H. Cheung, S.F. Leung and P.S. Wong. Water Technol. (in press) (1995).

Environmental Research Forum Vols. 5-6 (1996) pp. 45-54
© *1996 Transtec Publications, Switzerland*

Nutrient Circulation with Innovative Toilets in Multi-Family Dwellings

L. Jarlöv

Dalarna Research Institute, P.O. Box 743, S-79129 Falun, Sweden

Keywords: Sustainable Systems, Infrastructure Systems, Multi-Family Dwellings, Full Scale Experiment, Nutrient Circulation, Innovative Toilet, Urine Separating Toilet, Human Urine, Action Research, Mould Promoting Substances

Abstract

My perspective is an over-all perspective from an architect-researcher who is trying to start realizing sustainable systems within our existing urban housing areas. One part of this is nutrient circulation with urine separating toilets in multi-family dwellings. Trials have for a too long time been concentrated upon one-family houses and rural areas. We have got a lot of knowledge and experiences from "eco-villages" and now it is time to implement these experiences in a larger scale. Full scale experiments for me is the best way to get further and to identify the worst lacks of knowledge and the development of technics that is necessary.

As an action researcher I work in order to mobilize the interest of housing companies, municipal employees, farmers and tenants by information about actual research and by trying to establish contacts between different actors working with different parts of the infrastructure system. In Borlänge, a town with 50 000 inhabitants, a social housing company now is planning to start an experiment with urine separating toilets in one staircase with six flats in a two-storeyed house. In Nynäshamn, a town with 22 000 inhabitants, a similar but bigger project is planned, probably in 50 flats. Together with other groups of researchers I will follow and study these two examples, of which I myself have initiated the one in Borlänge.

To get as much knowledge as possible out of these experiments, studies on different matters are designed - attitudes and reactions of the tenants, the interest of the farmers, storing, transports, methods for spreading, effects on the quality and quantity of the crops, economy etc. Also the attitudes of different professionals is an important subject.

I hope that by the time of the conference in Waedenswil in September at least one of the full scale experiments has started according to the plans. But you never know. The infrastructure systems are very rigid. Then the interest must be focused on the obstacles.

1. Introduction

This paper concerns a trial to install urine separation toilets in existing multi-family houses which is planned in cooperation between a social housing company, Stora Tunabyggen, Dalarna Research Institute and the local authority of Borlänge, a Swedish town with about 50000 inhabitants. According to the original plans the trial should already have started but owing to unexpectedly high costs and financial problems it has been delayed and now is supposed to begin in the autumn 1995.

In Sweden today there is a number of new built dwellings with different types of innovative toilets where urine and sometimes also faeces are taken charge of and utilized as fertilizer in agriculture. These houses however are only an exceedingly small part of the total building stock. In order to change the unsustainable treatment of the resources within the housing stock evidently we have to set about the existing big scale residential areas. This is the starting-point of my interest in sewage systems for multi-family dwellings.

I shall from the very beginning declare that I am an amateur as sewage researcher. I am educated as an architect and scientist within the field of architecture and physical planning. My interest in sewage matters is a part of my interest in renewal of the built environment as a whole. The starting-point is my conviction that there is much in our present infrastructure system and in our present life style and culture that is not long-term sustinable and that it is urgent to begin to change the direction if it shall be possible to solve the global environmental problems. Since we in our part of the world are supposed to have a very high standard of living and are standing out as a model for many developing countries it is in a global perspective important what we are doing even if some of the problems, as for example lack of water and food, not seem to be specially urgent for us at the moment.

In Sweden the environmental conciousness has grown enormously in recent years. In most municipalities there are Agenda 21-actions going on and a great number of municipalities have declared themselves as "eco-municipalities" [1]. With regard to the built environment, however, we have remained in the small scale and mostly concerning new buildings. We have not really dared to approach the challenge of making the huge existing tall scale building stock sustainable.

As a researching architect I have for lots of years worked with renewal of problematic residential areas. The main point in my work has been on social and psychological matters as for example the meaning of home and tenants right to decide upon renovation measures in their own flats and on development efforts to change monotonous dwelling areas into comprehensive living neighbourhoods. Since the social problems are those most urgent today they are in focus when renewal is discussed. In fact, however, the work with those problems often leads to rebuilding measures. Besides, in Sweden, every now and then there is suggested that huge public investments might be concentrated on rebuilding the housing stock, particularly for labour market reasons. Finally, former or later, the housing stock has to be rebuilt. It would be miserable if we today invest a great amount of money in rebuilding acitivities, which are permanenting the present unsustainable infrastructure. That is why it is urgent to gather as much knowledge as possible about sustainable systems suitable for multi-family dwellings.

From this point of view I addressed the director of a social housing company in Borlänge in order to get him interested to bring about a full scale trial with a sustainable sewage system in multi-family dwellings. He was very positive and suggested an old housing area, Hushagen, as experimental area since the pipes there are very old, from around 1930, and anyhow ought to be renewed. The district, which has 250 flats, was built between 1880 and 1900 and has been renovated many times. The buildings are mostly two-storeyed wooden houses with flats of different sizes.

2. Why a full scale trial with urine separating toilets?

During the discussions about sustainable sewage systems it soon appeared that there are different "schools" among researchers, technicians and farmers, which recommend different solutions for sustainable sewage systems. It also appeared that all suggested systems are on the stage of development and that there is a great need for development of knowledge within all parts of the wandering of the nutrients from the dwelling to the farmer and back. We, however, relatively soon chose to try urine separation in Hushagen for the following reasons.

- There are in Sweden today a lot of different urine separating toilets on the market [2]. More and more people are interested in the system and choose it for one-family houses. A growing number of municipalities begin to discuss the possibility to install urine separating toilets even in multi- family houses, but it is not yet realized anywhere. Since there still are great shortages of knowledge about the function of the toilets, the technics for storage, transport and spreading, the quality of the urine under different conditions, infectious matters and effects of the plants etc it is important that studies in all these fields are set about.

-The supposed advantages of urine as nutrient compared with a mix of urine and faeces depend on the assumption that the urine is sterile while the faeces contain different kinds of infectious matters. This presume, however, that a total separation really has taken place. In practice it is difficult to obtain a hundred percent separation. Experimental studies are required to understand the connection between the extent of separation and the growth of microbes. There is also a need for experimental studies of the connection between the extent of separation, the technical design and the human behaviour. Probably persons who install separation toilets on their own initiative are more actively interested to get the system functioning than tenants who live in multi-family dwellings and are more or less forced to accept an innovative system.

- Full scale trials are necessary to carry the knowledge ahead. A full scale trial is a way to faster make visible problems and obstacles, which otherways will appear later. On the other hand it can also show that feared problems and obstacles have been overrated.

3. The intended trial

According to the actual time-table urine separating toilets to begin with are planned to be installed in one staircase with six small flats, in the autumn 1995. The faeces will for the present been taken care of by the existing sewage system.

In an examination paper at Luleå Technical Highschool the prerequisites for installing urine separation in the whole residential quarter of Hushagen has been studied [3]. In this study also local treatment of the gray water and the faeces by sedimentation and infiltration is suggested. The sludge was suggested to be composted together with the organic garbage from the households and then used by the tenants for gardening.

Then a proposal for an introductory trial with urine separating toilets in two houses with totally 18 flats has been pre-projected and cost estimated. This work has been done by BPA, the biggest firm for water and sewage engineering in Sweden in cooperation with SBUF, the development fund of the Swedish building branch. SBUF has emphasized their interest in thorough rebuilding activities in order to get work opportunities for the branch.

The estimated costs were 4250000 Skr (about 594405 dollars), that is about 263000 Skr (27466 dollars) per toilet! 333000 Skr (46573 dollars) were "costs for the commissioner of the building project" (that is costs for preparing work, financial and administration costs and costs for building permission etc), 445000 Skr (60839 dollars) were "costs for the working place" (that is sheds for the workers, telephone, toilets etc), 350000 Skr (48951 dollars) were costs for excavation and refilling, 670000 Skr (93706 dollars) were costs for inside building jobs, 350000 Skr (48951 dollars) were costs for inside installations and 260000 Skr (36363 dollars) were costs for the evaluation.

To try to understand the reasons for these high costs is a research task in itself, which I have not yet been able to spend any time on; I just have to notice that it was impossible to finance the project in that shape. Of course I cannot help wondering about the opposite interests of the different actors - on the one hand the building branch, desiring as much work as possible and consequently also as much money as possible, and on the other hand the environmental movement and some researchers, trying to show that it is economically advantageous to change the sewage system so that the nutrients can be utilized. I will go on examining this subject.

In order not to stop the process the project has been changed and the ambitions have been reduced to only one staircase with six flats. This makes it impossible to follow the previous plan to compare two different types of toilets. Now the type that can be installed with the smallest operations in the bathroom is chosen.The actual house has a cellar, where the storage tank for urine can be placed, which will reduce the costs for excavation.

The toilet has two flushing systems, one for urine and one for faeces. The urine is flushed with 0,1-0,2 litre of water and the faeces with 5 litres. The pipes for urine will be placed inside the existing wastewater pipes.

The costs now is estimated to 710000 Skr (99300 dollars), that is 88750Skr (12412 dollars) per toilet - also a rather expensive toilet shift! We are anyhow eager to fullfill the trial since it has been prepared for a long time and there are farmers in the neighbourhood and tenants who are very positive and interested.

4. Can urine separating toilets be a good solution for multi-family houses?

Since my trial concerns urine separating I will only discuss this type of innovative toilet systems in this paper. In fact urine separating is only one part of the water and sewage system. Stormwater, gray water, faeces fraction and eventually organic garbage from the households are other components which have to be included in the comprehensive solution.

How should a suitable solution for Hushagen be constructed? In my perspective the criterion for a good sewage solution for Hushagen and for all other residential quarters is a solution which is long-term sustainable and is contributing to a sustainable development both locally and globally in accordance with the Agenda 21-agreement which Sweden has signed. A sustainable solution for sewage system ought to have the following components:

- It is convenient and acceptable to the users in every part of the system.
- It is safe from a sanitary point of view.
- The nutrients are utilized in the best way.
- The use of water is minimal.
- It is energy- and resource-effective even in an overall perspective.

We have not yet sufficient knowledge about any of these components.

Convenient and acceptable

We do not yet know if urine separating toilets will be considered convenient and acceptable by tenants in multi-family dwellings, but we intend to examine the matter in Hushagen and Nynäshamn. We also do not know how the system is considered by caretakers, foodbreeders, consumers - we first must examine it. These studies probably will lead to reconsideration of present technology and to development of new technics followed by further studies from the users point of view.

Sanitary matters

Studies of growth of microbes in urine solutions with different concentrations of faeces are going on and more studies are needed concerning the security for consumers and for workers who are treating the human sewage in the different parts of the cirkulation chain [4].

Utilization of the nutrients

Concerning the best utilization of the nutrients in the sewage there are different "schools". It is important to discern what is possible to examine with existing scientific methods and what has to be classified among ethical standpoints. The chemical qualities of the urine under different conditions are studied with traditional scientific methods. These studies have for example shown the importance of keeping the urine in total air density in all parts of the treatment. Examinations of the effects of urine on different plants and in different stages of the growing process also are going on [4]. Much more has to be done.

Concerning standpoints which are difficult to study with traditional scientific methods there are for example among biodynamic farmers the opinion that the human urine contents vital substances which means that one-sided use of human urine in food production in the long run risks to have a degenerating effect on man. Advocates for other "schools" tend to put off such thoughts as superstition. I think, however, that a big scale change to utilizing human urine as nutrient is a profound cultural transition where all aspects have to be considered thoroughly. The argument that we earlier in history always have used our excrements in agriculture is not valid today when agriculture has a quite different character with huge unites and mono-cultures. Perhaps we ought to find a suitable mix between humanurine (-faeces), nitrogen-fixing plants and manure from animals to be able to use the human excrements in the best way. Perhaps composting all the excrements together with other organic material is a possibility. There also are arguments for caution in going *from* artificial fertilizers, which documentedly are impoverishing the soil, *to* only urine which is as one-sided a nutrient and also is lacking mould promoting substances.

Minimal use of water

In fact this component ought to be put under the next title, effective resource use, but I choose to point it out as a specially important resource matter. In regions with lack of water as for example Nynäshamn, saving water is a self-evident aim with urine separating. In Borlänge, on the contrary, there is no noticeable lack of water today. Concerning the urge for saving water there are also different "schools". Many people consider lack of water as a marginal problem for Swedish conditions. In a long-term perspective, however, it is apparent that the bigger the flow of water is the more energy and chemicals are needed for transports and cleaning. I have earlier mentioned another argument for water saving technics - our technology as a model for developing countries. Our possibilities to export are also worth mentioning, if we in our technologically advanced countries develop and ourselves try and refine water-saving sewage technics, since lack of water surely will be a growing problem in great parts of the world in the future. This ought to be an urgent matter to put ahead before the approaching UN Habitat-conference, which very much will treat sanitary problems in the big cities around the world.

Energy and resource effectivity

This component concerns both short-term and long-term judgements and resources, measured in monetary terms as well as in raw material and energy terms. It is quite clear that one reach different judgements when looking in a short-term monetary-economic perspective compared to a long-term resource-economic perspective [5]. It is necessary to put forward and analyse these two perspectives in order to be able to act according to the long-term perspective at the same time as we are practically tied to the local short-term perspective. From the results of such analyses it will be able to see which measures and investments are profitable or reasonable on the local level and which have to be submitted to the national environmental politics.

The present big gap between the short-term monetary-economic and the long-term resource-economic profit makes it very difficult to start full-scale studies concerning alternative technics. Since the existing big systems are both technically and financially constructed according to an unsustainable way of thinking which for example presumes access to cheap fossil energy it is impossible in the introductory phase of a system shift to make alternative systems profitable from the point of view of business economics. Such trials must rely on financial support. In our project we hope to get such support from SBUF.

It is not possible today to answer the question whether urine separating might be an economically profitable solution for Hushagen. Since it is not yet time for ordinary renovation of the houses all operations in the bathrooms caused by the installation of the urine separation system will be concidered as extra costs. As long as the trial only concerns a few flats or houses the general costs referring to for example the commissioner of the building project or to the working place will be disproportionately high. As long as the technics for transport, storage and spreading are not fully developed it is impossible for the houseowner to get paid for the urine even if the farmers in the neighbourhood are willing to recieve it. Studies from Nynäshamn show that farmers want the urine but are not willing to pay for it. The real cost diminishing through water saving is not enough to compensate for the extra costs for changing the toilet system in Hushagen. Urine separation is not a profitable affair under these circumstances. But it might surely be a good business when it is time to make the ordinary renovation of the flats, provided that the knowledge and the technics are developed at that time so that the nutrients in the urine can fetch a value on the market.

Then it might as well be possible that some other system has turned out to be superior. It might for example be vacuum toilets where the mixed excrements together with minced organic garbage from the kitchens are being retted in a bio-gas reactor, where both the gas and the rests from the retting will be utilized. This, however, provides that the necessary backbringing of mould promoting substances to the farmland will be solved in another way than through faeces and organic garbage from the kitchens, for example through plants and manure from animals.

Whatever the solution for a sustainable sewage system might be - it will probably be a lot of different solutions suitable for different places with different conditions - it surely will be cheaper than the present system. A presumption is, however, that the knowledge develop and it will only do so if we work keenly on it.

5. Cooperation with other project and researcher

During the time that the preparation for the trial in Hushagen has been going on plans for a couple of similar projects have started all over Sweden. By cooperation with other research groups engaged in these projects it might be possible to force the development of knowledge. Our project has above all a thorough cooperation with the Institution for System Ecology at the University of Stockholm [6]. The research group there is working together with a social housing company in Nynäshamn, which intends to install urine separation toilets in about 50 flats in multi-store houses in a residential quarter named Heimdal in the autumn 1995. Also here very high costs have delayed the project but not stopped it.

Through a certain program for research and development, "Organic Waste as Resource for Plant Nutrients" I have been invited to be part of a net-work of researchers working with different aspects of the problems of urine separation [4]. Owing to the delay of the start of the trial in Hushagen, we missed the possibility to be practical example for experimental studies made by other research groups. Now they had to choose a residential area with one-family houses, as there not yet exist any trial with urine separating toilets in multi-family houses. So now we have to design our own evaluation and make studies of certain interesting parameters together with the research group in Nynäshamn. We are specially going to study how the separation is functioning and how the tenants act and feel in relation to the innovative technics.

Within a newly constituted research group, " Culture and Recycling", a study of the attitudes and behaviour of the caretakers of the houses and of the nurses in the daycares in relation to sustainable technics, and how the recycling systems are affecting their daily work, will be made in Borlänge.

6. The continuation of the project

My method, action research, implies both studies of the presumptions for the trial, efforts to get it started and initiating and in certain cases also realizing evaluation and following up studies. It is quite natural that the project as a whole is depending on a lot of actors and of different unforseenable factors. I have chosen to put the main point in my work so far partly in the efforts getting the trial started and partly in the establishing of contacts with researchers within all the relevant knowledge fields, especially within water- and sewage-engineering and agriculture but also medicine, system ecology, economy and social and behaviour sciences. The result of the work up to now is summarized in an application paper to SBUF 1995-01-13.

The intention is to take up the discussions again with the tenants association and to talk thoroughly with the tenants in the actual flats. These talks will function both as information about the new toilet system and as interviews to catch their reactions, fears and expectations. The talks will be repeated within two months. At the same time analysis of the separated fraction shall be made and later on also studies of the storage, the transports and the spreading on farmland. This will be done in contact with the mentioned net-work. Parallell studies will be made in Nynäshamn.

As a part of the project in Hushagen we are discussing extended garden activities with interested tenants. By means of stimulating and facilitating such "recycling activities", which they themselves have individual or common interest in, we hope that their knowledge and positive cooperation in the innovative toilet system will grow.

7. The main conclusions up to now

If urine separation sewage system shall be an economically possible solution for existing multi-family dwellings it seems as if at least to presumptions must be carried out:

1. The installation of urine separation toilets, pipes, tanks etc must be made in connection with ordinary renovation when the bathrooms shall be renovated anyhow.

2. There must be users of the urine who are willing to pay for it.
If the urine shall become so interesting to the farmers that they want to pay for it knowledge, experiences and technics concerning all parts of the treatment chain and acceptans by the users are required. There is also a need for understanding of cultural and ethical subjects which are vital for people´s acceptance both of using innovative toilet and of consumption of food fertilized with human excrements.

Finally I will emphasize the importance of discussing the recirculation of nutrients to the farmland in form of human urine within a totality where recirculation of mould promoting substances as well is an important goal.

References

[1] A. Helmrot, G. Jonsson, Ö. Eriksson, Ringar på vatten - Va-verken och Agenda 21, VAV, VA-forsk, Report 1995-01 (1995)

[2] P. Svensson, Nordiska erfarenheter av källsorterade avloppssystem. Luleå Technical Highschool, Dept of Rest-product Technics, 1993:117 E (1993)

[3] U. Larsson, H. Rudholm, Lokal avloppshantering med urinsortering. Luleå Technical Highschool (1994) Dept of Rest-product Technics, 1994:122 E (1994)

[4] H. Jönsson, Source separated human urine. Application and research program 1995-05-01, Dept of Agr. Technics, University of Agriculture, Uppsala (1995)

[5] P. Söderbaum,. Ekologisk ekonomi. Stockholm (1993)

[6] A-M. Jansson, Kretsloppsanpassning i stadsbygd. Application and research program 1994-11-30, Inst for System Ecology, University of Stockholm (1994)

Environmental Research Forum Vols. 5-6 (1996) pp. 55-64
© *1996 Transtec Publications, Switzerland*

Graywater Reuse in Households
– Experience from Germany –

E. Nolde

Technische Universität Berlin, Fachgebiet Hygiene,
Amrumer Strasse 32, D-13353 Berlin, Germany

Keywords: Graywater, Service Water, Quality Standards, Hygiene, Decentralized Wastewater Treatment Plants, Soil Filter, Rotating Biological Contactor, Water Saving

Abstract

In Germany, waterborne diseases are no more of a major concern due to the indisputable supply of high quality drinking water. The installation of a second decentralized water supply net, with non-drinking water quality, hit against strong protest and opposition from the drinking water suppliers and the public health departments. However, acceptance is growing after it has been shown that graywater, following a biological treatment and an eventual UV-disinfection, fulfilled the quality requirements set in cooperation with the Public Health Department. Ecologically-motivated citizens, economically-dealing hotel managers and housing societies, as well as progressive water suppliers see to it that at the present time more graywater plants go into operation in Germany.

Introduction

Although Germany is generally considered a water-rich land, there exist regions with more than 1200 mm yearly drainage, while other regions exhibit less than 100 mm a year [1]. Therefore, ecological water concepts should always be planned to fit locally-governing conditions for both drinking water supply and wastewater disposal.

In the years between 1950 and 1980, the drinking water consumption in Germany rose from 85 to 139 liters per person and day as a result of the increasing hygienic needs and comfort, as well as an increasingly growing number of one- and two-person households. As a result, water supply problems, quantitative as well as qualitative, developed regionally, above all in densely populated urban areas, but also in rural areas. For example, cities like Frankfurt and Hamburg, which covered their rising drinking water consumption through increased groundwater withdrawal, created diverse ecological damage as a result of the sinking of the groundwater level.

Since 1980, interest in the implementation of drinking water saving measures and in the multiple use of water rose in Germany, in addition to the incorporation of stormwater in the domestic water supply.

The average drinking water consumption in Germany is calculated as 144 l per person and day [2]. Investigations [3] have shown that, in households where tenants have been informed on economic dealing with water, water bills for individual apartments are practiced and water-saving fittings have been installed (6-l flush toilets, flow regulators, two-grip fittings, water counters), the daily drinking water consumption could be reduced to 100 l per person.

A high level of environmental awareness and/or the expectation in the long term of an inevitable strong rise in the water costs (Berlin 1988: 2,72 DM/m³; Berlin 1995: 7,00 DM/m³; Darmstadt 1995: 13,64 DM/m³), may account for the fact that many people are showing more interest for service water use in households and in industry. Alone through the use of service water for toilet flushing, a reduction in the water consumption of about 30% can be achieved (30-50 l per person and day).

Background

In Germany, 98% of the population is connected to the drinking water supply and 90% to the sewerage system. Of these, 78% are connected to a biological and an advanced wastewater treatment plant [4]. Generally, connection to and use of a municipal wastewater treatment plant is obligatory as is the connection to the drinking water net, although self-supply through own extraction wells is possible. For this reason, it is difficult to realize decentralized water concepts, especially in urban areas.

After existing central service water nets have been abandoned for economical reasons, decentralized service water use from storm- and graywater systems have been both propagated and opposed since the end of the 70's. The positions taken in this respect varied from full support and approval, for example, the use of untreated graywater for toilet flushing, to strict refusal based on the demand for a drinking water quality for basically all uses in households as well as in public institutions.

While drinking water must fulfil the strictest quality standards of the drinking water decree, be it a municipal drinking water supply or a privately-owned plant, different effluent requirements are set for municipal wastewater treatment plants dependent on plant size, whereby, the disinfection of treated municipal wastewater is not dictated and practiced only in rare cases.

For the use of service water in buildings, there exist no officially established technical as well as hygienic/microbiological quality requirements.

Within the scope of a research project financed by the Berlin Senate Department for Building and Housing, general quality requirements for service water have been set that were later supplemented with measurable parameters as listed in Tables 1 and 2.

Table 1. General quality requirements for service water in buildings [5].

Hygienic Safety	Water which is to be used for toilet flushing, need not fulfil the strictest criteria of the drinking water decree. The use of service water in households should not be a source of hazard. By keeping to quality requirements as defined in the EU-Guidelines for recreational waters [6], a potential hazard is almost excluded.
No Comfort Loss	Smell irritations, depositions on sanitary facilities as well as clogging of sanitary fittings and equipments should be excluded.
Environmental Tolerance	The use of chemicals (esp. chlorine-containing disinfectants) and/or a higher energy expenditure are principally rejected.
Acceptable Operation Costs	The running operation costs for energy, maintenance, control and supervision should be levelled out by savings made in the costs for drinking water supply and wastewater disposal.

The criteria presented in Table 2 are more strictly formulated than those that apply for the large modern municipal wastewater treatment plants. Opponents of graywater reuse were convinced that these values can not be held to in practice and hoped that the chapter "service water" will be finally closed in Germany. Hügin et al. [7] have reported that the effluent from a pilot plant for graywater treatment in Wolfsburg was not good enough to be reused even for toilet flushing.

Table 2. Qualitative and quantitative parameters for service water in buildings [8].

Quality Aim	Assessment Criteria
Hygienically/ microbiologically indisputable	Total coliform bacteria: 0 / 0.01 ml (< 100/ml) Faecal coliform bacteria: 0 / 0.1 ml (< 10/ml) *P. aeruginosa*: 0 / 1.0 ml (< 1/ml)
Low BOD[1]	BOD_7 < 5 mg/l, to guarantee that the graywater is extensively cleaned.
Colorless and clear	UV-Transm. $_{254\ nm\ in\ 1\ cm\ cuvette}$: min. 60 %.
Oxygen-rich	> 50 % Saturation, for possibility of longer storage.
Almost free from suspended material; almost odorless; non-putrescible within 5 days	For proper functioning of sanitary fittings with no comfort loss for the user.

[1]   For technical reasons, other parameters such as TOC, if necessary COD, can be also considered
    instead as a suitable parameter.

Drinking Water Use and Graywater Origin

Table 3 shows the different uses for drinking water sorted according to quality requirement. It becomes clear that only 2% is used as drinking water, the rest 98% serves the purpose of cleanliness and hygiene, whereby, 45 l (33%) alone of highest quality drinking water are poured into the toilet, although for this purpose use of lowest water quality is just sufficient.

Table 3. Quantities and different uses of drinking water [2].

Liter Per Person and Day	Purpose of Use	Quality Requirement *
3	Drinking and cooking	Highest quality
10	Dishwashing	High quality
57	Bathing, showering	High quality
19	Laundry	High quality
7	Cleaning	Low quality
3	Garden	Low quality
45	Toilet flushing	Lowest quality

* Author's own opinion

In this context, **Graywater** is defined as household wastewater with the exclusion of toilet (faecal) wastewater. The respective amounts and inflow concentrations are strongly dependent on the individual behaviour of the user, on water consumption as well as on the dosed cleansing agents containing surfactants. Table 4 shows the different origins of graywater in households.

Table 4. Origin of graywater in households [8].

Graywater Origin	Remarks
Shower and bath	least polluted graywater
Hand-washing basins	poses usually no problems
Washing machines	increases treatment expenditure and is usually not necessary to cover the need - increased pollution load is of little problem - it is recommendable if installation costs for graywater collection system could be reduced
Kitchen wastewater	should rather not be used for drinking water substitution due to: - high (BOD) pollution load - increased maintenance expenditure

Graywater Quality

Graywater composition is subjected to wide fluctuations which have to be taken into consideration during the sampling procedure. Single random samples, taken on same week days and same time, deliver little meaningful results. The concentration of a single graywater source could vary widely. For example, COD concentrations measured in graywater from showers varied between 30 and over 1000 mg/l. In practice, systems which also treat kitchen wastewater require more maintenance. Apart from the high BOD load contributed by food, it is the combination of hair and fat which form depositions that are difficult to remove.

It has been observed that load fluctuations in larger graywater plants are, as expected, less marked than in smaller ones. In Table 5, values of different parameters for different graywater qualities in Berlin are presented. The hygienic/microbiological results show that, in bath tubs, showers and washing machines high faecal bacterial concentrations are also measured and that, a risk-free service water use is only possible after an extensive graywater treatment followed by disinfection.

Table 5. Different graywater qualities measured in Berlin plants [9, 10].

Parameter	Unit	Highly-polluted	Medium-polluted	Low-polluted
COD	mg/l	500 - 600	250 - 430	100 - 200
BOD_7	mg/l	280 - 360	150 - 250	50 - 100
N_{total}	mg/l	8 - 18	---	5 - 10
P_{total}	mg/l	2.5 - 4.5	---	0.2 - 0.6
Faecal coliforms	ml^{-1}	$10^2 - 10^6$	$10^4 - 10^6$	$10^{-1} - 10^1$
Total coliforms	ml^{-1}	$10^3 - 10^7$	$10^4 - 10^6$	$10^2 - 10^3$
Total counts (CFU)	ml^{-1}	$10^6 - 10^7$	$10^6 - 10^7$	$10^5 - 10^6$

Highly-polluted: household graywater excluding faecal wastewater, in households with water-saving measures; BOD measured as BOD_5

Medium-polluted: household graywater excluding faecal and kitchen wastewater, largely from washing machines contaminated with faecal bacteria from baby diapers

Low-polluted: household graywater excluding faecal and kitchen wastewater; mainly from bath and shower and a small part from washing machines

Up to this date, a precise plant dimensioning is still difficult due to the little available data and experimental values. During plant planning and dimensioning, a distinction must always be made between the origin of the respective graywater and the intended purpose of use of the service water.

Investigated Graywater Plants

The use of graywater in Germany has been primarily considered from the aspect of water saving. Based on that, many graywater plants were initially built as a storage tank in which slightly polluted wastewater, namely from bath tubs and showers, was passed over a surface filter and stored almost untreated. Such storage tanks have been built since 1987 mostly in one-family houses and privately-owned apartments. However, graywater plants have been also installed in larger places such as in multiple-family houses or in a hotel, either in the basement, in the bathroom under the ceiling or, recently, as the so-called "ecological tub" under the bath tub. As long as shower and bath wastewater is continuously aerated and stored temporarily, no problems would emerge in the first few weeks of operation. However, if the storage tank is not continuously cleaned, odors become gradually noticeable, aeration facilities become clogged and later the flush toilets will stop functioning properly.

No single graywater plant is known to the author undergoing no extensive wastewater treatment, and thus functioning properly for a longer period of time and with little maintenance. In order to reduce the maintenance expenditure as well as odor disturbances, disinfectants have been also used in some cases to impart to the water a breath of freshness. However, saving on drinking water should not be achieved through the addition of chemicals which exert a stress on the environment.

A graywater plant which, besides saving on drinking water, should also bring with it a relief on the wastewater side, was completed in 1987 as a 644 m² large plant-covered treatment plant in the so-called Block 6 in Berlin city centre. It deals with a horizontal-flow soil filter consisting largely of coarse granular soil material. The soil filter was planned to initially treat the household graywater (about 16 m³/d) from 200 persons, and later the total domestic wastewater, including faecal wastewater, which, however, could not be achieved.

With an average BOD_5 graywater inflow concentration of 233 mg/l, measured as sedimented sample, it was possible under a normal operation to clean daily with the above soil filter about 3000 l of graywater to a rest BOD_5 of < 5 mg/l. Through a series of experiments in which treated graywater, that once passed through the soil filter, was repeatedly aerated artificially, the amount of cleaned graywater could be doubled. Still, the actual daily service water requirement for toilet flushing and garden watering of between 7,500 to 10,000 l could be only partly covered. Teiner [9] showed that the aerobic degradation of the above graywater, investigated in laboratory experiments, was not less efficient than that of the normal municipal wastewater. The hygienic/microbiological quality requirements were attained due to the long detention time of the graywater as well as the good flow characteristics of the soil filter [11].

That the success of graywater treatment is related to the structure of the soil filter and the type of flow, shows the in 1987 constructed soil filter in Hamburg-Allermöhe, by the same planer of the Berlin plant, where at the beginning a similarly bad operation was recorded. Here, the total graywater, including kitchen wastewater, was treated in a horizontal filter, while faecal wastes were composted. Following the rebuilding in July 92 of a vertical soil filter with intermittent flow, better effluent concentrations were achieved in the 260 m² expanded plant for 130 residents [12]. The claim that graywater is more difficult to clean than domestic faecal wastewater is not correct, as demonstrated by the investigations in Hamburg-Allermöhe and other plants in Berlin.

Since 1989, further graywater pilot plants were constructed in Berlin which fulfil since 3 years the above development aims following a long period of optimization [10]. Diverse investigation results are available from a rotating biological contactor (RBC) found in a 15 m² basement treating the graywater from 70 people (Manteuffelstr. 40/41), and a 20 m² plant-covered vertical soil filter for 15 to max. 30 persons (Oranienstr. 5) in Berlin-Kreuzberg.

Figure 1 shows the principle of graywater treatment in the above two systems. Both variations, the multiple-stage biological treatment by means of the RBC and the vertical soil filter, extensively clean the graywater from showers, bath tubs and washing machines to the full satisfaction of the residents.

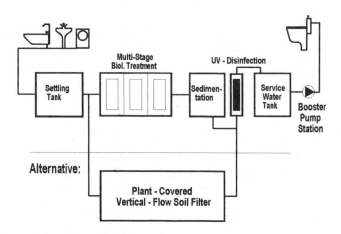

Figure 1. Recommended concept for graywater treatment for purpose of reuse as service water for toilet flushing.

A comparison between the RBC operation and the plant-covered vertical soil filter in Berlin-Kreuzberg is shown in Table 6.

Table 6. Operation parameters and requirements for the RBC and the soil filter in Berlin-Kreuzberg.

Critierion	Multi-Stage Rotating Biological Contactor	Plant-Covered Soil Filter
Connected persons	ca. 70	15 (max. 20 - 30)
Place	in basement	in garden; partly in basement
Place requirement per person	ca. 0.2 m²	ca. 1 m² (outdoors)
Cleaning capacity	3 m³/d	1 m³/d
BOD$_7$ of service water	< 5 mg/l	< 5 mg/l
Hygiene criteria	fulfilled	fulfilled
Odor of service water	almost odorless	odorless
Suspended material	very little	not detectable
Energy consumption (total)	ca. 3 kWh/m³	ca. 1.5 kWh/m³
Maintenance	< 1 h/week	< 1.5 h/month

The following four figures show results of investigations on service water from the RBC in Manteuffelstr. 40/41 which always demonstrated BOD$_7$ concentrations of < 5 mg/l (Fig. 2-5).

Figure 2 shows that the quantities of graywater inflow and service water need demonstrate similar curves. Measured as UV-transmission, service water is much better than treated municipal wastewater and shows only a little difference from drinking water (Fig. 3). Figure 4 shows that the concentrations of total and faecal coliforms in service water are much lower than the control values set in agreement with the Public Health Department.

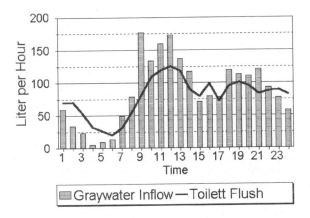

Figure 2. Daily course of graywater accumulation, toilet water consumption and the available service water (calculated from 4-week mean values).

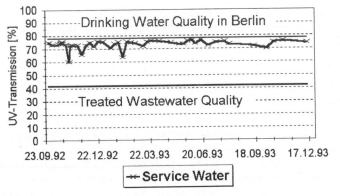

Figure 3. UV-Transmission of service water from the graywater treatment plant (for comparison: Berlin drinking water and treated municipal wastewater from Berlin -Ruhleben).

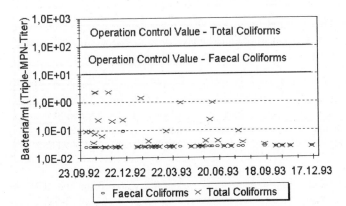

Figure 4. Concentrations of total and faecal coliforms in service water tank of the graywater treatment plant .

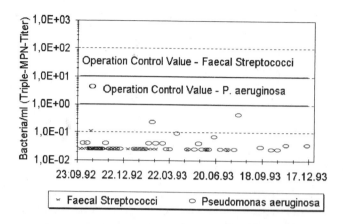

Figure 5. Concentrations of faecal streptococci and *P. aeruginosa* in service water of the graywater treatment plant.

Similarly, Figure 5 shows that the concentrations of faecal streptococci and *P. aeruginosa* in service water are much lower than the control values set in agreement with the Public Health Department for the Berlin pilot projects. The above hygienic/microbiological investigations clearly demonstrate that these strict requirements could be actually held in practice.

As a conclusion for the Berlin plants, in Tables 1 and 2 listed aims can be only achieved in connection with an extensive biological treatment and a final disinfection of the treated graywater.

If it is intended in households to substitute only water for toilet flushing with service water, it is usually sufficient to collect and treat only shower and bath wastewater. If a higher need exists, one can additionally treat the wastewater from hand-washing basins and washing machines.

Other Graywater Plants

Beginning of 95, a graywater experimental park for 48 apartments went into operation in Hannover. This project was in cooperation with the local water supply and wastewater disposal authorities and the German BauBeCon [13], under the technical and scientific supervision of the Department of Hygiene of the Technical University of Berlin. Initial investigations on four different plant types confirmed the research results of the Berlin plants. Systems with a plant-covered soil filter or a two-stage RBC achieved good results, whereas, systems which simply employed a so-called pre-treatment, caused problems and had to be partly shut down.

Still in this year a series of plants are expected to go into operation. Among others, a graywater treatment plant for 24 accommodation units in a multiple-family house in Kassel, as well as for a hotel with 60 beds in Schmölln and a luxury hotel with 400 beds in Offenbach.

Projects are known of, which instead of a biological treatment, plan to use a chemical/physical process, namely microfiltration and flotation. The advantage such systems may have is in being compact. However, as a disadvantage, they presumably have a higher energy expenditure and a larger volume of formed sludge. In this case, the graywater constituents would not be mineralized but left over for the municipal wastewater treatment, and this has very little to do with recycling.

Future Prospects

Lately, inquiries from the private sector, who sees a higher need for a graywater than a stormwater system, piled up, as well as from the industry who spots, still with hesitance, a future market for graywater.

On behalf of the Berlin Senate Department for Building and Housing, a leaflet for service water use in buildings has been worked out by the Department of Hygiene at the TU of Berlin [8], in which the local specialized authorities and representatives from other federal states also participated. The aim of this leaflet, which is to be published this year, is to provide help and orientation to interested clients, planers and operators who plan to realize service water use.

In this leaflet, regulations as well as the necessary requirements are described. A permanent carrying out of the quality requirements listed in the leaflet and which are based on several years' experience in Berlin, offers guarantee that the user need not fear either a hygienic risk or a loss of comfort.

In the open discussion on drinking water saving measures and service water use, the representatives of the water supply and wastewater disposal concerns continuously claim that water prices involve mainly fixed costs. They argue that if drinking water consumption sinks, the cubic meter price will inevitably rise for all consumers. However, it has to be said that for all consumers who realize drinking water saving measures and service water use, both drinking- and wastewater annual expenditure will sink inspite of the rise in the prices. Higher annual water costs would have to be paid only by those who did not implement these water-saving measures.

If a region rocks very close to its natural limits regarding its water supply and wastewater disposal plans, further plants can only be developed with very high capital expenditure (limit costs), or what the experiences in Hamburg and Frankfurt showed, taking into account the ecological impact.

The principles of "least cost planning" should be also applied to the water sector after showing positive experience within the energy sector. Graywater systems are appropriate in relieving the strain on the water cycle, whereby, the real application finds place first there, where a new construction or a total renovation of buildings is planned, or where an owner of a building intends to invest in a graywater treatment plant at an apartment level.

Another aspect, which finds until today little attention, is that graywater is nutrient-poor compared to municipal wastewater and as seen from the investigation results, it can be decentrally extensively treated with very little expenditure. Such a treated graywater may be possibly used for room cleaning, garden watering or for laundry, whereby, service water which is not needed in household can be decentrally collected together with stormwater for purpose of infiltration.

If it were practically possible to separate the nutrients (N, P) which are mainly found in urine or faeces and to compost them with other compostable wastes, likewise decentrally, to be recycled in the agricultural sector, it may also be possible in the future to do without the expensive canalization of the conventional wastewater treatment systems.

Many meaningful concepts exist for faecal waste disposal nowadays [14]. Some systems have apparently even proved their effectiveness in practice on a small scale. However, in order to realize the in Table 1 listed quality requirements for these systems, also in mutiple-storey buildings, a lot of work and development is still needed.

With this, the way from WASTEWATER TREATMENT can go through WASTEWATER RECYCLING (reuse of domestic graywater) to an integrated WASTE RECYCLING.

Only through an unprejudiced, cooperative and open collaboration of the different specialized disciplines and the people concerned, as well as the responsible departments and administrations, is there a chance to do the conference motto "RECYCLING THE RESOURCE" justice.

Acknowledgement
The Land Berlin represented by the Berlin Senate Department for Building and Housing, the Federal Ministry of Building and the "Berlin Forschung" of the Free University of Berlin for the financial support of the graywater projects.

References

[1] **Anonymous:** Regen-Grenze: reicher Westen - armer Osten; Faltkarte in GEO Nr. 6 (1993)

[2] **Anonymus:** Am meisten für Sauberkeit und Hygiene, Korrespondenz Abwasser, 42. Jahrgang, Nr. 5, S. 682, (1995).

[3] **Sommer, H.:** Trinkwasser-Spartechnik: Modellvorhaben zu Möglichkeiten und Strategien der wassersparenden Installation im komplexen Wohnungsbau; Herausgeber: Senatsverwaltung für Bau- und Wohnungsbau, Ref. IV Ö, Württembergische Straße 6-10, 10707 Berlin, (1994).

[4] **Umweltbundesamt:** Daten zur Umwelt 1992/93 Erich Schmidt-Verlag Berlin; S.319 ff (1994).

[5] **Nolde, E.; Dott, W.:** Grauwasserprojekte im Block 103 in Berlin-Kreuzberg; Kurzfassung Endbericht; Auftrag der Senatsverwaltung für Bau- und Wohnungswesen Berlin (1992).

[6] **Anonymous:** Richtlinie des Rates der Europäischen Gemeinschaft vom 8.12.1975 über die Qualität der Badegewässer (76/160/EWG-Amtsblatt der EG vom 5.2.1976-L31/1).

[7] **Hügin, D.; Wolffson; C., Kayser, R.:** Probleme und Grenzen der Nutzung von Wasch- und Badewasser zur WC-Spülung. Forschungsbericht 10202611, durchgeführt im Auftrag des Umweltbundesamtes (1991).

[8] **Nolde, E.; Dott, W.:** Merkblatt: Betriebswassernutzung in Gebäuden; Auftrag der Senatsverwaltung für Bau- und Wohnungswesen Abt. IV Ö, Berlin, Veröffentlichung in Vorbereitung, (1995).

[9] **Teiner, A.; Hegemann, W.:** Teilprojekt C2 - Abwasserreinigung und Regenwasserkonzept, unveröffentlicht in: Integriertes Wasserkonzept in Berlin-Kreuzberg, Projektdokumentation und Ergebnisse der Forschungsphase II (1990 -1993); Auftrag des Senators für Bau- und Wohnungswesen Berlin (1993).

[10] **Nolde, E.:** Betriebswassernutzung im Haushalt durch Aufbereitung von Grauwasser, WWT 1/95, S. 17 - 25, (1995).

[11] **Nolde, E.; Christen, S. und Dott, W.:** Teilprojekt D - Hygiene, unveröffentlicht in: Integriertes Wasserkonzept in Berlin-Kreuzberg, Projektdokumentation und Ergebnisse der Forschungsphase II (1990 -1993); Auftrag des Senators für Bau- und Wohnungswesen Berlin (1993).

[12] **Bahlo, K.; Wach, G., F.:** Ökologisches Bauen in Hamburg-Allermöhe, Hrsg.: Bundesminister für Raumordnung, Bauwesen und Städtebau Bonn, (1994).

[13] **Anonymus:** Grauwasseranlagen - Versuchspark-Hägewiesen, DI Deutsche BauBeCon AG, Schützenallee 3; 30519 Hannover (1995).

[14] **Lorenz-Ladener, C.:** Kompost-Toiletten: Wege zur sinnvollen Fäkalien-Entsorgung, Staufen bei Freiburg, Ökobuch (1991).

Environmental Research Forum Vols. 5-6 (1996) pp. 65-72
© *1996 Transtec Publications, Switzerland*

Wastewater Treatment in Estonia: Problems and Opportunities

J. Tenson

Senior Scientist, Pärnu Marine Biological Station, 2 Lootsi St., EE-3600 Pärnu, Estonia

Keywords: Constructed Wetland, Recycling, Tropical Plants

ABSTRACT

The problems Estonia currently faces in the field of wastewater treatment are basically the same as those of Eastern Europe or Russia: out-dated ideology, low standards of water treatment equipment and lack of funding as to reconstruction of the existing water treatment facilities. However, some success in this field can be noticed in Estonia. Making use of foreign know-how and financial support, a number of treatment facilities have been modernised (*e.g.* Pärnu, Paide, Kilingi-Nõmme). Still, the completion of the biggest plants in Tallinn and Tartu is far too slow yet. Due to the agriculture in Estonia currently being in deep depression, the waste disposal of agricultural origin has tremendously decreased. However, as a number of new small farms are established in the country, rural areas of Estonia need a lot of small waste water treatment plants, those working on the natural methods would be especially suitable for those purposes. The Stensund Conference (1991) resulted in distribution of new methods of water treatment both by introducing the ideas thereof in printed matters and on the conferences and workshops later on. There is a Centre for Ecological Engineering in Tartu which being based on the friendly support mainly from Switzerland and Germany, has proved itself to be a successful one. The co-operation between Stensund Ecological Centre and Pärnu Marine Biological Centre has contributed a lot to the successful experiments to apply several tropical plants (*e.g. Eichhornia, Azolla, Pistia*) in the water treatment plants during the summer season. In Pärnu, we have also launched a project on waste-free management of the beach complex successfully, including the beach, the port, the waste water treatment plant and the constructed wetland.

If we compare the current situation in waste water treatment in Estonia with that of 1991, during the last Ecological Engineering Conference in Stensund, it is significant that the situation has improved a lot. A number of laws concerning environmental issues have been adopted by the government. The latest among them, *the Law of Sustainable Development* was adopted on February 22, 1995. International co-operation in this field has also essentially improved, the most important landmarks being as follows:

- Participation in the CCB (Coalition Clean Baltic) for working out and accomplishing the Baltic Sea activities under the Baltic Sea Action Plan. The programme is extensive and includes various aspects of environmental activities, accomplished via Helsinki Committee, on state or organisational levels.

- The Blue Flag Campaign has also been launched successfully, assisting enormously in creating favourable conditions on our beaches.

- As a result of co-operation of Estonian, Swiss and German scientists and politicians, Centre for Ecological Engineering Tartu was founded on November 1, 1992. The main lines of activities of the Centre are as follows:

  (1)   Exchange of information in different ways between Estonia and other   countries;

  (2)   Training and publishing;

  (3)   Conferences;

  (4)   Launching research and application projects in the field of ecotechnology;

  (5)   Implementing ecological engineering solutions (building up constructed wetlands).

The seminar on *Constructed wetlands: implementation and perspectives in Estonia* arranged in July 1995 with experts from Norway, Sweden, Germany, Switzerland and Estonia participating can be considered as a good example of the Centre's successful work.

- On September 12, 1995, the Information and Consultancy Centre for Clean Production, which is financially supported by USAID and the World Environment Centre in co-operation with the Estonian Ministry of Environment, was opened in Tallinn. The main task of the Centre is supporting people with information concerning simpler ways of decreasing the pollution of environment and inexpensive but useful methods for Estonian enterprises.

- Several projects were launched already in 1993 and by now, significant results have been obtained, even though the technological basis is rather weak.

- The enthusiastic activities of the Estonian Green Movement propagating environmentally friendly technology especially among the youth should also be mentioned. They cleaned one of the small rivers in Hiiumaa Island in 1995.

- Estonian Green Movement, Coalition Clean Baltic, Stensund Ecological Centre and Pärnu Marine Biological Station have launched several training and research projects in Pärnu. As a result of such activities, the Development Plan of Environmental Economy for Pärnu based on the principles of sustainable development and favourable environmental technology is being worked out (see below).

As there are a lot of environmental problems in Estonia, it is impossible to concentrate on the limited number of such problems, *e.g.* ecological engineering, instead, success can only be gained by solving all these problems simultaneously.

First and foremost, it seems to be vital to intensify the environmental training, especially introduction of environmentally friendly methods of management. As to waste water treatment, the problems we face in Estonia are quite similar to those of Eastern Europe and Russia where on the way of changing the socialist system into the capitalist system the following shortcomings were taken along:

1. Outdated ideology - the technology of waste water treatment was determined in Moscow, changing thereof proved to be practically impossible.

2. It was only possible to use the non-efficient and unreliable equipment produced within the socialist system, materials to be used were prescribed for the draftsmen. Taking into consideration the difficult economic situation, it was awfully difficult to build the waste water treatment facilities for Pärnu in 1975 - 1991. At the same time, the production of waste was strongly concentrated, *e.g.* piggeries for 30,000 species worked practically without any waste treatment facilities. The penalties and taxes for pollution were very low.

After regaining independence, Estonia faced the dilemma: whether the existing waste treatment plants should be demounted or the new ones based on the western technology should be built. Numerous experts flooded in, huge amounts were written off as aid, unfortunately this aid actually meant clearing the outdated equipment out of the western warehouses. Getting tired of waiting, we decided to do something by ourselves, so we started to modernise our existing facilities little by little using the scarce means we possessed - and it seems that was the right decision. Although Estonia is but a small country, it is a country of multiform nature and varied economy, therefore the environmental problems are also manifold. As to agriculture, the pollution has dropped significantly due to the former big farms being split up. But quite soon, a number of small waste treatment plants suitable for small farms are needed due to the upspring of small households in rural areas. General information is now available for any farmer as the Ministry of Environment has published the brochures to teach farmers proper ways of using and treating water. However, more detailed information is still missing - information concerning, eg., agricultural machinery, maintenance of wetlands, planting different crops, etc. The increase in interest of country folks in environmental issues is significant.

Wetlands have been established and monitored in several places (*e.g.* sand-plant-filter in Põlva, seminatural wetland in Koopsi), production of energy forest has been experimented on the bases of waste water (willow-plantation in Aarike and Väike-Maarja). Constructed wetlans are especially popular among farmers. There is interest to use wetlands for polishing the wastewater from small villages. There are over 740 oxidation ponds constructed during 70 and 80ies in Estonia.

Several ecovillages are currently under the process of planning or construction. These are the good examples worth following. Near Pärnu, in the Halinga Parish, water closets with separation of urine are under construction for both the nursery home and local people (for 60 people in total). The waste water is treated in common treatment plants and then directed through the wetland into the river. A similar project shall be launched in Otepää in South-Estonia. Both objects are useful for both improving the conditions of the rivers and educating people as to environment protection. Both project are financed by Swedish churches. CCB programme on rehabilitation of rivers carries on successfully. The feasibility study necessary for restoration of the Selja River has been accomplished. In several small towns (*e.g.* Kilingi-Nõmme) and villages (*e.g.* Kõrgessaare in Hiiumaa) new, quite modern water treatment facilities have been built, the latter being funded by the means collected under Stockholm Water Action Programme.

Unfortunately, accomplishment of the two biggest waste water treatment plants in Tallinn (population: 450 thousand inhabitants) and Tartu (108 thousand inhabitants) carries on slowly indeed. Utilisation of the residue and sludge is one of the more general problems for Estonia. Settlement of environmental problems in Pärnu, the town with 52 thousand inhabitants, should be reviewed separately as it might serve as a good example for other towns. In Pärnu, a conception of general environmental problems was adopted as early as 1989. It served as a basis for pilot programmes in 1989 - 1990 and 1993 - 1995, the main goal there of - stopping direct discharge of waste water in the river and the bay - has now been gained; the operation of the waste water treatment facilities has significantly improved (see Table 1). While compiling the new programme for environmental protection for the period of 1996 - 2000, it became obvious that the principles of handling the matter should be changed - instead of environmental protection, environmental management should be aimed at. Fortunately enough, Pärnu had all the necessary prerequisites for such a change:

1. Water quality is inspected in Pärnu according to a definite programme;

2. The experiments carried out in co-operation with the Stensund Ecological Centre on application of the tropical plants on waste water treatment facilities were a success (see Figure 1);

3. A workshop on the use of wetlands supplied the public with the information on the necessity of application of new methods in waste water treatment.

Month	BOD IN	BOD OUT	SS IN	SS OUT	P IN	P OUT	N IN	N OUT
January	344	11	294	26	5.1	1.9	25.0	8.6
February	363	10	179	23	5.5	1.2	37.7	9.9
March	318	12	204	54	5.8	2.1	40.4	9.5
April	409	13	187	53	7.0	.6	45.7	5.8
May	438	13	259	19	7.6	.6	35.8	5.8
June	524	15	281	21	11.3	.6	39.7	3.5
July	560	14	273	15	8.9	.4	37.3	3.3
August	542	19	309	17	7.7	2.5	32.3	8.2
September	582	17	370	31	9.7	.7	34.9	4.9
October	470	17	253	20	9.2	1.5	25.5	5.8
November	552	18	278	23	9.8	.4	41.1	7.7
December	558	15	292	24	7.5	1.0	26.2	5.1
Year	472	14	265	27	7.9	1.1	35.1	6.5

Table 1. The datas of water quality in Pärnu sewage treatment plant

Figure 1. Tropical plants on Pärnu wastewater treatment plant

However, Pärnu currently faces two major problems:

1.   The beach is in desperate need for sand for regeneration;

2.   The commercial port of Pärnu needs dredging.

Taking into consideration all the necessities and current possibilities, a project on waste-free management of the beach complex was worked out, including the beach, the port, the waste water treatment plant and the constructed wetland (see Figure 2). The sand removed in the course of the dredging shall be used for improving the condition of the beach, other bottom sediment shall be used for landfill, some of it, still, to be used for creating additional sludge collectors, and a part of it going to the soil factory together with the solid waste and sludge from the waste water treatment plant. Even the works carried out up till now have managed to improve the condition of the ecosystem of the Pärnu bay (see Figure 3). After the port facilities are launched in full, we might expect the improvement of the ecosystem of the bay to the level of 1960s. Therefore, we can draw a conclusion that, in addition to all the problems, we can also see the clearly positive and optimistic opportunities.

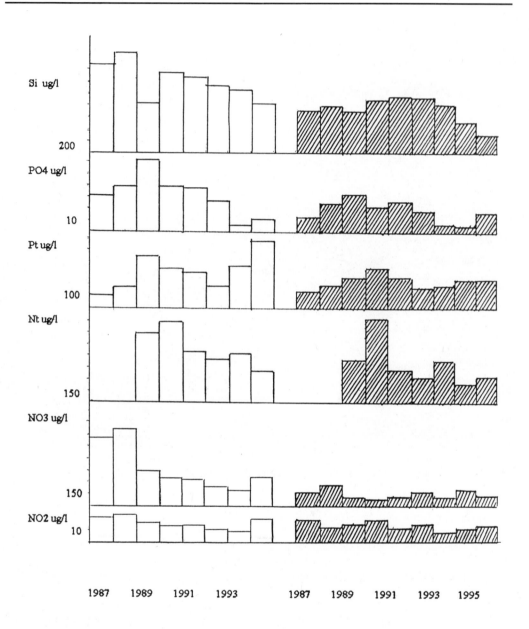

Figure 2. The content of nutrients in Pärnu Bay

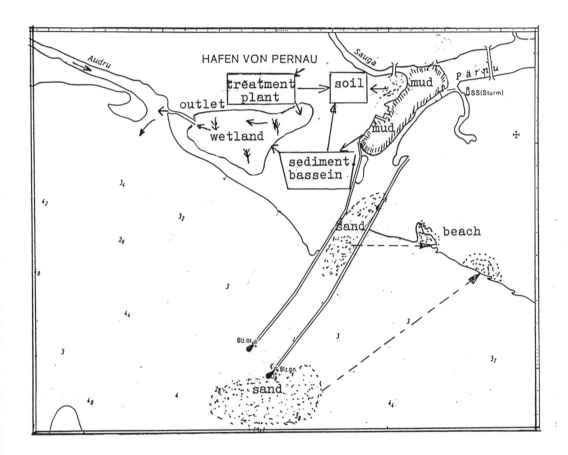

Figure 3. Management of the Pärnu Beach complex

Environmental Research Forum Vols. 5-6 (1996) pp. 73-82
© *1996 Transtec Publications, Switzerland*

Basics of Systems Analysis for Wastewater Treatment

T.A. Larsen

Swiss Federal Institute for Aquatic Science and Technology (EAWAG) and
Swiss Federal Institute of Technology (ETH),
CH-8600 Dübendorf, Switzerland

Keywords: Systems Analysis, Wastewater Treatment, Ecological Engineering, Tracer Experiments, Conservation Principles, Schattweid

Abstract: At the moment, ecological engineering suffers from a lack of tradition of systems analysis. This leads to inefficient work and loss of credibility, especially amongst more traditional environmental process engineers. With the use of traditional engineering methods of analysis, this situation can be improved.

Rules and methods for analysis of systems relevant to wastewater treatment are presented. These include the definition of system boundaries, application of mass balances (dynamic and steady state) and the use of conservation equations (mass, charge, etc.). One example of a technology of ecological engineering has been chosen to illustrate the theory.

The supporters of alternative wastewater treatment plants rarely discuss the energy and space requirements for these plants. Systems analysis makes it easier to discuss such difficult subjects in a rational way and will help ecological engineers coming up with more integrated solutions.

Introduction

The general public understands ecological engineering as the 'green' approach to wastewater treatment. Wastewater treatment in self designing ecosystems based on solar energy is a more official definition given in [1]. In practical life, ecological engineering often means compost toilets, buried sand filters and wetlands in the country side. Actual recycling of resources like wastewater fed aquaculture is most typically found in China [2]. Traditional environmental process engineers are primarily interested in the treatment of wastewater from larger urban areas. Wastewater treatment in non-urban areas is a special field where technologies similar to the ones employed in ecological engineering are used.

From the point of view of process engineering, ecological engineers in many cases do not supply the relevant information. On the basis of one system - the combined aerobic/anoxic sand filter and denitrifying wetland at the eco-center Schattweid in Switzerland - some of the reasons are discussed why environmental process engineers often may not understand the logic of ecological engineers. The analysis is based on [3, 4, 5].

Fig. 1: The principle of wastewater flow at the eco-center Schattweid. Feces are separately composted and used as garden fertilizer [5].

The choice of Schattweid as an example was made for reasons of efficiency: The documentation of the plant is readily available. The criticism is not specially directed towards this group of ecological engineers and I do not claim that traditional environmental engineers never make mistakes! The principle of the Schattweid wastewater treatment system is roughly outlined in Fig. 1.

Systems Analysis in Environmental Process Engineering

Systems analysis may be described as the science of applying natural laws and known or assumed causal relationships to well defined parts (systems) of the world. Scientists, including ecological engineers [6], will often use three-dimensional tensor notation for describing their systems whereas the tradition of environmental process engineering has limited itself to one-dimensional mathematics. There are obvious reasons for this difference: In the field of wastewater treatment, we have a well defined direction of interesting changes (input / output analysis) whereas this is not the case in natural systems. The use of one-dimensional mathematics implies that we will analyze all systems as if no gradients of state variables would exist perpendicular to the direction of interest. This has the advantage that analytical solutions to simple differential equations exist. The simplified equations have allowed engineers to come up with analytical solutions to their problems way before the invention of the personal computer. The tradition of simplified analysis still allows the engineer to perform valuable analyses on the basis of rather limited knowledge.

The tradition of systems analysis amongst process engineers of wastewater treatment is based on one single differential equation (Eq. 1)

$$\frac{dM_A}{dt} = J_A + P_A \qquad\qquad \text{Eq. 1}$$

where M_A (with the dimension [M]) is the total mass of component A in the system, J_A [MT^{-1}] is the net transport rate of component A *into* the system and P_A [MT^{-1}] is the net *production* rate of component A within the system. t [T] denotes the time.

In words, Eq. 1 expresses that the total storage rate of a component A in a system only depends of the net transport rate of A into the system and the net production rate of A within the system. It should be obvious from the definitions that transport out of the system as well as consumption will be negative. It should also be noticed that depletion is negative storage.

From this description it is also clear that we have to define a system (system boundaries) before we can use Eq. 1. System boundaries should generally be chosen in a way which allows us to answer the question of interest as precisely as possible.

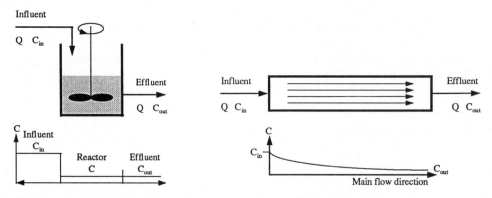

Fig. 2: Schematic representation of a completely mixed and a plug flow reactor.

Most introductory textbooks on process engineering will describe the standard developments of Eq. 1 in one-dimensional mathematics for different ideal reactors and systems. The classical textbook on this type of analysis is 'Chemical Reaction Engineering' [7], but for the education of environmental process engineers, locally re-written versions especially suited for the purposes of wastewater treatment are often used (e.g.[8, 9]). Two simple flow patterns, illustrated by the ideally mixed reactor and the plug flow reactor, and their typical behavior, are illustrated in Fig. 2. The mathematical descriptions of such reactors can be found in any introductory textbook on process engineering.

The most important aspect of Eq. 1 is the distinction between transport and transformation. This distinction is the basis of the engineering understanding of the world and guides many experimental studies in the field of process engineering. The transport properties of a given reactor[1] are described based on an in situ tracer experiment whereas the transformation processes may be studied separately in batch or controlled small scale reactor experiments.

Experimental Investigation of Transport Properties

Not only process engineers, also ecological engineers use the method of tracer curves for investigating the transport properties of their reactors. The tracer experiments in the system of Schattweid are reported in [4]. Here I will only comment on the tracer curve from the sand filter. This reactor type is probably dominated by the existence of different flow patterns as illustrated in Fig. 3.

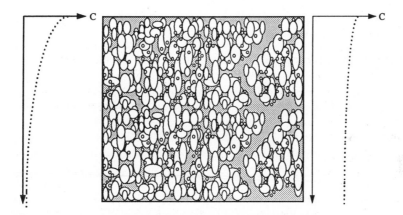

Fig. 3: In a filter, the existence of different flow patterns is the most important aspect of the transport properties. Preferential flow (to the right) results in only little contact to the surfaces with microbial growth. Consequently, the water passing rapidly through the filter may contain relatively high concentrations of pollutants in the outlet. Water trickling slowly through the filter in a more complicated way (to the left) will have much more contact to surfaces with microorganisms, resulting in better degradation and filtration of pollutants.

In Fig. 4, the responses of some ideal reactors to the most frequently used tracer experiment are shown. The modeling of a tracer experiment is based on Eq. 1. For a tracer, $P = 0$, which means that only the transport processes influence the response in the effluent. For the ideally mixed reactor as

---

[1] Reactor is here used in a broad sence. Of course the word has been chosen by engineers working with built reactors in a chemical factory. Systems analysis, however, may be used equally in anthropogenic systems and ecosystems.

well as the series of ideally mixed reactors, the computation is trivial. For the plug flow reactor, we do not even need to perform a computation. The shape of the tracer response is probably the best description of the transport properties in a plug flow reactor.

The interpretation of tracer experiments is based on statistical methods; the mean residence time and the variance are the key parameters obtained from such studies (See Box 1).

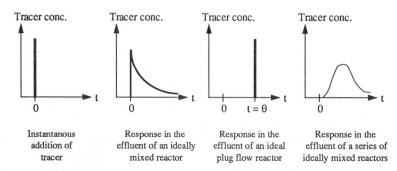

Tracer conc.	Tracer conc.	Tracer conc.	Tracer conc.
Instantanous addition of tracer	Response in the effluent of an ideally mixed reactor	Response in the effluent of an ideal plug flow reactor	Response in the effluent of a series of ideally mixed reactors

Fig. 4: The response in the effluent of ideal reactors on an instantaneous addition of tracer in the influent.

Box 1 Outline of the statistical treatment of tracer curve data.

For an instantaneous addition of tracer at time t=0, the measured function S_{tracer} (t) in the effluent is transformed into the probability density function $f(\tau)$:

$$f(\tau) = \frac{Q_{effluent} \cdot S_{tracer}(t)}{M_{tracer}}$$

The function $f(\tau)$ expresses the probability density that a molecule of water present in the inlet at time t=0 will reach the effluent at time $t=\tau$. It is obvious that the probability function should integrate to 1 for $t \to \infty$ (eventually all tracer will be found in the effluent). This treatment allows us to use the classical statistical means for describing the curve (1., 2. and higher moments). The first moment, τ_m, is equal to the mean residence time. The second central moment, σ^2, relative to τ_m^2 indicates the degree of mixing or dispersion in the reactor (for a plug flow reactor $\sigma^2/\tau_m^2 = 0$, for the ideally mixed reactor, $\sigma^2/\tau_m^2 = 1$).

In Fig. 5, tracer data from the sand filter in Schattweid [4] is transformed into the probability density function $f(\tau)$. The relevant data for understanding the system is found in Table 1. The effluent flow of the sand filter was not monitored, but since the time constants of the filter are large compared to the fluctuations of the hydraulic loading, it is possible to analyze the data based on the assumption of continuous flow. The model of total dispersion fits the form of the data reasonably well and indicates a minimum mean hydraulic residence time of 3 days. The total filter volume is 6 m³. Assuming a porosity of 33%, a pore volume corresponding to approximately 2 m³ is estimated. A mean residence time of 3 days for an average flow of 500 l/day corresponds to a water volume in the filter of 1.5 m³. According to this rough estimation, on the average more than half of the pore volume in the filter is filled with water, indicating that the filter volume is well used. In Table 1, data from Schattweid is compared to data from a laboratory sand filter [10].

The above example shows that even the use of very simple models makes it possible to start discussing the phenomena experienced in different types of treatment plants. Experiments are time consuming and the maximum of information should be extracted from the results obtained. With a little more knowledge of the traditional process engineering methods, ecological engineers could more rapidly develop their technologies and be more competitive on the market. Unfortunately, a more detailed discussion of the interesting results of the tracer experiment goes beyond the scope of this paper.

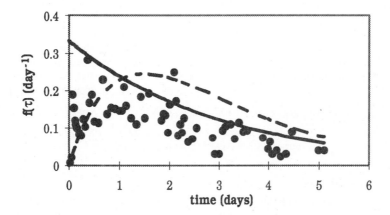

Fig. 5: Tracer curve from the sand filter in Schattweid. Experimental results are taken from [4] and transformed into the probability density curve. The full line indicates a model of full dispersion (one ideally mixed reactor) and the broken line a model of two ideally mixed reactors in series. For both models a mean residence time of 3 days and a constant wastewater flow of 500 l/day are assumed. The better explanation of the experimental data is obtained with the assumption of full dispersion.

Table 1: Comparison of data from the sand filter in Schattweid with data from a laboratory sand filter. Data from [5, 10].

Parameter	Units	Schattweid	Lab filter (1)	Lab filter (2)
Filter material diameter	mm	1-3	0.85**	0.85**
Filter depth	m	0.5	1	1
Loading	liter m^{-2} d^{-1}	42	120	120
Number of flushes	day^{-1}	5	12	3
Loading per flush	liter m^{-2} $flush^{-1}$	8.4	10	40
Mean hydr. residence time	days	3*	0.8	0.7
Moisture content	% of total filter volume	25	10	8
Retained water per unit area	liter m^{-2}	126	96	84

* The mean residence time is an estimated minimum and includes the volume of the distribution tank before the sand filter.

** The indicated value is a d_{50} value. The distribution coefficient $d_{10}/d_{60} = 2.1$.

Transformations

For the transformation processes in the treatment plants of Schattweid, it is possible to obtain valuable information already with simple input-output analyses based on conservation principles. Of special importance for wastewater treatment is the conservation principle for a theoretical COD which allows us to use conservation laws on non-specific organic matter [11]. An outline of the theory behind the conservation of theoretical COD is found in Box 2.

Box 2   Outline of the theory behind the conservation of theoretical COD.

The notion of the analytical COD is well established in the field of wastewater treatment. It is also well-known that under anoxic conditions, organic matter will be oxidized by nitrate rather than oxygen. All components in wastewater may be rated in oxidizing or reducing capacity relative to the oxidizing capacity of oxygen. The only requirement is that for every element, a state of 'zero' capacity must be defined. One logical possibility is to chose the states according to the response of raw wastewater to the analytical method of COD. In raw wastewater, nitrogen is primarily available as ammonium which is not oxidized under the conditions used in the analysis. The same is the case for sulfate and phosphate. From the theory of redox processes it is clear that oxidizing and reducing capacities are conserved. Following the above logic, the theoretical COD of the compounds relevant for the considerations made in this paper are the following:

O_2: $-1 g_{COD}$ g^{-1}_{oxygen}, NH_4^+ : 0 g_{COD} g^{-1}_N, NO_3^-: -4.57 g_{COD} g^{-1}_N, $N_2 = -1.71$ g_{COD} g^{-1}_N

(You may want to check that the oxidation of ammonium to nitrate takes 4.57 g_{O2} g^{-1}_N and that $(4.57-1.71)$ g_{COD} are oxidized when 1 g of nitrogen is reduced from nitrate to nitrogen gas).

The Sand Filter in Schattweid

In the sand filter of Schattweid, nitrification and denitrification takes place simultaneously. This means that the transfer of oxygen to some parts of the reactor must be limiting. It has the advantage that some of the COD of the wastewater is used for denitrification. In the case of Schattweid, the sand filter is followed by a wetland with additional nitrification, therefore the limitation of the nitrifiers is not critical. However, in order to understand such systems, a discussion of the oxygen limitation is relevant. It is specifically mentioned [5] that the sand filter may also be used in combination with normal toilets. This might give rise to increased organic concentrations of COD in the inlet to the sand filter with the risk of decreasing nitrification activity. Such problems have been seen in other sand filters treating normal toilet wastewater [10]. For comparison with other filter types, the transfer of oxygen may be computed. From the data available on the sand filter in Schattweid, a balance of the theoretical COD can be performed (Table 2). The numbers are based on average concentrations given in [5]. It is assumed that all nitrogen removed is denitrified - the error due to biomass formation is negligible. From the conclusions drawn by the authors, I must assume that the numbers given are flow proportional averages. The numbers in italics are my own computations.

Table 2: Balance of theoretical COD for the sand filter of Schattweid.

	Organic Matter $g_{COD}m^{-3}$	NO_3^--N g_N m^{-3}	g_{COD} m^{-3}	N_2 g_N m^{-3}	g_{COD} m^{-3}	Total g_{COD} m^{-3}
Inlet	320	0	*0*	0	*0*	*320*
Outlet	37	55	*-250*	31	*-50*	*-263*

According to the wastewater analyses, no other compounds contribute to the COD balance. Consequently, the difference in COD is made up by the oxygen transferred to the filter. In this case, the average oxygen transfer is approximately 580 g per m^3 of wastewater treated. An important means of transferring oxygen in a sand filter is the forced ventilation due to the intermittent loading. In a buried sand filter, oxygen is introduced due to forced ventilation. Ignoring possible other mechanisms of aeration, this means that about 1.8 m^3 of air per m^3 of water are transferred to the filter (1 m^3 of air contains about 320 g of oxygen). For other applications, the aeration may have to be improved whereas for the present application, the oxygen limitations seem to have advantages.

Denitrification in the filter is not only interesting from the point of view of organic matter, also alkalinity is of interest. In drinking water, practically all alkalinity exists in the form of HCO_3^-, I will therefore argue about alkalinity in terms of HCO_3^-. The following equations contain the relevant information on alkalinity changes in the processes of nitrification and denitrification in domestic wastewater:

Hydrolysis of urea: $\quad 0.5\,CO_2 + 0.5\,CO(NH_2)_2 + 1.5\,H_2O \Rightarrow NH_4^+ + HCO_3^-$

Nitrification: $\qquad NH_4^+ + 2\,O_2 + 2\,HCO_3^- \qquad\qquad \Rightarrow NO_3^- + 3\,H_2O + 2\,CO_2$

Denitrification: $\qquad NO_3^- + \text{organic matter} \qquad\qquad \Rightarrow 0.5\,N_2 + HCO_3^- \pm xH_2O \pm yCO_2$

In the overall reaction, hydrolysis of urea and nitrification of one mole of nitrogen consumes one mole of bicarbonate. If nitrate is denitrified to nitrogen gas, the bicarbonate is regained. In the case of Schattweid, on the average 95 $g_N m^{-3}$ of ammonia are produced by hydrolysis, 86 $g_N m^{-3}$ are nitrified, 31 $g_N m^{-3}$ are further denitrified. Without the denitrification, the net consumption of HCO_3^- would be about 5.5 moles m^{-3} of wastewater treated. With denitrification, the net consumption is only about 3.5 moles m^{-3}. A typical alkalinity of Swiss drinking water is 3-6 moles m^{-3}, indicating that the question of alkalinity is relevant. The filter material may contain alkalinity in the form of $CaCO_3$ which might slowly be leaking out of the filter. Measurements of alkalinity and the relevant computations would reveal important information concerning the stability of the process under different regimes of operation.

The Importance of the Storage Term: the Denitrifying Wetland of Schattweid.

Very often, one implicitly assumes that the storage term in Eq. 1 is equal to zero: Some material enters the reactor, part of it is degraded and the rest leaves the reactor again. In many situations, this *steady state assumption* (dM/dt = 0) is permissible. In a reactor with constant loading, soluble components like ammonia and oxygen will not accumulate over longer time periods. However, as we all know, loading is not always constant, not all compounds are soluble and in some cases, short time periods may be very important (e.g. in a situation of peak loading).

In the wetland, the possibility of negative storage (depletion) of organic matter is extremely important for the evaluation of the technology. In wastewater treatment, denitrification is normally limited by the amount of organic matter available. Again COD conservation determines how much organic matter is needed. Since in a wetland, no sludge is produced, the difference in COD of nitrate and nitrogen gas determines the amount of organic matter needed for denitrification (2.86 g of COD for denitrifcation of 1 g of NO_3^--N). The question of interest is whether this organic matter is produced in the wetland or whether the humus of the soil is used for denitrification. Table 3 and Table 4 give a number of estimates indicating that the question is relevant. I miss a discussion of this important matter in the publications from Schattweid which I have seen until now.

Table 3: COD demand for denitrification in the wetland of Schattweid [5].

Denitrification in Schattweid wetland	COD-demand for denitrification in Schattweid	Standard area of wetland	Actual area of Schattweid wetland	Actual COD demand denitrification
kg_N person^{-1} year^{-1}	kg_{COD} person year^{-1}	m^2 person^{-1}	m^2 person^{-1}	kg_{COD} m^{-2} year^{-1}
1.5	4.3	2.75	4.8	0.9

From Table 3 and Table 4 it is clear that the production of organic matter for the required denitrification is large enough. The question is whether this organic matter is available for the denitrifying organisms? It is also clear that the organic matter content in the soil would be large enough to support denitrification for many years. However, it would not be very wise to use valuable humus reserves for denitrifying purposes. In Schattweid, plant material is removed from the wetland. However, I have seen no comments on the risk of depleting organic material in the soil material.

Table 4: Estimated contents of organic matter and net primary production on different land types. After [12].

	Organic Matter Content kg_{DM}/m^2 *	Net Primary Production $kg_{DM}/(m^2$ year)*
swamps and marshes	30	2.5 (temp. zone)
cultivated land	10	1.2 (96% of temp. zone)
		11 (4 % of temp. zone)

* For wastewater treatment, we would use conversion factors of around 1 g COD/g DM. The conversion factor for primary production is probably higher, a reasonable estimate would be 1.5 g COD/g organic dry matter.

General Evaluation of the Schattweid Approach

The methods used above for analyzing the data from the two treatment facilities in Schattweid may also be used for a general evaluation of a technology. The methods themselves are not very demanding, only the choice of system boundaries may be critical. The point is that such evaluations are rarely made, neither by ecological engineers, nor by environmental process engineers. For the overall balances, I have found no relevant data in the publications, making it very difficult to come up with the interesting computations. I will restrict myself to a more qualitative discussion of a few points concerning recycling of nutrients, sludge quality, transport and energy.

Ecological engineers typically criticize the more traditional wastewater treatment on the following points which are also found in [5]:

- Heavy metals in sludge
- Pollution due to transport (of sludge)
- Poisonous residuals from incineration of sludge
- No recycling of nutrients in modern cities
- High costs for required denitrification

I more or less agree with this analysis: These are important deficiencies of our present wastewater treatment systems. However, in my opinion it is wishful thinking that the solutions offered by ecological engineers may seriously deal with this sort of problems. I would like to emphasize that this is not explicitly claimed by the authors of [5]. Still, when a given technology is presented as more ecological in comparison to other technologies, the advantages should be pointed out more precisely.

The first four points all relate to the difficulties of recycling sludge to agriculture. As it is the case for conventional wastewater treatment, the recycling of nutrients in Schattweid reduces to the recycling of phosphorous. The main problem of phosphorous recycling in Switzerland is the net-import of phosphorous in animal fodder and other food products. Only in the second place come the problems of the quality of sludge. Neither of these problems is solved by the Schattweid solution.

It is easy to agree that the problem of heavy metals should be controlled at the source, but the Schattweid solution does not suggest any solutions to this problem. The problem of industrial wastewater is solved by not having any industrial wastewater. The problem of heavy metals from private installations is not commented on. I guess that if ecological engineers start measuring the content of heavy metal in their own wastewater, at least in some places they might be in for a surprise: Heavy metals are no longer only a problem of industry.

It is of course easy to avoid transport of sludge and incineration if one has enough garden for disposing of ones own sludge. Although the data on this topic are missing, I assume that this is the case in Schattweid. In order to dispose of the phosphorous in sludge from 1 person equivalent, more than 400 m^2 of agricultural land are needed [17]. The exact area needed in Schattweid depends on the intensity of the gardening activity. It should be clear that the 4-5 m^2 of wetland per person equivalent do not contribute in any significant way to the recycling of nutrients, the main significance of the plants in the wetland would be the possible production of organic matter for denitrification. If the garden in Schattweid is in the order of 2000 m^2, local recycling of the phosphorous is possible. One must realize however that the wetland retains the phosphorous for years and then makes it all available in the moment of reconstruction.

It is claimed that the wetland solution could decrease the costs of denitrification. Without arguments on land prices and on the household of organic matter in denitrifying wetlands, this claim will not be taken serious. I accept that the solution could be interesting for very small treatment plants in the country side, but the question is whether it is necessary that these treatment plants denitrify? According to the authors, another advantage of denitrifying wetlands would be that they are biotopes for plants and animals. Personally, I have never seen any convincing arguments that biotopes heavily polluted with wastewater should have any ecological advantage over the same area of land left as natural ecosystem. For storm water retention the situation may be somewhat different, but this is not the question discussed here.

Another interesting question of ecological engineering technologies is the energy demand. It is often claimed that the technologies of ecological engineering are less energy consuming than the traditional ones, but I have never seen any computations actually making the comparison. Table 5 indicates the importance of wastewater treatment for the energy household in a western industrialized country. In Switzerland, large efforts are made to reduce the energy consumption of traditional wastewater treatment plants [16]. An analysis of the energy demand of compost toilets is given in [17]. It is a question whether the relatively small energy consumption for wastewater treatment justify the large amounts of space needed for the wastewater treatment technology suggested by ecological engineers.

Table 5: A range of approximate energy demands in Switzerland.

Wastewater treatment [13]	Compost toilets [17]	Hot water in households [14]	Total electricity demand [15]	Total energy demand [15]
10 W/person*	15 W/person	100 W/person	800 W/person	4000 W/person

* electricity

Conclusions

An analysis of the readily available publications on the Schattweid wastewater treatment system has revealed a number of deficiencies in the published interpretation of experimental results. The analysis of the transport properties does not make use of mathematical modeling of the results. Simple analyses on the basis of conservation principles of main interest to the evaluation of the technologies are not published. No overall analysis of the nutrient recycling system is performed. Hardly any connection between the general criticism of the traditional wastewater treatment technologies and the technologies presented can be established. The points where there might be an advantage of the Schattweid system in comparison with other small scale wastewater treatment technologies are not substantiated with any computations. The lacking information may be available in some longer reports. However, such reports will hardly be read by traditional environmental process engineers. If the publications presented in engineering journals would contain the relevant information, there are good chances that the professionals of wastewater treatment would better understand the logic of ecological engineers.

References

[1] W.J Mitsch and S.E. Jörgensen, Introduction to Ecological Engineering. In: W.J. Mitsch and S.E. Jörgensen (eds.) Ecological Engineering. An Introduction to Ecotechnology (1989).

[2] C. Etnier and B. Guterstam, Ecological engineering for wastewater treatment. Proceedings of the International Conference, March 24-28, 1991. Stensund Folk College, Sweden (1991).

[3] J. Heeb and B. Züst, Sand and plant filter systems. In: C. Etnier and B. Guterstam, Ecological engineering for wastewater treatment. Proceedings of the International Conference, March 24-28, 1991. Stensund Folk College, Sweden (1991).

[4] T. Yunping, A. Schönborn and B. Züst, Retention time of five waste water treatment plants (soil filters) in Switzerland. Zentrum für angewandte ökologie schattweid. CH-6114 Steinhuserberg, Switzerland (1993).

[5] A. Schönborn and B. Züst, GWA, **74** (8), 674 (1994).

[6] S.E. Jörgensen, Fundamentals of Ecological Modelling. Developments in Environmental Modelling 9. Elsevier Science Publishers B.V., The Netherlands (1986).

[7] O. Levenspiel, Chemical Reaction Engineering. John Wiley & Sons (1962).

[8] P. Harremoës, M. Henze, E. Arvin and E. Dahi, Teoretisk Vandhygiejne. Laboratoriet for teknisk Hygiejne. DTH. Polyteknisk Forlag, Denmark (1989) (In Danish).

[9] W. Gujer, Mathematische Beschreibung von technischen Systemen. Institut für Hydromechanik und Wasserwirtschaft, ETH-Z. CH-8093 Zürich-Hönggerberg (1990).

[10] M. Boller, A. Schwager, J. Eugster and V. Mottier, Wat. Sci. Tech., **28** (10), 99 (1993).

[11] W. Gujer and T.A. Larsen, Wat. Sci. Tech., **31** (2), 257 (1995).

[12] B. Bolin, E.T. Degens, P. Duvigneaud and S. Kempe, The Global Biogeochemical Carbon Cycle. In: B. Bolin, E.T. Degens, S. Kempe and P. Ketner (eds) The Global Carbon Cycle (1979).

[13] Tiefbauamt der Stadt Zürich, Stadtentwässerung, Jahresbericht 1993 (1993).

[14] E. Höliner: Warmwasserverbrauch in Wohnbauten. IWSA Tagung Basel, 6.-7. Februar 1990. IWSA-Kommision "Internationale Wasserstatistik" (1990).

[15] Bundesamt für Energiewirtschaft und Schweiz. Nationalkomitee der Welt-Energie-Rates: Schweizerische Gesamtenergiestatistik 1992. Bulletin SEV/VSE, **84**, Heft 12 (1993).

[16] A. Baumgartner, B. Kobel, H. Kutil, E. A. Müller, P. Stähli and R. Thommen, Energie in ARA: Musteranalysen. Materialien zu PACER. Bundesamt für Konjunkturfragen. 724.239.1d. CH-3000 Bern, Switzerland (1994).

[17] W. Gujer, Comparison of the Performance of Alternative Urban Sanitation Concepts. This volume (1995).

Acknowledgments

I would like to thank Andreas Schönborn and Johannes Heeb, zentrum für angewandte ökologie, Schattweid, Switzerland, for initiating this publication. I hope that it will contribute in a constructive way to the discussion of the methods and goals of ecological engineering.

Environmental Research Forum Vols. 5-6 (1996) pp. 83-98
© *1996 Transtec Publications, Switzerland*

Health (Pathogen) Considerations Regarding the Use of Human Waste in Aquaculture

M. Strauss

Swiss Federal Institute for Environmental Science & Technology (EAWAG),
Department of Water & Sanitation in Developing Countries (SANDEC),
Ueberlandstrasse 133, CH-8600 Dübendorf, Switzerland

Keywords: Wastewater, Wastewater Treatment, Aquaculture, Fish, Hygienic Quality, Pathogens, Pathogen Survival, Health Risks, Guidelines, Standards

Abstract

The author reviews the potential health risks and current epidemiological evidence for actual risks from pathogen transmission through wastewater aquaculture. The groups at risk, their exposure and the types of risk are defined. A distinction is made between potential and actual risks. The main factors determining pathogen transmission and survival are discussed. Water quality standards required for aquaculture are compared with the microbial quality of natural surface waters from which fish are harvested without restriction. Since wastewater treatment is an important technical tool to reduce pathogen transmission risks, several treatment processes are rated with respect to their pathogen elimination potential. Adequate wastewater treatment is needed for reasons of public health protection as well as for social acceptance of wastewater-based aquaculture. Possible legal approaches to minimise the pathogen transmission risks are discussed and examples of existing guidelines and standards relating to waste-fed aquaculture and aquatic food products are presented.

THE CALCUTTA AND MUNICH WASTEWATER-FED AQUACULTURAL SYSTEMS - HEALTH CONSIDERATIONS

Both the Calcutta raw wastewater-fed fish pond system - known as the "Calcutta Wetlands" - and the Munich treated wastewater-fed fisheries are frequently cited as showpieces of wastewater-fed fish production. In **India**, there are over 130 sewage-fed fisheries covering a total area of 12,000 ha. The **Calcutta Wetlands**, which have been operational since about 1930, are the world's largest wastewater fisheries, comprising an area of approx. 4,000 ha and providing employment for 4,000 families [1]. Various types of carps and *Oreochromis mossambicus* (Tilapia) are grown in these ponds. The catch from the ponds meets about 10% of Calcutta's market demands for consumable fish. The fish ponds are batch-fed with untreated sewage at low organic loading rates of 6-22 kg BOD/ha·day which correspond to maturation pond loading rates in warm climates. These rates have become established through decades of fish pond practice and are aimed at maintaining adequate fish environments. The low loadings lead to a rather safe hygienic quality of the pond water; with total coliform concentrations reportedly amounting to 10^2-10^3/100 ml [2]. This corresponds to a faecal coliform[1] level of 10^1-10^2/100 ml and would be equivalent to the quality of many surface waters from where fish are caught without restriction (see Table 2 below). If these values are confirmed and represent a long-term pond quality average, the potential health risks for fish consumers from pathogens contained in this waste-fed fisheries scheme could be judged minimal. Since all fish are cooked prior to consumption, the health risk is further reduced. The bulk of the pathogens entering the ponds with the wastewater settles in the pond sediments. This poses an occupational risk for the fishermen who wade in the ponds to harvest the fish. The effective health protection measure for the fishermen and their families is proper personal hygiene. Epidemiological investigations would be required to establish actual (not only potential) health risks.

---

[1] Faecal coliforms are specific indicators of faecal contamination, whereas total coliforms are not.

From the end of last century until the 1950's, wastewater-based fish production was rather widespread in **Germany**. Around 90 schemes existed at that time, and were either fed with mechanically-treated wastewater diluted with fresh water, secondary effluent, or with effluents from sewage fields. The aim then was to treat or polish wastewater and, at the same time, produce fish. The **Munich fishponds** is the only system still in use today. It covers an area of 230 ha and receives the diluted effluent from an activated sludge plant originally designed to treat sewage from 500,000 population equivalent [3]. Tertiary wastewater treatment has reportedly been added in recent years. The necessary dilution with water from the river Isar may therefore have been reduced from a 1:3 ratio (effluent:water) in 1989 to currently 1:1. The ponds, stocked with common carp, are only operated in the months of March/April to October/November since temperatures in winter are too low for substantial fish production coupled with wastewater polishing. The annual harvest of 100-150 tons is reportedly controlled by the health authorities prior to being sold to restaurants or large institutions. The total coliform concentration in the pond water was reported as 3×10^2 - 1.5×10^4/100 ml (cited in [3]). This may be taken as equivalent to a faecal coliform concentration of 10^1-10^3/100 ml; i.e., levels which are considered safe for the fish flesh to remain free from pathogens (see also the chapter on microbiological studies with fish). No pathogen-related health data for fish is available, however, fish quality is reported to comply with the standards imposed by the German Federal Bureau of Health [3]. Heavy metals and aromatic hydrocarbon concentrations in the fish ponds appear to satisfy government standards. Depuration[2] studies conducted with fish from the Munich ponds showed that depuration lasting for up to one year was effective for Cd and Pb but not for Hg and PCBs (cited in [3]).

In Germany (except in Munich), wastewater-fed fishery was abandoned due to increased land and operational costs, in addition to the requirement of alternative treatment processes over the winter months when fish production has to be suspended. The Munich ponds, though reportedly associated with very high operation and maintenance costs, will not be abandoned since they have been declared a European bird sanctuary and also serve as a recreational area.

THE RISK DIMENSION - WHO IS AT RISK ?

Pathogen and Indicator Levels in Faeces, Wastewaters and Rivers

Enteric bacteria, which form the intestinal flora, are shed daily in numbers ranging from 10^5 - 10^{10}/g of stool through the excreta of healthy individuals. Among them is the faecal coliform group with *Escherichia coli* as their main representative. Faecal coliforms (FC) are the most widely used indicator bacteria for faecal contamination, as their excreted load is similar or larger than that of pathogenic organisms, and their survival time in the environment longer than that of excreted bacteria and viruses.

Infected persons, exhibiting disease symptoms or asymptomatic carriers, shed huge loads of the respective pathogenic organism - for viruses up to 10^{11}/gram of faeces, for *Salmonella typhi* up to 10^8/gram, for *Entamoeba hystolitica* up to 10^5/gram, and for *Ascaris lumbricoides* eggs up to 10^4/gram. In water-borne waste disposal systems, part of the pathogen load will eventually reach the sewage treatment works. Concentrations in the raw wastewater are dependent on the number of individuals infected within a given population, as well as on the per capita wastewater generation. Virus concentrations in wastewater are often higher in summer than in winter. Pathogens reaching the treatment plant, while continuing to die-off "naturally", will be eliminated from the wastewater stream by scavenging mainly and accumulate in the treatment plant sludge. However, due to the often high numbers contained in the raw wastewater, considerable concentrations might still be found in the effluents, even though 99 % or more might be eliminated (see also the section on wastewater treatment below). Table 1 contains indicator and pathogen concentrations in faeces and wastewater.

---

[2] Depuration is the placing of the fish in clean water to allow cleansing from pathogens or contaminants.

Table 1   Indicator and Pathogen Concentrations in Faeces and Wastewater

Organism	In faeces[1] [no./g]	In wastewater	Country	References
Indicator bacteria				
• Escherichia coli	$10^7 - 10^9$	$10^7 - 10^8$/100 ml		
Pathogens				
• Salmonella	10^8	$10^{3.4}$/100 ml	Peru	[6]
		$\leq 10^4$/100 ml	U.S.A.	[7], [8]
• Enteric viruses	$\leq 10^{11}$	$10^2 - \leq 10^5$/l	U.S.A.	[9], [5]
		$10^2 - 10^4$/l	Spain	[5]
• Protozoa: Giardia lamblia cysts	10^5	50 - 10^4/100 ml		[8]
• Helminth eggs	$10^3 - 10^4$	≤ 40/l	Germany	[10]
		≤ 10/l	France	[10]
		≤ 2000/l	Russia	[10]
		≤ 300/l	Jordan	[11]
[1] Based on Feachem et al. [4] and L. Schwartzbrod [5]				

Concentrations in the receiving water bodies depend on whether and through which processes the wastewater is treated prior to discharge, on the dilution to which the discharged wastewater is subjected, and on the rate of further pathogen die-off in the receiving water.

Table 2   Faecal Coliform Concentrations in Rivers

Faecal coli-forms/100 ml	No. of Rivers per Region [1]			
	North America	Latin America	Europe	Asia & the Pacific
< 10	8	0	1	1
10-100	4	1	3	2
100-1000	8	10	9	14
1000-10,000	3	9	11	10
10,000-100,000	0	2	7	2
> 100,000	0	2	0	3
100-500	**Switzerland** River **Rhine** upstream of Basle [2]			
10^4 near bank < 10^3 midstream	**India** River Ganges at Patna (Bihar), approx. 2/3 downstream from its source [3]			
$10^4 - 10^5$	A hypothetical river just downstream from a wastewater treatment plant discharge (FC in raw wastewater = 10^7-10^8/100 ml; plant removal = 99 %; **dilution factor = 10**)			
$10^3 - 10^4$	A hypothetical river just downstream from a wastewater treatment plant discharge (FC in raw wastewater = 10^7-10^8/100 ml; plant removal = 99 %; **dilution factor = 100**)			
[1] [12] [2] [13] [3] [14]				

Table 2 shows faecal coliform (FC) concentrations in rivers. It also includes a hypothetical case where a ten and hundred-fold dilution, respectively, has been presumed for the effluent of a secondary (mechanical-biological) wastewater treatment.

The microbiological quality of rivers has been cited here for reasons of comparison between the hygienic quality of natural surface waters, which are traditionally harvested for freshwater fish, and that of artificial wastewater-fed aquacultural systems further discussed below. Fish caught from rivers and sold in markets are generally considered safe for consumption and not rejected for aesthetic reasons, even though levels of faecal indicator bacteria in the waters from which they were caught might be high and excreted pathogens might be detected, too.

Health Risks from Human Pathogens in Aquaculture

Pathogens and Their Transmission; Risk Definitions

There are about 30 excreted infections of public health importance, and many of these are of specific importance in excreta and wastewater use schemes. However, the aquacultural (or for this purpose also the agricultural) use of excreta and wastewater may lead to an actual public health risk only if all of the following prerequisites concur:

(a) If an infective dose of an excreted pathogen reaches the pond or a natural water body, or if the pathogen multiplies in an intermediate host residing in the pond or in the aquatic environment to form an infective dose;

(b) If this infective dose reaches a human host through contacts or consumption of the aquacultural products;

(c) If this host becomes infected; and

(d) if this infection causes disease or further transmission.

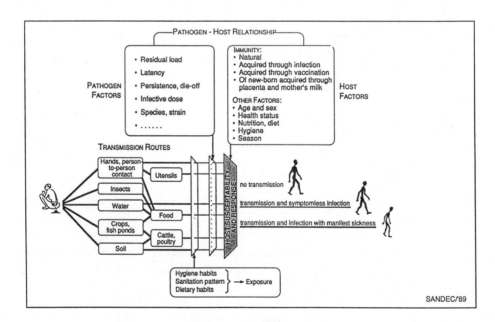

Fig. 1 The Pathogen-Host Relationship and Possible Transmission Routes for Excreta-Related Infections

(a), (b) and (c) constitute the **potential risk** and (d) the **actual risk** to public health. If (d) does not occur, the risks to public health remain potential only. The potential risk is assessable by microbial investigations and inferences, whereas the actual risk is assessed through epidemiological (health impact) studies. The sequence of events required for an actual health risk to take place is illustrated in Fig. 1 together with the pathogen-host properties and interactions that influence each step in the sequence. If the sequence is interrupted at any given point, the potential risks may no longer constitute an actual risk. This is the rationale behind the various methods of public health protection discussed below.

The actual public health risks occurring through waste use in aquaculture can be divided into three main categories; i.e., those affecting consumers of the aquatic products grown in wastewater-fed waters (**consumer risk**), those affecting the operators of the aquacultural system who might become exposed to treated and/or diluted wastewaters (**operators' risk**), and those who handle and process the products (e.g. fish) (**workers' risk**). Depending on the kind of health protection measures used and enforced, all, none or only selected risks may occur.

Of the four categories of excreted pathogens - viz. **viruses, bacteria, protozoa**, and **helminth eggs** (cited in ascending order of size), all may play a role in aquaculture. Viruses, bacteria and protozoa may be carried **passively** on the scales and in the inner organs of fish, crustacea, shellfish, or plants. Nowadays, helminthic pathogens play a minor role in industrialised countries. Many helminths are, however, endemic in most tropical, less industrialised countries, and have, via waste-fed aquaculture, various potential transmission patterns. *Ascaris* (roundworms), the most common helminthic pathogen in developing countries, may be carried passively by fish and other aquacultural products although its major transmission foci are soils and waste-fertilised crops. *Schistosoma*, a typical water-related disease of stagnant waters in the tropics, is transmitted through contact with water containing the pathogens that are taken up and later released by the compulsory snail hosts. *Clonorchis*, the Chinese liver fluke, endemic in certain regions of China and SE-Asia, and other specific tapeworms and flukes, are transmittable via **fish as compulsory intermediate hosts**. Fig. 2 illustrates the transmission cycles of passively carried pathogens, such as viruses, bacteria, and protozoa, and of helminths which use fish as intermediate hosts.

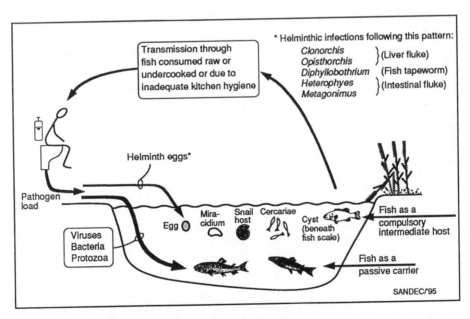

Fig. 2 Transmission of Excreted Infections Through Fish

There are currently over 100 viruses known to be faecally excreted by man. Public health importance of **viral infections** is based mainly on their low infective dose (< 100). Hepatitis A and polioviruses are epidemiologically the most important excreta-related viruses in temperate zones. Of the numerous types of ***Salmonella***, which are widespread world-wide, both humans and animals act as reservoirs (except for *Salmonella typhi* which is exclusively a pathogen of man). Food poisoning is often caused by *Salmonella*. These phenomena and the fact that the majority of infections follow an asymptomatic course or remain unreported, renders salmonelloses epidemiologically important. For the reasons outlined above, relevant investigations on pathogen survival and transmission through aquaculture (besides the use of indicator bacteria such as *E. coli*) have focused on enteric viruses and *Salmonella* mainly (see chapter on microbial studies).

Factors Determining the Transmission of Excreta-Related Infections and Their Importance in Human Waste Use in Aquaculture or Agriculture

Table 3 lists the factors and their role for the transmission of enteric infections through the use of excreta or wastewater in aquaculture or agriculture. In aquacultural use, pathogenic organisms may accumulate in the intestines and tissues of fish, crustacea and shellfish.

Table 3   Factors Determining the Relative Importance of Excreta and Wastewater Use in Transmitting Excreta-Related Infections

Pathogen	Survival in the environment	Infective dose	Immunity	Predominant routes of infection	Latency	Importance of pathogen transmission through fish, shellfish or crustacea
Helminths	Long	Low	None or Little	Soil, crops	Long	Fish as compulsory intermediate hosts for liver flukes and fish tapeworms
Protozoa	Short	Low to Medium	None or little	Person-to-person, food, water	Zero	No published data
Bacteria	Short to Medium	Medium to High	Short to Medium	Person-to-person, food, water	Zero	Yes
Viruses	Medium	Low	Long	Person-to-person	Zero	Yes

Die-off or **survival** of excreted pathogens is an important factor when considering the public health dimension of waste-fed aquaculture. In principle, all pathogens die off exponentially upon excretion, however, at different rates. Exceptions are pathogens whose intermediate stages multiply in compulsory intermediate hosts, such as the miracidia of *Clonorchis* or *Schistosoma* which multiply in aquatic snails and are later released into the water body. Table 4 lists survival periods in faecal sludges and wastewater for temperate and tropical climates [4, 15, 10].

The main **factors** influencing die-off are **temperature, dryness** and **UV-light**. Die-off rates increase in proportion to the level or intensity of these variables. NH_3 (ammonia) is known to be bactericidal. Therefore, bacteria die-off is enhanced at pH $\geq$ 8.5-9 due to an increase in the NH_3/NH_4 ratio.

Another important factor is the **infective dose** of pathogens or the dose required for a human host to contract the disease. For helminths, protozoa (*G. lamblia*, amoeba) and viruses, the infective dose is low (< 10^2). For bacteria, it is medium (~ 10^4) to high (> 10^6). This influences the choice and effectiveness of health protection strategies in reducing transmission risks from the various kinds of pathogens. Manifestation of the disease differs among the pathogens: with viruses, protozoa and bacteria, an infected person will either become ill or not, depending on whether the infection is symptomatic. With helminths, however, an infected person will exhibit various degrees of disease intensities depending on the number of worms it carries in its intestines. Thus, when considering an excreta or wastewater treatment strategy, a reduction of worm eggs by e.g. 80-90% already has a major positive public health effect. With viruses, bacteria and protozoa, higher degrees of removal are required; i.e., 99.9-99.999% or 3-5 log cycles.

Latency is the period between the excretion of a pathogen in the human stool and its becoming infective again. It is zero for protozoa, bacteria and viruses which. Therefore, short transmission routes such as person-to-person contacts normally play a proportionally more important role for these pathogens than longer transmission paths such as via soil or crops. .For helminthic infections, the latency lasts for several weeks. This renders helminths potentially important in the longer transmission paths such as through reuse in agriculture or aquaculture.

Table 4   Pathogen Survival in Faecal Sludge and in Wastewater

| Organism | Average Survival Time at Ambient Temperature in Wet Faecal Sludge and Wastewater | | | |
| | In **temperate** climates (10-15 °C) [days] | | In **tropical** climates (20-30 °C) [days] | |
	Wet f. sludge	Wastewater	Wet f. sludge	Wastewater
• **Viruses**	< 100	> 50	< 20	< 50
• **Bacteria:**				
-**Salmonellae**	< 100	< 150	< 30	< 30
-**Cholera**	< 30	< 30	< 5	< 5
-**Fecal coliforms**	< 150	< 120	< 50	< 30
• **Protozoa:**	< 30	> 50	< 15	< 15
• **Helminths:**				
-**Ascaris eggs**	**2-3 years**	(accum. in sludge)	**10-12 months**	(accum. in sludge)
-**Tapeworm eggs**	**12 months**	(accum. in sludge)	**6 months**	(accum. in sludge)

Based on the above, a ranking of pathogens with respect to their relative importance in creating a health risk in waste-fed aquaculture is proposed in Table 5. Such a ranking has previously been established for the situation typical of developing countries [16]. There a large number of enteric infections are endemic, numerous parallel transmission routes exist, and the socio-economic status and level of hygiene are typically low. The risk of disease transmission through waste-fed aquaculture in industrial societies might be of greater epidemiological significance than in developing countries, since the prevalence and incidence of these infections have been drastically reduced in the past decades. Helminthic diseases have been practically eliminated. Of the enteric protozoal diseases, *Giardia lamblia* remains of importance also in temperate climates with prevalences ranging from 2-5 %.

**Table 5   Relative Health Risks from Untreated Wastewater
and Excreta Use in Agriculture and Aquaculture**

Type of pathogen	Relative amount of excess frequency of infection or disease through reuse [1]	
	In developing countries [16]	**In industrialised regions**
1. **Intestinal nematodes** (roundworms): *Ascaris, Trichuris,* hookworm, e.g.	High (in agriculture primarily)	Least
2. **Bacteria**: *Salmonella, Shigella,* bacterial diarrhoea (e.g. cholera, typhoid)	Medium (in agriculture primarily)	Medium
3. **Viruses**: Hepatitis A, polio, viral diarrhoea	Least	High (particularly in shellfish and crustacea)
4. **Trematodes (flukes) and cestodes (tapeworms)** : • schistosomiasis • chlonorchiasis • taeniasis	From high to nil, depending upon the particular waste utilisation practice and local circumstances	Least

[1] Excess frequency is the risk attributable to the particular route of transmission occurring over and above the frequency already existing in the particular community

The epidemiology of e.g. hepatitis A has been thoroughly investigated. Nowadays, infection occurs at a later age than previously. In industrial societies, the proportion of persons having acquired hepatitis A immunity is also steadily decreasing. A similar shift has occurred with the epidemiological pattern of poliomyelitis [4]. Table 5 shows the relative importance of the various types of pathogens in the use of untreated excreta or wastewater, as established by WHO [16] for developing countries and as proposed for industrial societies with high socio-economic levels and hygiene standards. The information in Table 5 relates to the use of untreated wastes. Risks can be reduced or eliminated if proper health protection measures, such as waste treatment, good (hygienic) manufacturing practices in fish processing, and adequate personal hygiene are adopted. Adequate cooking destroys pathogens except possibly in shellfish (see chapter below) and proper conservation of food prevents the growth of pathogens .

RELEVANT MICROBIOLOGICAL STUDIES

Virus Studies

Analysis of hepatitis A virus in environmental samples such as in wastewater or in fish is extremely difficult. Evidence of its occurrence is indirect; i.e., based on the isolation of the virus from infected persons and inferred from the occurrence of enteroviruses, such as polio-, echo- and coxsackie viruses or other enteric viruses. Feachem et al. [4] have extensively reviewed literature on virus uptake by molluscs, such as oysters, mussels, cockles, and clams, and by crustacea, such as crabs, lobsters, shrimps, or prawns. All bivalve molluscs are filter feeders and concentrate pathogenic organisms from the water in their tissues. Oysters may filter up to 1,500 litres of sea water per day in looking for food. This may lead to >100 times higher concentrations than in the surrounding waters. Crustacea, too, accumulate enteric viruses if they live in pathogen contaminated water or if they feed on contaminated molluscs. Shellfish grown in polluted waters pose a public health risk particularly when they are eaten raw - a common habit in many societies. However, a residual risk remains even after cooking: 7-20 % of viruses were found to survive stewing, frying, steaming or boiling (cited in [4]).

When common carp was exposed to water containing $7x10^7$ bacteriophages/l [3] for 7 days, phages were recovered from all fish organs, including muscles. Phage levels in the digestive tract was higher than in water by a factor of 6. In water containing $3x10^5$ phages/l, a virus concentration which may already be considered very high, none of the organs except the digestive tract contained phages after 30 days of exposure [17].

Viruses cannot multiply outside living cells, and it has been found that human viruses do not multiply in fish or shellfish. Virus survival is generally prolonged at low temperatures and with the presence of particulate matter, however, die-off rates vary considerably according to the type and strain of virus. Viruses appear to survive longer within shellfish than within contaminated water. Enteric viruses in oysters were found to survive refrigeration for one to four months. At freezing temperatures, 10-35 % of the viruses survived for one month or more.

Depuration of viruses from fish and shellfish is often proposed or practised prior to their consumption. Depuration from shellfish in clean water is a mechanical process and takes place at a rate of 2-3 orders of magnitude within 2-4 days [4, 18]. Depuration therefore would have to last from a few days to a few weeks depending on the accumulated virus concentration in the shellfish. Depuration works best if the shellfish are at high feeding activity with temperatures around 20 °C and salinity > 18 ppt.

Bacterial Studies

Several studies regarding the bacterial contamination of fish reared in sewage-fed fish ponds have been performed and are reported in the literature [19, 20, 21, 22]. An extensive review has been published by Edwards [23]. Muscle tissue, as the edible part of the fish, is relatively safe from the invasion of bacteria (and viruses). Invasion occurs only if the fish grow in water containing high levels of microorganisms, among them pathogens. Buras et al. [19] report that when fish accumulate bacteria beyond a given threshold concentration (approx. 10^4 tot. bacteria or SPC/g or ml of *Oreochromis aureus* (tilapia) or *Cyprinus carpio*, (common carp), their capability to cope with high concentrations of particles diminished and bacteria began to invade muscle tissues and other organs. Bacterial concentrations in pond waters, which corresponded to the threshold levels in fish, amounted between 1.0 and $5.0x10^6$ SPC/100 ml.

CEPIS [20] investigated the hygienic quality of *Oreochromis niloticus* (Nile tilapia) in fish ponds fed with maturation pond effluent of approx. 10^4 faecal coliforms/100 ml. FC concentrations within the fish ponds ranged from 10^2 to $3.3x10^3$ /100 ml. The fish were allowed to grow for 4-5 months. *Salmonella* recovery from the fish fillets were always negative. Total coliform (TC) levels ranged from 10^1 to $2x10^3$/g per fish fillet and may indicate FC levels amounting to 10^0 - 10^2/g. Van den Heever and Frey [22] compared the hygienic quality of catfish (*Claris gariepinus*)) grown in a wastewater treatment plant effluent containing $6.3x10^3$/100 ml faecal coliforms and $7.2x10^3$/100 ml coliphages, with fish grown in an impounded reservoir ($1.3x10^2$ FC/100 ml and $2.6x10^3$/100 ml, respectively). Most of the water samples from the wastewater treatment plant (pond scheme) were *Salmonella* positive, however, only a small proportion of the reservoir water contained bacteria. No faecal coliforms, *Salmonella* or coliphages were isolated either from the fillets of wastewater-fed fish or from the fillets of fish grown in the "natural" water.

Depuration from fish, where bacteria and other pathogens mainly accumulate in the digestive tract, appears to be variably effective. It is dependent on the microbial load of the fish prior to depuration, as well as on depuration time. Muscle tissue appears to depurate more slowly than other organs, if at all. Periods of several weeks might be required [19].

---

[3] Bacteriophages are viruses of bacteria and are used as indicators for viral pathogens.

Microbiological evidence acquired to date suggests that there is little likelihood of enteric organisms, including pathogens, invading edible fish tissues if the FC concentration in fish pond water is < 10^3 per 100 ml [19]. However, even at lower microbial levels in the pond water, the fish usually accumulate high load of microorganisms in the digestive tract and in the intraperitoneal fluid. The public health risk of waste-fed fish may therefore be associated mainly with the potential cross-contamination of the fish fillets through high loads of microorganisms contained in other organs of the fish. Persons processing the fish may become foci of disease transmission unless they adhere to good personal and institutional hygiene.

EPIDEMIOLOGICAL EVIDENCE

Blum and Feachem [24] reviewed the epidemiological evidence for disease transmission associated with the aquacultural use of excreta and wastewater. There is clear epidemiological evidence for the transmission of certain trematode (fluke) diseases, principally *Clonorchis* (oriental liver fluke), *Fasciolopsis* (giant intestinal fluke), and for *Schistosoma* (bilharzia). A two-year study on health assessment of nightsoil use and treatment carried out in several locations in China from 1988-90, among them Jiujian in Jiangxi Province, revealed a 10% *Clonorchis* prevalence in pond workers and a 35% *Fasciolopsis* and *Schistosoma* prevalence [25]. Control groups showed zero prevalence for these pathogens. Blumenthal, Abisudjak and Bennett [26] investigated the risk of diarrhoeal disease associated with excreta fertilisation of fishponds in Indonesia. The study encompassed a total of 2000 exposed and control families. Results indicate that children under 5 years are at increased risk through pond water contacts. Further to this, individuals consuming fish from the excreta-fertilised ponds but otherwise not having any pond water exposure, appear to be at an increased diarrhoeal risk, too. There is circumstantial evidence from several hepatitis A outbreaks reported from the United States that infection was caused by the consumption of shellfish grown in polluted waters (several authors cited in [4]). Passive carriage of pathogens by fish or shellfish grown in waste-fed ponds must therefore be considered an important potential transmission route.

WASTEWATER TREATMENT

In **conventional** sewage treatment plants, comprising sedimentation followed by biological treatment (trickling filters or activated sludge), the removal of excreted viruses, bacteria and parasites (protozoa and helminth eggs) is primarily by discrete settling (helminth eggs), by entrapment in settleable flocks, or by adsorption onto settling particles (bacteria and viruses). Most microorganisms thus concentrate in the treatment plant sludge where they survive for varying periods of time depending on the sludge treatment applied. Virus adsorption by the biological slime in trickling filters, and elimination by predation, are additional factors in pathogen elimination. Retention times in conventional sewage treatment plants ($\leq$ 24 hours) are too short for natural die-off to play a significant role in pathogen reduction.

The effectiveness of **rapid sand filtration** in removing excreted pathogens depends on the kind of pathogen. While helminth eggs and protozoal cysts are likely to be completely eliminated due to their relatively large size (5-70 μm), the removal of bacteria and viruses is less reliable. Bacteria are from 1-10 μm and viruses from 0.001-0.1 μm in size. Fig. 3 below illustrates the % elimination of particles by deep-bed filtration as a function of particle size [27]. The diagram reveals that the removal of bacteria size particles by filtration of activated sludge effluent is 20-30 % more efficient if the process is preceded by $FeCl_3$ coagulation (contact filtration) than with non-coagulated wastewater. Bacteria-size particles are removed by 1-2 log cycles. For virus-size particles ($\leq$ 0.1 μm), elimination must be assumed at << 50 %; i.e., 1 log cycle or less.

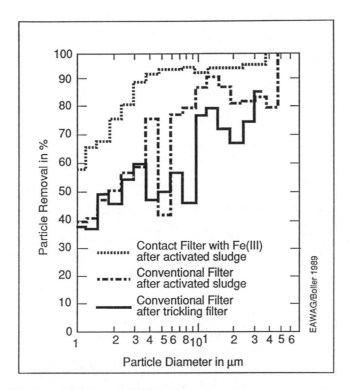

Fig. 3 Particle Removal in Deep-Bed Filters as a Function of Particle Size and of Pre-Filtration Coagulation (Boller 1989)

Faecal coliform (FC) and coliphage elimination at the Zurich-Werdhoelzli advanced wastewater treatment plant was investigated in 1988 and 1989 [28]. The plant comprises activated sludge treatment (including nitrification) with simultaneous phosphorus precipitation, secondary sedimentation and rapid sand filtration of coagulated secondary effluent (contact filtration). Table 6 shows FC and coliphage removals by the activated sludge and filter treatment. Higher removal efficiencies were observed for bacteria than for viruses. Virus removal by coagulation-filtration was investigated also by the County Sanitation Districts of Los Angeles County [9]. In a system comprising coagulation, sedimentation and filtration, a so-called "complete treatment", poliovirus removal was 1.3-1.5 $\log_{10}$ cycles. In contact filtration, removals were about 90 % (1 $\log_{10}$ cycle).

Table 6   Faecal Coliform (FC) and Coliphage Removal by Activated Sludge and Contact Filtration at the Zurich-Werdhoelzli Wastewater Treatment Plant

Sampling site	Faecal coliforms		Coliphages	
	Concentration [no./100 ml]	% removal in treatment step	Concentration [PFU/l]	% removal in treatment step
Act. sludge - Influent - Settled effluent	1.5×10^7 9×10^4	99 (2 $\log_{10}$ cycles)	1.5×10^6 6.5×10^4	96 (~2 $\log_{10}$ cycles)
Contact Filter - Effluent	1.2×10^4	87 (~2 $\log_{10}$ cycles)	4×10^4	38 (~1.5 $\log_{10}$ cycle)

Waste stabilisation ponds are very effective in pathogen reduction. They are particularly important in tropical and subtropical areas for the production of effluent qualities suitable for the safe use in agriculture or aquaculture. Non-aerated ponds in these climates require a gross surface area of approx. 3 m^2 per capita. They are also widely used in temperate and cold climates in Europe though primarily in schemes for less than 10,000 inhabitant equivalents where they require a gross area of 12-14 m^2 per capita. However, there they are designed mainly to produce effluents complying with environmental discharge standards. WHO [16] in its health guidelines on wastewater use in agriculture and aquaculture recommends a faecal coliform (FC) limit of $\leq$ 10^3/100 ml (geometric mean) for irrigation of vegetables consumed uncooked. Wastewater used in fish or macrophyte ponds should have a maximum of 10^3-10^4 FC/100 ml and an absence of eggs of flukes and tapeworms. In tropical and subtropical climates, the bacterial criterion can be satisfied with a total of 20-30 days retention. Bacteria and virus removal and inactivation is achieved by entrapment in and adsorption onto settleable particles and flocs, by the algae-induced pH increase, by the effect of UV-light, time and temperature. pH levels between > 8.5 and 9, which are typically reached in ponds in the afternoon, are mainly detrimental to bacteria. To protect irrigation workers, wastewaters should contain less than one intestinal nematode (such as *Ascaris* or hookworm) per litre (arithmetic mean). Complete helminth egg removal is achieved within a 10-14 days retention time; i.e., by about half of the required pond surface to meet the bacterial guidelines. According to CEPIS (1990), effluents containing $\leq 10^3$ FC/100 ml exhibit very few if any *Salmonella*. There is evidence that viruses are very likely also not found in such effluents.

Table 7   **Expected Removals of Excreted Microorganisms in Various Wastewater Treatment Systems (Based on [29] and Table 6 Above)**

Treatment Process	Removal in $\log_{10}$ units of[1]			
	Viruses	Bacteria	Cysts	Helminth eggs
• Primary sedimentation, plain	0-1	0-1	0-1	0-2
• Primary sedimentation, coagulated	0-1	1-2	0-1	1-3
• Activated sludge + secondary sedimentation	0-1	0-2	0-1	0-2
• Trickling filter + sediment'n	0-1	0-2	0-1	0-2
• Contact filtration of sec. effluent (see also Table 6)	0-1	1-2	1-3	1-2
• Waste stabilisation ponds	1-4	1-6	1-4	1-3
• Chlorination or ozonation[2]	0-4	2-6	0-3	0-1

[1] The higher numbers in the ranges indicated in the table relate to well-designed, non-overloaded and well operated treatment units   [2] not recommended

Faecal coliform removals in a **planted soil filter** plant treating domestic wastewater and in a plant treating wastewater originating from a cheese factory were observed to range from 1-3 $\log_{10}$ cycles (90-99.9 %) [30]. Hydraulic retention times in the plants investigated were approx. 16 and 48 hours, respectively.

Table 7 summarises the range of pathogen removal efficiencies observed and reported in literature for various sewage treatment processes.

MICROBIAL GUIDELINES AND STANDARDS

What Type of Barrier for Health Protection ?

There are various **barriers** to protect the health of consumers and workers from excreted pathogen transmission through wastewater-fed aquaculture. They are listed according to the path sequence of the aquacultural product:

a   Minimising the level of excreted pathogens in the water where fish or plants are grown; i.e., prescribing the **microbiological water quality**

b   Maintaining an adequate personal and **institutional hygiene** during transport, handling, processing, and marketing of the products

c   Treatment for **conservation** of raw, unprocessed products prior to sale and consumption, or **cooking** prior to consumption

Engineered wastewater-fed aquaculture is to the author's knowledge not practised yet on a commercial scale neither in Europe nor in North America (with the exception of the Munich fisheries described in the introductory chapter). For the Munich fisheries, the hygienic quality of the harvested fish is being controlled prior to sale. This is equivalent to setting a **type a barrier**. Wherever the introduction of wastewater-fed aquaculture would be considered, minimum standards for pond water quality or equivalent treatment should be made a legal requirement (**barrier type a**). Evidence from microbiological studies carried out to date would suffice to provide the scientific basis for setting suitable standards. Such regulations would be a basic and necessary condition to attain social acceptance of wastewater-fed aquaculture.

Faecal contamination both from humans and animals is omnipresent, and any fish, whether grown in or harvested from natural or aquacultural water bodies, will carry faecal indicator organisms or be potentially contaminated by excreted pathogens. Most regulatory agencies therefore do not stipulate microbiological standards for freshly caught fish, shellfish or crustacea, but specify and enforce standards for processed products. In this way; i.e., by **barriers of type b** and **c**, control can be exerted over food processing and marketing.

Guidelines

WHO [16], the Joint FAO/WHO Codex Alimentarius Commission (CAC) [31], the International Commission on Microbiological Specifications for Food (ICMSF) (a body of the International Union of Microbiological Societies) [32], and the International Consultative Group on Food Irradiation [33] (composed of FAO, the Int. Atomic Energy Agency and WHO) have formulated **guidelines** which either directly relate to wastewater-fed aquaculture, to the hygienic quality of fish, shellfish or crustacea, or to the hygiene requirements in food handling and processing.

The WHO guidelines [16] formulated for wastewater use in aquaculture are based mainly on studies and experimental wastewater-fed schemes in Israel, U.S.A., and Peru from where most of the then available data originated. Wastewater used in aquaculture should be free from flukes (trematodes) and tapeworms (cestodes) and contain FC levels of $\leq 10^3\text{-}10^4/100$ ml. ICMSF [32] and the Int. Consultative Group [33] have formulated guidelines for the microbial quality of fish products. Both organisations stipulate "acceptable" and "rejectable" levels. Table 8 lists the proposed limits.

CAC [31] has published a draft code of hygiene practice for aquacultural products. The code prescribes the hygienic practices required in the production, handling, transport, and processing of these products. For the microbiological quality of wastewater-based aquaculture in warm climates, CAC refers to the above WHO guidelines and stipulates that "more data are needed before faster expansion of this practice can be generally recommended".

Table 8 Microbial Guidelines for the Quality of Fish and Crustacea [1] as Formulated by ICMSF [32] and the FAO/IAEA/WHO Consultative Group [33]

Indicator or pathogen	ICMSF[2]		Consult. Group[3]	
	Acceptable level no./g	Rejectable level no./g	Acceptable level no./g	Rejectable level no./g
• SPC	500,000	1,000,000	500,000	10,000,000
• *E. coli*	11	500	--	--
• *Salmonella*	0	--	--	--

[1] Guideline values apply to the weight unit of the edible part of the product. Not more than three out of five samples are allowed to fall between these two levels.
[2] The stipulated values apply to fresh, frozen and cold-smoked fish
[3] The stipulated values apply to fish and crustacea

Standards

Through a limited search comprising the Commission of the European Community (CEC), Swedish authorities and the Swiss Federal Health Department, the author has attempted to collate information on regulations relating to the hygienic quality of fish, shellfish and crustacea, as well as of water in which they grow or are being produced. In its ordinance relating to the quality of freshwater from which fish are harvested [34], **CEC** lists physico-chemical parameters including heavy metals, yet no microbiological criteria are included. In 1991, CEC introduced an ordinance establishing the water quality criteria in natural environments and in aquaculture from where shellfish are harvested [35]. The ordinance stipulates that *Salmonella* and *Vibrio* should not be detectable in such waters. Moreover, four classes of water and levels of restriction have been defined as shown in Table 9.

Table 9 Classification of Shellfish Waters as per Ordinance of the Commission of the European Community [35]

Class	Faecal coliform limit	Measures or restrictions
1	< 300/100 ml	None; direct sale of shellfish allowed
2	< 6,000/100 ml in 90 % of samples	Sale allowed only after depuration or thermal treatment
3	< 60,000/100 ml	Depuration for at least two months or thermal treatment required
4	> 60,000/100 ml	Harvesting prohibited

Swedish regulatory agencies have drafted microbial quality standards for fish, shellfish and crustacea, as well as for processed products. The proposed rejectable level for raw and frozen packed fish fillet amounts to 10^5/g SPC and 10^1/g *E. coli* [36].

The **Swiss** ordinance regulating the microbiological quality of food products, the "hygiene ordinance" [37] was revised and enacted in June 1995. No standards have been established for fresh fish, however, the regulations stipulate tolerance levels in smoked fish which are 10^6 total aerobic bacteria per g, 10^3 enterobacteriaceae (enteric bacteria) per g, and 10^3 staphylococcus aureus per g.

CONCLUSIONS

The pathogen transmission risk through wastewater-fed aquaculture represents a controllable risk; i.e., pathogen loads can be reduced to acceptable levels and adequate measures adopted. A zero risk may not be attained, however, it would neither be necessary nor feasible due to the omnipresence of excreted pathogens and for economic reasons. Microbial requirements for waste-fed aquacultural schemes should be compatible with background levels in natural waters, since the harvesting of fish and other aquatic animals is generally unrestricted and socially accepted. Wastewater treatment, adequate institutional hygiene during handling and processing of the aquacultural products, adequate personal hygiene, food conservation treatment, and cooking are suitable health protection measures. Through them, public health can be reliably protected. The quality of the products from waste-fed aquaculture, such as fish, shellfish, and crustacea, would have to comply with the hygiene regulations for food grown under "natural" conditions.

REFERENCES

[1]   Edwards, P. and Pullin, R.S.V., eds. (1990). Wastewater-Fed Aquaculture. *Proceedings, International Seminar on Wastewater Recycling and Reuse for Aquaculture.* Calcutta, 6-9 December.

[2]   Strauss, M. and Blumenthal, U.J. (1990). *Human Waste Use in Agriculture and Aquaculture - Utilization Practices and Health Perspectives.* IRCWD Report 08/90. EAWAG/SANDEC, Duebendorf, Switzerland.

[3]   Prein, M. (1990). Wastewater-Fed Fish Culture in Germany. In: *Wastewater-Fed Aquaculture,* Proceedings of the International Seminar on Wastewater Reclamation and Reuse for Aquaculture, Calcutta, India, 6-9 December, 1988.

[4]   Feachem, R.G., Bradley, D.J., Garelick, H., Mara, D.D. (1983). *Sanitation and Disease - Health Aspects of Excreta and Wastewater Management.* John Wiley & Sons.

[5]   Schwartzbrod, L. (1991). Virus et milieux hydriques. In: *Virologie des milieux hydriques.* Schwartzbrod, L., ed. Tech. et Doc., Paris (in French).

[6]   CEPIS (1990). *Health Risk Evaluation Due to Wastewater Use in Agriculture.* Panamerican Centre for Sanitary Engineering and Environmental Sciences, Lima, Perú.

[7]   Akin, E., Jakubowski, W., Lucas, J.B., Pahren, H.R. (1978). Health Hazards Associated with Wastewater Effluents and Sludge: Microbiological Considerations. In: *Proceedings of the Conference on Risk Assessment and Health Effects of Land Application if Municipal Wastewater and Sludges,* Sagig and Sorber, eds.. Center for Applied Research and Technology, The University of Texas at San Antonio, San Antonio, Texas 78285.

[8]   U.S. Environmental Protection Agency (1992). *Guidelines for Water Reuse.* EPA/625/R-92/004. Technology Transfer Manual.

[9]   Asano, T. and Mujeriego, R. (1998). Ptretreatment for Wastewater Reclamation and Reuse. In: *Pretreatment in Chemical Water and Wastewater Treatment.* Springer, Berlin and Heidelberg.

[10]   Schwartzbrod, J. (1994, 1995). Personal communications.

[11]   Al Salem, S.S. and Tarazi, H.M. (1989). Wastewater Reuse and Helminths Infestation in Jordan: A Case Study. In: *Regional Seminar on Reuse of Treated Effluents.* WHO/CEHA, Amman, Jordan.

[12]   World Health Organization (WHO)/United Nations Environment Programme (UNEP) (1987). *Global Pollution and Health. Results of Health-Related Environmental Monitoring..* GEMS, Global Environmental Monitoring System. Geneva.

[13]   Cantonal Laboratory Zurich (1995). Personal communication.

[14]   Rao, S.N., Chaubey, R., Srinivasan, K.V. (1990). Ganga Water Quality in Bihar. *Indian Journal of Environmental Health,* **32**, 2, 393-400.

[15]   Strauss, M. (1985). *Health Aspects of Nightsoil and Sludge Use in Agriculture and Aquaculture - Part III::* Pathogen Survival. IRCWD Report no. 04/85 (International Reference Centre for Waste Disposal, CH-8600 Duebendorf, Switzerland).

[16]   World Health Organization (WHO) (1989). *Health Guidelines for the Use of Wastewater in Agriculture and Aquaculture.* Report of a WHO Scientific Group, WHO Technical Report Series 778.

[17] Buras, N., Hepher, B., Sandbank, E. (1982). *Public Health Aspects of Fish Culture in Wastewater*. IDRC, International Development Research Centre, Ottawa, Canada.

[18] Mann, R., Vaughn, J.M., Landry, E.F., Taylor, R.E. (1979). *Uptake of Heavy Metals, Organic Trace Contaminants and Viruses by the Japanese Oyster, Crassostrea Gigas, Grown in a Waste Recycling Aquaculture System*. Woods Hole Oceanographic Institution Technical Report WHOI-79-50.

[19] Buras, N., Duek, L., Niv, S., Hepher, B., Sandbank, E. (1987). Microbiological Aspects of Fish Grown in Treated Wastewater. *Water Research*, **21**, 1, 1-10.

[20] CEPIS (1991). *Reuso en Acuicultura de las Aguas Residuales Tratadas en Las Lagunas de Estabilización de San Juan*. Centro Panamericano de Ingeniería Sanitaria y Ciencias del Ambiente, Lima, Perú (in Spanish)

[21] Hejkal, T.W., Gerba, C.P., Henderson, S., Freeze, M. (1983). Bacteriological, Virological and Chemical Evaluation of a Wastewater-Aquaculture System. *Water Research*, **17**, 1749-1755.

[22] Van den Heever, D.J. and Frey, B.J. (1994). Microbiological Quality of the Catfish (*Clarias gariepinus*) Kept in Treated Wastewater and Natural Dam Water. *Water SA*, **20**, 2, 113-118.

[23] Edwards, P. (1992). *Reuse of Human Wastes in Aquaculture - A Technical Review*. UNDP-World Bank Water & Sanitation Program.The World Bank, Washington, D.C., U.S.A..

[24] Blum, D. and Feachem, R.G. (1985). *Health Aspects of Nightsoil and Sludge Use in Agriculture and Aquaculture - Part III: An Epidemiological Perspective*. IRCWD Report No. 05/85 (International Reference Centre for Waste Disposal, CH-8600 Duebendorf, Switzerland).

[25] Niu, S. and Ling, B. (1991). *Health Assessment of Nightsoil and Wastewater Reuse in Agriculture and Aquaculture* (Final Report, WHO Technical Services Agreement WP/ICP7RUD7001/RB/88-310). Chinese Academy of Preventive Medicine, Institute of Environmental Health and Engineering.

[26] Blumenthal, U.J., Abisudjak, B., Bennett, S. (1991). The Risk of Diarroeal Disease Associated with the Use of Excreta in Aquaculture in Indonesia, and Evaluation of Microbiological Guidelines for Use of Human Wastes in Aquaculture. In: *Proceedings, The Third Conference of the International Society for Environmental Epidemiology*, Jerusalem, 11-15 August, 1991.

[27] Boller, M. (1989). Verfahrenstechnik der Chemischen Phosphor-Elimination. *Chem.-Ing.-Tech.*, MS 1810/89 (in German).

[28] Wehrli, M. (1995). Personal communication.

[29] Mara. D.D. and Cairncross, S. (1989). *Guidelines for the Safe Use of Wastewater and Excreta in Agriculture and Aquaculture*. (WHO Geneva).

[30] Züst, B. (1995). Personal communication. Zentrum für Angewandte Oekologie, Schattweid, Switzerland.

[31] FAO/WHO Codex Alimentarius Commission (CAC) (1994). *Proposed Draft Code of Hygiene Practice for the Products of Aquaculture*. Codex Committee on Fish and Fishery Products, 21st Session, 2-6 May, Bergen, Norqay.

[32] ICMSF (1995). Sampling for Microbiological Analysis: Principles and Specific Applications. In: *Micro-Organisms in Food* (chapt. 2). The International Commission on Microbiological Specifications for Food. Blackwell Scientific Publications.

[33] International Consultative Group on Food Irradiation (1989). *Consultation on Microbiological Criteria for Foods to be Further Processed Including Irradiation*, 29 May - 2 June 1989, WHO, Geneva.

[34] European Commission (1978). Directive 78/659/CEE. Published in: Journal officiel des Communautés européennes, no. L 222/1, 14 August 1978.

[35] European Commission (1991). *Directive 91/492/CEE*. Personal communication by H.Belvèze, Législation Vétérinaire et Zootechnique, Direction Générale de l'Agriculture.

[36] Norrberg (1995). Personal communication.

[37] Swiss Federal Department of Health (1995). *Verordnung über die Hygienisch-Mikrobiologischen Anforderungen an Lebensmittel, Gebrauchsgegenstände, Räume, Einrichtungen und Personal*. (Hygiene ordinance). In German.

Environmental Research Forum Vols. 5-6 (1996) pp. 99-104
© *1996 Transtec Publications, Switzerland*

The Future of Ecological Engineering

B. Guterstam[1] and C. Etnier[2]

[1] Stensund Ecological Center, S-619 91 Trosa, Sweden

[2] Department of Economics and Social Sciences, Agricultural University of Norway,
P.O. Box 5033, N-1432 Åas, Norway

Keywords: Ecological Engineering, Nitrogen, Phosphorus, Recycling, Wastewater Systems

Abstract

Ecological engineering is a revolutionary influence on wastewater treatment. Design is developing in two different directions, large-scale systems for cities and smaller systems for individual neighborhoods or even houses. Both types are enhanced by an increasing consciousness of the importance of beauty. Continued development will be fastest through bold experiments at the local level, done by ecologists and engineers working together. Gradually, the concept of wastewater treatment will be replaced by the more general field of resource management.

The people assembled for this second international conference on ecological engineering for wastewater treatment have gathered to discuss the end of the throw-away attitude towards wastewater and its resources. Most of our future work will be concentrated on the techniques and benefits of recycling. Ecologically engineered systems are, however, only a minor part of both the infrastructure and the planning in most of the countries with temperate climates that this conference is primarily directed towards. Before ending the conference, we would like to give a report from the future of this young discipline.

What is ecological engineering?

What is this ecological engineering, or ecotechnology, whose future we are discussing? Examples and definitions can be found in Mitsch and Jørgensen [1], Etnier and Guterstam [2], and Straskraba [3]. We can sum them up and state that ecological engineering uses the resource management principles of ecosystems. In reality technical and ecological solutions cooperate. Maybe also political decisions should be included in the process?

In our own perspective, we only know that wastewater needs to make a comeback as a resource. This means that it should not be destroyed by mixing everything. It means that we must use the principles of recycling and management. Technical solutions must consider the long-term effects of natural laws, like those of thermodynamics, and the concept of limited resources (such as phosphorus), ecosystem diversity and primary productivity.

During this conference we have heard about everything from bacteria for biogas production to fish farming in large-scale systems. Just as ecological engineering requires cooperation with sanitary engineers, it can also be used in cooperation with urban hydrologists, fermentation technologists, and organic farmers, to name a few. We want to consider ecological engineering for wastewater treatment as generally as possible, without trying to exclude too much, since what we are really looking for is the best *feasible* solution. There are many systems which take their inspiration from ecological engineering without living up to all of its goals. The range of possibilities for ecological engineering is also different from what is achievable in 10 or 50 years. One key criterion for these systems: Keep track of your phosphorus and nitrogen *and keep them in a form you can reuse.*

"To the necessary alliance between ecologists and engineers"

This is the dedication to the proceedings of the first conference in this series, held at Stensund in 1991. As ecologists and engineers now work closer together, it is important that we speak the same language. Ecological engineering has its roots in systems ecology, which is a discipline working with the functional parts of nature that we call ecosystems. Investigations focus on the ecosystem metabolism, i.e. the flows of energy and materials managed by the diverse composition of organisms. Systems ecologists know that nature is very complex and that their studies can only lead to a simplified picture of reality.

An ecological engineer takes a wide responsibility for her or his solution to a practical problem. Howard T. Odum is the ecologist who not only lay the groundwork for and coined the phrase "ecological engineering," but also developed the science of systems ecology. According to him, an ecological engineer is an expert who at the same time has an holistic mind, i.e. he or she finds technical solutions to fit both nature and society. Of course, this is difficult and many of us know that the social side of the task can stop a project in its infancy.

Ecological engineering uses the processes of nature to organize the human relationship to such natural resources as fuels, nutrients, fields, fisheries, etc. This requires both theoretical and practical skills. Ecological engineers tend to put together nature in constructed ecosystems like wastewater-fed aquacultures, wetlands, integrated farms and even residential areas. Helpful tools are models and computers. The solutions delivered may be described using data on flows, populations dynamics, productivity, etc.

As ecologists, we were surprised at this conference to hear an engineer criticize ecological engineering for the lack of systems thinking. It is a sign of how little we still know about each other. Hopefully more joint projects and meetings will improve that understanding.

Tough years to come

For the future, we can see tough years to come for ecological engineering for wastewater treatment in Europe. As ecological engineering becomes more competitive with established solutions, it is also being fought by singe-minded defenders of old concepts.

For example, phosphate-containing detergents are forbidden both in Norway and in Switzerland. Did you know that phosphate-containing detergents have less negative impact on the environment than phosphate-free alternatives? This is the conclusion of a study called *The Swedish phosphate report* [4]. This investigation is sponsored by the chemical industry, written by consultants after *anonymously* interviewing 17 European wastewater scientists about what they *believe* will be the impact of these detergents on the aquatic ecosystem. It is quite curious that anonymous interviews are used to build the case for phosphate use, since the standard basis for assessing environmental impact in this century is measured results reported in signed papers.

The aim of the study seems to be to influence decision makers, especially in Norway and Switzerland, to change their minds, and allow the use of our limited plant nutrient phosphorus as a washing agent. The report recommends the Swedish wastewater treatment concept be implemented in the rest of Europe.

The Swedish wastewater establishment have recently concluded that their wastewater treatment systems also are the best solutions for the recycling of plant nutrients into food production [5]. Their conclusion about ecologically engineered solutions for wastewater treatment is that they are more expensive and less efficient. Their use in Sweden can only become complimentary. There is no place left for the visionary thinking of wastewater-fed aquacultures in Stockholm, or urine separation in the municipalities of Borlänge, Tanum or Trosa. There will be little money left over for research and development of serious projects in this field. We must state that a future for ecological engineering for wastewater treatment in Sweden is not seriously desired by conventional interests.

On the other side, the concept of ecological engineering, which is by definition sustainable, is promoted by people who do not give up that easily. To implement well-functioning systems of ecological engineering, the strategy must be continued efforts to seek alliances with holistically thinking wastewater engineers and to find working solutions with new systems and retrofitting of existing systems.

Reasons to implement ecological engineering

Population growth and urbanization has led to the need for wastewater treatment. The methods of industrial agriculture which have made urbanization possible squander nitrogen and phosphorus, both polluting water more and using up fossil fuel and fossil phosphorus rapidly. And human waste, concentrated far from agricultural areas, becomes a disposal problem. The same is true for animal manure on feed lots.

To bring order in this mismanagement of natural resources, we have use for a new terminology. Keeping to the end of the pipe, a new name is needed which focuses on the resource emitted from society. Why not rename "wastewater" to "recycling water." This would give the treatment plants the new task to become something like "fertilizer reclamation factories" or "nutrient processing centers."

Changes in Sweden

Understanding the state of the art and the vision of ecological engineering for wastewater treatment, together with wastewater treatment's history, can give us some clues about the future. We will take the history of Sweden, with which we are best acquainted.

Technically, the wastewater situation was already solved in the 1970's with about 80% of the urban population connected to advanced wastewater treatment plants, which include a three-step procedure with chemical phosphorus precipitation. For nitrogen, about 75% was still discharged and lost.

In the middle of the 1980's, it was found that most Swedish coastal waters were overloaded with nitrogen, and since then a large effort has been made to remove 50% of incoming nitrogen. One problem at a time has been solved concerning bacteria, BOD, phosphorus, and nitrogen. Unfortunately there is a lack of an holistic approach, where both water pollution and resource recycling could be addressed.

Another sign of the lack of holistic thinking is the relative lack of attention to nitrogen discharges from agriculture and emissions from traffic and other combustion, which constitute the bulk of the anthropogenic nitrogen to the Baltic. If Sweden is representative, no more than 10% [6] comes from wastewater.

The Stensund Wastewater Aquaculture was conceived in the beginning of the 1980's, and was finally constructed to serve the folk college in 1989 [7]. During this period, environmentalism had become fashionable in Sweden, and some people could envision a recycling society. The project has been overwhelmed by an interested and supporting public. Tens of thousands of visitors have come to the wastewater aquaculture. They include more than 25% of Sweden's 286 municipalities with their wastewater experts. Other important groups are students, school children, politicians, business people, and governmental agencies, both from Sweden and abroad. For them, holistic approaches to recycling the wastewater resources have become interesting. Many other people and organizations have begun promoting nutrient recycling from wastewater during this time. Interest in the Stensund project has reflected a latent public interest in new views of wastewater management. We can see a tension here between the public interest and the skepticism on higher levels to new technologies.

Growing understanding of the possibilities for ecological engineering

In 1991, the year of the first conference at Stensund, we weren't sure what the potential of ecologically engineered systems was. Since then, we have seen that human waste has great potential to replace those nutrients taken from the land in growing our food. This ought not to have been a surprise.A very interesting development has been urine separation as a viable alternative for capturing most of the nutrients in an urban setting. There is also a variety of natural and constructed land and wetland systems taken into operation for wastewater treatment.

At the Mid Sweden University in Östersund, the new three-year curriculum of Ecotechnics does a good job of meeting the educational needs of ecological engineers [8].

Also, the interest from regulatory authorities has increased since then, and not just in Sweden and the USA, whose authorities had representatives attending the 1991 conference. This time we have representatives from three governmental authorities and two municipalities participating.

Trends in development

Innovative solutions for wastewater treatment are appearing in different forms and on different scales. Here are some of the trends we see:

Small-scale, local solutions. At the extreme, the house is an independent watershed, because of its rainwater collection, and treatment plant, relying on minimal inputs of clean drinking water. In an urban area, such a system needs to be connected with nutrient transport back out to a farm.

Large-scale solutions for urban areas. This is a reality in Asia as in Calcutta, Hanoi or cities in China. In Europe, the Munich wastewater-fed aquaculture serves as a historical monument, where it was shown all this century that it can be feasible to use ecological engineering on a large scale. Maybe we can find tools for a sustainable Europe here [9]?

Better marriage of ecological engineering with ecological design. As buildings and neighborhoods are designed with a thought towards their taking care of their own water needs, the ecology will be more skillfully incorporated into the design and the buildings will be more beautiful and easier to live in [10, 11]. The barrier between the inside and the outside will be broken down, and the inside of our homes will perhaps resemble the African savannah our species evolved to live in [12].

Boldness and innovation. There is a paradigmatic revolution afoot and, as usual, the new ideas are coming from outside the establishment. The goals of the wastewater system are changing from safe disposal to resource management, and there are good ideas on how to do this amongst ecologists and urban hydrologists. These people are bold enough to go onto turf traditionally reserved for others and try out their ideas. It also means mixing science and PR in a new way, setting up experiments that invite the public in to new ways of conceiving their waste at the same time they try to refine the technology. We try to do this at Stensund; examples of other organizations which successfully combine science and publicity for new ideas are Ocean Arks International in Massachusetts, USA, and Biosphere II, in the desert of Arizona.

Prudence, humility, and local options. The "necessary alliance" must be forged between the ecologists and the engineers. Engineers understand pumps and pipes and all the other hardware that is common to both traditionally and ecologically engineered systems. Also, even the full-scale experiments with new methods need to be small, as we learn to handle the new technology. The Borlänge and Trosa urine projects mentioned at this conference are good examples of that [13, 14].

There is the possibility that systems for recycling the wastewater nutrient will be installed too quickly. Gun-Britt Mårtensson, a leading politician in Sweden, has proposed gradually replacing all toilets in Sweden with urine-separating ones [15]. At this time, we know too little about such systems to do this. What are the problems with storage of the urine? How should the urine be applied to the fields? What goes wrong when many different users employ the same system, or when the system gets large loads? Do people destroy the urine's value by throwing toxic chemicals down the toilet? These and many more questions need to be answered by small, yet full-scale, experiments before a national strategy based on urine-separating systems can be adopted.

A whole system that functions well in a complete eco-cycle needs to be built and tested for some years, otherwise you could end up with something analogous to automobiles in the US or nuclear power—poorly thought-out, unsustainable, and producing much waste that nobody knows what to do with.

Develop the technology with people. Assuming we can solve the hygienic problems of reusing wastewater, we have wondered how people respond viscerally. It turns out that if they get to experience the systems themselves, and the systems are designed to be beautiful as well as functional, then the people are sold on them. Some even become very foolish in their expectations, like some visitors to Stensund's facility who want to and even do drink the water exiting the aquaculture, despite our warnings that it neither fulfills drinking quality standards nor is meant to.

Scenarios for the future

It is said that an Arab proverb states, "He who predicts the future lies, even if he is right." There is always the possibility that wastewater's laser beam or microchip hasn't been invented yet and will change everything in a way impossible to foresee. But we know something about the biology of ecocycles, and nature has been doing experiments for us for a couple billion years, so we should have a good overview of the possibilities.

As we gaze into our cloudy crystal ball, we anticipate developments in the following time frames and places:

1. *Countries with sewage systems, in the next 50 years.* Sewage treatment plants will continue to be used. Industrial water will become cleaner through closed and clean production strategies and separate treatment, so the sludge will be more useable in agriculture. Urine separation will be experimented with successfully in medium to large cities, with many different variations on the theme. Some cities will also use large-scale ecologically engineered systems. Green roofs, porous concrete and asphalt, and skillful use of ponds and infiltration sites will minimize stormwater runoff and its effects both on treatment plants and waterways. Economic incentives will be developed and used to make it as foolish economically to throw away nitrogen and phosphorus as it is ecologically.

2. *Countries with sewage systems, 50-150 years from now.* The field of wastewater treatment will gradually be replaced by general resource management. Large-scale sewage systems will be slowly abandoned, going over reuse and treatment of the water in the house or neighborhood, with either local reuse of the nutrients in urban farming or systematic transport of the nutrients back to the countryside. Urban sprawl will have ceased, with more small towns built if the population continues to increase. This will increase the edge effect and make it easier to recirculate nutrients from the city to the countryside.

3. *Countries without sewage systems, in the next 50 years.* Large-scale investments in dinosaur systems will continue for a while, at the same time that experiments are performed with ecologically engineered systems. After a while, the economic and environmental advantages of these systems will become clear even to bankers and economists and they will become the preferred alternative. In 50 years, these countries may have a more advanced water infrastructure than the ones that currently have sewage systems.

Getting there

The best way to make a transition to a sustainable management of recycling water is to follow the poetically worded advice of a man unfortunately more known for his oppressive politics than openness to ecological engineering: "Let a hundred flowers bloom." Experiments in different social and climatic conditions, with different waters, will lead to a tool kit full of useful systems that can be combined as necessary.

And finally, the "necessary alliance" needs to be strengthened. It gives us hope, that the Swiss Federal Institute of Technology (ETH) and the Swiss Federal Institute for Environmental Science and Technology (EAWAG) have been active in planning this conference, which is taking place at the Engineering School of Wädenswil.

References

[1] W.J . Mitsch and S.E. Jørgensen. *Ecological Engineering: An Introduction to Ecotechnology*, Wiley-Interscience, New York (1989).

[2] C. Etnier and B. Guterstam, *Ecological Engineering for Wastewater Treatment*, Proceedings of the International Conference 24-28 March 1991, Stensund Folk College. Gothenburg, Sweden: Bokskogen, 365 pp. (1991).

[3] M. Straskraba, Ecotechnology as a new means for environmental management. *Ecological Engineering*, 2, 311-331 (1993).

[4] R. Wilson and B. Jones,*The Swedish Phosphate Report: A Delphi Study to Compare the Environmental Impacts of Deregent Builders in Nordic Countries and the Implications for the Sustainable Management of Freshwater Resources in Europe*,London: Landbank Environmental Research and Consulting, ISBN 0 9525639 0 8 (1995).

[5] VAV, *VA-teknik idag och i morgon - hygien, funktion och uthållighet*. Swedish Water and Wastewater Association (VAV; in Swedish; 1995).

[6] Swedish E.P.A., *Monitor 1988* (1988).

[7] B. Guterstam, Demonstrating ecological engineering for wastewater treatment ina Nordic climate using aquaculture principles in a greenhouse mesocosm. *Ecological Engineering - The Journal of Ecotechnology*, (in press).

[8] Ecotechnics, Curriculum 120 points., Mid Sweden University, Östersund, Sweden (1993).

[9] P. Edwards and R.S.V. Pullin (Eds.), *Wastewater-fed aquaculture. Proceedings of the international seminar on wastewater reclamation and reuse for aquaculture, Calcutta, India, 6-9 December 1988*, xxix+296 pp. Environmental Sanitation Information Center, Asian Institute of Technology, Bangkok, Thailand (1990).

[10] N. J. Todd and J. Todd, *Bioshelters, Ocean Arks City Farming: Ecology as the Basis of Design*, San Francisco: Sierra Club Books (1984).

[11] N. J. Todd and J. Todd, *From Eco-cities to Living Machines: Principles of Ecological Design*, Berkeley: North Atlantic Books (1994).

[12] M. Fredriksson and B. Warne, *På akacians villkor - att bygga och bo i samklang med naturen*. Warne Förlag (In Swedish; 1993).

[13] Trosa Municipality, *Reconnecting the City and the Countryside: Preventing Wastewater and Developing Sustainable Agriculture through Urine Recycling in Trosa Community*, Technical Description, EU Life proposal No. LIFE95/S/A13/S/320/ÖMSV (1995).

[14] L. Jarlöv, "Nutrient circulation with innovative toilets in multi-family dwellings." This volume (1996).

[15] Å. Askensten, "Förslag om jättelikt ROT-projekt," *VVS-Forum* 4, (in Swedish; 1994).

Workshop Presentations

Environmental Research Forum Vols. 5-6 (1996) pp. 107-110
© *1996 Transtec Publications, Switzerland*

Workshop 1:

Aquacultures for Wastewater Treatment

Summary and Conclusion by the Chairperson

G. Dave

Department of Zoophysiology, University of Göteborg,
Medicinaregatan 18, S-413 90 Göteborg, Sweden

Abstract

The primary objective of this paper is to address 6 questions on wastewater aquaculture. These are: 1. What is the state of the art of wastewater aquaculture? 2. Can aquaculture contribute to a sustainable use of wastewater? 3. Are there inherent dangers in this technology? 4. What should future research concentrate on? 5. What role can wastewater aquaculture play in wastewater treatment systems in the future? 6. What is the potential for wastewater treatment and food production in temperate climates? These questions were discussed at the 2nd International Conference on Ecological Engineering, and this paper is an attempt to provide very short answers to these very complex questions.

Introduction

The aquaculture workshop session of the 2nd International Conference on Ecological Engineering comprised 12 oral papers presented by researchers from 10 countries. Thus, the session was truly international and the speakers and the other participants contributed by suggestions and comments on the presentations and in the discussion on the questions given below. However, like the situation at most international conferences the time schedule was tight, and most discussions had to be interrupted before complete maturation. In connection with the 2nd International Conference on Ecological Engineering there was also an E-mail Conference for those who could not attend and some questions were brought up in relation to the papers presented. The answers were provided by the author and as a chairperson I had no role in coordination of the E-mail conference.

This paper is not a synthesis of the papers presented. This might have been worthwhile because the papers represented a variety of topics, and it may be difficult to put them all into a common perspective. However, in my opinion it is the responsibility of the author(s) to put their paper into its appropriate scientific perspective. Therefore, I leave this task to the author(s) and refer the reader to the actual paper(s) of interest.

Therefore, the primary objective of this paper is to address 6 questions on wastewater aquaculture raised by the organizing committee for the 2nd International Conference on Ecological Engineering. These questions (1-6) are given in the headings below.

1. What is the state of the art of wastewater aquaculture?

Wastewater aquaculture is an old technique which have been used extensively in both Asia and Europe. Recycling the resource, the major theme of this conference, is the objective for wastewater aquaculture. The technology to be used for accomplishing the task is a important element in ecological engineering. Wastewater aquaculture is still practiced extensively in Asia, but in Europe this technique has been replaced by activated sludge sewage treatment technology. If wastewater aquaculture shall expand in Europe new insights and elements must be incorporated into it. Therefore, the state of the art for wastewater aquaculture in Asia is fairly advanced, while in Europe it is at a very early stage of development.

The state of the art is usually described by reviewing the literature. This has been done quite recently by Edwards and Pullin ([1], [2]). Several reviews were also published in the proceedings from the 1st International Conference on Ecological Engineering ([4]). The next logic steps would be to organize a workshop with a limited number of invited experts in order to produce a state of the art report. Major ingredients in an evaluation of the state of the art are quality of handbooks, educational programs and scientific organization including conferences, journals and societies.

Excellent reviews on wastewater aquaculture are available in [2] and in [4], and there is probably no pressing need to review this topic right now. However, the overview of available courses for students on the topic of wastewater aquaculture ([3]) needs to be updated.

2. Can aquaculture contribute to a sustainable use of wastewater?

The answer to this question is a straight yes. Examples in support of this statement are numerous. Wastewater aquaculture provides a more direct recycling system for nutrients into food than the activated sludge treatment system and chemical flocculation practiced in the Western world today.

However, a more direct recycling may also impose greater risks for the consumer of the products. Furthermore, the present sewage treatment systems have replaced many aquaculture treatment systems that were operating in the past. Rather than criticizing the alternatives we should try to understand why aquaculture systems were replaced by the activated sludge treatment system. Rather than forget, we should try to learn as much as possible from the past experiences with aerated lagoons, fish ponds and wetlands before we construct new ones. In science there is always a risk that old findings be forgotten, and that we repeat the same tests and experiments that were made in the past. Education is the key to avoid repeating the same mistakes over and over again. This is true in other areas and it is probably also true in wastewater aquaculture. Therefore, educational programs on aquaculture and wastewater treatment are needed.

3. Are there inherent dangers in this technology?

Inherent dangers or risks connected with the use of wastewater aquaculture must be assessed in relation to prevailing alternatives. The alternatives are discharge into natural surface waters either directly or after sewage treatment.

Risk is the product of the probability of occurrence of a non-desirable event and the severity of this event (consequence) to the target. The target is often synonymous with humans, i.e. the consumer of the product or the treated water for recreational and/or household purposes (drinking water). But the natural environment, the receiving body of water is also a target, even if it has received less attention.

Wastewater aquaculture is the management of a hypereuthrophic aquatic environment. This can be an extremely productive system, but it can also be a stinking disaster if not managed properly. It is the responsibility of the manager to direct its production into edible components like fish. Key processes in the successful management of this system is photosynthesis/respiration and recycling of nutrients. Key parameters are the concentration of dissolved oxygen, pH, temperature and the concentrations of ammonia and nitrite ([7]). In a given situation many of these processes and parameters are interrelated and it may be sufficient to monitor only one or two of these, or to use other parameters like water clarity and fish behavior (surfacing and gasping for air) to adjust addition of wastewater and initiate forced aeration.

In order to determine the risk of an operation, if it is a industry, a sewage treatment plant or a wastewater aquaculture, it is essential that the operation is functioning. First when it is operating within its predefined limits is it possible to estimate the risk. Like all processes which discharge into the environment the immediate hazard is that production goes wrong. The long-term hazard is the total amount of toxicants discharged over time, and the risk is a function of the process and it environmental location.

The risk for the consumer is of two types and related to hygiene and toxicity. Disinfection of human pathogens is related to temperature and time of treatment and the health (pathogen) aspects of wastewater use was treated by Strauss at this conference (Strauss, this volume). The risk for food chain transfer of toxic chemicals should not be ignored, but it should neither be overemphasized. Known pathogens as well as known environmental contaminants must be measured in food products originating from wastewater aquaculture operations as well as from other sources including the wild environment. Therefore, the risk for the consumer should neither be neglected nor exaggerated, but determined by the same procedures as used for other food products and the same regulations and standards should be applicable to all food products irrespective of it origin.

4. What should future research concentrate on?

Future research should concentrate on real problems and not on anticipated, potential or constructed problems. Wastewater aquaculture is an applied science, which has to deal with certain specific problems. Some of these problems may not be covered by the basic sciences, but much of the knowledge needed in wastewater aquaculture can probably be gained from sciences dealing with general aquaculture and wastewater treatment.

However, in the application of principles and theories from other disciplines, certain gaps in knowledge may be revealed, and research may be needed to investigate unforeseen cause-effect relationships or to develop new culturing regimes or remedial actions. Experience and lack of knowledge of fundamental importance for wastewater aquaculture must guide the selection on what research that should receive funding. This implies that the research that is needed in one country is not necessarily the same that is needed in another country. There may be site specific needs related to the wastewater quality and climate and also species specific needs relating to the cultured species.

5. What role can wastewater aquaculture play in wastewater treatment systems in the future?

Wastewater aquaculture is a well established technique in many tropical and subtropical parts of the world today, especially in China and India. Historical aspects related to population density and availability of food have probably affected the social acceptance of fish produced by wastewater aquaculture. Today there is a difference in acceptance between ethnic groups. In a certain area there is also a difference in acceptance within the population. Therefore, in tropical and subtropical countries the future role of wastewater aquaculture is more of a social, public and political matter than a technological one. The technique is available, even if it certainly can be improved.

In temperate climates of the Western world wastewater aquaculture has been an important treatment technique rather than a food production reality, and today most wastewater fish pond aquaculture operations have been replaced by activated sludge treatment plants. Major reasons for abandoning wastewater aquaculture are climatic constraints (short growing season), areal limitation (cost of land is high in urban areas), and social acceptance of fish raised in human excreta. Today the situation is slowly changing and there is an increasing awareness among the administrators, the politicians and the public about the need for recycling of materials in society. This will certainly affect public acceptance for products cultured in wastewater. However, the health authorities and the public needs to know that the food is safe to eat. This has to be and can also be proven by conventional techniques for food analysis. Therefore, in the western world the general attitude towards recycling is positive, but the general attitude towards food production is critical.

6. What is the potential for wastewater treatment and food production in temperate climates?

Wastewater aquaculture has not been successful in temperate climates for reasons mentioned above. The winter season of temperate climates has been a constraint for a year-round operation of aquaculture and agriculture production in general, not only wastewater aquaculture production. There are two ways to overcome the climatic constraint. One is to change the climate by encapsulation of the treatment system as practiced at Stensund aquaculture and Ocean Arks, i.e. to use greenhouse technology ([5]). The other is to store the waste during the winter season. This is not practically feasible if all wastewater has to be stored, but it may be feasible for a fraction of it. This is especially so for the urine fraction, which contains the major part of the dissolved nutrients (N and P). Urine may then be used in agriculture and/or aquaculture during the growing season and replace the demand for inorganic fertilizers. The potential for this technique is enormous because it can save large amounts of energy. The production of 1 kg of nitrogen in fertilizers uses energy equivalent to 2 liters of oil. At least part of this energy may in stead be spent on greenhouse technology to produce high value crops like vegetables at the site of utilization, and save further energy used for transporting fresh food by air.

Item	Total Swedish cost (SK)	Estimated annual per capita cost (US$)
Buy of fertilizers	1,700,000,000	33
Nitrogen leakage	500,000,000	10
Nitrogen purification of wastewater	1,000,000,000	19
Phosporous purification of wastewater	1,250,000,000	24
Water for waste transport	150,000,000	3
Total sum:	4,800,000,000	89

Table 1. Estimated costs for not recycling the nutrients in domestic wastewater in Sweden (from [6]).

The annual per capita cost of not recycling the nutrients in domestic wastewater was recently estimated in Sweden by Schönbeck and Westberg [6]. The annual per capita cost was 89 US$ (Table 1). Thus, the economic potential for wastewater treatment and food production in temperate climates is very high, but the technology needed for accomplishment of this task is not developed so far. Therefore, the potential is very high but the probability to realize this potential in the near future is very low.

References

[1] Edwards, P. 1992. Reuse of human wastes in aquaculture: a technical review. Water and Sanitation Report No. 2, UNDP-World Bank Water and Sanitation Program, The World Bank, Washington D.C., 350 pp.

[2] Edwards, P. and R.S.V. Pullin (eds). 1990. Wastewater-fed aquaculture, Proceedings of the International Seminar on Wastewater Reclamation and reuse for aquaculture, Calcutta, India, 6-9 December 1988, xxix+296 pp. Environmental Sanitation Information Center, Asian Institute of Technology, Bangkok, Thailand.

[3] Edwards, P., R.S.V. Pullin and J.A. Gartner. 1988. Research and education for the development of integrated crop-livestock-farming systems in the tropics. ICLARM Studies and reviews 16, 53 pp.

[4] Etnier, C. and B. Guterstam (eds). 1991. Ecological engineering for wastewater treatment. Proc. Int. Conf., March 20-24, 1991, Stensund Folk College, Sweden. 365 pp.

[5] Guterstam, B. and J. Todd. 1990. Ecological engineering for wastewater treatment and its application in New England and Sweden. Ambio, 19:173-175.

[6] Schönbeck, A. and L. Westberg. 1995. Overfed water - undernourished land. Länsstyrelsen i Göteborg och Bohus Län, Göteborg, Sweden. (In Swedish)

[7] Timmons, M.B. and T.M. Losordo (eds). 1994. Aquaculture water reuse systems: engineering design and management. Developments in Aquaculture and Fisheries Science, 27, Elsevier, 333 pp.

Environmental Research Forum Vols. 5-6 (1996) pp. 111-116
© *1996 Transtec Publications, Switzerland*

Recycling the Resource from Greenhouse Cultures

B. Züst[1], J.B. Bächtiger[2] and J. Staudenmann[2]

[1] Centre for Applied Ecology Schattweid, CH-6114 Steinhuserberg, Switzerland

[2] School of Engineering Wädenswil, CH-8820 Wädenswil, Switzerland

Keywords: Aquaculture, Nutrient Solutions, Greenhouse Cultures, Plant Production, Fish Production, Plankton Production, Wastewater Treatment, Reuse and Recycling, Ecological Engineering, Indigenous Species, Protein Production

Abstract

The present report gives an overview over the first three years of an indoor aquaculture project at the School of Engineering at Wädenswil in Switzerland. The nutrients of culture solutions from soil-independent vegetable production are at the base of a constructed food chain: plants for consumption (*Nasturtium*) and algae and plants for plancton (*Daphnia*) or fish (*Tilapia mossambica* and *Ctenopharyngodon idella; Rutilus rutilus, Scardinius erythrophtalamus*) feed. The aim of the project is to gain experience in cleaning the water with indigenous species (making use of the nutirents) and producing in temperate climate.

1. Introduction

At the School of Engineering at Wädenswil (ISW) several gardening professions are taught in several departments. Horticulturists are introduced to different horticultural techniques, among others to the soil-independent production of vegetables. In 1993, a new, computer-monitored greenhouse was completed for this purpose. Culture solutions, having provided nutrients to growing vegetables (tomatoes, cucumbers, peppers), still contain nutrient salts easily accessible to plants. There are four ways to deal with depleated culture solutions:
- regeneration for reuse (add depleated salts)
- disposal with or without treatment (into sewerage system)
- use for irrigation
- use in aquaculture

Collegues of the department of ecological techniques within the School took up the challenge to explore meaningful ways for making use of and recycle the nutrient solutions of the horticulturists with aquaculture technique. At the same time the aquaculture system is a center of teaching and research mainly for post graduate students . For this purpose, the School of Engineering at Wädenswil joined forces with the specialists of low technology wastewater treatment of the Center for Applied Ecology Schattweid. Two external specialists joined the team and the Swiss Confederation subsidized these studies during the years 1993 - 1995.

The present aquaculture project is still in its beginnings. Few quanitative results are available as the analysis of some experiments is still in process.

2. Aims of the project

There are three main reasons for studying aquaculture at the ISW:
- *recycle the resource* in wastewater of soil-independent culture
- *educational and demonstration purposes* for students in ecological techniques

- *research and development* in aquaculture:
  e.g. is aquaculture a meaningful technique to recycle nutrients in temperate climate?
       can it be run with indigenous plant and animal species?
       which is the minimum, which is the maximum size of an indoor aquaculture system?
       and what is the energy consumption of an indoor aquaculture system?

On a more general base, there are several trades and industries producing wastewater with a high content of nutrients suitable for an aquaculture system. Several case studies focussing on such topics were dealt with in the course of this conference (see this volume).

3. Conditions for the operation of an aquaculture system at the School of Engineering at Waedenswil

The aquaculture plant is located in a non-heated glasshouse (next to the vegetable cultures), but in winter the temperature never falls below freezing. This space is needed for other school-purposes once a year with the consequence that the system has to be flexible so as to be easily taken down. This means that any biological balance acheived after a certain time is upset and has to be regained when the system is rebuilt.

All persons involved with the aquaculture plant work part time (20 - 70 %); thus, reliable coordination and communication are extremely important.

4. Operation in 1994

The **1994 goals** were to

- gain first practical experiences with aquaculture organisms and systems such as flow characteristics, nutrient demands and uptake, nutrient concentrations at different locations of the system, the suitability of plant and animal species and the bassins
- optimize the hydraulics of the system and then connect it to the computer-monitored vegetable cultures.

During 1994, our aquaculture-system was composed of three parallel subsystems: In each subsystem, the water (rain water or nutrient solution) passed three plant bassins in sequence followed by an aquarium, a settling bassin, a pump and a sand filter from where it was either directed back to the greenhouse to be reconditioned or discarded.

Plants used: *Carex acutiformis, Juncus inflexus, Elodea canadensis, Lemna minor* (one species per bassin)

Fish involved: *Tilapia mossambica* and *Ctenopharyngodon idella* (one species per bassin)

Conclusions

The surface, respectively the volume, of the aquaculture system was too small for the amount of nutrients introduced and hence the facility was overloaded. The system was also too small for the quantity of water treated (600 l/d). On the other hand this short retention time of the water offered the advantage for the fishes not to have the risk of intoxication with own wastes.

Due to hydraulic problems with joints, pumps and leakages no representative results (chemical analyses) could be obtained; however the experience was important for future planning. Plant growth was satisfying. The choice of plants and animals had to be revised: plants have to be able to remove nutrients and should be appreciated by the fish; fish should grow well and not need heating; an aquaculture with several plant and fish species (polyculture) might be more stable and at the same time nutrients can be used more completely; the plant-production should be in accordance with the amount of fish food needed.

5. Operation in 1995

5.1. Goals
- Apply the aquaculture-experiences gained during 1994
- Aim for good efficiency in wastewater treatment (e.g. nutrient uptake)
- Work with indigenous species (Central Europe)
- Aim at a balance between plant production and food demand of the herbivorous fish
- Collect data which can be used for statistical analysis
- Obtain more precise results regarding design criteria for such aquaculture systems

5.2. Experiments in 1995
In 1995, the aquaculture system was devided in three main sections which were each studied by a student (see Fig. 1):
- *section for nutrient uptake and plant production* (section 1; Anuschka Heeb)
- *section for production of zooplancton and submersed plants* as additional fish food and polishing step for wastewater treatment (section 2;Philipp Staufer)
- *section for plant consumption and fish production* (section 3; Karin Scheurer).

The surface for plant production (N and P), the amount of fishes plus the volumes of the fish tanks (3) and, the algae and plancton bassins (2 and 5) were dimensioned in such a way that in each section the amount of foodstuff necessary for the following section was produced.

For practical reasons (more security and more flexibility for each investigator), the three sections were run individually, but clearly with the intention to join them in the future.

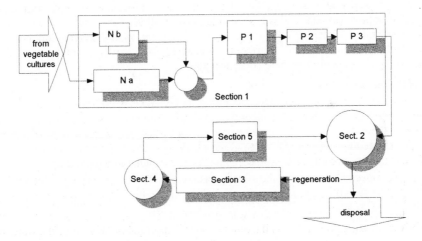

Legend: 1 Section for plant production and water cleaning
            Na Nasturtium officinale: culture on corrugated plastic sheet
            Nb Nasturtium officinale: cascade culture
            P1-3 polycultures of acquatic macrophytes
       2 Section for production of zooplancton and submersed plants
       3 Section for plant consumption and fish production
       4 Sand filters
       5 Algae and plancton bassins

Fig. 1 Principle scheme of the pilot plant 1995

5.2.1. Section for plant production and water cleaning
(see also Heeb, A et al.; this volume)

Preliminary **experiments** (February - June 1995)
Different concentrations of nutrient solutions and harvesting methods were tested. The results were confirmed in the main experiment (see below).

Main experiment (July - October 1995)
Different indigenous species were exposed to different nutrient solutions at two concentrations and their growth and purification efficiency was recorded (emerse plants with substrate *Alisma, Butomus, Callitriche, Sagittaria,* emerse plant without substrate *Nasturtium,* floating plants *Azolla, Lemna, Salvinia,* and submerse plants *Ceratiphyllum, Elodea, Myriophyllum, Stratiotes sp.*).

Important variations in *purification efficiency and growth* were observed between different species; growth of submerse plants was hindered by algae which competed for light and nutrients. For plant production a polyculture of the floating plants *Azolla, Lemna, Salvinia* and emerse plants *Alisma, Butomus, Sagittaria* was used.

Polyculture proved to be suitable as long as the water body was kept free from algae owing to shading by the floating plants.

Two different *harvesting methods* were applied to polycultures while growth and purification efficiency of the polycultures were compared. Harvesting took place 4 times at intervalls of 10 days.

- „*Large harvest*": all but 5 leaves from rooted plants were harvested and - for the floating plants - 75 % of the surface was cleared
- „*Small harvest*": all but 10 leaves from rooted plants were harvested and - for the floating plants - 50 % of the surface was cleared.

A clear effect of the two harvesting methods has been observed for biomass production: *Lemna* grew fastest after large harvests and was soon able to cover 50 % of the water-surface, whereas *Azolla* was predominant after "small harvest": it then took up 60 % of the water-surface. In terms of purification efficiency, no significant difference was noted. Both harvesting methods resulted in a clear decrease of nitrate and phosphate concentrations.

Results and Conclusions
Nasturtium officinale showed good nutrient removal ability and was used for human consumption. *Lemna* and *Salvinia natans* showed also good purification efficiency and growth when grown individually. All other species tested failed to qualify for further use because of poor growth (emerse species needing substrate for rooting), unsatisfying nutrient removal rates (*Azolla sp.*) or both (all submerse species unless combined with a large zooplancton population).

Harvesting to remove nutrients from the system was possible for selected species but not for emerse species. Luxury consumption was not observed. Polyculture in the selected combiantion was not succsessful.

5.2.2. Section for plancton production
(see also Stauffer, P et al.; this volume)
The phytoplancton-zooplancton system is an important part of aquatic ecosystems and a basic food source in natural rivers and lakes. Phytoplancton is a major transformer of anorganic nutrients into organic material. These organisms are highly productive and are able to adapt rapidly to changing environmental conditions.

Preliminary experiments
Planctonic algae (Chlorophytae) and Daphnids were kept in continuous culture. The uptake of nitrate and phosphate as well as the growth of the Daphnids were registered.

Results: The growth of algae was inconsistent, the reduction of nutrients was quite variable, and nitrite and ammonium concentrations once reached toxic levels and killed the Daphnids.

Main experiment and results

Wastewater from the fish tanks was pretreated in sand-filters to mineralize organic nutrients and make them accessible to phytoplanctonic algae. Following multiplication of the algae, they were fed into the polyculture described below.

An attempt was made to increase the stability of the zooplancton-system by creating a rich polyculture with submerse vascular plants (*Potamogeton, Elodea, Ceratophyllum*) different planctonic crustatceans (*Daphnia, Cyclops, Ostracoda, Bosmina*), water snails (*Lymnaea, Planorbis, Anisus, Radix*), insect larvae, water mites, rotatoria and others.

The development of phytoplancton, zooplancton and biomass production in general was evaluated qualitatively. Nutrient contents at the inlet were compared to the ones at the outlet. Initially the zooplancton (mainly *Daphnia* sp.) developped well; with an increasing phosphate content, however, algae grew so much that *Daphia* died but *Eudiaptomus* took over. The submerse vascular plant grew well. Towards the end of the experiment (October) *Potamogeton* started its winter rest, the emerse parts decayed.

Such systems are found to be effective in removing nutrients from the water but only at low nutrient concentrations.

5.2.3. Section for fish production and nutrient uptake (=plant consumption)
(see also Scheurer, K et al.; this volume)

Goals
- Work with indigenous fish species adapted to nutrient-rich water and able to feed exclusively or predominantly on plants
- Grow fish suitable for human consumption
- Determine the amount of food necessary per fish per day
- Compare the amount of the ingested feed with groth rates
- Find out whether animal feed influences the intake of plant feed (quantity and species prefered)
- Examine the influence of the type of feed on the water quality
- Determine the influence of polyculture on the feeding habits

Preliminary experiment: fish production

Which one(s) of four indigenous fish species (*Cyprinus carpio* (carp), *Rutilus rutilus* (roach), *Scardinius erythrophthalmus* (rudd), *Tinca tinca* (tench)) are best adapted for the present aquaculture system?
Duration of experiment: May - June 1995.
Results and conclusion: Only *Scardinius* and *Tinca* adapted well to life in capticvity. Rudd was observed to feed on *Elodea, Hippuris, Callitriche* and *Lemna*, whereas tench fed on *Lemna* only.

Design criteria for main experiment

Fish are the crucial factor in the aquaculture system: In 8 bassins (1'100 l each) with 10 rudds and 5 tenches each; body-weight per fish 7-40 and 20-50 gr, respectively (total ca 2 - 3 kg fish) and daily food need of 50 - 100 % of body-weight, the estimated necessary surface for plant production is between 40-80 m^2.

Quantity of water in circulation: either the ammonium is nitrified in the fish bassins or about 800 l water/day have to be in circulation/renovation.

Main experiment

In 4 bassins the fish polyculture (10 *Scardinius* together with 5 *Tinca*) are fed exclusively on plants whereas in the other 4 bassins feed was both vegetal and animal (additional 5-25% of total fish weight per basin and day with mosquito larvae). Duration of experiment: July - October 1995. Parameters to be controlled: growth of the fish, feed taken, water quality. As the experiment still is going on, only preliminary results are available.

Results

While the weight of fishes (especially Tench) fed with both vegetal and animal feed slightly increased, exclusively vegetal fed species lost weight. With increasing amount of fed mosquito larvae, less plants were eaten. At supply animal feed of 15% of fish weight per day, hardly any plant material was taken anymore.

5.2.4. Sandfilters

The purpose of the sand-filters is twofold: - eliminate possible pathogenic germs and mineralize organic nutrients. In 1995, they were not included in a scientific experiment.

6. Open questions

The predominant question is one of finances: are there public or private bodies that, firstly, recognize the time-span necessary to obtain answers to aquaculture questions, that, secondly, judge such answers menaingful and important and, thirdly, are prepared to allocate the necessary funds for such studies?

In addition, scientific questions are abundant. Some of them were discussed during the case study devoted to the Wädenswil aquaculture plant (chairperson Teresa Ozimek; see this volume)

7. Acknowledgements

Stephan Biro, (CH-4106 Ettingen), Ursin Ginsig, (ISW, CH-8820 Wädenswil) Anuschka Heeb, Institut für Umweltnaturwissenschaften der Universität, CH-8057 Zürich), Karin Scheurer, (Zoologisches Museum der Universität, CH 8057 Zürich), Regula Treichler, Philipp Staufer (both ISW, CH 8820 Wädenswil) all participated, assisted, worked, helped in the projekt: We greatly appreciate their contributions.

Environmental Research Forum Vols. 5-6 (1996) pp. 117-122
© *1996 Transtec Publications, Switzerland*

Pisciculture and Biomagnification of Cadmium in a Wastewater Ecosystem

V.S. Govindan

Centre for Environmental Studies, Anna University, Madras - 600 025, India

Keywords: Pisciculture, Wastewater System, Biomass, Cadmium, Bioaccumulation, Biomagnification

Abstract: Waste stabilization ponds in tropics are efficient in removing nutrients, BOD and COD. The effluents are high in quality, rich in oxygen and free from odour. The effluent, rich in phytoplankton, zooplankton and benthic organisms could be used profitably for pisciculture. Polyculture system is more beneficial than the monoculture system. The associated toxicity problems of cadmium for effluent reuse were a function of its concentration in the effluent and the diet as well.

INTRODUCTION

Indiscriminate discharges of untreated sewage into water courses deplete dissolved oxygen in the receiving water and pollute surface and ground waters. To control pollution and reuse wastewater components, sewage from a village or town can be treated in a waste stabilization pond followed by reuse for pisciculture and agriculture [1]. Although reuse techniques are gaining importance and popularity, the toxicity problems associated with wastewater reuse cannot be ignored. The wastewater in waste stabilization pond is treated by the symbiotic action of aerobic bacteria and algae. The design aspects of waste stabilization pond, the physico-chemical components of effluents, biomass productivity, seasonal changes and succession of biota in waste stabilization pond had been investigated [2,3]. However, the information available on heavy metals and their impact on phytoplankton and animal life in a waste stabilization pond ecosystem is scanty [4]. Hence an attempt has been made in this study to assess the treatability performance of waste stabilization pond for pollution control, reuse of the treated effluent for pisciculture and explore the chemical profiles of waste stabilization ponds with special reference to the seasonal variation of cadmium in raw sewage, effluent and in different trophic levels to establish the interrelationship between the biotic and abiotic components. Such an ecologically balanced system is well suited for tropical villages in developing countries.

MATERIALS AND METHODS

The Guindy Engineering College Campus' sewage, about 0.34 million litres/day was treated in waste stabilization ponds, single pond (150 x 150 x 1.5m) and ponds arranged in series (Fig.1) in which fish were grown in Ponds II and III. Single pond supported rich growth of *Tilapia mossambica*. Raw sewage, effluents and sediments of single pond collected weekly were analysed for cadmium. The seasonal succession of plankton in this pond was studied as was the bioaccumulation of cadmium in phytoplankton, zooplankton, chironomid larvae and fish. The fish were dissected and the different tissues and organs as well as the whole fish were analysed for cadmium accumulation as per Standard Methods [5]. The influent and effluent samples collected weekly from series ponds were also analysed for chemical and biological components.

The growth pattern of *Cirrhina mrigala, Cyprinus carpio, Labeo fimbriatus* and *L. rohita* in a polyculture system and yield of fish were studied in Ponds II and III. The growth pattern and yield of *C. carpio* in a monoculture system were also studied. The average weight and length of fingerlings at the time of stocking were 1.0 g and 3.5 cm, respectively. Observation on the growth of fish was made periodically by catching 20 to 30 fish using a cast net and releasing them into the pond after measurement.

RESULTS AND DISCUSSION

The results of the physico-chemical and biological analysis of raw sewage and effluents are given in Table 1. The temperature during the peiod of study varied from 26 to 33°C. The high pH values recorded in the single pond and Pond III were due to intense photosynthetic activity as evidenced by high D.O (29 mg/l) recorded at 14.30 h. During the period of study, raw sewage did not contain any D.O. However, the mean D.O. level of effluents varied from 6.9 to 8.7 mg/l. The maximum D.O. recorded was 29 mg/l at 14.30 h and the minmum D.O. values were at the predawn

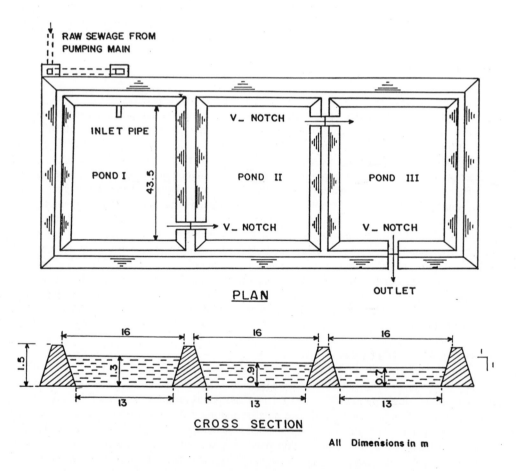

FIG 1. PLAN AND CROSS SECTION OF SERIES PONDS

hour, the extreme range illustrating the intensity of photosynthesis. The intense biological activity and oxidation of ammonia into nitrite and nitrate when the influent flowed through the series of ponds resulted in low ammonia nitrogen in Pond III. The per cent BOD reduction was 79 in the single pond and 80.3 in the series ponds. It was also noted that high concentrations of organic matter in wastewaters were associated with lower BOD and COD removals than were attained with low concentrations of organic loads. Higher BOD and COD reductions were also recorded during summer months rather than the monsoon and winter months. Phosphate removal was better in the series ponds than in the single pond. The sludge samples from the ponds had higher concentration of cadmium (on a weight per g sludge or per ml water). The order of cadmium concentration in the sludges was single Pond = Pond I < Pond II < Pond III. The bacterial reduction was higher in the series ponds than the single pond. Nawaz Tariz and Ahmed [6] stated that a longer detention time resulting in better bacterial removal. Presumably the higher pH resulting from photosynthetic activity of algae rather than detention time, reduces the bacterial load in the ponds.

Table 1. Characteristics of raw sewage and effluents (mean values)

Parameter	Raw sewage	Effluents from			
		Single Pond	Pond I	Pond II	Pond III
Temperature (°C)	28.2	29.8	29.8	29.9	30.0
pH	6.6	8.4	7.5	8.4	8.8
D.O. (mg/l)	Nil	8.6	6.9	7.2	8.7
BOD (mg/l)	178.0	36.0	78.0	55.0	35.0
COD (mg/l)	441.0	229.0	210.0	140.0	85.0
Ammonia N. (mg/l)	15.6	4.5	9.0	4.1	1.0
Nitrate N. (mg/l)	Nil	2.5	0.6	2.4	3.7
Phosphate (mg/l)	17.0	8.5	12.0	8.5	5.5
Cadmium (mg/l)	0.24	0.22	0.22	0.21	0.19
Phytoplankton/ml	-	272×10^3	163×10^3	207×10^3	269×10^3
Algal yield (mg/l)	-	140.0	104.0	131.0	169.0
Fecal Coliforms/100 ml	103×10^7	64×10^3	123×10^5	96×10^4	26×10^3
Fecal Streptococcus/100 ml	38×10^5	44×10^2	170×10^4	63×10^3	24×10^2
BOD load (kg/day)	-	28.6	28.6	12.4	8.7
Detention time (day)	-	12.0	5.0	3.5	2.7

The effluents supported about 115 taxa belonging to Chlorophyta (67 spp.) Bacillariophyta (14 spp.), Euglenophyta (10 spp.) and Cyanophyta (24 spp.). Continuous exposure of ponds to sunshine and nutrient-rich effluents may be responsible for such species diversity. Single pond supported a luxuriant mixed algal population consisting of green, blue green algae and diatoms, of which *Chlorella, Scenedesmus, Ankistrodesmus* and *Pediastrum* species occurred profusely. Volvocales,

Eugleneae and Desmids were predominantly present in Pond I. *Closteriopsis* was recorded only in Pond I. Pond II supported a luxuriant mixed algal population of *Chlorella, Crucigenia, Navicula, Nitzchia, Pandorina, Scenedesmus, Arthrospira, Merismopedia, Microcystis, Oscillatoria* and *Phormidium* species. *Korshikoviella limnectica* was observed only in this pond. *Microcystis aeruginosa* appeared in blooms in Pond III throughout the period of study. Clearly, there was a succession of algae from Volvocales, Eugleneae and Desmids in Pond I to a mixed algal population comprises of green algae, diatoms and blue green algae in Pond II which finally gave way to the dominance of blue green algae in Pond III as the BOD in the effluent steadily decreased.

During the period of study *Daphnia, Cyclops* and *Cypris* were recorded predominantly in single pond. *Moina, Bosminia, Filinia* and *Diaptomus* occurred rarely. *Daphnia* and *Cyclops* population reached a maximum in June and July when the algal population started declining. When the population of *Daphnia* and *Cyclops* declined in monsoon (September - November) *Cypris* appeared predominantly. Probably a declining algal population and the non-availability of food limited *Daphnia* and *Cyclops* population in monsoon. Cowgill [7] in his studies on crustaceans has recorded that algae were the major diet for zooplankton. During the winter (December - March) *Moina* and *Bosminia* reappeared. *Brachionus* was seen only in winter. *Daphnia, Cyclops, Filinia* and *Cypris* also occurred commonly. Presumably the winter temperature favoured the zooplankton diversity. The occurrence of *Daphnia* and *Cyclops* throughout the period of study indicated the ubiquitous nature of the species.

FISH PRODUCTION

The average weight of *Cyprinus carpio* in monoculture system was 620 g at six months and the total fish production was 7700 kg/ha/yr. The average weight and length of *C. carpio* at 205 days in polyculture system were 656 g and 32.7 cm in Pond II and 831 g and 35 cm in Pond III, respectively. The maximum recorded weight was 1002 g in Pond III. The average weight and length of *Labeo fimbriatus* were 273 g and 28 cm in Pond II and 299 g and 29 cm in Pond III, respectively. The average weight and length of *Cirrhina mrigala* were 612 g and 39 cm in Pond II and 532 g and 35.7 cm in Pond III, respectively. In 205 days *L. rohita* attained a maximum weight of 298 g in Pond II and 321 g in Pond III. The average length of this fish in Pond II and III was 29.7 and 30.4 cm, respectively. The growth of *C. mrigala* was better in Pond II. The total yield of fish obtained in polyculture system was estimated at 4,245 kg/ha/yr in Pond II and 7,300 kg/ha/yr in Pond III (Table 2). Since no supplemental feeds were provided, the entire fish biomass might have resulted from the organic constituents and plankton present in the ponds. The mean algal yields obtained from the effluents of the waste stabilization ponds ranged from 50 to 70 t/ha/yr. The conversion factor of algal biomass to fish (F/A) biomass varied from 0.71 to 1.45 in Pond II and from 0.16 to 1.24 in Pond III.

Table 2. Fish yields in Ponds II and III

Species	Pond II (kg)	Pond III (kg)	Total yield (kg/ha/yr)
Cyprinus carpio	65.640	166.280	6352.0
Cirrhina mrigala	61.200	53.250	3123.0
Labeo rohita	14.875	32.063	1285.0
Labeo fimbriatus	13.670	14.955	785.0
Total	155.385	266.548	11545.0

HEAVY METALS

Cadmium profile in waste stabilization ponds during summer (April-August), monsoon (September-November) and winter (December-March) as well as in tissues and organs of fish are presented in Table 3. A maximum cadmium of 200 µg/l in raw sewage and 150 µg/l in effluent was recorded in June in single pond and the corresponding minimum values were 100 µg/l and 5.4 µg/l in September. Interestingly, the minimum cadmium level of 5.4 µg/l of the effluent in September coincided with the maximum cadmium level of 2.82 µg/g in the sediment and the maximum level of 150 µg/l in the effluent in June coincided with the minimum level of 1.42 µg/g in the sediment in June. The high cadmium levels recorded in the effluent during the summer months may be due to the release of cadmium from the sediment by the intense biological activity at higher temperature. During the monsoon, the cadmium in the raw sewage got adsorbed onto the surface of the particulate matter which settled and increased the cadmium content in the sediment and thereby cadmium level in the effluent was low.

Table 3. Cadmium profile in single pond and tissues/organs of fish

Parameter	Summer		Monsoon		Winter	
	Range	Mean	Range	Mean	Range	Mean
Raw sewage (µg/l)	130.00-200.00	166.00	100.00-124.00	114.70	130.00-160.00	147.50
Effluent (µg/l)	7.00-150.00	89.00	5.40-9.00	7.33	14.00-120.00	83.50
Sediment	1.42-2.43	1.75	2.12-2.82	2.39	1.43-2.00	1.69
Algae	69.50-94.00	78.90	80.60-95.00	83.70	90.00-104.20	88.50
Crustaceans	70.00-109.40	89.10	76.50-102.40	91.90	74.80-116.40	103.50
Chironomid larvae	75.00-89.50	80.00	95.80-103.00	98.40	76.00-85.00	81.40
Fish	78.00-122.30	83.00	76.70-88.00	84.00	82.00-115.30	99.70
Gill	50.00-72.00	68.00	48.00-68.00	66.00	49.00-70.00	64.00
Gut	38.00-54.00	50.00	42.00-56.00	54.00	48.00-60.00	58.00
Muscles	12.00-18.00	16.00	12.00-16.00	14.00	10.00-15.00	12.00
Liver	70.00-89.00	84.00	74.00-90.00	86.00	72.00-94.00	88.00
Brain	9.00-14.00	11.00	8.00-12.00	9.00	8.00-12.00	10.00
Kidney	65.00-78.00	74.00	70.00-96.00	84.00	72.00-94.00	90.00

All values except sewage and effluent are in µg/g.

When the cadmium level in the effluent was maximum (89 µg/l) in summer, algae (79 µg/g) and crustaceans (89 µg/g) recorded the minimum cadmium in their body. This may be due to the non-availability of free cadmium in the wastewater for algal absorption which resulted in low cadmium content in the body of algae and crustaceans. Sprague [8] also reported that algae absorbed only freely available cadmium ions. Further, when the algae registered higher levels of cadmium (104 µg/g) in their body in December, the crustaceans also showed higher levels of cadmium (116 µg/g) in their body which could be assumed that the cadmium rich algal diet consumed by crustaceans resulted in the maximum accumulation of cadmium in their body. The cadmium level in the body of chironomid larvae depended on the cadmium level in the sediments. The higher concentration of cadmium recorded in chironomid larvae (103 µg/g) in September coincided with the maximum concentration of cadmium in the sediments in September (2.82 µg/g). Chironomid larvae being bottom dwellers bioaccumulated higher levels of cadmium through feed. The cadmium level in the whole body of *T. mossambica* ranged from 77 to 122 µg/g.

During the period of study the different tissues and organs of fish were also analysed for cadmium profile. The mean cadmium (µg/g) recorded in tissues/organs of fish was 61 in the gills, 48 in the gut, 14 in the muscles, 86 in the liver, 12 in the brain and 84 in the kidneys. The cadmium accumulation in guts, liver and kidney of fish depended on the cadmium level in the diet of fish. The cadmium level in brain and muscles was relatively low and insignificant. This study clearly indicated that the whole fish analysis recorded relatively low cadmium content but real amplification takes place when different tissues/organs of fish were analysed individually for cadmium accumulation. When amplification of any element through a food chain is discussed, whole body analysis of fish is considered. However, the amplification occurring in different organs and tissues may be masked due to the whole weight of the fish. The real magnification of the toxicant can be known only if different organs and tissues are analysed for the particular toxicant.

SUMMARY

Waste stabilisation ponds in tropical climate are efficient in removing BOD, COD and nutrients. The effluents are high in quality, well stabilized, rich in oxygen and free from odour. The effluent supports luxuriant phytoplankton, zooplankton and benthic organisms. There is continuous cycling of nutrients and heavy metals from the sediment to effluent and vice versa. The bioaccumulation of cadmium in phytoplankton, zooplankton and chironomid larvae varied seasonally. When the concentration of cadmium in algae increases, the concentration of cadmium in crustaceans (consumers) also increases. The concentration of cadmium in chironomid larvae is closely related to the concentration of cadmium in the sediments as is to be expected. The whole fish analysis recorded relatively low cadmium content but real amplification takes place when different tissues/organs of fish were analysed individually for cadmium accumulation. However the bioaccumulation of cadmium in fish depended not only on the concentration of cadmium in the diet but also the concentration of cadmium in the aquatic phase. The pattern of accumulation of cadmium in gut, liver and kidney of fish clearly indicates that it is related to the availability of cadmium in the diet. The cadmium accumulation in gill of fish depends on the cadmium level in the wastewater. The cadmium accumulation in brain and muscles was relatively low and insignificant.

REFERENCES

[1] V.S. Govindan, Arch. Hydrobiol. 28, 95-104 (1987).
[2] V.S. Govindan, J. Biological wastes. 30, 169-179 (1989).
[3] M.G. McGarry and G. Tongkasame, J. Wat. Pollut. Control Fed. 42, 824-835 (1971).
[4] V.S. Govindan and Geetha Ramesh, J. Wat. and Wastewater International. 9, 48-51 (1994).
[5] Standard Methods, Am. Publ. Hlth. Asso. Am. Wat. Wks. Asso., and Wat. Pollut. Control Fed. New York (1989).
[6] J.M. Nawaz Tariq and K. Ahmed, Asian Environ. J. Environ. Sci. Tech. 6, 15-19 (1985).
[7] U.M. Cowgill, Wat. Res. 21, 1453-1462 (1987).
[8] J.B. Sprague, Cadmium in the aquatic environment, John Wiley and Sons Inc. USA (1987).

Environmental Research Forum Vols. 5-6 (1996) pp. 123-130

Mussel Aquaculture for Nutrient Recycling and Wastewater Treatment: Physiological Abilities and Limits in Cold Environment

A.V. Gudimov

Murmansk Marine Biological Institute, 17 Vladimirskaya Street, Murmansk, 183010, Russia

Keywords: Mussels, Physiological Abilities, Powerful Natural Biofilter, Wastewater Treatment, Limits, Cold Environment, Food Supply, Recycling

Abstract

In marine and estuarine coastal areas mussels can filter enormous volumes of water. Great productivity of filtration apparatus coupled with high efficiency of particulate matter retention determine such bivalves as powerful natural biofilters. Physiological features of common mussels (*Mytilus edulis* L.), their worldwide distribution and high resistance to pollution have founded the basis for use them in wastewater treatment. However, there are not much practical attempts and experimental evidences to recycle wastewater nutrients by mussel aquaculture in temperate and cold environment.

At the same time blue mussels are not non-selective, mechanical and continuous feeders. The results of our experiments have shown to exhibit the cyclical changes of feeding, oxygen consumption and shell moving activity correlated with tidal, diurnal, monthly and annual periods. Although low temperature and other factors undoubtedly influence the activity and growth of mussels, our data both on growth rates and physiological responses support the hypothesis that, in cold environment, availability of food can be the principal determinant of the mussel metabolic and growth rates. Besides, in several experiments have been shown the negative significant effects of inorganic suspended matter and some toxicants on energy balance, behaviour and growth of common mussels on the East Murman. In effect, these parameters set the upper limit to growth rate of mussels in wastewater aquaculture in northern conditions.

Introduction

Aquaculture and ecological engineering are based in a way on the same use of natural processes and beneficial recycling of nutrients in nature. The farming of aquatic organisms includes certain manipulations that use and alter the natural life cycles of fish, molluscs, crustaceans and aquatic plants. Cultivation results in a greater stability of supply and price, which increases economic efficiency and attracts capital investment.

Unlike aquaculture, the basic principle of ecological engineering is ecology, while the guideline of aquaculture is a production processes for the human direct benefit. The object of ecological engineering concerning wastewater treatment "is not only one part of the ecosystem but also the whole, and not only immediate interests but long term benefits."[1]

Using wastewater to aquaculture has a long tradition in China and other Asian areas. Chinese macrophyte- and microphyte-fish culture for direct and indirect utilization of wastewater are two major kinds of ecologically engineered treatment systems at the freshwater. At the same time, half of the yields from aquaculture are from coastal sea water. Marine bivalve molluscs, especially oysters and mussels, have been cultivated for centuries. The common mussels, *Mytilus edulis* and *M. galloprovincialis* are widely acknowledged from European aquaculture.

It should be pointed out that production processes in such seston-eating bivalves are accompanied by efficient cleaning up the ambient water due to filtration activity of mussels. Ecological engineering in mussel aquaculture should be able to apply the natural self-purification abilities of mussels for treatment of wastewater. This partnership with nature involves the design of this natural environment using quantitative approaches and basing our approaches on basic science [2]. The evaluation of the mussel aquaculture potential for wastewater treatment in terms of biological activity and physiological processes rates demands certain attention as well as its fitness to the ecologically sound recycling of wastewater resources in temperate and cold environment.

Physiological abilities

Ecotechnology involves the theoretical and applied investigations in ecology to the designing of sustainable ecosystems . Choice of the biological species for ecological engineering is dependent on their physiological abilities in wastewater utilization and recycling.

Filtration. Filtration abilities of seston-eating bivalves principally determine clams as powerful natural biofilters.   The filtration rate (=clearance rate) reflects the level of feeding activity and is definited as the volume of the sea water cleared of particles in unit time. These filter- feeders process large volumes of the surrounding water and efficiently retain small suspended particles [3].   Cirri of the bivalve gill act as filters that retain suspended particles in the through current.   At low particle concentrations in the water the gills transport water at constant high rates and efficiently retain suspended particles of size down to a few µm. Vahl [4] demonstrated 80-100 % retention by *Mytilus edulis* of all particles in the range 2-8 µm. diameter. A comparison between filtration intensity of the most active freshwater filtrators and marine bivalves has shown the highest values of filtration rate in mussels (Table 1).   This is evidently the result of adaptation to lower concentrations of suspended matter in the sea water.

Table 1. Filtration intensity of bivalves per g of unit wet (dry) weight tissues.

Species	Filtration intensity, ml/h	Reference
Dreissena and Unio	85	Alimov, 1981
Mytilus edulis	386	Voskresensky, 1948
Mytilus edulis	(3900)	Vahl, 1973
Mytilus edulis	(3200)	This paper
Arctica islandica	180	Winter, 1969
Modiolus modiolus	240	Winter, 1969

The maximum filtration rate is observed at very low suspension concentrations quantified in the natural environment. Filtration rate in natural sea water amounts to 2-4 l per h for a mussel 4-5 cm in length [5]. Increasing food concentration results in a pronounced reduction in filtration rate. In contrast to the assimilation rate which is practically independent of food concentration, and to the amount of food consumed (ingestion rate) which increases slightly.

Edible part of ingested particulate matter is digested with following discharge of dissolved organic matter (DOM) and faeces. With an increase in food concentration the loss of energy in form of faeces increases drastically and the amount of energy available for growth decreases.

Biodeposition. At still higher food levels, finally a part of the algal cells filtered-out is accumulated to pseudofaeces, and the filtration and ingestion rate are reduced to a remarkable degree [6,7]. Already at suspension concentration twice exceeding the natural one the filtration activity declines and pseudofaeces appear [8]. Their amount increases with increase of suspension concentration, and some of the algae filtered out of suspension are rejected. Substances agglutinated in pseudofaeces and faeces appeared to be the most nutritious food for detritophages, in particular of polychaetes and Chironomid larvae [9].   Excretions of living molluscs are not only included into the detritus food web, but also influence greatly formation of primary production.

Stimulating of primary production. Metabolites excreted by mussels, i.e. dissolved organic matter (DOM) possess a certain biological activity in relation to producers. Thus, in regions of mussel cultivation and above their natural populations numbers of bacterioplankton cells are 4-6 times and of phytoplankton cells 10-30 times as higher than in control areas. Concentration of DOM, carbohydrates, and silicon containing substances in water off mussel banks exceeds their control values 2.5-3 times. Experimental studies of Galkina [10] have shown that the Barents sea mussels excrete about 12 % of their ration in the form of DOM. Excretion in mussels cultivated on the rafts was 1.6 times higher than that in natural populations.

Evidently DOM excreted by mussels influence phytoplankton in the native state, but unlike DOM of sea birds excretions they are not the source of biogens formed in destruction process. While concentrations of chlorophyll "a" practically do not increase assimilation activity of cells increased to a considerable extent, it varies with season and has the highest values in spring at a deficit of biogens in the coastal water of the East Murman. Moreover as it has been confirmed experimentally, mussel excretions control bacterial populations development. Thus, mussels promote recycle some of their nutrients for the benefit of the ecosystem.

Recycling. In marine and estuarine coastal areas mussels can filter enormous volumes of water. Great productivity of filtration apparatus coupled with high efficiency of particulate matter retention determine such bivalves as powerful natural biofilters [11]. Mussels consume relatively large amounts of seston, feeding directly on the primary producers, phytoplankton and bacteria, and non-living organic material. Unlike herbivorous cultivated filter-feeding fish mussels are low in the food web. For each stage of the food chain an animal is removed from the source of primary production, there is an energy loss of some 80-90% for herbivores and 90% for carnivores in converting food into flesh [12]. Therefore, while the predatory fish are relatively wasteful of the basic production, mussels are more efficient converters and so lend themselves well to cultivation, producing greater yields per unit area than species at higher trophic levels [5]. In the summer and autumn (during 5-6 months in temperate climate) 20-30 % of filtered organic matter may be fixed as meat weight by *Mytilus edulis* [13]. The net effect of mussel aquaculture will always be a reduction of nutrients in the area.

Other abilities. The water purification processes also involves any other ecologically sound features of mussels to solving environmental problems:

* Blue mussels are able to concentrate and deposit oil residuals, extracted from ambient water. Mussel tolerance to oil pollution allows to consider a possibility of using them in an integrated biological purification of sea water from oil products. In the process of filtration mussels accumulate oil in thin mucous brown "cords" of pseudofaeces .

*The lack of capacity to regulate internal concentration of elements leads to accumulation of heavy metals and other different anthropogenic chemicals in bivalve tissues to a marked degree over sea water levels.

* The shading of phytoplankton owing to much seston in water is removed through mussel clearance activity.

* Mussels promote the circulation water in the water bodies.

* Continuous quality control of water flowing in and out for purification cycles is possible by bivalves owing to their bioaccumulation features and monitoring of shell moving activity.

* The mussel's sessile habit also makes it suitable to cultivation.

Physiological features of common mussels (*Mytilus edulis* L.), their worldwide distribution and high resistance to pollution found the basis for wastewater treatment in coastal areas. However, there are a few practical attempts and experimental evidences [14] to recycle wastewater nutrients by mussel aquaculture in temperate and cold environment. Some factors are able to limit the efficiency of water purification by this natural biofilter in nothern climate.

Limits

Various factors affect physiological activity and growth of mussels in temperate and cold waters, that growth is rapid during the spring and summer and slight or absent during the colder winter months. The results obtained by many authors on physological responses of bivalves very seldom include any experimental evidences on influences of each factor in natural fluctuated conditions.

It is apparent that growth is regulated according to the food supply and maximum rates of growth is reached only at the highest algal concentration. As a consequence of this the relationship between the growth of an animal and its metabolic activity is also affected by ration.

1 . Availability of food

Metabolic Rate. It should be noted that the metabolic rate (the oxygen consumption) depends on the amount of available food [15,16]. Above the certain threshold concentration of food particles, oxygen consumption increases from a low "standard" rate to a higher "routine" rate during prolonged feeding. The stable routine rate is established when the ingested ration exceeds the certain maintenance level [15,5]. Over a considerable range of further ration increase, there appears to be practically insignificant effect of ration on metabolic rate. From this point it is obvious that the dependence of routine rate of oxygen consumption upon food availability indicates the lack of food supply.

The results of our experiments have been estimated from this reason. Oxygen uptake, filtration rate and shell moving activity in *Mytilus edulis* were established by author for cold environment of the Barents sea (at the East Murman area). The experiments were carried out in running-water aquaria with constant sea water flow. Despite constant water flow with natural seston, the respiration rate varied in the course of an experiment. Variations in oxygen consumption rate are correlated with diurnal or tidal rhythms. The rate of oxygen consumption by mussels as a function of tidal phases is presented in Fig. 1.

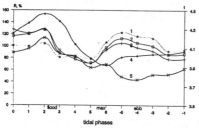

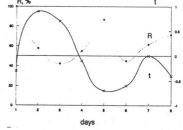

Fig. 1. Respiration rate (R, in % of fixed level) of mussels in relation to body size(1-4) and temperature(5): curves - moving average of 3 terms.

Fig. 3. Oxygen uptake by mussels and mean temperature of water in a week.

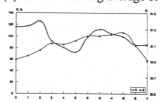

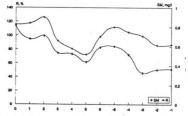

Fig. 2. Mean oxygen uptake by mussels and salinity changes in tidal cycle.

Fig. 4. The correlation of oxygen consumption rate of mussels and seston amount.

A possible relationship might be expected between oxygen consumption and temperature. However, superposition of these temperature and salinity tidal changes in May and February on the observed respiration pattern for the mussels had shown no statistically valuable correlations (Fig. 1, 2 and 3).

These relationships reflect the marked increase in metabolic rate that occurs with increase in the seston amount (Fig. 4), partly accompanied by a less pronounced increase in the water temperature. Certainly, in experiments performed under natulal fluctuated conditions the effects of seston availability and physical factors (e.g. temperature, salinity) could not be distinguished with accuracy, because many environmental influences as well as some physiological responses are closely linked. The cumulative effect of the factors might also be expected in the respiration patterns.

Our results confirm the relationship between shell moving activity and filtration rate that was found by various authors in different lamellibranchiate bivalves [17,6,18,19]. Factors which might synchronize the observed physiological responses in *Mytilus edulis,* if it does represent a daily or tidal exogenous rhythm, include temperature, light and food.

The continuous recording of moving activity over the period exceeded half a year revealed a distinct daily and tidal fluctuation in shell gape size for all mussels . Results of the present experiments indicated that moving activity in *Mytilus edulis* was quantitatively affected by changes in concentration of seston (Fig. 5). Variations in moving activity were most predictable and were not correlated significantly with any others of environmental variables tested.

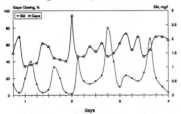   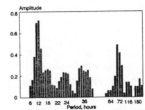

Fig. 5. Shell gape closing and concentration   Fig. 6. Periods of mussel shell movements
of suspended matter as a function of time.   in the power spectra.

Both local tidal sequence and diurnal cycles exerted a significant influence on oxygen consumption, feeding and behavior of mussels. Time series analysis of valve movements showed that mussels exhibited both 12-h and 24-h periodicity of activity (Fig. 6). Power spectra often contained multiple peaks, denoting the occurence of more than one component of periodicity. A close correlation between the phases of shell movements, the filtration rate and the amount of food indicate distinct feeding cycle in *Mytilus.*

Growth. Food supply is probably the most important factor in determining activity and growth rate of mussels for northern environment. This relationship, however, varies to a high degree in relation to the season of the year [5].

At the coastal water of the Barents and the White seas the most favourable period for mussel growth is very short and lasts from April till September. Within this productive growth period of the year the energy input is available for growth and reproduction. The organic seston during the months under consideration is formed by blooms of phytoplankton, the concentration of which does not exceed 300×10^6 cells/m^3. It is thus suggested that at the East Murman on the whole even in summer the suboptimal conditions for mussel growth take place and variations in growth rates occur according to local conditions.

During the long cold and dark winter the concentration of phytoplankton decreases drastically untill its almost full absence. The quality of seston as food for mussels changes for lesser content of proteins, while organic detritus supplies a large part of the mussels diet. At this low particle concentrations in the water metabolic reserves are utilized until more favourable conditions arise. The growth rate changes accordingly the food supply. Even mussels grown under apparently identical conditions can exhibit widely different rates. The results obtained by the authors [20] have shown that in the floating capron cages growth rate of adult mussels were from 5 to 10 times higher than in sublittoral zone of the Murman coast with the suboptimal food conditions.

2. Temperature

The lack of food supply could be significant, in winter months at least, but low temperature also is an important factor in controlling metabolic and growth rate. On the other hand *Mytilus* occurs under extreme arctic conditions, where in long cold winters the temperatures of water along the coastline are often at or below 0 °C. In intertidal zone mussels sustain the temperatures at below minus 10-12 °C during exposure to air. It is common knowledge that, in general, metabolic rate increases with increasing temperature.

The results obtained by author [21] on routine oxygen consumption rate was expressed as function of the dry of soft tissues of mussels by a ordinary power function :

$$R = 0.186 \ W^{0.702} \qquad \text{in July, at 7.0 °C} \qquad\qquad (1)$$
$$R = 0.081 \ W^{0.702} \qquad \text{in November, at 4.0 °C} \qquad\qquad (2)$$

Comparable results were found by Bayne, Thompson and Widdows [5] in *Mytilus edulis* from the North sea. In order to make possible such a comparison, it is necessary to transform these equations by Q_{10} coefficient to unit temperature at 15 °C, considering also b- values:

$$R = 0.339 \ W^{0.702} \qquad \text{Bayne et al., 1976, summer, routine} \qquad (3)$$
$$R = 0.341 \ W^{0.702} \qquad \text{Gudimov, 1989, summer, routine} \qquad (4)$$
$$R = 0.263 \ W^{0.72} \qquad \text{Bayne et al., 1976, winter, standard} \qquad (5)$$
$$R = 0.207 \ W^{0.72} \qquad \text{Gudimov, 1989, winter, routine} \qquad (6)$$

The equations indicate that metabolic rate in mussels in summer at the East Murman is lower than that from the North sea principally due to the effect of temperature.It is obvious, that the winter level of oxygen uptake by mussels in the Barents sea is close to standard rate of that in mussels from the North sea. From this comparison it is clear, that there appears to be distinct effect of ration on the metabolic rate of mussels.

Thus, it is suggested that availability of food set the upper limit of metabolic and growth rate of mussels in cold environment.

3. Suspended matter.

In coastal water the tidal currents exert a significant influence on the growth and energetics in *Mytilus edulis* by presence of suspended matter. Besides, the maximal growth rate in mussels was reached only at the highest algae concentration in the addition of 5 mg silt/l [22]. In shallow waters it is often the case that the concentration of suspended inorganic matter exceeds the concentration of particulate organic matter.

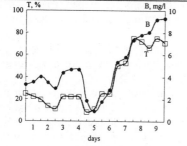

Fig. 7. Effect of inorganic suspended matter on shell closure time in *Mytilus edulis.*

On the basis of the our data obtained for the oxygen consumption rate and shell moving activity in *Mytilus edulis* the linear correlation between the concentration of inorganic suspended matter and diurnal shell closure time (Correlation coefficients = 0.89-0.9) has been found out. The reliable increase (p < 0.01) of average daily duration of the shell closure over 6 mg concentration of dry clay powder (the bentonit) per litre has become evident (Fig. 7). In the aquaria with clay suspension added the standard oxygen consumption rate for starved mussels has increased up to 140 mg dry clay per litre. The values of active metabolic rate in mussels fed by algae decreased at all clay concentrations.

That's why a significantly increased isolation of mussels from environment begins above threshold concentration of suspended matter with value higher than 6 mg/l. Aerobic respiration decreases at benthonit concentration higher than 180 mg/l, the scope of anaerobic metabolism increases which, entails energy expenditures of inner energy resources. and exhaustion of an organism.

4. Toxic effects.

The limitations of recycling in wastewater treatment are affected by the problems of toxic pollution effects. The untreated mixed wastewaters is not suitable for aquaculture in fishponds [23]. At the same time domestic and industrial wastes are often mixed into the water bodies. As a consequence of this the nutrient recycling concept stresses detoxification as a first stage of wastewater treatment.

It is common knowledge, that many bivalves possess a high pollution resistance. Owing to functioning of detoxification system [24] molluscs retain vital activity, although their growth slows down and even stops.

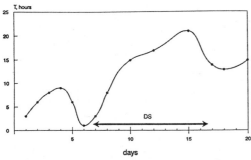
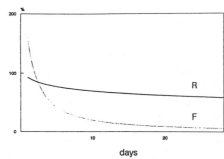

Fig. 8. Time course of shell closure duration (T) in mussels at the 0.001 % concentration of drilling solution (DS-exposure time)

Fig. 9. Time course of oxygen consumption(R) and filtration rate (F) of mussels at 1% DS, in % to control.

The negative significant effects of some drilling fluids on energy balance and behavior of common mussels at the East Murman have been shown in our several long duration tests [25].   The highest sensitivity of mussels behavioral responses to changes in chemical water composition was found. The Barents sea mussels had significant differences from the control by gape size-index even at 0.001%   concentration of drilling fluids (Fig. 8.). The differences from control in oxygen consumption and filtration rate of mussels were observed only at 1% concentration of drilling solution (Fig. 9.). These results allow to use the shell moving activity of mussels for bioassay and monitoring of water quality 'in situ'.

Conclusions

Wastewater treatment by   mussels aquaculture is relatively new application of ecological engineering. It is obvious, the filter-feeding mussel was born as a   natural powerful biofilter. Filtration capacity and high efficiency of suspended matter retention coupled with high resistance to pollution and worldwide distribution attract immediate attention to mussel aquaculture for solving environmental problems in an ecologically sound way.

Mussel populations meet the basic concepts of ecological engineering (by Mitsch,[2]):

* Mussel ecosystems are able to sustain themselves through self-design.

* Mussel aquaculture is most adaptable to the human needs for the benefit of the natural ecosystem.

These "living machines"(Todd.91) not only promote recycling of nutrients into bioproduction but greatly reduce environmental stress.

Metabolites excreted by mussels, cause stimulating effects on primary production and appears to be control of the microorganisms development.

Bivalves can be the first as well as the last stage of purification process. For instance, on pretreatment stage of pond fish culture mussels are able to protect one throughout accumulation and monitor of the harmful components. They can finally purify water from DOM, particulate matter and phytoplankton.

In general, mussel aquaculture possess the mechanical, biological, and chemical treatment of the wastewater.

The physiological ability of mussels to accomplish engineering goals is dependent, in large part, on the environmental conditions during the treating process. There are several major factors limiting the physiological activity and growth rate of mussels in cold environment.

Food supply is probably the most important factor in determining activity and growth rate of mussels for northern environment. The lack of food supply could be more significant, than even low temperature in controlling metabolic and growth rate, at least during half a year at the East Murman. If food is scarce, then growth is retarded regardless of all other conditions.

Under the circumstances, the wastewater contained considerable amounts of organic matter will provide the basis for improvement of food supply. It is suggested, that this recycling of wastewater organic matter would also increase the possibility of future yields and reduce water pollution significantly at the same northern conditions.

As concerns the effects of toxic pollutants and of inorganic suspended matter, it is limiting through the inflow of wastewater in right doses.

Recycling potential of bivalve aquaculture should be applied to the treatment of wastewater both in coastal and fresh waters.

References

1. Yan,J.S., and S.J. Ma, Ecol.Eng.for Wastw.Treat. Proc. of Intern.Conf., Stendsund, March 24-28,1991,80-94 (1991)
2. Mitsch, W.J. and S.E. Jorgensen, Ecol.Eng.: An Introd. to Ecotechnology, 472 (1989).
3. Jorgensen, C.B.,Mar.Biol. V 61, 277-282 (1981)
4. Vahl, O. Ophelia. V 10,10-25 (1972).
5. Bayne, B.L., R.J.Thompson, J. Widdows. In: Mar.mussels: their ecol. and physiol., 494 (1976).
6. Winter,J.E.,Mar. Biol. V 4, 87-135 (1969).
7. Winter, J.E. Haliotis. N 7,71-87 (1976).
8. Alimov, A.F. Funct.ecol.presnov.dvustvor.moll.,L,Nauka,248 (1981)
9. Isvekova, E.I., A.A. Lvova-Katchanova. Pol. arch. hydrob.V 49, 203-210 (1972)
10. Galkina, V.N. Ph. thesis. 23 (1985)
11. Voskresensky, K.A. Tr. Gos. oceanogr.in-ta, V 6(18),55-120 (1948)
12. Odum, E.P. Fundamentals of Ecology. (1971)
13. Boje, R. Kieler Meeresforsch. B 21, 81-100 (1965).
14. Kautsky N. & C. Folke. Proc. Intern. Conf., Stensund,320 (1991)
15.Thompson, R.J. & B.L. Bayne. Mar. Biol.V 27,317-326 (1974).
16. Gudimov, A.V. Preprint, Apatity,36 (1994).
17. Theede H. Kieler Meeresforsch. Bd 19, Hf.1,20-41 (1963).
18. Salanki, J., F.Lukacsovics and L.Hipipi. J. Annal. Biol. Tihany. V 41, 69 (1974).
19. Davenport, J. & A.D. Woolmington, J.Exp. Mar. Biol. Ecol. V 62,55-67 (1982).
20. Gudimov, A.V. & V.E. Kostylev, Symp. Onthogeny of Mar. Inverteb. Vladivostok, 23 (1988).
21. Gudimov, A.V. In: Feeding relat. bottom inverteb... Apatity,96-103 (1989).
22.Kiorboe, T., F. Mohlenberg, O.Nohr, Mar. Biol. V 61, 283 (1981).
23. Li, S. Proc. Intern. Conf., Stensund, 216 (1991)
24. Mauri, M., Symbioses. V 17, N 2, 89-92 (1985).
25. Gudimov, A.V. In: Arctic seas: bioind. envir... Apatity, 36-44(1993).

Environmental Research Forum Vols. 5-6 (1996) pp. 131-136
© *1996 Transtec Publications, Switzerland*

Evaluation of the Effect of a Conveyor Production Strategy on Lettuce and Basil Productivity and Phosphorus Removal from Aquaculture Wastewater

P.R. Adler[1], S.T. Summerfelt[2], D.M. Glenn[1] and F. Takeda[1]

[1] USDA-ARS, 45 Wiltshire Road, Kearneysville, WV 25430, U.S.A.
[2] The Conservation Fund's Freshwater Institute, PO Box 1746, Shepherdstown, WV 25443, U.S.A.

Keywords: Crop Production, Ecological Engineering, Effluent Treatment, Nitrogen, Phosphorus

Abstract

Plants cannot remove nutrients to low levels in water without a reduction in productivity; that has been the predominant paradigm regarding use of cash crops to clean effluents. Since greenhouse space is expensive, maintaining maximum productivity is critical to sustaining a profitable operation. With the conveyor production strategy, plants are positioned near the inlet when they are young and moved progressively towards the outlet as they mature. Greenhouse studies demonstrated that luxury consumption of nutrients by young plants made it possible for them to maintain maximum productivity to harvest maturity while reducing phosphorus to less than 0.01 mg/L.

Introduction

Nutrient cycling is fundamental to sustainability and as a principle that drives ecological engineering design, it creates greater efficiency and productivity [1,2]. The application of ecological engineering principles to water pollution control can reduce water treatment costs. The decline of federal subsidies for municipal wastewater treatment in the United States is driving the search for lower cost alternatives [3,4]. Confounding this quest for lower cost systems is the move towards more stringent nutrient discharge regulations. The increase in effluent pretreatment at the industrial source reduces toxic heavy metals in the municipal wastewater stream thereby allowing its use as a water and nutrient source for irrigation of food crops. Other effluents suitable for clean up by food crops are the vast point source effluents generated by agricultural production facilities [5], particularly the aquaculture industry. Aquaculture facilities generate large volumes of low nutrient water. Since nutrient mass loading is the critical factor contributing to environmental degradation, effluent treatment is important. Aquacultural effluents are difficult to treat because they contain relatively dilute nutrients (< 1 mg/L phosphorus).

Thin-film technology is a crop production system in which plants grow in water with nutrients flowing continuously as a thin-film over the root system. Water flow across the roots decreases the stagnant boundary layer surrounding each root, which in turn enhances the mass transfer of nutrients and permits crops to maintain high productivity at steady-state phosphorus levels below 0.300 mg/L [6,7,8]. However, as the phosphorus concentration is reduced in water, influx rates are too low to sustain maximum productivity. Plants are able to absorb and store much higher levels of nutrients than are required by metabolism, a phenomenon called luxury consumption. To maintain high plant productivity, the nutrient influx rate must be such that the nutrient concentration within the plant does not drop below the critical deficiency level, that concentration necessary to maintain maximum plant growth. Due to luxury consumption, maximum productivity can be sustained even as nutrients are depleted to C_{min}, the concentration where net influx is zero, if the mean influx rate over the entire plant's growth cycle is high enough to maintain the nutrient concentration above the critical deficiency level [9].

The conveyor crop production strategy enables plants to store phosphorus early in their growth cycle. This reservoir of phosphorus is available for plant growth as phosphorus levels in the water decrease and influx can no longer meet current demands. With the conveyor production strategy, lettuce or basil seedlings are introduced at intervals near the inlet of a thin-film system and sequentially moved progressively towards the outlet as they mature and where they are then harvested. At the front end of the thin-film troughs where nutrient concentrations are high, young plants absorb and store nutrients in excess of their immediate needs. Luxury consumption of nutrients during this early growth phase sustain the plants later when they are moved down towards the outlet where nutrient concentrations in solution are too low for absorption kinetics to meet their growth needs.

Materials and Methods

'Ostinata' lettuce and sweet basil were seeded into Oasis cubes (Smithers-Oasis, Kent, OH). The cubes were watered with a complete nutrient solution [(mM): 3 $Ca(NO_3)_2$ and 4 KNO_3, 1 KH_2PO_4, 2 $MgSO_4$, and micronutrients (mg/L): Fe as $FeSO_4$ (2.5) and DTPA (2.5), B as H_3BO_3 (0.5), Mn as $MnSO_4$ (1.0), Zn as $ZnSO_4$ (0.05), Cu as $CuSO_4$ (0.02), and Mo as $(NH_4)_6Mo_7O_{24}$ (0.01); adjusted to pH 6.0 with KOH]. At 8 and 12 days, the lettuce and basil seedlings, respectively, were placed into thin-film troughs and the above solution was recirculated. At 16 and 20 days, lettuce and basil, respectively, were moved to a nonrecirculating thin-film system configured with the conveyor production sequence. Rainbow trout wastewater was pumped with peristaltic pumps (model no. 7520-35 Cole Parmer Instrument Co., Chicago, IL) at a constant flow rate of 250 and 300 ml/min for basil and lettuce, respectively. The wastewater was from the rainbow trout system of The Conservation Fund's Freshwater Institute and on average contained the following macronutrients (mg/L): NO_3 (15), P (0.6), K (5), Ca (55), Mg (20), and S (9), at pH 7.2. To maximize phosphorus removal, the following nutrients were added to make phosphorus the most limiting nutrient: 0.1 mg/L Fe (as Seq 138), 130 µg/ml Mn (as $MnCl_2$ in a 2.5 mm film of 1% agar on the trough bottom), 0.005 mg/L Mo (as $(NH_4)_6Mo_7O_{24}$), and 14 mg/L K (as K_2SO_4). The troughs, roughly 10 x 366 cm (Genova Products, Inc., Davison, MI), were covered with 1.6 mm PVC having 3.2 cm holes evenly spaced at 17.5 cm, and planted with 20 seedlings. Plants were grown from the beginning of June through mid-July in a greenhouse kept at 27-34°C during the day and at 21-24°C during the night.

Six connected troughs, each 366 cm long, formed the foundation for the conveyor production scheme. With this production strategy, the rate of biomass production per unit area, hydraulic loading rate, and effluent phosphorus concentrations were relatively constant. Lettuce and basil were harvested every 4 days; the lettuce and basil life cycle was 40 and 44 days, respectively. Each of the 6 sections represented 4 days in the system, so both lettuce and basil were in the system for 24 days. The plant rotation was as follows: plants were harvested at the outlet end of the system, the plants in the remaining 5 sections were move down one position, and 16 day old lettuce or 20 day old basil plants were set into the system at the inlet end (Fig. 1). This cycle was repeated 5 times to move a given set of plants completely through the system to harvest.

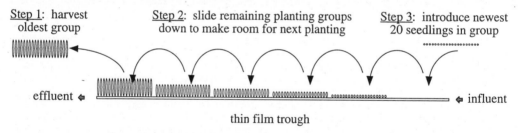

Figure 1. Conveyor crop production schematic.

Ion Analysis

Ion analysis in both tissue and water samples were performed as previously described by Adler et al [10].

Results and Discussion

After steady-state planting and harvesting was achieved, the conveyor production strategy maintained plant productivity and health while removing 99% of the dissolved phosphorus and 60% of the nitrate from the flow (Fig. 2). Basil and lettuce removed phosphorus to 3 µg/L and less than 1 µg/L, respectively, from an influent concentration of greater than 500 µg/L. Rate of phosphorus removal was greater than 60 mg P/m²·d and nitrate removal was 980 mg N/m²·d for lettuce.

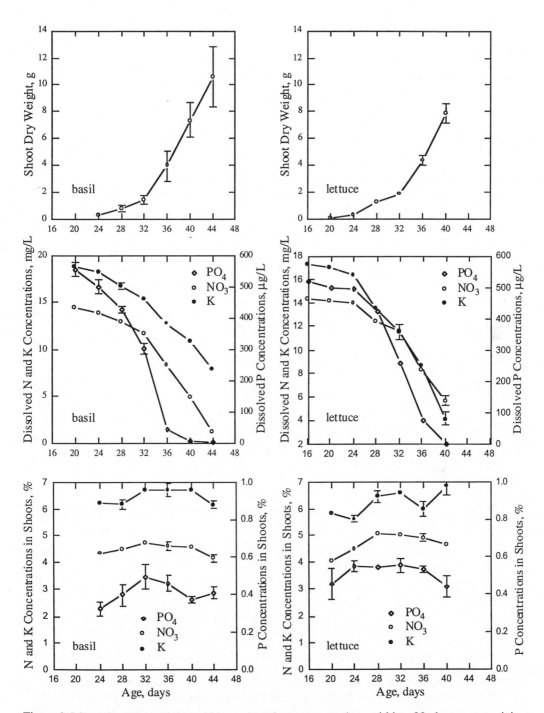

Figure 2. Mean shoot dry weight and biomass nutrient concentrations within a 20 plant group as it is conveyed down the trough; And, dissolved nutrient concentrations in the flow leaving a 20 plant segment according to plant age (or placement down the thin-film trough) during the conveyor production experiment.

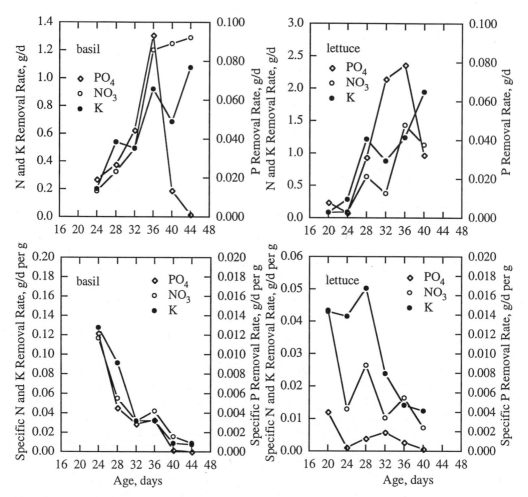

Figure 3. Nutrient removal rates (g nutrient per day and g nutrient per day per g biomass) for each 20 plant section (0.56 m²).

Basil removed phosphorus faster than lettuce (Fig. 3). Nutrient uptake rates were greatest, however, when dissolved nutrient concentrations were highest (Fig. 3).

Cellular nutrient concentrations were sufficient to sustain growth even after nutrients within the flow were limiting (Fig. 2), as indicated by the large decrease in nutrient uptake rates per unit of plant biomass (Fig. 3). In general, if the nutrient being removed was not limiting, removal followed growth, whereas when nutrients were limiting, removal depends on a combination of growth and nutrient concentration in the effluent. Nutrient storage through luxury consumption allowed the conveyor production strategy to remove phosphorus to very low levels without a reduction in productivity or quality. By positioning plants towards the inlet when they were younger, more nutrients were transferred and stored. These nutrients were available for use when the plants had been moved towards the end of the thin-film trough and less dissolved nutrients were available.

Basil and lettuce were both healthy and their growth did not appear to be limited. About 85% of growth occurred during the last 12 days of the 40-44 day production cycle (Fig. 2). Productivity averaged approximately 12 and 16 g /m²·d for lettuce and basil, respectively. In a model being developed, the authors estimated that with a growth rate of about 12 g /m^2·d, the maximum phosphorus removal rate would be about 100 mg P/m^2·d.

Conclusions

With these recent improvements, new opportunities are emerging to reduce costs of tertiary treatment of municipal and animal effluents through the application of ecological engineering principles. The use of thin-film technology and a conveyor production strategy allowed for production of high value crops while meeting stringent phosphorus discharge regulations. Under this tertiary water treatment scheme, water treatment costs are reduced through use of a less expensive technology, improved use of greenhouse space, and production of a high value crop. These greenhouse studies demonstrated that the conveyor production system produced phosphorus removal to consistently less than 0.01 mg/L without a reduction in crop productivity or quality.

Acknowledgments

We would like to thank Dan Bullock and Kevin Webb for their invaluable technical assistance.

References

[1] Mitsch, W.J. and S.E. Jørgensen (eds.). 1989. Ecological Engineering: An Introduction to Ecotechnology. John Wiley & Sons, New York, NY.

[2] Etnier, C. and B. Guterstam (eds.). 1991. Ecological Emgineering for Wastewater Treatment. Proc. Int. Conf. at Stensund Folk College, Sweden. March 24-28, 1991. Bokskogen, Gothenburg, Sweden.

[3] Jewell, W.J. 1991. Wastewater treatment using combined anaerobic filtration and hydroponics. p.37-52. *In* R. Isaacson (ed.) Methane from Communiity Wastes. Elsevier Applied Science, New York, NY.

[4] Jewell, W.J. 1994. Resource-recovery wastewater treatment. American Scientist 82:366-375.

[5] Sharpley, A.N., S.C. Chapra, R. Wedepohl, J.T. Sims, T.C. Daniel, and K.R. Reddy, 1994. Managing agricultural phosphorus for protection of surface waters: issues and options. J. Environ. Qual., 23:437-451.

[6] Asher, C.J. and J.F. Loneragan, 1967. Response of plants to phosphate concentration in solution culture: I. Growth and phosphate content. Soil Science, 103:225-233.

[7] Asher, C.J. and P.G. Ozanne, 1967. Growth and potassium content of plants in solution cultures maintained at constant potassium concentrations. Soil Science, 103:155-161.

[8] Barber, S.A., 1984. Soil Nutrient Bioavailability: A Mechanistic Approach. John Wiley and Sons, New York, NY, 398 pp.

[9] Adler, P.R. 1995. Modeling the effect of water-flow pattern on phosphorus removal and constructed wetland productivity. Proc. Int. Symp. Water Quality Modeling, Orlando, FL. ASAE Publ. 5-95. St. Joseph, MI. p. 468-474.

[10] Adler, P.R., S.T. Summerfelt, D.M. Glenn, and F. Takeda. 1994. Creation of wetlands capable of generating oligotrophic water. Proc. 67th Water Environ. Fed. Annual Conf., Chicago, IL. 8:37-44.

Environmental Research Forum Vols. 5-6 (1996) pp. 137-144
© *1996 Transtec Publications, Switzerland*

Toxicity of Human Urine, its Main Nitrogen Excretory and Degradation Products to *Daphnia magna*

M. Adamsson[1] and G. Dave[2]

[1] Department of Applied Environmental Science,
Medicinaregatan 20A, S-413 90 Göteborg, Sweden

[2] Department of Zoophysiology, Medicinaregatan 18, S-413 90 Göteborg, Sweden

Keywords: *Daphnia magna*, Toxicity, Human Urine, Ammonia, Nitrate, Nitrite, Creatinine, Uric Acid, Storage

Abstract

The discovery of nitrogen as a major cause of eutrophication, especially in the marine environment, has intensified the research on nitrogen removal from sewage water. About eighty percent of the nitrogen in domestic sewage water originates from human urine. One alternative for nitrogen reduction technology in sewage treatment, is to separate the urine from fecal matter, and use the urine as a fertilizer, either in agriculture or in aquaculture. In temperate regions urine has to be stored during the winter season.

The objectives of this study were to determine the toxicity to *Daphnia magna* of human urine, its major nitrogen excretory products (urea, uric acid, ammonia and creatinine) and their degradation products (nitrite and nitrate) and to determine how storage affects the chemical composition of these nitrogen compounds in urine and the resulting effect on urine toxicity to *D.magna*. The 48-h EC50 for *D.magna* of fresh morning urine from ten persons ranged from 2.8 to 12.5 percent by volume (mean ± SD: 5.4 ± 2.9). The 48h EC50s (expressed as mM of nitrogen) for the nitrogen compounds were 679 for urea, 78 for nitrate, 49 for creatinine, 2.4 for uric acid, 1.5 for ammonia and 0.6 for nitrite, at pH between 8.2-8.5. The toxicity of urine increased approximately five times during storage in 32 days at 20°C. The major cause of this increase was conversion of urea to ammonia, which was accompanied by an increase in pH from 5.7 to 9.7, making the ammonia even more toxic. Therefore, the use of human urine as a fertilizer in aquaculture and probably also in agriculture must consider the potential toxicity of ammonia.

1. INTRODUCTION

The primary objectives of sewage treatment are reduction of human pathogens (HP), oxygen demand (BOD,COD), suspended solids and nutrients, mainly phosphorous (P). In addition some reduction of toxic compounds and nitrogen (N) may take place. More recent innovations in sewage treatment include nitrogen reduction by nitrification and denitrification processes. The identification of nitrogen as a limiting nutrient for marine algae [9, 15] has intensified the research of N-reduction techniques, but so far the reduction efficiency (in percent) for N have been considerably lower than that for HP, BOD and P. Furthermore, the increasing demand for recycling of natural resources and the limited resources of fresh water in many areas has increased the interest in using domestic waste water for irrigation and aquaculture [11].

A considerable part of the nutrients in domestic wastewater originates from human urine [20]. Separation of urine from fecal matter has been proposed as an alternative to N-and P-reduction by sewage treatment plant technology and/or wetland construction. In temperate climates this would require a storage capacity for urine in the winter season of 0.5 to 1 m^3 per person, which is technically feasible.

The composition of human urine reflects the physiological state of the body, and variations in its composition constitutes the basis for its use in medical diagnostics [6]. Average values for its major constituents are given in Table 1.

Table 1. Chemical composition of fresh human urine.

Parameter, unit	Mean Literature Value (range)	Analyzed concentration
Density , kg L⁻¹	1.015 (1.001-1.028)	nd
Dissolved solids, g L⁻¹	40 (36-47)	nd
Urea, mM N	687 (467-1149)	654
Ammonia, mM N	36 (15-65)	34
Creatinine, mM N	42 (27-56)	27
Uric acid, mM N	3.3 (1.4-5.0)	3
Nitrate, mM N	-	0.9
Nitrite, mM N	-	<0.1
Total Nitrogen, mM N	819(650-1500)	719
Sodium, mM	150 (95-310)	nd
Potassium, mM	70 (40-100)	nd
Calcium, mM	5.9 (<10)	nd
Magnesium, mM	5.4 (2.5-8.3)	nd
Chloride, mM	135 (80-270)	nd
Phosphate, mM	20.0	nd
Sulphate, mM	36 (33-41)	nd
pH	4.5-8.2	5.5 (4.8-6.6)

Literature values from Geigy Scientific Tables [8] and Altman and Dittmer [3]. A man with 70 kg body weight and a daily urine production of 1.0L was used to convert excreted amounts into concentrations. Analyzed values were those determined on fresh morning urine samples from 10 healthy persons used in the present study.

Among the nitrogen compounds, urea is dominating, constituting about 90% of the N from a person on a high protein diet, and between 50 and 60% when on a low protein diet. Other nitrogen excretory products in human urine are ammonia, creatinine and uric acid. Total urinary N is roughly 1.25 times the urea content [7, 14]. After the urine is voided, the urea will decompose into ammonia and carbondioxide. The natural nitrogen cycling processes will determine the further fate of ammonia into nitrite, nitrate, nitrous oxide and/or nitrogen gas.

The objectives of this study were to investigate the toxicity of human urine and its nitrogen excretory products to *Daphnia magna (Cladocera, Crustacea)* and to study the effects of extended storage on urine toxicity. *D.magna* was used because it is an important part of an aquatic food chain used in an aquacultural, ecological wastewater treatment [11, 12] and it has also been shown to be a suitable organism in toxicity studies [5]

2.MATERIALS AND METHODS
2.1 Test materials
Fresh morning urine was collected from ten healthy persons (five men and five women) of 13 to 53 years of age. Each sample was tested separately for acute toxicty to *D. magna* as described below to determine individual variability. In order to study the effect of storage an equal volume (225 mL) from each sample (N=10) was pooled into a 5 L glass beaker, kept at $20 \pm 2°C$ and stirred gently with a magnetic stirrer for 1 h six times daily. A 32 ml sample was
taken immediately (0 d) and after 2, 4, 8, 16 and 32 days. These samples was tested for acute toxicity to *D.magna* as described below. Additionally a 60 mL sample was taken and stored at -20°C for subsequent chemical analysis. All pure chemicals (toxicants) used for toxicity tests were of analytical grade. The toxicants tested were ammonium sulphate, sodium nitrate, sodium nitrite, urea, uric acid and creatinine (2,6,8-trihydropurine).
2.2 Toxicity tests
Toxicity tests with *Daphnia magna* were made according to ISO [13]. Neonates for testing (6-24 h old) were obtained from our laboratory culture. The dilution water for culture and toxicity tests was well water adjusted to a hardness of 250 mg L⁻¹ as $CaCO_3$ and pH 8.1 by the aid of ISO [13] stock solutions. Ambient conditions were $20 \pm 2°C$ and a photoperiod of 16 h light and 8 h darkness. All exposures were made in Petri-dishes (id 10 cm) with 50 mL test

solutions and 20 daphnids, using a 0.5 dilution factor. Dissolved oxygen and pH was recorded at start and after terminating exposure. Immobility was recorded after 24 and 48 h and EC50s were determined by probit analysis or moving average according to [16]. All toxicity tests with the pure chemicals were repeated and a reference toxicant test with potassium dichromate was made in parallell with all toxicity tests.

2.3 Chemical analysis

After storage at -20°C for a maximum period of 3 months, the water samples of stored urine (taken after 0, 2, 4, 8, 16 and 32 d of storage) were brought to 20°C. A subsample was diluted to 1% with deionized water prior to chemical analysis of ammonia, nitrate, nitrite and total nitrogen by an certified (accredited) laboratory (Waterworks laboratory, Göteborg). Non-diluted samples were sent for analysis of creatinine and uric acid by clinical methods at the laboratory of a local hospital (Östra sjukhuset, Göteborg). The urea concentration was calculated from the total nitrogen minus the other measured nitrogen chemicals (ammonia, nitrite, nitrate, creatinine and uric acid).

2.4 Statistical treatment

Toxic concentrations (24- and 48-h EC50s) for urine were expressed as percent by volume, while those for the pure nitrogen compounds were expressed as mM N (millimoles of nitrogen
L^{-1}), considering that ammonium sulphate and urea contains 2 N, creatinine contains 3 N and uric acid 4 N per molecule. Correlation tests was made by a statistical software package, Crunch Software Corp., Oakland, CA, USA.

In order to determine the toxic contribution of each nitrogen compound in fresh and stored urine the additive toxic unit (TU) approach was used. The toxic unit required to immobilize 50% of the test-organisms for each analyzed nitrogen compound after each time of storage (0-32 d) and each exposure time (24 and 48 h) was determined by the equation (1):

$$\text{Toxic unit} = \frac{\text{EC50 for urine (\%) x concentration of compound in 100 \% urine (mM-N)}}{100 \text{ x EC50 for compound (mM-N)}} \quad (1)$$

This calculation of TU was made for each analyzed compound (ammonia, creatinine, nitrate and uric acid), the calculated urea concentration and also from calculated un-ionized ammonia-nitrogen, according to the following equation (2). The pKa at 20° C for ammonia is 9.4 [2].

$$\text{Per cent un-ionized ammonia} = \frac{100}{1 + \text{antilog (pKa - pH)}} \quad (2)$$

If one assumes an additive toxicity for the studied compounds, then sum of all calculated TUs equal to 1 means that the entire effect on the *D.magna* exposed could be accounted for by the chemicals in the sample. A TU equal or close to 1 for a certain chemical in the sample also means that the major part of the toxicity is due to this chemical.

3.RESULTS

3.1 Toxicity of nitrogen excretory products

Toxicity tests with the pure nitrogen excretory and degradation products showed that nitrite, ammonia and uric acid were the most toxic compounds on a nitrogen molar basis, with mean values of 2.6 to 5.1(24-h EC50s) and 0.6 to 2.4 mM N (48-h EC50s), respectively.
The other compounds (nitrate, creatinine and urea) were one to two orders of magnitude less toxic (Table 2).

Table 2. Acute toxicity to *Daphnia magna* of human urine nitrogen excretory products and their degradation products. Dissolved oxygen (DO) was about 100% ASV and pH ranged between 6.3 and 8.5.
All EC50s are expressed as mM of N. Mean EC50s shown were used in the calculation of Toxic Units (TUs).

Chemical	test	24-h EC50	48-h EC50
Urea, mM-N	1	699 (532-1065)	699 (532-1065)
	2	719 (532-1065)	659 (532-1065)
	Mean	**709**	**679**
Creatinine, mM-N	1	≥ 191	55 (43-68)
	2	154 (122-199)	42 (11-56)
	Mean	**173**	**49**
Uric acid, mM-N	1	3.6 (2.7-4.6)	2.3 (1.8-2.9)
	2	6.5	2.5(1.7-3.8)
	Mean	**5.1**	**2.4**
Ammonia,mM-N	1	3.4 (2.5-4.9)	1.9(1.5-2.5)
	2	1.8 (1.2-2.4)	1.0(0.8-1.2)
	Mean	**2.6**	**1.5**
Free Ammonia, mM-N[1]	1	0.31	0.17
	2	0.16	0.09
	Mean	**0.23**	0.14
Nitrate, mM-N	1	85 (66-114)	73(51-124)
	2	93 (71-141)	82(35-141)
	Mean	**89**	**78**
Nitrite, mM-N	1	nd	nd
	2	2.7 (2.0-3.6)	0.6(0.4-0.8)
	Mean	**2.7 (2.0-3.6)**	**0.6(0.4-0.8)**

[1] Calculated from the total ammonia concentration above, considering that test pH was 8.4 and the pKa value at 20° C is 9.4.

The 24-h EC50 for the reference toxicant ranged from 1.06 to 2.83 mg $K_2Cr_2O_7$ L^{-1} (mean value: 1.9 ± 0.5). Some of these reference toxicant values were higher than accepted by ISO [13], (accepted between 0.9-2.0 mg/L as 24 h EC50 for $K_2Cr_2O_7$), but they were within the range obtained at our laboratory and also at other laboratories using this method.

3.2 Individual variation in urine toxicity
Twenty four hour EC50s for morning urine samples from 10 persons ranged from 9.7 to $\geq$ 32%. Corresponding values for 48 h exposures ranged from 2.8 to 12% (Table 3). Dissolved oxygen and pH measured after 48 h ranged from 22 to 72% ASV and 7.2 to 8.7. Mean values for 24 and 48 h EC50s were 16 and 5.4%, respectively. No correlation between toxicity and pH was found.

Table 3. Individual variation in acute toxicity to *Daphnia magna* of fresh morning urine from ten persons.
pH ranged from 7.2-8.7 no correlation with toxicity was found.

Parameter	Mean	SD	Min	Max
24-h EC50, % volume	16,3	7,5	9,7	32
48-h EC50, % volume	5,4	2,9	2,8	12,5

3.3 Effect of urine storage
Dissolved oxygen and pH in the stored urine ranged between 22 and 98 % ASV and 5.7 and 9.8, respectively. Dissolved oxygen and pH in the toxicity tests (at EC50 value) ranged between 46 and 98 % ASV and 7.8 and 8.9, respectively. The effect of urine storage at 20°C on the concentrations of nitrogen compounds ,toxicity, dissolved oxygen and pH, is shown in Tables 4 and 5.

Table 4. Effect of storage of human urine on the concentration of nitrogen excretory products, their degradation products, pH and dissolved oxygen.

Parameter (unit)	Period of storage at 20° C (days)					
Time (days)	0	2	4	8	16	32
Ammonia-N, mM-N[1]	34	183	432	483	334	566
Creatinine-N, mM-N	27	26	26	28	23	20
Uric acid-N, mM-N	3	1,5	1	0.8	0.6	0.6
Nitrate-N, mM-N	0.9	0.2	0.6	0.2	0.2	0.2
Nitrite-N, mM-N	<0.1	<0.1	<0.1	<0.1	<0.1	<0.1
Urea, mM-N[2]	654	550	234	244	256	117
Total N, mM-N	719	761	694	756	614	704
pH	5.7	9.1	9.6	9.8	9.4	9.7
Dissolved oxygen (% ASV)	75	55	22	82	98	68

[1] Measured as total ammonia (ammonia+ammonium)

[2] The concentration of urea was estimated by calculation of the difference between the concentration of total nitrogen and the concentrations of the other analyzed nitrogen compounds (ammonia, creatinine, uric-acid, nitrate and nitrite) on a nitrogen molar concentration basis.

Table 5. The acute toxicity of human urine to *Daphnia magna* during storage at 20° C.

Parameter, unit	Period of storage at 20° C					
	0	2	4	8	16	32
24-h EC50, % by volume	4.8	1.8	0.4	0.4	0.3	0.3
48-h EC50, % by volume	1.6	1.0	0.4	0.4	0.2	0.3
pH[1]	7.8	8.9	8.8	8.8	8.6	8.7
Oxygen,% ASV[1]	46	89	98	96	98	86

[1] pH and dissolved oxygen at the EC50 value.

Measured concentrations of total nitrogen ranged from 614 to 761 mM without any tendancy towards decreasing or increasing concentrations (Figure 1). The observed variations in measured total N are within the analytical variability for this method, ±10% [18, 19]. The major change in urine composition and toxicity took place within
4 d, when urea was converted to ammonia and a concomitant increase in pH and toxicity. During the first 4 d the increase in ammonia concentration was 398 mM N, or 12.7 times. The conversion of urea to ammonia was the major alteration in urine nitrogen compound composition taking place. No major changes in the other analyzed nitrogen compounds were found. The increase in urine toxicity during the first 4 d of storage was 12 times for the 24-h EC50.

Figure 1. Total Nitrogen in human urine after storage in 20 ± 1°C in 32 days.

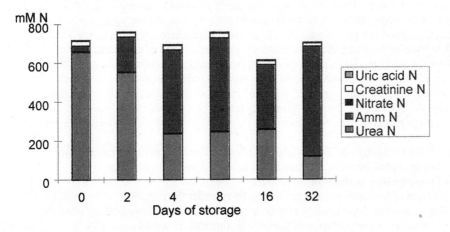

3.5 Toxic Units
Calculated toxic units for the different compounds is shown in Table 6. Ammonia measured (ammonia+ ammonium) was the dominating fraction even without taking pH into account. The "free" ammonia is calculated from total ammonia (ammonia) and different pH at the EC50 values, according to equation (2).

Table 6 . Toxic units for nitrogen compounds in human urine during storage for 0 to 32 days.
A. 24 h exposure

Compound	Period of exposure					
	0	2	4	8	16	32
Ammonia	0.63	1.27	0.66	0.74	0.39	0.65
Urea	0.04	0.01	<0,01	<0,01	<0,01	<0,01
Creatinine	<0,01	<0,01	<0,01	<0,01	<0,01	<0,01
Uric acid	0.03	<0,01	<0,01	<0,01	<0,01	<0,01
Nitrate	<0,01	<0,01	<0,01	<0,01	<0,01	<0,01
Nitrite	<0,01	<0,01	<0,01	<0,01	<0,01	<0,01
Sum	<0.71	> 1.0	<0.66	<0.75	<0.40	<0.66
Free Ammonia	0.18	3.44	1.51	1.69	0.60	1.23

B. 48 h exposure

Compound	Period of exposure					
	0	2	4	8	16	32
Ammonia	0.36	1.22	1.15	1.29	0.45	1.13
Urea	0.02	0.01	<0,01	<0,01	<0,01	<0,01
Creatinine	0,01	0,01	<0,01	<0,01	<0,01	<0,01
Uric acid	0.02	0,01	<0,01	<0,01	<0,01	<0,01
Nitrate	<0,01	<0,01	<0,01	<0,01	<0,01	<0,01
Nitrite	<0,01	<0,01	<0,01	<0,01	<0,01	<0,01
Sum	<0.42	< 1.26	<1.16	<1.30	<0.46	<1.14
Free Ammonia	0.10	3.14	2.48	2.78	0.65	2.01

4. DISCUSSION
4.1 Variations in urine toxicity from different persons
The physiological state of the body will affect the chemical composition of its urine. Concentrations of all components are affected by the water balance of the body through hormonal regulatory systems affecting, thirst, intestinal uptake and kidney function [10]. These physiological control systems are designed to regulate primarily the blood volume and the osmotic pressure of the blood plasma, but also the concentrations of ions important for the proper functioning of the cells, like calcium, hydrogen, phosphate, magnesium, potassium and sodium [6]. Urine is the major external result of this regulation of the body fluids, which is called homeostasis. The diurnal variations in urine volume and composition can be considerable. Therefore, urine excretory patterns are usually expressed as content of the components excreted per day, rather than as concentrations in single voided urine samples. These temporal variations in urine concentrations can to some extent be counteracted by the use of morning urine samples, which is practiced in routine medical examinations. The same approach was used in this study in order to eliminate short-term variations and determine individual variability rather than diurnal variability.

Considering these precautions to reduce temporal variability the individual range in urine 24-and 48-h EC50s for *D.magna* was 3- to 4-fold and the CV (Coefficient of Variation) was about 50%. No correlation between measured pH values and toxicity was found.
4.2 Toxicity identification evaluation
An evaluation of toxicants present in urine must consider the possibility of several toxicants acting in concert. Joint toxicity may be less than additive, strictly additive or more than additive. However, a consideration of these three alternatives would require a

considerable amount of data from controlled experiments with combinations of toxicants and defined mixtures based on the analyzed concentrations of plausible toxicants present in different samples of urine. Based on the chemical composition of human urine (Table 1) the present study was focused on the nitrogen excretory products and their degradation products.

D. magna is a freshwater organism with a fair tolerance to dilute seawater. In Scandinavia *D. magna* occurs in rock pools, which are small ponds close to the sea, with a salinity affected by seawater spray and seabird droppings. A study on the natural occurrence of three *Daphnia spp* in rock pools [17] have shown that the salinity tolerance of *D. magna* is higher than for *D. pulex* and *D. longispina*. *D. magna* was commonly found at salinities from 0.5 to 3.4 g L^{-1} (per mill) and occasinally from 0.03 to 12.5 g L^{-1}. Forty eight hour survival of our laboratory clone of *D. magna* (>48 h) in our dilution water [13] was not affected by addition of NaCl up to 1.6 g L^{-1} (or 0.97g $Cl^- L^{-1}$) or by addition of natural sea water salt up to 2.8 g L^{-1} (or 1.7g $Cl^- L^{-1}$)(Dave, unpublished). The concentration of chloride in human urine is 4.8 (2.8-9)g L^{-1} (Table 1) and the highest concentration of salt tested in this study was 32% urine (1.5 g Cl^-). Therefore, salt alone was ruled out as a plausible limiting toxicant in human urine for *D.magna*. During storage of urine the increase in toxicity was accompanied by a decrease in the concentration of urea and an increase in the concentration of ammonia, suggesting that ammonia was the limiting toxicant.

All calculated TUs are presented in table 6, with separate TUs for 24- and 48-h exposures. The sum of all TUs approximated 1, indicating that all major toxicants were taken into account. The dominating TU was that for ammonia on all occasions. The un-ionized ammonia, which is more toxic then the ionized form, gave very high TUs, confirming that pH plays an important role.

The lower TU values for the fresh urine (storage time 0) could be due to the fast initial conversion of urea to ammonia, and the use of initial analyzed concentrations in the calculations of all TUs, meaning that the actual exposure concentration of ammonia during the *D. magna* toxicity tests were higher than accounted for in the TU calculation. A similar explanation may be valid for the 16 d storage time, because of the lower analyzed values for both ammonia and total nitrogen compared to that observed after 8 and 32 d. Therefore, the major conclusion is that ammonia was the dominating toxic agent in both fresh and stored urine.

The toxicity of ammonia is highly dependent upon ambient pH, being more toxic at high compared to low pH [2, 4]. The control of pH is, therefore, as important as the control of ammonia concentrations. In waters fertilized by manure which supports high densities of microalgae, a photosynthesis-driven elevation of pH may result in massive kills of zooplankton and fish due to ammonia intoxication. This was recently shown to be the case at a wastewater treatment facility using aquacultural principles [1]. Thus the toxic potential of human urine due to ammonia must be considered carefully in the use of urine as a fertilizer in aquacultures.

5. CONCLUSIONS

1. Urine toxicity to *D.magna* from different persons showed a variation of ± 50% with a mean 24-h EC50 of 16% and a mean 48-h EC50 of 5.4%.

2. Urine toxicity to *D.magna* increased during storage (one month) from 4.8 % to 0.3 % for 24-h EC50s and from 1.6 % to 0.3 % for 48-h EC50s.

3. Urine toxicity of fresh as well as old urine was mainly caused by ammonia. Therefore, a careful control of ammonia concentrations is essential in ecological wastewater treatments.

Acknowledgement — This research has been supported by the Swedish Council for Building Research, The Futura Foundation and the Public Water Works in Göteborg. Technical assistance was given by Eva Nilsson and Birgitta Vallander. The authors are grateful for the assistance offered by these organizations and persons involved.

REFERENCES

[1] Adamsson, M., Dave, G., Forsberg, L. Guterstam, B., (1995). Toxicity identification and evaluation at the Stensund wastewater Aquaculture, Sweden, in manuscript.

[2] Alabaster and Lloyd, (1982). Water quality criteria for freshwater fish., second edition, Agriculture Organization of the United Nations,Butterworth Scientific

[3] Altman, P.L. and Dittmer, D.S. (1974). Biological handbooks. Biology Data Book, Second edition, Vol. III, Federation of American Science for experimental Biology, Bethesda, Maryland, USA.

[4] Ankley, G.T., Schubauer-Berigan, M.K., (1994). Background and overview of current sediment toxicity identification evaluation procedures. Paper presented at the 1st conference in Sediment Quality Asssessment, August 1994.

[5] Baudo, R., Giesy, J. and Muntau, H., (1990). Sediments: Chemistry and toxicity of inplace pollutants. Lewis Publ.

[6] Berne, R.M. & Levy, M.N. (Eds.), (1990). Principles of Physiology. Wolfe Publ. Ltd., 690 pp.

[7] Burge, J.C., Choban, P. McKnight, T., Kyler, M., Flancbaum, L., (1992). Urinary Ammonia plus urinary urea nitrogen as an estimate of total urinary nitrogen in patients receiving parenteral nutrition support. JPEN, vol. 17, pp.529-531.

[8] Geigy Scientific Tables, (1981), Units of measurement. Body fluids composition of the body nutrition.Editor: Cornelius Lentner, CIBA-Geigy Limited, Basle, Switzerland.

[9] Graneli, E., Wallström, K., Larsson, U.,Graneli, W. & Elmgren, R., (1990). Nutrient limitation of primary production in the Baltic Sea area. AMBIO, 19:142-151.

[10] Green, J.H., (1976). An introduction to human physiology, Fourth S.I. edition (rev 1987)Oxford university press, Hong Kong

[11] Guterstam,B and Etnier, C, (1991), Ecological engineering for wastewater treatment. Proceedings of the International Conference, March 24-28, 1991. Stensund Folk College, Sweden. Ed. Guterstam Etnier, Bokskogen, Tryckverksta′n, Limmared, Sweden

[12] Guterstam and Todd, (1990). Ecological engineering for wastewater treatment and its application in New England and Sweden. Ambio 3/1990, pp. 173-175

[13] ISO, (1982). Water quality - Determination of the inhibition of the mobility of *Daphnia magna* Straus (*Cladocera, Crustacea*), International Organization for Standardization, ISO 6341-1982. 15pp

[14] Konstantinides, F., Konstantinides, N., Li, J., et al., (1991).Urinary urea nitrogen: Too insensitive for calculating nitrogen balance studies in surgical clinical nutrition. JPEN 15:189-193.

[15] Kuparinen, J., Leonardsson, K., Mattila, J. & Wikner, J., (1994). Food web structure, carbon flow and trends in the Gulf of Bothnia, Baltic Sea. Vatten (Lund), 50:201-219. (In Swedish with abstract in English).

[16] Peltier, H.W. and Weber, C.I., (1985). Methods for measuring the acute toxicity of effluents to freshwater and marine organisms (3rd edition) EPA160014-851013, march 195, US EPA, Cincinnati; Ohio 45268, Apendice Eg pp.170-216.

[17] Ranta, E., (1979). Niche of *Daphnia* species in rock pools. Arch. Hydrobiol. 87:205-223.

[18] SIS, Standardiseringskommissionen i Sverige, (1991). Översikt SIS10. Utgåva 2. Vattenundersökningar Förteckning över svensk standard, Stockholm, Sweden (In Swedish).

[19] Swedish EPA (Naturvårdsverket) (1993) Naturvårdsverkets metodhandbok. Vatten., Naturvårdsverket, Solna, Sweden. (In swedish).

[20] Swedish EPA (Naturvårdsverket) (1995), Vad innehåller avlopp från hushåll, Rapport 4425, Naturvårdsverket, Stockholm, Sweden (In Swedish).

Environmental Research Forum Vols. 5-6 (1996) pp. 145-148
© 1996 Transtec Publications, Switzerland

A Case Study of Ecological Engineering for Treatment and Use of Wastewater from a Silk Reeling Mill

J.S. Yan and Y.S. Zhang

Nanjing Institute of Geography & Limnology, Academia Sinica, Nanjing, 210008, China

Keywords: Ecological Engineering, Wastewater, Reeling Mill, Beneficial Circulation, Sustainable Development

ABSTRACT

Wastewater from al silk reeling mill is treated an utilized with a series of components and techniques including adjusting the pH value to attain the isoelectric point; cultivation of the photosynthetic bacteria, water hyacinth, water spinach, *Azolla filicubides*, *Lolium multiforum*. This combination results in the purification rate of 95.7% TSS, 92.6% BOD_5, 96.6% COD_{Cr}, 86.6% TN, 79.8% NH_4^+-N, 90.5% TP and 88.7% PO_4^{3-}-P in order to promote the effective and beneficial circulation of the products and waste, a parallel connection of symbiosis networks is created by joining the fish culture ponds, biogas, germination tanks, geese and duck farm, and mulberry grove. This ecological engineering makes some contributions to sustanamble development.

1. INTRODUCTION

The township and village enterprises is the new raising and developing enterprises in China, since 1980s. Although they make great contribution to Chinese economic development, their pollution is a problem because most of their waste could not be treated well by normal traditional path and technologies of environmental engineering limited by deficient in investment and technological energy sources. It is a urgent task to study and seek the new paths and technology for treating waste water with low investment, less consumption of technological energy sources and operation. For this propose, we had engage in a series of studies about this aspect. This paper will show a case study of them.

2. SITE DESCRIPTION

Maodong Silk Reeling Mill ist the largest enterprise and the main pollution sources in Xinggon town, Danyang county. The annual discharge of wastewater from this enterprise is about 144'000 t wich contains about 10'630mg/l COD_{Cr}, 3'834mg/l BOD_5, 12.5-54.22 mg/l NH_4^+-N, 413 mg/l TIN 1'035mg/l TN, 80-90 mg/l PO_4^{3-}-P and 5.335-9.54g/l TSS including much fat and protein.

3. PURIFICATION OF WASTEWATER

This high concentration organic polluted wastewater can not be purified directly by most hydrophytes, because it exceeds the limit of tolerance of most species of aquatic plants It is necessary to pretreat the wastewater for aquaculture.

3.1. Recovery of Protein and Fat

much of protein and fat with nutrients in the wastewater are basically a resources. If unrecovered and unused, they will become the waste and organic pollution sources. It is used as the first step and pretreatment to means of adjusting the pH value to attain the isoelectric point. It is well known that the protein should coagulated easily when the pH value of the medium attains its isoelectric point. On the basis of this principle, at first, the pH value of the high concentration organic wastewater is adjusted to 4.5, to promote coagulation of protein and fat, by adding hydrochloric acid. Then, the coagulated solid is recovered by sedimentation and flotation. The average recovery rate of this coagulated solid from this wastewater is 30kg/m3 (range from 25 to 35 kg/m3), wich contains 18.5-26.2% of raw protein and 53.5-62.7% of fat. This results in the purification rate of 79-92%, 71-80%, and 92-95% for COD_{Cr}, BOD_5, and TSS respectively. After this treatment, COD_{Cr} decreases to 1'274 - 1'470mg/l from 6'890-15'680mg/l; BOD_5 to 590-792mg/l from 2'734mg/l; and TSS to 208-395mg/l from 2'712-8'643mg/l.

3.2. Conversion of some pollutants into usable resources

Although much organic pollutants have been removed from wastewater and the concentration of organic material and nutrients is decreased in the first link, it needs to purify further for cultivation of some aquatic plants, because the concentration of COD_{Cr}, is in excess of tolerant limitation of most aquatic plants still. Therfore the photosynthetic bacteria is introduced into the wastewater after adjusting the pH value to 6.0-7.0. This link results in COD_{Cr} further decreases to 80-142 mg/l from 1'274-1'4703mg/l, and it purification rate is more 90%. It is successful in increasing disease resistance, survival rate, growth rate and yield of fishes, chickens, hens, duck, geese, and pigs that the photosynthetic bacteria and its sludge were used as an additive for these animals.

The wastewater after treatment by these two link flows into a sewage channel which is about more 1'200m long and 4-6m wide. About 0.2 ha of water hyacinth was cultured and hydroponic cultiation in 0.2 ha water surface of water spinach (*ipomoea aquatik*) was cultured from May to October. Water spinach begin resting at 10°C, begins growth at 15°C, and its optimal growth temperature is 30°C to 35°C. It can normally grow in the environment with less than 800mg/l COD_{Cr}. In our static experiment; its purification rate is 100% for COD_{Cr} after 5 days from 74.4mg/l and 122.4mg/l to 0mg/l of COD_{Cr}, and in our dynamic experiment, its purification rate are 86.2%, 83.0% and 78.9% from 236.6mg/l, 482.5mg/l and 559.0mg/l to 32.6mg/l, 73.4mg/l and 117.6mg/l of COD_{Cr} after 6 days respectively. Its average purification rate in the sewage channel ist about 62% (59.7-72.9%) from 467.9mg/l to 142.6mg/l of COD_{Cr} during 18 days, as well as 37.8-50.0% for TN, 50% for TP respectively. Its turnover time ist about 4.4. Its production by harvest in rotation is about 45-51 t/ha during May to October [2]. All of these are lower than that of water hyacinth, but its price is higher than taht of water hyacinth in economy.

During October to May of next year, 0.2 ha of *Azolla filiculoides* and hydroponic cultivation in 0.2 ha of *Lolium multiform* was planted instead of water hyacinth and water spinach. In this sewage channel *Azolla filiculoides* can not tolerate the environment with higher concentration organic pollutant (COD_{Cr} is over 80mg/l), but it can grows in slight salinity (0.1%-0.5%) environment. It begins growth at 5°C, and optimal growth temperature is in 20°-25°C, decrease growth rate in more 25°C, and stops growth at more than 35°C. In our dynamic experiment, its average purification rate of TN is 37.6% during 7 days from 6.8mg/l to 4.24mg/l; 13.7% for NH_4^+-N during 8 days from 3.87mg/lto 2.99mg/l; 48.2% for COD_{Cr} during 9 days from 168mg/l to 87mg/l; and 56.0% for BOD_5 after 5 days from 78.9mg/lt to 43.5mg/l. Its efficency for decreasing nitrogen in the water is less, because it is a symbiont of a species of fern (Pteridophyte) with some nitrogen-fixing blue algae and has the capability for nitrogen fixation from air. Its roots system suspended in the surface layer of water is very short and can only provide and enlarge a few of matrices for microorganism which can decompose the organic material, so that its purification rate for COD_{Cr} and BOD_5 is less than that of water hyacith. But its advantage for treating the wastewater is that it can grow in winter and early spring when water hyacinth can not grow. Its purification rate for COD_{Cr}, TN, TP, and TSS in the wastewater channel are 49.9%, 52.7%, 43.2% and 50% respectively. Its annual production by harvest in rotation is about 45-60t/ha.

On the basis of our eight service, this combination of these lants results in that 16,145.8 kg of TSS, 13'729kg pf CPDCr, 7'725 kd of BOD_5, 1'196.2kg of TN, and 226.9kg of NH_4^+-N are removed out of decreased in more 50'000t of wastewater flowed through this sewage channel during May to October 1988. Their pruification rate were 95.3-96.4%, 90.0-93.6%, 91.7-97.2%, 68.9-94.4%, and 58.7-97.1% respectively. While these pollutants are convertet into some vegetables and 125t of green fodder.

On the basis of the original wastewater, this combination of first, second and third links results in purification rate of 95.7% TSS, 92.6% COD_{Cr}, 96.6% BOD_5, 86.6% TN, 79.8% NH_4^+-N, 90.5% TP, and 88.7% PO_4^{3-}-P. respectively the discharged water after this treatment attains to the National Standard of Surface Water in China.

3.3 Promote Effective and Beneficial Circulation

A parallel connection of symbiosis networks is created by joining the fish culture ponds, biogas, germination tanks, geese and duck farm, and mulberry grove to promote effective and beneficial circulation of the products and waste in this system.

The organic material recovered from first link is used as a part of fine fodder for pig, geese, and duck to converge it into commodilty, and save much commercial food for these domestic animals. The experiment on using recovered organic material as the additive of mixed feeds (1:5) for duck shows that there is an about 10% increase of body weight in experimental groups as compared with that in control experimental gruop after 30 days. Water hyacinth, Azolla and rye produced from the purifying channel are harvested and used as the green fodder for duck, geese, and nerbivorous fishes. For instance 114t of water hyacinth had been used as a part food for culture fish by Daihu Fishery Farm and resulted in saving more 2.0t of fine commercial feeds and increasing 1.5t yield of fish [5]. This is not only utilization but also promote the transformation and transfer of these plants and avoid secondary pollution in this channel by their saprophytism and decomposition for utilization.

The excrement and residues of food from the pig, chicken, duck and geese are used as the raw materials for biogas generation system. The biogas is used as the energy sources. The solid slag from biogas fermentation, as a solid waste are used as a part of raw materials of media in the cultivation of edible fungi. The residue and detritus of edible fungus bed after harvesting the mushroom can be used as the medium mixed with other waste for earthworm breeding. The earthworm is a good quality bait for fish, and poultry feeds, and it can also be processed a kind of tradition Chinese medicine. The residue and detritus from earthworm bed and the bottom organic materials from the fish culture ponds including the excrement of fish is also a good fertilizers for mulberry grove to improve soil. The mulberry leaves are used as the fodder for silk worms, and the withered and fallen parts of mulberry grove are used as the food for fish. The silkworm cocoon is the raw material of silk reeling, and its waste, the chrysalises , is a fine food for duck, pig and som species of fish, thus making an effective and beneficial circulation of substance in a limited region.

4. DISCUSSION

This combination of series connections of food chains or web and parallel connections of different ecosystem to promote multi-step and multi layer utilization of substances including waste to promote the beneficial and effective circulation of some substances including waste not only promote economic beneficial, ecological environmental and human social advantage in this region are maximized [3, 4], but also contribute to sustainable development. There are three categories of valuable characteristics whose sustainability is desirable, production and yield; species diversity; and self-regulation [1]. This ecological engineering can increase the production and yield of many organisms, especially economic animals and plants, which is a prerequisite for man's existence; and can protect the species diversity by improve the environmental conditions also. The self-regulation of these artificial ecosystems set is increased by the increasing of biodiversity, and multi-step and mult-layer utilization of substance including waste.

REFERENCE

[1] Brink, B.T., 1991: The AMOEBA aprroach as a useful tool f or establishing sustainable development. In Onno Kuik & Harmen Verbruggen (Eds), in Searche of Indicators of Sustainable Development Kluwer Academic Pub, Netherlands 1991, pp 71-88.

[2] Chen, shupai et al 1991: Treatment of wastewater from a silk reeling mail by hydroponic cultivation of water spinach. Environ Scj. 12(4): 47-57.

[3] Yan et al, 1992: Ecological techniques and their application with some case studies in China: Ecol. Eng. 1(1992) :261-285

[4] Yan et al., 1993: Advance of ecological engineering in China "Ecological Engineering" 3 (1993: 193-216

[5] Li Weixing et al, 1991: Study on the purification of organic wastewater discharge by township and village enterprise mainly with water hyachinth. Rural Eco-Environment No4, 1991: pp 54-58.

Environmental Research Forum Vols. 5-6 (1996) pp. 149-154
© *1996 Transtec Publications, Switzerland*

Culture of Common Carp *Cyprinus carpio* in Dairy Wastewaters: A Feasibility Study

R. Nagendran

Centre for Environmental Studies, Anna University, Madras - 600 025, India

Keywords: *Cyprinus carpio*, Dairy Wastewater, Feeding, Food Conversion, Energy Budget, Economics of Culture

ABSTRACT

The potential of aquaculture for economic and efficient reuse of wastewaters is being recognized over the world. This concept is attracting the attention of the developing countries in view of the increasing demand for cheaper protein-rich food resource. Pisciculture has been recognized as a means of economic handling and disposal of wastewaters. While studies on the use of domestic wastewaters for aquaculture are readily available, extension of similar attempts to industrial effluents appear to be wanting. The present paper reports on the feasibility of culturing the common cyprinid, *Cyprinus carpio* in wastewaters from a dairy. Composite wastewater from a milk dairy was collected and analysed in the laboratory using standard methods. The slightly acidic effluent exhibited higher chemical oxygen demand (COD) and biochemical oxygen demand (BOD). The nutrient concentrations were low.

Individuals of *Cyprinus carpio* weighing 450 to 680 mg were used for the study. The fish were exposed to different concentrations of the effluents and the LC_{50} at 96 h was computed. The LC_{50} was found to be 25%. The fish were reared in freshwater, 2.5 and 5.0% of the effluent for 30 days and were fed on *ad libitum* diet of tubifex worms. At the end of the study period the food intake, growth and food conversion efficiency were computed. Fish exposed to the wastewater exhibited considerable increase in growth. The conversion efficiency exhibited by fish reared in 2.5 and 5.0% were 8.80 ± 0.55 and $6.63 \pm 0.78\%$, respectively. Based on the data collected during the study, a model energy budget for *Cyprinus carpio* has been developed and the economics of the proposal has been discussed.

INTRODUCTION

Food production from agricultural resources has not been meeting the ever increasing demand owing to substantial growth of human population. In this context, aquaculture is being projected as a viable and very often a profitable supplement to agriculture. The scientific and technological advances made during the last three decades have resulted in enormous increases in the harvest of fish, shrimps and other aquatic organisms. This trend would make a headway in meeting the protein requirements of rural poor, especially in the developing countries like India.

Growing industrialization has been showing its negative impacts on aquaculture as in the case of agriculture. Indiscriminate disposal of industrial effluents into freshwater bodies is rendering most of them useless for aquaculture. This situation has attracted the attention of aquaculturists and scientists resulting in pollution related studies on freshwater resources and organism of economic importance. An effective and more often a challenging option which is being projected of late, has been the use (or reuse as the case may be!) of polluted water itself for aquacultural purposes.

Reports on the characterization of water courses receiving effluents from different industries and their effects on respiratory physiology [1], haematology [2], feeding behaviour [3,4] etc., of aquatic animals are readily available. However, studies dealing with the impacts of pollutants and polluted water conditions on the culture prospects of aquatic organisms are meagre. Literature on this aspect is mostly restricted to domestic wastewaters [5,6] and effluents from Kraft mill [7], textile units [4] and distilleries [8]. A preliminary qualitative study on the dairy effluents indicated the possibility of employing "aquaculture" as a means of economic handling (reuse) and disposal of this wastewater. Hence the present study.

MATERIALS AND METHODS

Untreated composite wastewater from a large-scale dairy was collected at the first common outlet which drains the effluents to the disposal point and transported to the laboratory in a 5 l plastic container. It was stored under refrigeration to minimize bacterial degradation.

Fingerlings of *Cyprinus carpio* used for the present study were collected from a local fish farm and stocked in the laboratory in plastic tubs. The fishes were acclimated to the laboratory conditions for 15 days. Individuals showing signs of disease or injury were removed and healthy ones were used for experiments. The fishes were fed *Tubifex tubifex* during the maintenance as well as experimental period. The tubifex worms were collected from a local fish feed dealer and stocked in the laboratory.

Selected physico-chemical parameters of the dairy wastewaters were determined employing the methods detailed in Standard Methods [9]. Three samples were analysed during the study. LC_{50} value for fish exposed to the dairy wastewater was computed following the method detailed in Nagendran and Katre [10].

For the present feeding experiments, fingerlings of *Cyprinus carpio* in the weight range of 450 to 680 mg were selected from the stock. All feeding experiments were carried out for a period of 30 days in glass aquaria (29.5 x 17.5 x 23 cm; capacity = 10l) containing 5 l of tap water (control) or a specific concentration of dairy wastewater (i.e., 2.5 and 5.0% wastewater). For each experiment, eight replicates were maintained and 10 individuals were reared in each tank. Throughout the experimental period the fish were fed *ad lib* diet of *Tubifex tubifex* worms. The medium was continuously aerated and the medium was replaced once in seven days. The food intake, growth and conversion efficiency of the fish were calculated as detailed in Nagendran [3]. The efficiency of culture of *C. carpio* was calculated after the method of Onodera [11]. Crude protein, carbohydrate, fat and ash contents of the feed and fish were determined by methods detailed in [12] to [15], respectively. The energy content of the feed and the fish was computed using the physiological fuel values and the proximate biochemical composition of feed/fish [16].

RESULTS AND DISCUSSION

A. Physico-chemical characteristics of the dairy wastewater

The physico-chemical characteristics of the dairy wastewater are presented in table 1. The wastewater had a pH of 6.4 and its total solid content was 2720 mg/l. The alkalinity of the wastewater was 391 mg/l as $CaCO_3$. The wastewater exhibited a considerably high BOD (1079 mg/l) and COD (1926 mg/l) values. The nutrient concentrations were markedly low (total phosphate - 6.9 mg/l; total nitrate - 53.0 mg/l). While the present values of physico chemical parameters are comparable to those reported earlier, the nutrient concentrations are significantly lower [17].

Table 1.   **Physico-chemical characteristics of dairy effluent (each value represents the average of three samples - triplicates in each sample were analysed)**

S.No.	Parameters	Values
1.	pH	6.4
2.	Temperature (°C)	30.2
3.	Total solids (mg/l)	2720.0
4.	Total dissolved solids (mg/l)	862.0
5.	Total suspended solids (mg/l)	1858.0
6.	Alkalinity (mg/l) as $CaCO_3$	391.0
7.	BOD (mg/l)	1079.0
8.	COD (mg/l)	1926.0
9.	Total phosphate (mg/l)	6.9
10.	Total nitrate (mg/l)	53.0

B. Dose response of *Cyprinus carpio* exposed to dairy wastewaters

While no mortality was recorded in control series, the fish exposed to various concentrations of the dairy wastewater responded differently. Test individuals died more rapidly at a concentration of 30 and 35% wastewater and instantaneously beyond this concentration. A concentration of 25% dairy wastewater represented the 96 h LC_{50} for *Cyprinus carpio* (weight range - 555 to 638 mg). Interestingly, the present value is higher than that obtained for the same fish exposed to distillery effluent (less than 2% [8]), and copper containing wastewater (800 ppb [18]). Similarly, *C.carpio* appears to be more resistant and tolerant to pollutants as compared to minor cyprinids such as *Puntius* spp.[3] and *Barbus stigma*[19]. The present observations bring to light the chemical - physiological compatibility between the dairy wastewater and *C. carpio*. A profitable trade - off between these two factors may help evolve a suitable strategy for the culture of *C. carpio* in moderately strong dairy wastewaters.

C. Feeding energetics of *Cyprinus carpio* exposed to dairy wastewaters

Table 2.   **Feeding energetics of *Cyprinus carpio* exposed to dairy wastewaters (each value represents the average of 80 individuals; ± = SD)**

Concentration of wastewater (%)	Initial weight of fish (mg)	Final weight of fish (mg)	Food intake (mg/g live fish/day)	Growth (mg/live fish/day)	Growth rate (mg/g live fish/day)	Conversion efficiency (%)
Control	587.26 ± 82.36	1456.81 ± 152.79	305.72 ± 18.03	869.54 ± 72.58	49.74 ± 3.44	16.30 ± 1.33
2.5	780.45 ± 48.74	1606.15 ± 81.99	404.59 ± 8.36	832.45 ± 46.63	35.66 ± 1.91	8.8 ± 0.55
5.0	712.10 ± 78.12	1270.11 ± 126.43	395.82 ± 11.55	558.00 ± 68.10	26.23 ± 2.90	6.63 ± 0.78

Table 2 summarizes the feeding energetics of *Cyprinus carpio* reared in freshwater and selected concentrations of dairy waste water.

From the table, it is evident that *Cyprinus carpio* in the weight range of 587.26 ± 82.36 mg consumed on an average 305.72 ± 18.03 mg of Tubifex worms / g live fish/day. In contrast to this a much lower feeding rate of 108 mg/g live fish/day for the same fish reared in freshwater has been reported earlier [8]. It is well known that food intake in fishes is controlled by several factors including age [20], density [21], quality of food [22] etc. The lower feeding rate cited above for *C.carpio* could be attributed to one or more of these factors. The growth rate of the fish recorded presently was 49.74 ± 3.44 mg/g live fish/day. This value is considerably higher than those observed in minor cyprinids such as *Rasbora daniconius, Nuria danrica* and *Puntius stigma* [10]. The gross conversion efficiency (K_1) of *C.carpio* reared in freshwater was 16.30 ± 1.33%. This value is comparable to that obtained for other cyprinids reared in freshwater and fed on tubificids [3]. Interestingly, cyprinids fed on other food/feed items [8,19] and species belonging to other families such as *Mystus vittatus* [23,24], *Lepidocephalichthys thermalis* [25], *Oreochromis mossambicus* [26,27] exhibit comparable conversion efficiencies. The present observations indicate different trophic strategies adopted by different species to share and co-exist in a common habitat.

From Table 2 it is also evident that the food intake of *Cyprinus carpio* exposed to dairy wastewater increased significantly compared to those reared in freshwater. Though such higher food intake by fish exposed to pollutants is not uncommon and considerable supporting literature is readily available [8,13,19], reduction in food intake on exposure to polluted waters have also been reported in a few fish species [3,4,25,27]. The increased food intake observed presently could be attributed to the elevated metabolic rate and differential energy channelisation by the fish to overcome the initial physiological shock as well as sustained maintenance on continuous exposure to polluted media. The initial behavioural responses exhibited by *Cyprinus carpio* bear testimony to this (Nagendran, unpublished data). Discussing on this aspect, Webb [28] has reported that "the ability of fish to increase food consumption when under stress leads to unanswered question of voluntary appetite control". Despite considerable increase in food intake, both the rate of growth and the efficiency of food conversion were markedly low in fish reared in the selected concentrations of the wastewater. The reduced growth may be attributed to one or more of the following : (i) accumulation of metabolites in the body (ii) increased mitochondrial metabolism (iii) general metabolic stress and (iv) higher energy demand for body maintenance. Investigations on these aspects are in progress and lack of specific experimental data render further discussions prohibitive. While the food intake was not influenced by the concentration of the wastewater, growth rate and food conversion decreased with increase in concentration.

D. Economics of the culture of *Cyprinus carpio* in dairy wastewater

From the data presented in Table 2, the 'Housekeeping' of *Cyprinus carpio* in different media can be established.

For example, for fish reared in freshwater,

Total increase in biomass in 30 days = 69.5g

In 30 days the Specific Growth Rate (SGR - wt. increase/initial weight/day) = 69.5/47.0/30 = 0.049 g/day.

Total amount of food given = 430.87 / 47.0 / 30 = 0.305 g/day.

Therefore, a daily ration of tubifex worms equivalent to 30.55% of initial biomass produced a daily weight increment of fish flesh equivalent to 16.08% of the given food during 30 days. The 'Housekeeping' can now be suggested as follows. In 50 days, the necessary amount of food will be equal to 0.305 x 50 = 15.25 times the initial biomass, while the daily increment accumulates to yield 0.305 x 50 x 0.0160 = 2.44 times the initial biomass. In other words, 1 kg of initially stocked *Cyprinus carpio* in the weight range of 587.26 ± 82.36 mg may need 15.25 kg of tubifex worm as food to produce 2.44 kg of new flesh. Similar calculations have been made for fish reared in wastewater and the balance sheet for the culture of *Cyprinus carpio* thus obtained has been presented in Table 3.

Table 3.   Balance sheet of *Cyprinus carpio* culture

Concentration of wastewater	Initial biomass (g)	Food intake (% of initial biomass)	Increment of flesh (% of total food intake)	Total food required by an initial stock of 1 kg to produce 1 kg of new flesh in 50 days (kg)
Control	47.00	30.55	16.08	6.21
2.5%	62.44	40.45	8.70	11.55
5.0%	56.96	39.57	6.56	15.33

Though the amount of food required to produce 1 kg of new flesh from an initial stock of 1 kg in 50 days is double or a little more than double the amount as compared to the control fish, the culture prospects appear to be bright. This view is strengthened by the fact that the fish culture can be practised by simple dilution of untreated dairy wastewater. This does not require any skilled manipulation and minor variations in the factor of dilution may not exert excess physiological pressure on *Cyprinus carpio* owing to the moderate strength of the wastewater. Even when the wastewater exhibits higher organic contents, the fish culture can be resorted to at the polishing stage of treatment, if not earlier.

E. Energy profile of *Cyprinus carpio*

Based on the biochemical composition of tubificid worms and *Cyprinus carpio*, their energy contents were calculated using the physiological values of 4.1, 5.7 and 9.5 cal/g for carbohydrates, proteins and lipids, respectively. The major contributor for the energy content for tubifex and fish was protein followed by lipids and carbohydrates. The initial and final energy contents of the experimental fish are presented in Table 4.

Table 4.   Energy content of *Tubifex tubifex* and *Cyprinus carpio* (values calculated from % dry matter of carbohydrate, protein and lipid fractions)

Feed/Fish	Energy content (cal/100g)		Total gain or loss of energy (%)
	Initial	Final	
Tubifex worms	489.09	-	-
Control fish	489.09	512.01	+6.65
Fish reared in 2.5% dairy wastewater	489.09	447.25	-6.83
Fish reared in 5.0% dairy wastewater	489.09	313.43	-34.66

While fish reared in freshwater registered a gain of 6.65% over its initial energy content during the 30 day period, fish reared in wastewaters indicated considerable loss. The loss of energy in fish reared in 5% wastewater registered a remarkable loss of 34.66%. The present observations appear to support the reasons attributed for marked effects of the wastewater on the rate of growth and food conversion, as discussed in section C above. Further investigations to correlate metabolic effects of wastewater and the energy profile of *Cyprinus carpio* are in progress.

REFERENCES

[1]   M.A.Haniffa and M.Porchelvi, J. Environ. Biol. **6**, 17-23 (1985).
[2]   M.A.Haniffa and S.Maryvijayarani, Indian J. Exp. Biol. **27**, 476-478 (1989).
[3]   R.Nagendran, Ph.D. Thesis (1980).
[4]   M.A.Haniffa and A.G.Murugesan, Proc. Natl. Symp. Environ. Biol. 212 (1992).
[5]   B.B.Sundaresan, S.Muthusamy and V.S.Govindan, Proc. San. Dev. APHA. 1-18 (1977).
[6]   V.S.Govindan, J. Biol. Waste. **30**, 169-179 (1989).
[7]   R.H.Ellis, Natl. Council for Air and Stream Bulletin. **210**, 1-15 (1967).
[8]   M.A.Haniffa and K.Balakrishnan, Poll. Res. **7**, 103-106 (1988).
[9]   Standard Methods, APHA, AWWA, WPCF (1971).
[10]   R.Negendran and S.Katre, Indian J. Exp. Biol. **17**. 270-273 (1979).
[11]   K.Onodera, FAO 62/6, 3.
[12]   B.H.Lowry, M.J.Rosenbrough, A.L. Ferry and R.J. Randall, J. Biol. Chem. **193**, 265-275 (1951).
[13]   J.H.Roe, J. Bio. Chem. **212**, 335-343 (1955).
[14]   M.J.Gurr and A.T.James, The Lipid Biochemistry (1957).
[15]   R.T.Paine, Ecology **45**, 384-387 (1964).
[16]   W.H.Peterson and F.M.Strong, General Biochemistry (1957).
[17]   V.S.Govindan, Ph.D. Thesis (1982).
[18]   R.Rehwoldt, G.Bida and B.Nerrie, Bull. Env. Cont. Tox. **6(5)** 445-448 (1971).
[19]   M.A.Haniffa and S.Sundaravadhanam, Life. Sci. Adv. **2**, 143-146 (1983).
[20]   Menghe Li and R.T.Lovell, Aquaculture, **103**, 153-163 (1992).
[21]   E.H.Jorgensen, J.S. Christiansen and M.Jobling, Aquaculture, **110**, 191-204 (1993).
[22]   S.Arunachalam, E.Vivekanandan and S.Palanichamy, Comp. Physiol. Ecol. **10(3)** 101-104 (1985).
[23]   S.Palanichamy, S.Arunachalam, S.A.Ali and P.Bhaskaran, Proc. Second Asian Fish. Forum (1990).
[24]   S.Arunachalam, S.Palanichamy, M.Vasanthi and P.Bhaskaran, Ibid. (1990).
[25]   S.Palanichamy, P.Bhaskaran and Balasubramanian, Proc. Symp. Pest. Resid. Env. Pollution. 97-102 (1986).
[26]   P.Bhaskaran, S.Palanichamy, and S.Arunachalam, J. Ecobiol. **1(3)**, 203-204 (1989).
[27]   M.A.Haniffa and M.Jaceentha, Indian J. Environ. Hlth. **30**, 228-233 (1988).
[28]   P.W.Webb, Ecology of Freshwater Fish Production, 184-214 (1978).

Environmental Research Forum Vols. 5-6 (1996) pp. 155-160
© *1996 Transtec Publications, Switzerland*

Wastewater-Fed Aquaculture in Germany: A Summary[1,2]

M. Prein

International Center for Living Aquatic Resources Management (ICLARM),
MCPO Box 2631, 0718 Makati, Metro Manila, Philippines

Keywords: Municipal Sewage Treatment, Dairy Wastes, Household Effluents, Recycling, Fish Culture, Constructed Wetlands, Biological Treatment

Abstract

For over a century, household and municipal wastewaters have been recycled into fish biomass in Central Europe. The main reasons were (1) utilisation of otherwise un-used nutrients, (2) the production of protein, (3) the recovery of costs for treatment facilities, and (4) a more environmentally friendly disposal of otherwise polluting wastewaters. A variety of different systems were either only tested experimentally or actually implemented for numerous years. The simplest forms received only mechanically pretreated wastewater (household, dairy, brewery, municipal), others acted as polishing ponds for effluents from treatment systems ranging from constructed wetlands to biological treatment systems. Scales of implemented systems ranged from single backyard fishponds, over units comprising several ponds, to sophisticated designs with 30 large ponds managed by a designated team of operators. Nearly all systems have ceased to operate, with the largest ever built system in Munich being the notable exception. The main characteristics of the systems, the problems encountered and the reasons for their decline are pointed out, together with potentials and considerations for the implementation of similar, adapted systems in developing countries.

HISTORY AND DEVELOPMENT

Since the Middle Ages, household and village latrines were emptied into ponds stocked with fish. Castle moats were known for their fish productivity, and these had latrines built above them [2,3]. The addition of such wastewaters into ponds provided a more hygienic form of waste disposal than was common at the time, namely the direct discharge into the ground or onto the streets. The increased productivity of the ponds was a welcome added benefit, aside from their main purpose as a water reservoir for fire extinguishing.

In the 19th century, wastewaters from the cities polluted streams and rivers to an extent that treatment systems were called for, invented and installed. For quality control, effluents from these systems were fed into small ponds with trout as indicator organisms.

---

[1] This is a condensed version of an earlier, comprehensive account [1].
[2] ICLARM Contribution No. 1165

In 1887, experiments with fish culture in effluents from constructed wetlands (sewage fields) were initiated in Berlin [4]. In 1897 a proposal was patented for a three-stage aquaculture system comprising a microorganism/algal pond, followed by a second zooplankton pond and a final fish pond [5]. The experiments proved that effluents could be used safely and beneficially for fish production. This system was later implemented on a large scale in Berlin, Bielefeld and Dortmund.

In Munich experiments began in 1902, which led to successful implementations of fishpond systems which received a variety of wastewaters [3]. Most had a mechanical, sludge-settling pretreatment facility. In 1908, it was suggested that the entire wastewaters of the city of Munich could be treated with this wastewater-fishpond method. In the late 1920s, the largest wastewater-fed aquaculture system ever built in Europe was implemented in Munich, designed to treat the wastewater of its population of 500,000. The system, termed Hofer's wastewater fishpond method, was adopted in a number of other towns and cities, such as Amberg, Strasbourg, and Vienna [1, 3, 6, 7].

Aside from towns and cities, the wastewater-fed fishpond systems were installed individually for sanatoria, hospitals, prisons, military barracks, breweries, slaughterhouses, dairies, textile and paper factories and train stations. More than 100 are documented in the literature and it can be assumed that a far greater number were actually installed and operated. Today, the only system in operation is that in Munich, where the original form of management has been altered from a raw wastewater treatment facility to that of polishing ponds receiving the effluents of the recently modernized three-stage biological treatment process [1, 7, 8].

DIFFERENT SYSTEM DESIGNS

All systems were constructed in close proximity to the sources of origin. Two major types can be distinguished. The first are wastewater fishponds in the narrow sense, receiving raw or mechanically pretreated wastewater. The second are polishing fishponds, receiving effluents from more elaborate treatment systems. The waters had low amounts of suspended solids but had high amounts of nutrients. The second type can be further classified into those receiving effluents from (a) sewage fields or constructed wetlands, and (b) biological treatment systems such as various types of filters or activated sludge processes [1].

OPERATIONAL CHARACTERISTICS

Water Quality

An important aspect is that the fishpond environment must be adjusted to enable production of healthy fish suitable for human consumption, but at the same time achieve a high level of water treatment. Loading rates of wastewater to the ponds ranged from 200 to 2000 persons·ha^{-1}·day^{-1}. In most cases, wastewaters were diluted with fresh surface water in a ratio of 1:4 to 1:9 depending on quality. Ponds were operated as a flow-through system. Water temperatures ranged from 5 C in the spring and autumn to over 25 C in summer. Reductions in nutrients, COD and BOD ranged from 20 to over 90 %. Bacterial counts were reduced by 72 to 99 % [1]. Treatment efficiencies (total dry matter) were determined to be around 40% for the wastewater-fed fishpond systems in Strasbourg and Munich.

Fish Production

The main species cultured was the common carp (*Cyprinus carpio*) which were stocked as one-year-old fish weighing 100 to 150 g at a density of 1000 to 1400 per hectare and fattened within a single season to a market weight of around 500 g. This is considered to be extensive fish culture, requiring large pond areas at low productivity (around 500 kg·ha^{-1}·7 months^{-1}). Fish relied entirely on the natural food productivity of the ponds, consisting mainly of chironomid larvae in the upper layers of the pond mud. In some cases additional food was provided to the fish, but this constituted an additional cost. Several other fish species (e.g. tench and trout) were tried but with limited success, mainly due to their unsuitability to the environmental and operational features of the wastewater-fed fishpond.

Public Health Aspects

Effluents from wastewater-fed fishponds were released into surface waters such as streams or rivers, from which the dilution water was usually taken. It was generally agreed that the quality was much better than if the wastewater had been directly released into the environment. Fish were monitored for pathogens etc., by government laboratories monitoring food quality. Human pathogens were never found in the fish muscle. As fish are generally processed under accepted hygienic standards (i.e. gutting, cleaning, cooking) wastewater-fed fish culture was not considered a health risk to the public. With industrialisation and agricultural intensification, potential health risks through contamination with heavy metals, PCBs and pesticides have been identified [9].

REASONS FOR TERMINATION

It is difficult to obtain official reports on the reasons for termination of wastewater-fed fishpond operations. However, accounts of the problems encountered during operation are available. Aside from the initial costs for the construction of the facilities themselves, operational costs were high due to the maintenance of the extensive areas and the need for professional fish keepers. Located on the urban fringes, land prices increased considerably as cities expanded. Water for dilution was becoming increasingly sparse and in some cases less available in the required quality.

Systems could operate only for part of the year, during which fish could be grown, i.e. from April to December. During the winter months when the ponds were left to dry, alternative systems had to be available. In the simplest form these were storage lakes or stabilisation ponds. Undersized, unmarketable fish had to be stored in special overwintering ponds, and some operations even had their own hatcheries and nursing ponds to produce fish fingerlings. Additional feeding was necessary in some cases to ensure marketability, adding to operational cost.

In the 1920s and 30s, when most systems were installed, there was a need for simple but effective technologies for wastewater treatment. The added benefit of fish production made wastewater-fed fishponds attractive. On the other hand, when compared to the overall fish production through aquaculture at the time, even the output from the Munich system at 100-150 $t\cdot y^{-1}$ was minor.

Social acceptance and marketability of the produced fish was occasionally cited as a problem, yet the rural population was accustomed to fish from village ponds and surface waters. In the cities, and this is well-documented for Munich, a public awareness and information campaign was implemented, stating the benefits of recycling wastes for nutrient recovery, and that the wastewater-fed fishponds only spatially confined a naturally occurring process. The products were considered safe for human consumption.

POTENTIALS AND CONSTRAINTS FOR FURTHER IMPLEMENTATION

In recent decades, experiments on the reuse of wastewater for aquaculture with simultaneous sewage treatment have been conducted in a number of developed countries, experiencing similar design constraints that led to the termination of the systems in Central Europe [8, 10, 11]. To date these renewed efforts have not led to such a widespread application as was once the case in Germany.

In developing countries, the potentials for application are considerable. Existing, partly traditional systems in Asia have been widely documented, but only few have been studied in depth [8, 11, 12, 13]. Recent renewed interest has led to studies in Lima, Peru, which were successful in respect to treatment, fish production, public health concerns, economics and social acceptance [14, 15, 16].

It should be considered that, in many countries, the reuse of human wastes for agriculture or aquaculture is prohibited, either legally or by social taboos. In reality, in spite of the existing prohibitions, systems still exist and in some cases the use is spreading. Public health concerns are of highest importance [17, 18, 19]. Very little is known about the true health risks associated with wastewater reuse for, and treatment through aquaculture and in-depth studies are urgently necessary, given its present rate of expansion.

A risk assessment on a case-by-case basis is necessary, depending on local factors and type of system planned, which can range from well-controlled municipal facilities all the way to uncontrolled backyard uses. Preliminary health guidelines have been issued [20, 21]. Local traditions, beliefs, requirements and constraints need to be considered. Approaches to appropriate technology development are often only successful with the participation of target groups in the development process itself [22, 23]. Solutions may require public awareness and information programs to alter behavioral and nutritional habits.

References

[1] M. Prein, p. 13-47 *in* P. Edwards, R.S.V. Pullin (eds.) Wastewater-fed aquaculture. Environmental Sanitation Information Center, Asian Institute of Technology, Bangkok (1990).
[2] H. Keller, Zentralbl. Bauverw. **38**, 351 (1918).
[3] B. Hofer, Münch. Med. Wochenschr. **52**, 2266 (1905).
[4] G. Oesten, Fisch.-Z. **7**, 565, 585 (1904).
[5] G. Oesten, Pat. Cert. No. 101706, Kaiserliches Patentamt, Munich (1897).
[6] D.C. Demoll, Handb. Binnenfisch. Mitteleurop. **6**, 222 (1926).
[7] P. Beck, p. 78-82 *In* Abwassertechnische Vereinigung e.V. (eds.) Lehr- und Handbuch der Abwassertechnik, Vol. IV. Wilhelm Ernst & Sohn, Berlin (1985).
[8] P. Edwards, Water and Sanitation Report 2, UNDP-World Bank Water and Sanitation Program, The World Bank, Washington, D.C. (1992).
[9] P. Wissmath, J. Klein, Arch. Lebensmittelhyg. **31**, 65-66 (1980).
[10] C. Etnier, B. Guterstam (eds.) Ecological engineering for wastewater treatment, Proceedings of the International Conference at Stensund Folk College, 24-28 March 1991, Bokskogen, Gothenburg and Stensund Folk College, Trosa, Sweden (1991).
[11] P. Edwards, R.S.V. Pullin (eds.) Wastewater-fed aquaculture. Environmental Sanitation Information Center, Asian Institute of Technology, Bangkok (1990).
[12] K. Ruddle, G. Zhong, Integrated agriculture-aquaculture in South China: the dike-pond system of the Zhujiang Delta. Cambridge University Press, Cambridge, UK (1988).
[13] D. Little, J. Muir, A guide to integrated warm water aquaculture, Institute of Aquaculture Publications, University of Stirling, Stirling, UK (1987).
[14] S. Cointreau, Applied Resource and Technology Unit (WUDAT) Tech.Note 3, Water Supply and Urban Development Department, The World Bank, Washington, D.C. (1987).
[15] C.B. Bartone, M.L. Esparza, C. Mayo, O. Rojas, T. Vitko, Monitoring and maintenance of treated water quality in the San Juan lagoons supporting aquaculture (UNDP/ World Bank/ GTZ Integrated Resource Recovery Project GLO/80/004): final report of phases I-II. Pan American Center for Sanitary Engineering (CEPIS/PAHO), Lima, Peru (1985).
[16] J. Moscoso, A. Florez, Reuso en aquicultura de las aguas residuales tratadas en las lagunas de estabilizacion de San Juan, seccion I: resumen ejecutivo. Centro Panamericano de Ingeneria Sanitaria y Ciencias del Ambiente (CEPIS), Lima, Peru (1991).
[17] R.S.V. Pullin, H. Rosenthal, J.L. Maclean. (eds.), ICLARM Conf. Proc. 31 (1993).
[18] I. Hespanhol, Health and technical aspects of the use of wastewater in agriculture and aquaculture. WHO, Geneva (1988).
[19] M. Strauss, U.J. Blumenthal, Human waste use in agriculture and aquaculture - utilization practices and health perspectives, IRCWD, Duebendorf, Switzerland (1990).
[20] WHO, WHO Tech. Rep. Ser. 778 (1989).
[21] WHO Study Group on the Control of Foodborne Trematode Infections, WHO Tech. Rep. Ser. 849, WHO, Geneva (1995).
[22] R. Chambers, Challenging the professions: frontiers for rural development, Intermediate Technology Publications, London (1993).
[23] J. Werner, Participatory development of agricultural innovations: procedures and methods of on-farm research. Schriftenreihe der GTZ No. 234, Eschborn, Germany (1993).

Environmental Research Forum Vols. 5-6 (1996) pp. 161-164
© *1996 Transtec Publications, Switzerland*

Wastewater Treatment of the Pig-Breeding Farm "Shirvinta" and Fish Rearing in them

A. Pekiukenas[1], V. Marazas[2] and Z. Siaurukaite[1]

[1] Institute for Ecology, Akademijos 2, LT-2021 Vilnius, Lithuania

[2] Institute for Immunbiology, Akademijos 2, LT-2021 Vilnius, Lithuania

Keywords: Wastewater Treatment, Pig-Breeding Farm, Dungwater, Ponds of Fishery, Phytoplankton, Zooplankton, Carp, Parasitic Helmints, Salmonellas, Coli-Form Bacteria

ABSTRACT

With the purpose of designing the installation and construction of the wastewater treatment in the pig-breeding farm "Shirvinta", the new method of sewage treatment was tested. 3 small ponds of fishery (0,1 ha area, 1000 m^3 water volume in each) were used in May-September. Every week dungwater was given to the first pond up to 10 % of its volume. The concentration of organic matter in dungwater according to the chemical consumption of oxygen was 2000-2500 mgO_2/l. In dungwater were 669+ 35 mg/l N, 57+ 9,1 mg/l P and 2904 18 mg/l K. The eggs and larvae of parasitic helminths, bacteria of Salmonella and the great amount of Coli-form bacteria were found in dungwater. In the first pond the rich vegetation of algae (phytoplankton) was observed. 24 species were ascertained, dominating Chlorella volgaris. Every week the water from the first pond was given to the second one. The second pond was very rich in small Crustacea (zooplankton). Its total biomass reached up to 500-600 g/m^3. Among 6 species found, Daphnia magna predominated. The water from the second pond was given to the third one in the same way. In June 3500 specimens of carp fingerlings were released into this pond. In October carp underyearlings reached the average weight of 50 g. The total fish production was 900 kg/ha. During this process was utilized approximately: 76,85 % N, 87,76 % P and 73,45 % K. Parasitic helminths and bacteria of Salmonella were not found, Coli-form bacteria numbers decreased up to the minimum. Hydro-chemical and veterinary-sanitary water indices corresponded to the standards (requirements) of fishery ponds. It can be utilized for the removal of manure in the pig-breeding farm. All that allowed to create the closed system of water utilization in the construction design. Besides, the secondary production - cheap and of a very good quality - fish restocking in inland waters was obtained.

In Litauen gibt es 33 grosse Schweinezuchtanlagen. Für Mistbeseitigung benutzen sie das Wasser. Eine Anlage, die 25'000 Schweine züchtet, hat jeden Tag 500-900 m^3 Abwässer. Die Entsorgung und Verwendung dieser Abwässer ist ein aktuelles Problem [5]. Sie haben viele organische und mineralische Substanzen und sind ein gutes Düngemittel. Aber bei der stetigen Bewässerung der Wiesen und Felder mit den ungeklärten Abwässern entsteht die Gefahr der Umweltverschmutzung. Im Acker reichern sich die Nitrite und Nitrate an. Sie geraten über die Pflanzen in den Viehorganismus und können verschiedene Erkrankungen verursachen [3]. Dazu gehören verschiedene pathogenen Mikroben und die Eier und Larven mancher parasitischen Helminthen in diesen Abwässern [2,6].

Für die Klärung der grossen Mengen der Abwässer der Vierzuchtanlagen wird ein weitverbreitetes biologisches Klärungsverfahren benutzt. Es ist auf der mikrobischen Wirkungsweise begründet [1,4]. In unseren Forschungen in der Schweinezuchtanlage "Schirwinta" haben wir neben mikrobischen Klärung die Mikroalgen (Phythoplankton), die winzigen Wasserkrebschen (Zooplankton) und die Fische benutzt. Die Angaben unseres Versuches sollten der Projektierung und dem Bau der Abwässerklärungeeinrichtungen dienen. In dieser Anlage sind zuerst die Abwässer. Die dicke Fraktion kompostiert man mit dem Torf. Die dünne Fraktion (Gülle) pumpt man in den Güllespeicher, von wo sie für die Begiessung der Wiesen benutzt werden.

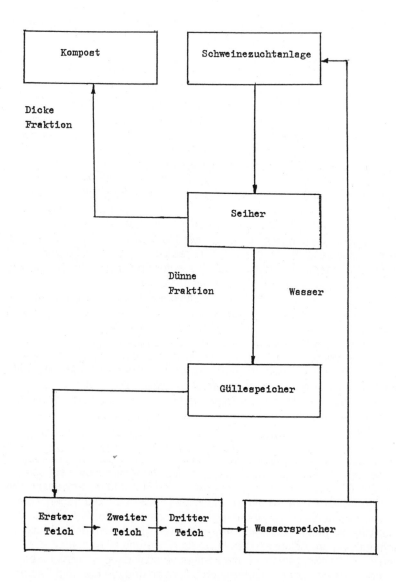

Abb.1 Das Schema der Abwasserklärung

In unserem Versuch benutzten wir 3 Teiche. Jeder Teich hatte 0,1 ha Flache, 1,0 m Tiefe und 1'000 m³ Wasservolumen Die Abwässerklärung wurde gemäss unserem Verfahren folgendermassen durchgeführt (Abb. 1): Bakterien und Algen (erster Teich) -> Zooplankton (zweiter Teich) -> Fische (dritter Teich).

Der Versuch spielte sich in den Monaten Mai bis September ab. Im Laufe dieser Periode gaben wir jede Woche in den ersten Teich bis 10 % von seinem Wasservolumen die Gülle aus dem Speicher. In diesem Teich entwickelten sich sehr üppig die Algen. Aus diesem Teich wurde schon teilweise geklärtes Wasser in derselben Mengen und zur gleichen Zeit in den zweiten Teich gegeben. Hier entwickelten sich sehr intensiv die kleinen Wasserkrebschen (Zooplankton), welche für ihre Ernährung die Algen verbrauchten. Das Wasser aus dem zweiten Teich wurde in derselben Ordnung in den dritten Teich gegeben, wo wir Ende Juni 3'500 St. Karpfenbrut ausgesetzt hatten.

Während des Versuches wurden jeden Tag im jeden Teich Wassertemperatur, Sauerstoffgehalt und Wasserdurchsichtigkeit registriert sowie jede Woche hydrochemische und hydrobiologische Untersuchungen durchgeführt. In den Monaten Juni und September wurden die veterinärmedizinisch-sanitären Untersuchungen vorgenommen.

Die Angaben der hydrobiologischen Forschungen erkärten dass die intensivste Entwicklung und Vermehrung der Algen im ersten Teich war. Hier wurden 24 Arten gefunden. Dominierende Arten: Chlorella vulgaris, Micractinium pussilium. Die grösste Menge der Algen wurde in der ersten Hälfte der Sommer bei der höchsten Wassertemperatur und Isolation festgestellt. Die grösste Anzahl der Algenzellen reichte bis $4-6.10^6$ Zellen/ml.

In dem zweiten Teich wurden 6 Arten des Zooplanktons gefunden. Dominierende Arten: Daphnia magna und Brachionus calicyflorus. Die gesamte Biomasse des Zooplanktons erreichte im zweiten und dritten Teich im Mai/Juni 500-600 g/m³. Im July verminderte sich das Zooplankton im dritten Teich wesentlich, da die Karpfen es für die Ernährung benutzten. Im dritten Teich sind die Karpfen ausgezeichnet gewachsen. Durchschnittliche Masse eines einsömmerigen Karpfen erreichte im Oktober bei der Abfischung 50 g, die gesamte Fischproduktion 900 kg/ha.

Die Konzentration der organischen Substanzen in der Gülle war gemäss der chemischen Sauerstoffverzehrung 2'000-2'500 mg O_2/l. Sie hatte 669 ± 35 mg/l N, 57 ± 9,1 mg/l P und 290 ± 18 mg/l K.

Ort der Probeentnahme	Monat	
	Juni	September
Abwässerseiher	49 500 000	250 000 000
Güllespeicher	5 800 000	6 000 000
Algenteich	3 700 000	4 000 000
Wasserkrebschenteich	600	1 000
Fischteich		

Tabelle 1, Coliformbakterienindex (Bakterienzahl in 100 ml Gülle, Wasser)

Während des Klärungsverlaufes aus den Abwässern wurden durchschnittlich 76,85 % N, 87,76 % P und 73,45 % K utilisiert. Im dritten Teich war BSV_5 durchschnittlich $6,7\pm 1,7$ mg O_2/l. Dies entspricht den Anforderungen für die Fischteiche.

Wir mussten auch die Resultate der Abwässerklärung in der veterinärmedizinischen-sanitären Hinsicht einschätzen. In der Gülle wurden die Salmonella- und die Coliformbskterien, die Eier Ascaridata, die Larven Oesophagostomum dentatum untersucht. Im Laufe des Versuches waren die Proben aus verschiedenen Orten unseres Abwasserklärungsysteme entnommen. Die Ergebnisse der Untersuchungen sind folgende: die Bakterien von Salmonella wurden noch im ersten Algenteich konstatiert, während im zweiten und dritten Teich schon keine mehr festgestellt wurden. Die Eier und Larven der parasitischen Helminthen wurden weder im Wasser noch im Boden aller 3 Teichen gefunden. Die Coliformbekterien verminderten sich stufenweise (Tab.I).

Gemäss den Angaben unserer hydrochemischen, hydrobiologischen und veterinärmedizinisch-sanitären Untersuchungen entsprach das Wasser in dem dritten Teich den Anforderungen für die Fischteiche. Es dürfte auch als technologisches Wasser für die Mistbeseitigung in der Schweinezuchtanlage benutzt werden.

Diese Ergebnisse dienten als Basis für die Vorbereitung des Projektes und für den Bau der geschlossenen Wasserbenutzungs- und Abwässerklärungsysteme in der Schweinezuchtanlage "Schirwinta".

LITERATURVERZEICHNIS

[1]   **Jurowskaja E.M.:** Mikrobiologitscheskoja otechistka promyschlennych stotschnych wod. Sdorowje. Kiew (1984), 160 S. (russisch).

[2]   **Klimow N.J., Cekaew M.G.:** Ratsionalnoje ispolsowanije semelnych i wodnych resursow. Otschistka newososoderschaschtschich stotschnych wod Tselinogradskoj oblasti. Alma-Ata (1984), S.161-165 (russisch).

[3]   **Koltypin J.A.:** Teoretitscheskije i technologitscheskije aspekty pererabotki olchodow schiwotnowodstwa s ispolsowanijem samknutych pischtschewych biologitecheskich sistem. Awtoreferat doktora biologitscheskich nauk. Moskwa (1980), S. 1-33 (russisch).

[4]   **Milto N.I., Karbanowitsch A.I.:** Mikrobiologitscheskaja charakteristika stokow swinowodischeskogo kompleksa. Ochrana okruschajutschey sredy, Minsk 3 (1984), S.28-32 (russisch).

[5]   **Morosow N.W.:** Problema obeswreschiwanija stotschnych wod schiwotnowodtscheskich kompleksow i wosmoschnyje puti jejo reschenija. Wodnyje resursy. Kiew 5 (1983), S.142-152 (russisch).

[6]   **Nikitin D.P.:** Krupnyje schiwotnowodtscheskije kompleksy i okruschajutschaja sreda. Moskwa (1980), 255 S. (russisch).

Environmental Research Forum Vols. 5-6 (1996) pp. 165-172
© *1996 Transtec Publications, Switzerland*

Reclamation of Wastewater for Aquaculture:
Trace Metal and Bacterial Contamination in Fish

Y. Liang[1], R.Y.H. Cheung[2] and M.H. Wong[1]

[1] Department of Biology and Center for Waste Recycling & Environmental Biotechnology,
Hong Kong Baptist University, Hong Kong

[2] Department of Biology and Chemistry, City University of Hong Kong, Hong Kong

Keywords: Wastewater, Reuse, Aquaculture, Trace Metals, Bacteria

ABSTRACT

The purpose of this study was to investigate the possibility of reclaiming organically polluted river water for aquaculture purpose. A series of ponds were constructed at Au Tau Substation of the Agriculture and Fisheries Department, Hong Kong Government. Water was abstracted from the Kam Tin River and passed through a serious of eight ponds in sequence. The first four ponds were used for water treatment and algal culture by providing sedimentation and aeration and the remaining four ponds were used for fish culture. Fries of six species of freshwater fish were stocked in the ponds in January, 1994. Big head (*Aristichthys nobilis*) and silver carp (*Hypophthalmichthys molitrix*) are filter-feeders which feed on zooplankton and phytoplankton, grass carp (*Ctenopharyngodon idellus*) is herbivore and black bass (*Micropterus salmoides*) are carnivors; while common carp (*Cyprinus carpio*) and tilapia (*Oreochromis mossambicus*) are omnivores, feeding on almost anything. Results indicated that fish growth was satisfactory. The levels of trace metals (Cu, Zn, Cd, Ni, Cr, Pb, Sb) and fecal coliforms in the fish flesh were within the safety limits set by the Public Health and Municipal Service Ordinance (Hong Kong) and International Commission on Microbiological Specification for Food accordingly.

INTRODUCTION

Wastewater reuse in aquaculture is receiving increasing scientific attention as its potential to recycle nutrients in productive, socially acceptable systems has been realized [1]. It is prevalent in China, Indonesia, India and Vietnam where wastewater and waste materials including human excreta, fecally polluted surface water and conventional sewage are used for fertilizing fish ponds [1, 2, 3, 4]. Public health is the most important concern in wastewater-fed fish and there is a lack of knowledge concerning the possibility of getting water borne diseases from the consumption of wastewater-fed fish [5]. Pathogenic microorganisms exist in wastewater-fed fishponds. Their survival, numbers, infective capability and the risk they might impose on fish farmers, fish handlers and consumers should be dealt with [6]. Bacteria had been recovered from various organs and muscles of fish exposed to waters containing bacteria [7, 8]. On the other hand, chemical (metals and organochlorines) contamination in wastewater-fed aquaculture should also be considered. Chemical contaminants such as trace metals and PCBs can be bioaccumulated in aquatic organisms directly from water (bioconcentration) or from contaminated food (biomagnification). As the degradation rate of these contaminants is often slow, their persistence in fish tissues is of great concern of human health. High concentrations of trace metals (such as Cu, Pb) and PCBs [9, 10, 11, 12] were detected in the tissues of fish fed with sludge. However, no significant difference in trace metal contents was detected in fish flesh reared in the oxidation ponds when compared with those in control ponds [13, 14].

The fishpond system constructed at Au Tau Substation, Agriculture and Fisheries Department, Hong Kong Government was used to study the feasibility of treating the polluted riverwater and at the same time using the treated water for fish culture. A laboratory study [6] simulating the on-going fishpond system was conducted. The results showed that the level of ammonia was reduced by aeration and the metals (Cu, Zn) were also removed from the riverwater. The algal products contained a high crude protein (42%, dry weight basis) which could be used to feed carp and tilapia.

In this study, we examined bacterial and trace metal contamination of different fish species stocked in this fishpond system. It is hoped that the results will yield some information related to the public health aspect of this wastewater-fed aquaculture system.

MATERIALS AND METHODS

Experimental site

The layout of the fishpond system is shown in Fig. 1. The polluted river water was intruduced into the fishpond C12. Then the river water was pretreated by passing through the ponds C11 to C9. Ponds C13-C16 were used for fish culture.

Experimental materials

Six species of freshwater fish were used for this study. The fish include: big head *(Aristichthys nobilis)*, silver carp *(Hypophthalmichthys molitrix)*, grass carp *(Ctenopharyngodon idellus)*, black bass *(Micropterus salmoides)*, common carp *(Cyprinus carpio)*, and tilapia *(Oreochromis mossambicus)*. The fish were stocked in ponds C13-C16 of the fishpond system in January, 1994 (Fig. 1). The feeding habitats of the fish are shown in Fig. 2.

Sample collection

Fish samples were collected at Au Tau substation between March and June 1995. The fish samples were put in plastic bags, and stored on ice during transportation. Micorbiological examinations were conducted within six hours after collection.

Microbiological examination of fish

Microbiological examination of fish flesh was conducted using the methods described by APHA [15] and AOAC [16]. Hydrophobic grid membrane filter method was used for detecting total aerobic count, total coliform and fecal coliform. The fish were washed by tap water and then the surface of the fish were swabbed by 75% alcohol and blot dried. After the skin of the fish were removed, about 50 g muscle tissue of fish were picked and put into a sterile plastic bag containing 150 mL sterile diluent and homogenized for 5 min. 10 mL of the slurries were then treated by 2 mL sterile 10% trypsin solution at 35-37°C for 20-30 min. Tryptic soy-fast green agar was used for total aerobic count examination. M-FC agar was used for the detection of total coliform and fecal coliform.

Trace metal analysis in fish flesh

Fish flesh was removed and freeze-dried to a constant weight. Moisture contents of the fish tissue were recorded as weight losses of the tissue before and after freeze-drying. 3 g of the dried fish flesh were digested in concentrated HNO_3 at room temperature overnight and then at 130°C for 5 h. 5 mL concentrated $HClO_4$ was then added into the digestion tube and kept reflux at 130°C for 2 h. After digestion, the samples were diluted and filtered to 25 mL volumetric flasks and diluted to the volume. The trace metals (Cu, Zn, Cd, Ni, Cr, Pb, Sb) were detected by an atomic absorption spectrophotometer. NBS 1566a (oyster tissue) was used as the standard reference material in the process of analysis.

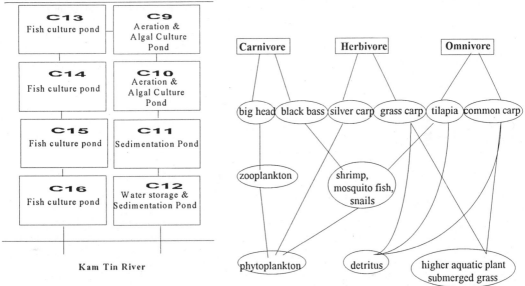

Fig. 1 Reclamation of polluted riverwater for aquaculture: the fishpond system experimental layout at Au Tau

Fig. 2. Food web developed at Au Tau fishpond system

RESULTS

Bacteria in fish flesh

Table 1 shows the bacterial contents in the fish flesh. The International Commission on Microbiological Specifications for Foods (ICMSF) [17] recommended microbiological limits for fresh and frozen fish using a three class sampling plan. In this plan, the lower limits for acceptable fish are : 1,000,000 CFU per gram for standard plate count (SPC) at 25°C, 4 per gram (MPN) for fecal coliform and 1,000 per gram for *Staphylococcus* count. The maximum limits for these sanitary indicators are 10,000,000, 400, and 2,000 per gram, respectively. For freshwater fish, we examined total coliforms instead of *Staphylococcus*. In Massachusetts State of the United States [22], the criterion for coliforms in raw meat was not to exceed 100 per gram, while not more than 100,000 per gram for aerobic plate count (APC). According to the standards described above, the levels of fecal coliforms in all of the fish were acceptable in the study. A hundred percent of the fish had APC not more than 100,000. As for coliform groups, 25 percent of big head, 20 percent of tilapia, grass carp and black bass exceeded the criterion, while silver carp and common carp were all within the limit.

Table 1. Bacterial content of wastewater cultured fish

	Sample No.	Coliform group (MPN/g)	APC (CFU/g)	Fecal coliform (MPN/g)
		Membrane Filter Colony Count per gram fish meat		
Tilapia	1	<4	140	<4
(*Oreochromis mossambicus*)	2	<4	550	<4
	3	60	560Est**	<4
	4	190	1,600	<5
	5	50	730	<5
Grass carp	1	22	2,900	<4
(*Ctenopharyngodon idellus*)	2	130	510	<4
	3	30	360	<4
	4	20	420	<4
	5	<4	<4Est	<4
Black bass	1	230	280	<4
(*Micropterus salmoides*)	2	<4	<4Est	<4
	3	8	100	<4
	4	<4	2Est	<4
	5	<4	64	<4
Silver carp	1	<4	<4Est	<4
(*Hypophthalmichthys*	2	10	20,000Est	<4
molitrix)	3	<4	<4Est	<4
	4	<4	<4Est	<4
	5	17	40,000	<4
Big head	1	<4	4Est	<4
(*Aristichthys nobilis*)	2	<4	<4Est	<4
	3	N.D.*	9,900	<4
	4	<4	2,400 Est	<4
	5	6,600	29,000	<4
Common carp	1	<4	44 Est	<4
(*Cyprinus carpio*)	2	<4	2,100	<4
	3	66	9,600	<4
	4	<4	1,000	<4
	5	<4	1,300	<4

N.D.*----Not detected.
Est**----Estimated value

Trace metals in fish flesh

Table 2 lists trace metal (Cu, Zn, Cd, Ni, Cr, Pb, Sb) concentrations and moisture content in fish flesh. The results were also compared with the limits set in Public Health Regulations for Food in Hong Kong [20] and median standards for edible tissues of fish in other countries [18]. The standards for trace metals in food are usually set on a wet weight basis. However, dried samples were used for the detection of trace metals in this experiment. Considering the moisture content in the fish flesh, the results showed that metal contents in flesh of the fish cultured in this system were far below the upper limits.

Table 2. Concentrations of trace metals in fish flesh

Species	Trace metal (µg/g dry weight)							Moisture content (%)
	Cu	Zn	Cd	Ni	Cr	Pb	Sb	
black bass (*Micropterus salmoides*)	1.68±0.246	23.5±1.92	0.0618±0.0101	<0.3	<0.4	<0.4	<1.0	78.5
big head (*Aristichthys nobilis*)	1.56±0.164	22.9±0.95	0.0854±0.0273	<0.3	<0.4	<0.4	<1.0	77.7
common carp (*Cyprinus carpio*)	1.51±0.244	29.1±5.57	0.0313±0.0105	<0.3	<0.4	<0.4	<1.0	76.9
silver carp (*Hypophthalmichthys molitrix*)	1.42±0.204	25.5±3.37	0.0715±0.0193	<0.3	<0.4	<0.4	<1.0	74.3
grass carp (*Ctenopharyngodon idellus*)	1.43±0.304	26.6±7.11	0.0650±0.0140	<0.3	<0.4	<0.4	<1.0	79.7
tilapia (*Oreochromis mossambicus*)	1.08±0.168	22.2±1.27	0.0817±0.0205	<0.3	<0.4	<0.4	<1.0	77.7
	Trace metal (µg/g wet weight)							
	Cu	Zn	Cd	Ni	Cr	Pb	Sb	
Public Health Regulation (Hong Kong)	-----	-----	2.0	-----	1.0	6.0	1.0	
Public Health Regulation (median standards in other countries)	20	45	0.3	-----	1.0	2.0	1.0	

DISCUSSION

The risk of bacterial contamination in fish flesh increases when pathogens in the pond water reach high numbers. The contaminated fish can transmit pathogens to its handlers and consumers. The World Health Organization had a tentative bacterial guideline for fishpond water which is below $10^3/100$ mL [19]. The potential for muscle infection by bacteria increases when fish are exposed to water containing fecal coliforms greater than 10^4-10^5/100 mL [19]. The environment in wastewater-fed fish ponds can be stressful for fish and stress can lower the threshold value of penetration of bacteria into the muscles of fish and increase their susceptibility to diseases [7]. Therefore, it is essential that bacterial contaminated wastewater should be treated prior to introduction into fishponds in order to produce uncontaminated and healthy fish [7]. On the other hand, pathogens can also be removed in the process of fish culture [5]. In the present fishpond system, fecal coliforms obtained in the fishpond water in May 1995 were between 6.5×10^3/100 mL and 2.3×10^4/100 mL [21]. Under such a condition, fish are at the marginal state of increased bacterial infection. Further studies should be conducted aiming at bacterial reduction in wastewater-fed systems.

As to trace metal contamination in aquatic environment, instances caused by Hg and Cd are by far the most documented. Many countries have public health regulations for trace metals in food. In Hong Kong, regulations have also been promulgated for trace metals including Sb, As, Cd, Cr, Hg, Pb and Sn in food. In this study, we examined Cd, Cu, Zn, Ni, Cr, Pb and Sb in the fish flesh. Other trace metals including As, Hg and Sn will be examined in the future. Cu, Zn and Pb concentrations in fish flesh were lower than in the algal product (Cu: 102, Zn: 339, Pb: 1.24 µg/g dry wt) obtained from the same ponds [6]. Ni and Cr were undetectable in water, fish flesh and algae [6]. It is likely that food and particulate are a much more important source of pollutants uptake than water [23]. The preliminary findings on trace metal contents of the different fish species indicated that they were within the safety limits set for human consumption. However, pollutants can be accumulated or transferred within the ecosystem, a more detailed investigation is being carried out on the trace metal contents in different trophic levels of the ecosystem. Further more, there is a lack of information related to trace organic contamination in fish cultivated in wastewater-fed systems. Further studies should be conducted along this line.

ACKNOWLEDGMENTS

The authors thank the Director of Agriculture and Fisheries Department for his permission in collecting samples. Financial support from The Biotechnology Research Institute is gratefully acknowledged.

REFERENCES

[1] P. Edwards, *In* H. Pullin, Rosenthal and J. L. Maclean (eds) Environmental and Aquaculture in Developing Countries. ICLARM Conf. Proc. **31**, (1993).

[2] Z. S. Zhang, *In* P. Edwards and R.S.V.Pullin (eds.) Wastewater-fed Aquaculture. Environmental Sanitation Information Center, Asian Institute of Technology, Bangkok (1990).

[3] P. Edwards, Water and Sanitation Report No. **2**. UNDP-World Bank Water and Sanitation Program. The World Bank, Washington, D. C. (1992).

[4] P. A. Tuan and V. Trac, *In* P. Edwards and R.S.V.Pullin (eds.) Wastewater-fed Aquaculture. Environmental Sanitation Information Center, Asian Institute of Technology, Bangkok (1990).

[5] P. Edwards, *In* P. Edwards and R.S.V. Pullin (eds.) Wastewater-fed Aquaculture, Environmental Sanitation Information Center, Asian Institute of Technology, Bangkok, Thailand (1990).

[6] M. H. Wong, R. Y. H. Cheung, S. F. Leung, and P. S. Wong, *In* Proc. of 4th Int. Conf. on Wetland systems for Water Pollution Control, Guangzhou, PRC, Nov. (1994).

[7] N. Buras, *In* R. S. V. Pullin, H. Rosenthal and J. L. Maclean (eds) Environmental and Aquaculture in Developing Countries, ICLARM Conf. Proc. **31**, 359 p (1993).

[8] N. Buras, L. Duek, S. Niv, B. Hepher and E. Sandbank, Water Research, **21**(1):1 (1987).

[9] M. H. Wong and S. H. Kwan, Toxicology Letters, **7**:367 (1981).

[10] M. H. Wong, Y. H. Cheung and W. M. Lau, Toxicology Letters, **12**:65 (1982).

[11] M. H. Wong and F. Y. Tam, Archives Hydrobiologie, **100**:423 (1984).

[12] M. S. Yang, W. F. Fong, C. T. Chan, V. W. D. Chui and M. H. Wong, Environmental Technology, **14**:1163 (1993).

[13] A. W. Sin and M. T. L. Chiu, Resources and Conservation, **13**:217 (1987).

[14] A. W. Sin and M. L. Chiu, Resources and Conservation, **13**:231 (1987).

[15] M. L. Speck, Compendium of Methods for the Microbiological Examination of Foods. 2nd edn. American Public Health Association, Inc., Washington, DC, U.S.A. (1984).

[16] AOAC, Official Methods of Analysis, 15th edn. Association of Official Analytical Chemists, U.S.A. (1990).

[17] International Commission on the Microbiological Specification of Foods, Microorganisms in Food. Vol. **2**, Sampling for Microbiological Analysis: Principles and specific applications. Toronto: Univ. Toronto Press (1982).

[18] C. C. Nauen, FAO Fish. Circ. **764**, (1983).

[19] WHO, WHO Technical Report Series **778**. World Health Organization, Geneva, Switzerland (1989).

[20] Hong Kong Government. Public Health and Municipal Services Ordinance, Cap.**132,** Food Adulteration (metallic contamination) Resolution, Hong Kong.

[21] AFD, Agriculture and Fisheries Department , Hong Kong Government, unpublished data.

[22] H. M. Wehr, Food Technology, **36** (9): 45, 92 (1982).

[23] P. A. Miller, K. R. Munkittrick, and D. G. Dixon, Canadian Journal of Fisheries and Aquatic Sciences, **49**:978 (1992).

Environmental Research Forum Vols. 5-6 (1996) pp. 173-176
© 1996 Transtec Publications, Switzerland

Workshop 2:

Recycling Challenges and Solutions

Summary and Conclusion by the Chairperson

W. Gujer

Swiss Federal Institute for Water Science and Technology (EAWAG)
and Swiss Federal Institute of Technology (ETH),
CH-8600 Dübendorf, Switzerland

Introduction

The workshop dealt with two groups of problems:

1. Reuse of nutrients in agricultural activities

    1.1 Reuse by irrigation with pretreated effluent
    1.2 Reuse of solids and sludge separated from wastewater

2. Technology and systems for urban sanitation and water reuse

    2.1 Solutions for treatment of separated wastestreams (Graywater, Urine...)
    2.2 Integral solutions for sanitary as well as storm wastewater in urban areas.

Reuse of nutrients in agricultural activities

Reuse by irrigation with pretreated effluent

The use of pretreated wastewater for irrigation and nutrient recycling is used to enhance productivity of energy forests in Sweden and Estonia. Today these activities are operated in pilot scale only. From India and China we had reports on the use of wastewater for irrigation of edible crops.

Problems arising:

- Large areas with monocultures may be subject to pests (insects, etc.). Today's experience with highly fertilized energy forests is limited to small pilot areas. Well fertilized biomass may be more nutritious and therefore more attractive for game.

- Irrigation and nutrient reuse is only possible during the vegetation period. A second route for wastewater treatment is required for winter operation. This frequently reduces overall efficiency of nutrient reuse.

- In China an alternative irrigation route was introduced to deal with wastewater during periods when rice cannot be irrigated. A third route was introduced for winter conditions.

- Wastewater quality may be a problem, its salt content may, depending on climate and soil conditions require that irrigation is operated skillfully or else the soil structure may suffer.

- Little is known on the fate of heavy metals or non-biodegradable organics. This problem must be solved with source control. Chemical analysis for on-site control of these chemicals are typically not available.

- When irrigation is also used for wastewater treatment, efficient pretreatment is lacking and even more undesired pollutants may accumulate in the irrigated land.

Reuse of nutrients in concentrated form (Sludge, stored urine)

Due to the unknown composition of sludge, agriculture is very reluctant to accept sludge from wastewater treatment plants as a substitute for fertilizer. One community in Sweden therefore is trying to fertilize its own energy forest with this material.

In Sweden a biological farmer is planning to use separated urine as a mineral fertilizer substitute. Early in the vegetation period he requires mineral nutrients for his crop. He hopes that urine will count as 'biological fertilizer'.

An example from Poland was introduced, where an artificial soil / humus for recultivation of mined areas shall be produced. The proposal is very site specific but it indicates that engineering frequently combines simultaneous solution of several problems.

The production of human urine in the diurnal cycle was quantified and it was identified, that 2/3 of the urine is typically shed in the home environment.

Problems arising:

- Since fertilizer is only used during the vegetation period all concentrated forms of nutrients (Urine, sludge, dried sludge, etc.) must be stored during the winter period.

- Heavy metals and non-biodegradable organics pose the same problems as with the use of wastewater for irrigation purposes.

- Little is known on the characteristics of separated urine and the effects that storage of urine will have with regard to nutrient loss.

Technology and systems for urban sanitation and water reuse

Treatment of separated wastestreams (Graywater, Urine...)

Proposed solutions for partial problems such as graywater treatment, urine separation and storage, blackwater treatment, etc. become more and more abundant. What is lacking is the full scale experience and information on day to day performance under non research conditions.

'High tech' solutions as well as nature based systems are available for graywater treatment.

Graywater reuse is one means to reduce freshwater requirements. However we have to realize that the chemical properties of water are changed by its repeated use. This might p.e. result in increased detergent requirements in all washing processes. Full scale experience where graywater may be recycled several times is still lacking.

Questions arising:

- Will recycled gray water be accepted by the general public?

- How are the chemical properties of recycled water affected by repeated reuse?

- Can decentralized technology be maintained at a high level of performance?

- Is the multiple installation of houses with double cold and warm water as well as draining systems error tolerant, economically feasible, sustainable, ...?

- What happens to virus and other pathogens in graywater treatment?

- Will the distribution systems for cold and warm treated graywater be infected by biofilms? Warm water, which is rich in organics has proven to cause allergies due to bacterial growth.

Integral solutions for sanitary as well as storm wastewater.

One report described a planned system where new housing for several hundred people should be equipped with decentralized nature based graywater treatment followed by groundwater infiltration. A vacuumsystem allows for blackwater collection and central anaerobic treatment as well as reuse of the treated blackwater with its nutrients in nearby agriculture. Storage over winter would be provided in concentrated form on the farms. Stormwater can be infiltrated to the ground. First evaluations show the proposed system to be favorable from the point of view of energy requirements and cost.

Systems Analysis

Many proposed systems focus on a fractional aspect of rural or urban sanitation. Frequently these systems are proposed with a narrow goal in mind (saving water, recycling nutrients, treating graywater, etc.). Only limited work is available which analyses the proposed systems from an integrated point of view. Systems analysis could help to identify feasible alternatives in an early stage of development.

Resources should not be seen with a narrow focus on nutrients only but rather with a broad overview over different aspects of resource requirements.

Question arising:

- What resources (Building material, energy, chemicals, personnel, etc.) are required for the construction and the operation of existing and proposed systems. Is the use of these resources sustainable?

Energy

Relative to the energy requirements of industrial economies only a small fraction is required for urban water management (drinking water, urban drainage, wastewater treatment). Therefore it was questioned whether energy requirements are relevant at all and energy consumption of the systems proposed by ecological engineers was not really a topic of prime interest. This is most surprising. A few numbers might identify the problem:

Urban water management in Zurich (Water supply, wastewater treatment) requires approx. 18 W / inhabitant. This is less than 2% of today's electricity consumption. If however the entire world population would require a comparable service quality, 50% of all electricity produced in the 320 atomic power plants in operation worldwide would be required to supply this electricity.

Protection of the atmosphere requires that in the year 2050 consumption of fossil fuel is reduced by 80%. Global warming could then be limited to 1-2°C. (Deutscher Bundestag: Enquête-Kommission "Schutz der Erdatmosphäre"). Today industrial nations produce 75% of the CO_2 - Emissions but comprise only 25% of the worlds population (Swiss Environmental Protection Agency 1994):

Neglecting further population growth we may conclude from these two statements, that industrial nations must reduce their fossil energy consumption (CO_2 Emission) by a factor of 15 in order to allow equal per capita consumption to developing countries! And this goal should be achieved in 50 years whereas phosphorus resources are expected to become scarce in 100 - 150 years. What is only 2% today might well be 30% in 50 years!

Our concern about recycling nutrients and especially phosphorus is important. However the best recycling strategies are in vain if we do not reduce fossil energy consumption: *Concerned engineers therefore should concentrate on developing systems and strategies which provide efficient service with the minimum utilization of energy: A principle to be learned from ecology.*

Environmental Research Forum Vols. 5-6 (1996) pp. 177-182
© *1996 Transtec Publications, Switzerland*

Nutrient Recycling and Crop Production by Utilizing Ecologically Balanced Soil-Vegetation Bio System

D. Chatterjee[1], A. Gangopadhyay[2] and P.B. Maity[1]

[1] Department of Chemistry, University of Kalyani, Pin - 741235, India

[2] Department of Civil Engineering, Jadavpur University, Calcutta - 70032, India

Keywords: Treated Effluent, Nutrient Recycling, Land Application

ABSTRACT

The present work is based on the utilization of available nutrients in the waste for raising useful crops, and renovation of those effluents for prevention of surface water pollution attributable to waste discharges or to recharge falling water tables. Understanding of the real nature of waste water is essential for its appropriate use. The quantity and quality of sewage in terms of organic load, nutrients (N, P, K) are to be vigorously monitored. A soil-vegetation bio-system with an adequate climatic condition is employed to remove the bulk of the nutrients and other contaminants. The performance of this developed system is tested in a simulated experimental bed to understand their suitability for irrigation purposes in crop production. Waste water application to soil-vegetation bio-system can provide an efficient treatment under strict conditions of control of the quality of land-disposed effluent, hydraulic load and method of application and land management.

INTRODUCTION

India has tradition in sewage farming particularly in the Eastern part of the country. In sewage farming the mostly suggested proposition is to utilize effluent from secondary treatment system of municipal waste water as irrigation water for crop production. This application of effluent to land will serve two-fold purposes. The crop will take up nutrients for their growth and the soil will also arrest a part of it thereby removing nutrients from the effluent to prevent surface water pollution and recharging of falling water tables. Researches have shown that such soil-plant systems can effectively treat waste water effluents [1, 2, 3]. The degree of renovation of effluent often depends upon the predominant N, P, K species and upon agronomic practice such as crop selection, irrigation loading and agricultural management [1, 4, 5].

The present work is based on the utilization of soil-vegetation bio-system as a "living filter" to remove the bulk of the nutrients from the nunicipal wastes for raising useful crops. The soil-vegetation bio-system can be viewed as ecologically balanced system. The experiments have two equally important aims: to study the resource potential of waste water, vis-a-vis, the selection of effective soil-plant system, and to investigate the performance of the developed system to treat the waste water.

In order to produce a potentially valuable crop, Hybrid Napier grass (used as fodder crop in several countries) has been harvested with the treated effluent from Ganguly Bagan Housing Scheme (GBHS, Phase-I). This has been a short term investigation which reveals that the efficiency of crop uptake is quite higher in beds where treated effluent has been used. This has led to further investigation as regards the potential utilisation of treated municipal waste water in rice production to understand the N, P, K uptake in producing a reasonable crop yield in India. The climatic and soil conditions followed by crop management practice are considerably important for plant growth and nutrient availability as well as the design and construction of the irrigation system particularly for the developing countries.

The different combination of soil-vegetation – fertilizer – climatic conditions as practised in crop production in Eastern India, is presented in Table 1 [6].

Table 1. Soil – vegetation – fertilizer – climatic conditions as practised in crop production in Eastern India.

Soil	Vegetation	Fert. input kg./hec. N : P : K	Climatic Zone	Moisture Index *	Temp. Index	PPt. Mean mm
Black Forest	Maize	100:60:60	Subtropical per humid	50	4.0 C- / 8 C	2500 / -2700
Terai	Wheat	80:60:60	Tropical per humid	50	5.5 C / -12 C	3200 / -3500
Old alluvial	Rice,oil / Oil seed	60:40:40 / 80:40:40	Tropical humid	20 / 50	5.5 C / -15 C	1700 / -1800
New alluvial	Rice / S.cane / Jute	60:40:40 / 150:100:100 / 60:40:40	Tropical moist humid	1-20	8 C / -25 C	1100 / -1200
Red Laterite	Cashwnut	20:40:40	Tropical dry sub humid	1-3	10 C / -30 C	1000 / -1100

$$* \quad \frac{PPT.PET}{PPT} \times 100$$

PPT = Mean precipitation per year.
PET = Mean evapotranspiration per year.

MATERIALS AND METHODS

In the first phase, the GBHS urban dwelling (10 acres) is the collection site for treated waste water. The origin of the waste water (900 cubic meter per day) is primarily from kitchen, bathroom and lavatory. The waste water is being subjected to secondary treatment with bio-filter and the plant effluent has been collected and used for experimental bed having appropriate soil-vegetation bio-system.

The experimental bed 8 ft (2.44 m) x 5 ft (1.52 m) with 4 ft (1.22 m) deep sandy loam soil (15% silt, 80% sand) is employed for vegetation. Hydraulic load about 1 in (2.5 cm) twice in a week in the harvesting season (June-October) and flooding type method of application are used for cultivation. Twine suction type lysimeters are employed for collection of leachates. Soil and leachates collected from 6 in. (15 cm), 2 ft. (60 cm) and 4 ft. (120 cm) depths.

Leachates are collected within 24 hrs. after each frequency of irrigation and soil samples are analysed before and after cessation of irrigation in the harvested season.

Throughout the study period, weekly waste water and leachate samples are analysed for pH, conductivity, total-N, NO3-N, total-K, total-P and BOD$_5$ (only in waste water). Weekly samples of leachate are also stored to obtain one sample per month for analysis of COD, sodium, chloride and filtrable orthophosphate. To study accumulation of organic carbon and organic matter, available-N, available-K, available-P of the soil are collected and analysed before and after each harvesting season.

The second phase of the present investigation has been based on reuse of treated Municipal waste water under Ganga Action Plan [7]. Most of the townships along the bank of river Ganga have secondary treatment plants (capacity ranging 10-20 mld) either by activated sludge process or biofilter unit. Simultaneously, an adequate portion (ranging between 30-40% of the load capacity) of waste water is also processed by stabilisation pond. Most of the waste waters are mixed in character due to the inflow of industrial load to the municipal sewerage/drainage system. Proper characterisation of waste water of a definite area is necessitated to know the potential of its utilisation as fertiliser and the possibility of presence of any substances causing hazard to the agricultural production. Collection of samples and analyses have been done in this phase during the period from June 1993 to December 1994. The ground water samples are also monitored with a view to knowing the constitutents originally present.

RESULTS AND DISCUSSION

Phase-I (GBHS community waste water) :

The soil-vegetation bio-system (living filter) studies carried out on the experimental bed (lysimeter) utilizing "local Soil-H. Napier grass - biofilter plant effluent". The quantities of irrigant alongwith nutrients applied and green weight harvested are presented in Table 2.

Table 2. Irrigant and nutrients applied and green weight harvested

Irrigant (hectare-metre)	Kg./hectare			
	N	P2 05	K2 0	green weight harvested
0.60*	65	23	97	11,765

* 0.60 hectare-metre in 3 months = 5 cm/week/hectare.

The efficiency of crop uptake (ECU) values for H.Napier grass as observed in the present investigation pertaining to N, P, K are shown in Table 3 and compared with values obtained by different researchers for Reed canary, Timothy, Smooth Broom and Timothy Alfalfa which are several common animal forage crops in U.S.A. [8].

Table 3. Efficiency of crop uptake (ECU) values in relation to nutrients observed in different studies

Constituent	ECU					
	Present Study* H.Nap. grass	Pennsylvania* Reed Canary	Cape Cod*			
			Tim.Alf.	Smo.Bro.	Tim.	
Nitrogen (N)	68	122	156	51	58	
Phosphorus(P)	59	63	42	14	18	
Potassium (K)	64	117	155	82	103	

* 5 cm of effluent weekly. $ECU(\%) = \dfrac{\text{Mass in harvested crop}}{\text{Mass applied in effluent}} \times 100$

The results of chemical analysis of leachate samples collected from 6 in. (15 cm), 24 in.(60 cm), 48 in. (120 cm) are presented in Table 4.

Table 4. Physico-chemical characteristics of leachate

Constituents	Leachate from 6 in.(15 cm)		Leachate from 24 in.(60 cm)		Leachate from 48 in.(120 cm)	
	range	average	range	average	range	average
pH	7.5–8.94	8.23	7.23–9.95	7.88	7.43–8.38	7.88
Conductivity	1.06–4.2	2.674	1.27–6.36	4.137	3.23–6.89	4.38
(ms cm^{-1})						
NO3-N(mgL^{-1})	0.59–10	3.17	1.2–16.0	6.13	1.45–13.2	5.84
Total-P(mgL^{-1})	*ND–0.41	0.089	*ND–1.39	0.243	*ND–0.15	0.038
Total-K(mgL^{-1})	1.6–4.75	2.89	2.25–4.75	3.07	2.3–3.8	3.01
COD (mgL^{-1})	22–63	44	a		a	

*ND = nondetectable.

Applied nitrogen have been found to leach out of the experimental bed. Thus, for maximization of the overall performance of the soil-vegetation bio-system the cover crop must have high nitrogen uptake through the root system. Pre and post irrigation soil chemical analysis results are presented in Table 5.

Table 5. Chemical analysis of pre and post irrigation soil - oven dry basis

Characteristic	0-6 in.(15cm)		6-24 in.(60cm)		24-48 in.(120cm)	
	BI	AI	BI	AI	BI	AI
Organic carbon%	0.228	0.510	–	–	0.222	0.560
Organic matter%	0.392	0.877	–	–	0.382	0.960
Available N, ppm	64.80	56.20	50.15	–	33.30	42.50
Available P205,ppm	41	48	30	27	27	53
Available K20, ppm		70	–	50	–	60

BI = before irrigation, AI = after irrigation.

The total nitrogen has also increased slightly, however, the available nitrogen has decreased at 6 and 24 in. depth, and has increased at the 4 ft depth. The decrease may be attributed to plant uptake and downward leaching, whereas the increase in the deep layer may be due to accumulation of nitrogen for being beyond the root zone of the cover crop. The available phosphorus has increased in the top 6 inches, as well as at the 4 feet depth probably due to greater renovation of waste water borne available phosphorus in the soil mantle than the plant uptake.

Phase-II (municipal waste water)

Mineral constituents and plant nutrients (N, P, K) available in the treated municipal waste water are separately shown in Table 6 and Table 7 respectively and GBHS treatment plant effluent are also mentioned for comparative study.

Table 6. Major mineral constituents in treated municipal waste water

Constituent (mgL)	Phase-I (GBHS)	Phase-II Conventional	Low cost
TDS*	3,046	426	416
Chloride	1,010	415	450
Sulphate	54	26	27
Boron	0.14	0.12	0.10
Sodium	543	69	67
Percent sodium**	61	52	51
SAR***	8.65	2.68	2.59

* TDS = Total dissolved solids.
** Percent sodium = $Na/(Ca + Mg + Na + K) \times 100 \; (mgL^{-1})$
*** SAR (Sodium adsorption ratio) = $Na/\sqrt{(Ca + Mg/2)}$.

Table 7. N, P, K values in treated municipal waste water

Constituent (mgL^{-1})	Phase-I (GBHS)	Phase-II Conventional	Low cost
Total-N	9.63	15.412	14.196
Total-P	1.50	3.18	2.68
Total-K	12.08	12.60	12.85

The values of TDS, sulphate, sodium in the treated municipal waste water are relatively low with respect to earlier observation (Phase-I). On the other hand the nutrients are on the higher side as compared to that of GBHS biofilter plant effluent. Additional analysis shows that the deleterious effect of high sodium content in effluents, as Phase-I is also absent. Moreover, SAR values of the effluents in the present analysis are found to be below 4 and this confirms their suitability for nutrient recycling because of

their less phyto-toxic character [9]. At the same time the boron
and sulphate concentrations in waste water are quite acceptable
for irrigated agriculture crop. All the above observations predict
that skilful irrigation management practice is not needed for regular production in agricultural crops, if the treated effluents
are used for irrigation purposes.

The cultivable land adjoining to the waste water treatment
plants under present investigation comprises of mostly new alluvial
soil and rice is the commonly irrigated agricultural crop in this
area. The Table 1 shows N, P, K (chemical fertilizer) values necessitated for rice production are in the ratio of 60:40:40 whereas
present experimental findings of N, P, K values in the treated waste
water shows the ratio being 60:12:57. Invariably it concluded that
treated waste water when recommended for irrigation in crop production, although the phosphorus content (in comparison to the chemical fertilizer normally used for the same purpose) is deficient as
in the Phase-I & II of the present study, will produce reasonable
yield.

CONCLUSION

The experimental results conclusively prove that both GBHS
biofilter plant effluent and treated municipal waste water have
potentiality to be used as irrigant with adequate nutrients content. The efficiency of crop uptake as revealed in Phase-I of the
present study is quite high in experimental bed where GBHS biofilter plant effluent has been used as irrigant under H.Napier vegetation cover. The treated municipal waste water, having higher
nutrients content than Phase-I can be viewed as more efficient
irrigant in crop production. The primary observation in Phase-I of
the study show that there is possibility of NO3-contamination of
the underground water table by leaching of the irrigation water
through soil-vegetation bio-system. This possibility may be overcome by selecting vegetation having higher nutrient uptake rate,
better crop management practice and optimisation of the loading
rate.

The Phase-II of the ongoing project is yet to be completed.
The rice has been selected as the cover crop in the experimental
bed and treated municipal waste water as the irrigant. New alluvial
soil has been used for sustaining the high yielding crop. It is to
be observed, whether the municipal waste water having deficiency in
phosphorous content, may be equally efficient to produce reasonable
crop yield.

REFERENCES

1 F.M. D'Itri (ed.), Land treatment of waste water. Ann Arb.(1982).
2 J.E. Hook, J. Environ. Qual. 8, 496 (1979).
3 D.R. Linden, J. Environ. Qual. 13,256 (1984).
4 L.T. Kardos (ed.), Renovation of waste water. Pen.St.Univ.(1973).
5 D.R. Linden, J. Environ. Qual. 10, 507 (1981).
6 National Agriculture Research Project, New Delhi, 1 (1989).
7 Ganga Action Plan Report, Min. of Env.& For, New Delhi (1988).
8 R.C. Loehr (ed.), Spray irrigation system. Ann Arb.Pub. (1977).
9 Water quality criteria, Fed.water poll.cont.Adm.W.Y.D.C.(1978).

Environmental Research Forum Vols. 5-6 (1996) pp. 183-188
© *1996 Transtec Publications, Switzerland*

Wastewater Irrigation of Energy Plantations

K. Hasselgren

Svalöv Community, S-268 80 Svalöv, Sweden

Keywords: Municipal Wastewater, Land Treatment, Slow Rate Treatment, Spray Irrigation, Soil-Plant System, Natural System, Nutrients, Nitrogen, Phosphorus, Energy Forestry, Willow, *Salix*, Kågeröd, Svalöv, Sweden

Abstract

Municipal secondary effluent is used as water and nutrient resources in fast-growing willow plantations meant for energy purposes. The 5-year project started 1992 and concerns mainly measurements of biomass growth and removal of various wastewater constituents in the soil-plant system. Wastewater application up to 12 mm/d during two growth periods has not negatively influenced on superficial groundwater quality. The willow plants develop better in parcells applied with wastewater than in control parcells.

Introduction

Cultivation of selected species of willows (*Salix* spp.) for energy purposes has rapidly increased in Sweden. To date, ca 15 000 hectares of *Salix* are established and during 1996 another 6-8 000 hectares are expected to be planted. The total cost for chip production from energy willow forestry by known technique amounts to 110-115 SEK/MWh which figure is comparable to the price of chipped residuals from conventional forestry. On-going R&D projects indicate that the costs will decrease in a few years time due to improvements within the areas of plant breeding and cultivation technology.

Fertilization is of great importance and corresponds to 15-20 % of the chip production cost. As complement or alternative to manufactured fertilizers recycling of various waste products from the society rich in nutrients, such as municipal or industrial wastewater, sewage sludge, ashes as well as leachate from sanitary landfills have been discussed [1].

Application of wastewater, undewatered sludge or landfill leachate would also partly or completely replace possible water requirements. A positive water effect on *Salix* growth was clearly shown in a landfill leachate irrigation project in the south of Sweden [2]. A good circulation of water in the root zone provides a uniform distribution of dissolved nutrients. This will create optimum conditions for plant nutrition. Thus plant energy requirement for development of roots searching for nutrients and water in deeper soil layers could be reduced in favour of the above-ground biomass production. Further, if the plant has to economize with water, the stomata openings will close to counteract transpiration. This will reduce the necessary conditions for an optimum photosynthesis.

Early small-scale tests with wastewater application to *Salix* stands planted in slilty loam lysimeters resulted in promising wastewater treatment potentials [3]. A willow-soil system may be defined as a natural physical-biological-chemical wastewater treatment reactor including the following main active parts and processes:

- *Soil particles* which filter suspended solids and fix dissolved components in the wastewater by adsorption, ion exchange or precipitation.
- *Macro- and microorganisms* which transform and stabilize organic substances and transform nutrients in applied wastewater.
- *Vegetation* which utilizes macro- and micronutrients in the wastewater for growth and maintains or increases the infiltration capacity in the soil.

Human wastes have been used on land since ancient times. The first land treatment facility documented in the literature was Bunslaw, Germany, where a sewage irrigation system commenced in 1531 and was in operation for over 300 years [4]. Many "sewage farms" existed in the latter half of the19th century and during the first few decades of our century. They were replaced gradually by in-plant alternatives starting round 1920 when the activated sludge method and other biological treatment processes were introduced.

In the United States, land treatment methods in general and especially irrigation applications have regained respect in recent years as cost-effective and competitive alternatives to conventional wastewater treatment processes. This is a result mainly originating from the pioneering and extensive work at The Penn State University in the 1960's and the early 1970's [5]. In Sweden a few wastewater irrigation facilities were established in the 1980's on the islands Gotland and Öland. Here, deficient water resources in agriculture are the main motive while fertilization and wastewater treatment are positive side effects.

In this paper are summarized a project description and some preliminary results concerning biomass growth and removal of some wastewater constituents in *Salix* plantations after wastewater irrigation at the Kågeröd Wastewater Treatment Plant (KWTP), Svalöv Community, in the south of Sweden.

Experimental procedures

The KWTP receives sewage from the village of Kågeröd with about 1500 inhabitants and from a large milk-powder factory corresponding to an organic load of 6-7000 persons. The wastewater is treated to a tertiary level according to the Swedish standard, *i.e.* chemical precipitation of secondary effluent for phosphorus removal. The stipulated discharge parameters are biochemical oxygen demand (BOD_7) and phosphorus (Total P) with maximum allowable values of 10 mg O_2/l and 0.3 mg P_{tot}/l, respectively, as monthly means.

During 1992 started a project concerning wastewater irrigation of energy forest parcells adjacent to the KWTP. Thus, secondary effluent is taken out from the activated sludge unit before the P removal step and applied to *Salix* plots at six different rates, *i.e.* 2, 4, 6, 8, 10 and 12 mm/d, respectively. Wastewater is applied during the growth period (May-October) using solid-set sprinklers. The seven plots are each 900 m^2 (30 m x 30 m) in size. One plot is not irrigated and thus kept as control.

Three commercial *Salix* clones, ORM, RAPP and ULV (mainly from osier willow material), developed especially for energy forestry purposes by the breeding company Svalöf Weibulls, are used in the study. Cuttings were planted in spring 1992 according to the standard double-row concept giving a plant density of ca 18 000 individuals per hectare. The soil is a silty clay with a silty top-soil layer of about 0.3 m. The field was earlier used for traditional crop production and fertilized by manufactured fertilizers. Centrally in each plot is installed a tube well for observation of groundwater fluctuations and sampling of superficial groundwater. Chemical analyses of applied wastewater, soil, groundwater and *Salix* material, mainly in terms of nutrients (fractions of N, P and K), organic material (BOD) and some heavy metals (Cu, Zn, Cd and Pb), are carried out within the project.

A first evaluation phase of the project will be completed during 1998. The project would hopefully answer questions such as: How much wastewater is required for optimum willow growth? Which resources in the wastewater are of concern? Can natural soil-plant treatment compete with conventional technology? Which are the alternatives during the cold period of the year?

Preliminary results and discussion

Application of wastewater and nutrients

In 1993 application of wastewater to the six plots amounted to between 239 mm (2 mm/d-plot) and 1439 mm (12 mm/d-plot) during 120 days from May to November. After the introductory year with some initial mechanical problems, the rates in 1994 reached the expected 335 mm for the 2 mm/d-plot and 2018 mm for the 12 mm/d-plot. The annual precipitation in the area averages 775 mm of which an amount of 396 mm falls during the irrigation period May-October. Although the irrigation volumes have exceeded the precipitation amounts and despite the relatively fine silty soil present, ponding on the horizontal soil surface (indicating potential surface run-off) has only occurred after some heavy storms and only in the plot receiving 12 mm wastewater per day. No damage to the plants was recorded.

Irrigation during 1993 resulted in macronutrient applications of 41-247 kg N/ha, 3-19 kg P/ha and 34-206 kg K/ha. The applied rates increased during 1994 as the irrigated volumes increased and amounted to 47-284 kg N/ha, 2-12 kg P/ha and 45-268 kg K/ha. Experiences gained from other cultivation tests indicate an N-concentration of 0.5-1.0 % in *Salix* stems on a dry matter (DM) basis at growth rates in the range of 15-20 tonnes of DM/ha and year, resulting in an N-requirement of 75-200 kg N/ha and year [6]. Thus, the selected application rates of nitrogen cover a wider range than the expected uptake rate. Early comprehensive laboratory studies of nutrient require-ments of *Salix* species resulted in the N-P-K-relationship of 100-13-65 for optimum growth and biomass yield [7]. The N-P-K relationships in applied secondary effluent 1993 and 1994 were 100-8-83 and 100-4-94, respectively. This means that if irrigation with this wastewater is designed to balance the crop uptake of nitrogen, which would be most likely, only 25-50 % of the P requirement would be met.

The phosphorus concentration is generally higher in ordinary secondary effluent. The activated sludge step at the KWTP is operated by alternating between aeration and non-aeration in one part of the unit in an attempt to test the potential for N removal by nitrification/denitrification. It is obvious that besides N reduction we also have conditions promoting biological P removal. The soil supply of phosphorus has been sufficient up to now and will be that also further on within the project period. On a long term basis and in a full scale situation, however, it will be necessary to change the operation procedure of the activated sludge unit to reach lower P removal rates. This should be sufficiently based on the knowledge of the content of phosphorus in raw wastewater resulting in an almost ideal relationship of macronutrients concerning the requirement of willow plants.

Plant growth

The plants were cut in December 1992 after the first growth period for stimulation of the shoot production the following year. Early growth data from measurements in autumn 1993 with 1-year-old shoots on 2-year-old roots indicate that already in the establishing phase irrigation with wastewater promotes the growth (Figure 1). The number of productive shoots increased with increased irrigation rates up to 4 or 6 mm/d. Of the three clones, ULV seems to respond best to wastewater. The year was relatively wet with large precipitation amounts during the growth period (26 % higher than the average). When also taken into account that the initial nutrient supply of the

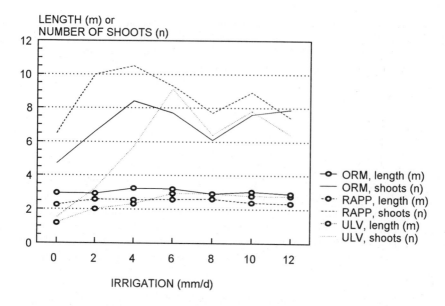

Figure 1. Growth data for *Salix* clones with 1-year-old shoots on 2-year-old roots measured in the winter 1993-94. Length of maximum shoot and number of productive shoots per plant.

soil was relatively high, the plant growth differences between the control and irrigated parcells most likely will increase the following seasons. In the winter 1995-96 harvest of the 3-year-old shoots is planned to take place when also the plants will be weighed for calculations of the dry matter production.

Reduction of nitrogen

Nitrogen is considered as one of the most important component in wastewater from a resource as well as a water pollution point of view. Test results so far indicate a high N removal capacity in this soil-plant system (Figure 2). Applied secondary effluent is sampled monthly for detailed calculations of applied amounts in a latter stage of the project. Here the samples are put together to quarterly values to balance the quarterly groundwater analyses. It should be pointed out that wastewater irrigation is carried out from May to November each year even though the N content in the wastewater during the whole year is presented here.

In superficial groundwater the N values from 1994 and further have been below 4 mg N/l more or less independently of the load. The depth of the unsaturated zone has varied between 0.2 and 1.3 m due to the time of the year as well as differences between the parcells. Regarding possible dilution aspects even though the groundwater samples are taken superficially, analyses of chloride are included in the sampling programme. Assuming that the Cl⁻-ions are not accumulated in the soil, the chloride concentration in wastewater and groundwater could be used as a reference parameter for assessments of removal of various wastewater constituents. As shown in Figure 3, the chloride concentrations in general increased as the application rates increased. This would imply that the presence of wastewater in superficial groundwater beneath the plots has increased with wastewater irrigation rates.

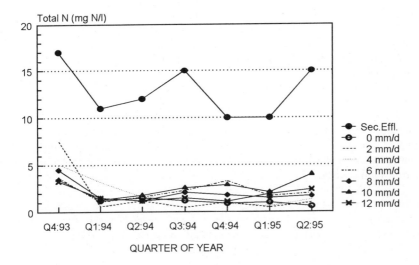

Figure 2. Concentrations of Total N in secondary effluent and quarterly samples from superficial groundwater in the test plots during the period October 1993-June 1995.

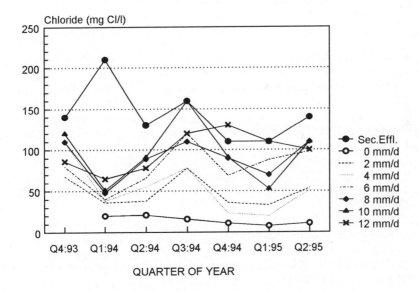

Figure 3. Concentrations of chloride in secondary effluent and in quarterly samples from superficial groundwater in the test plots during the period October 1993-June 1995.

Thus, when comparing the Figures 2 and 3, an early conclusion is that application of wastewater during the growth period, May-October, up to 12 mm/d has not so far dramatically affected groundwater quality concerning total nitrogen. It is assessed that the main N removal mechanisms are uptake of N by *Salix* and nitrification/denitrification in the soil. Ammonia volatilization is primarily depending on alkaline soil pH values. Since the soil pH is neutral or slightly acidic, the contribution from volatilization of NH_3 has likely been of minor importance.

Discharge of nitrogen from municipal wastewater to surface waters is a topical question. Today, larger sewage treatment works (> 10 000 person equivalents, PE, in terms of organic load) are being licenced with maxium discharge values in the range of 8-15 mg N/l. These values are based on the so called "best available technology" which to date has been variants of the conventional in-plant nitrification/denitrification concept.

The official approach to wastewater treatment in Sweden as elsewhere is to find sustainable alternatives to processes in use at present where the resources involved should be subjects to recycling and reuse in an environmentally safe manner. Based on project results obtained so far, the example described here seems to include the basic qualifications to fulfil this ambition.

References

[1]   Hasselgren, K (1988). "Sewage sludge recycling in energy forestry". Proceedings 1: 189-197. The 5th International Solid Wastes Conference, 11-16 September, 1988, Copenhagen, Denmark. International Solid Wastes Association. Academic Press, London, England.

[2]   Hasselgren, K (1992). "Soil-plant treatment systems". *In*: Landfilling of wastes - Leachate. Cossu, R; Christensen, T and Stegmann, R. (Eds.). Elsevier Applied Science, pp 361-380.

[3]   Hasselgren, K (1984). "Municipal wastewater reuse and treatment in energy cultivation." Proceedings 1:414-427. Water Reuse Symposium 3, 26-31 August, 1984, San Diego, California, USA. American Waterworks Association Research Foundation, Denver, Colorado, USA.

[4]   Reed, S C and Crites, R W (1984). "Handbook of land treatment systems for industrial and municipal wastes". Noyes Publications, NJ, USA. ISBN 0-8155-0991-X.

[5]   Sopper, W E and Kardos, L T (1973). "Recycling treated wastewater and sludge through forest and cropland". The Pennsylvania State University Press, University Park, PA, USA. ISBN 0-271-01159-9.

[6]   Perttu, K (1995). Personal communication.

[7]   Ericsson, T (1981). "Effects of varied nitrogen stress on growth and nutrition in three *Salix* clones." *Physiol. Plant.* **51**:423-429.

Environmental Research Forum Vols. 5-6 (1996) pp. 189-196
© *1996 Transtec Publications, Switzerland*

Energy Forest Use in Wastewater Treatment in Estonia

Ü. Mander[1], A. Koppel[2], V. Kuusemets[3] and T. Mauring[3]

[1] Institute of Geography, University of Tartu, 46 Vanemuise St., EE-2400 Tartu, Estonia

[2] Institute of Zoology and Botany, Estonian Academy of Sciences,
181 Riia St., EE-2400 Tartu, Estonia

[3] Centre for Ecological Engineering Tartu, 4 J.V. Jannseni St., EE-2400 Tartu, Estonia

Keywords: *Salix viminalis, Salix dasyclados, Alnus incana*, Riparian Buffer Zones, Nitrogen Fixation, Denitrification, Productivity, Biomass, N and P Removal, Phosphorus Retention

ABSTRACT

Biomass production values in the end of plantation establishment year were very low in natural soils, reaching 0.13 t ha^{-1}. In a willow plantation for municipal wastewater purification, however, the plants displayed extremely vigorous growth, reaching the height 2.8 m after 3.5 month growth in September 95. The preliminary estimation of the above-ground production was 2.4 and 0.7 t ha^{-1} in stems (bark and wood) and leaves, respectively. The average values of the purification efficiency of organic matter and total-N were 73 and 25 %. Retention showed relatively low values being 2.1, 0.5, and 0.06 g m^{-2} d^{-1} for organic matter, total-N and total-P, accordingly. Willow plantations can be used as effective and cost-efficient biological purification systems, however, the leaching of wastewater into groundwater could be avoided
Despite the significantly higher N and P loading in the grey alder test site downhill from the slurry application field relative to the less loaded area (0.4-4.3 and 2.0-62.1 mg N l^{-1}, accordingly; 0.6-7.09 and 0.42-1.05 mg P l^{-1}, respectively) the output concentrations were at the comparable level (0.4-3.0 mg N l^{-1}, and 0.08 to 0.65 mg P l^{-1}). Denitrification value within the grey alder forest in the less loaded test site varied between 12 and 21 µg N m^{-2} hr^{-1}. In the loaded site, it was 3-14 µg N m^{-2} hr^{-1}. In adjacent wet meadow and set-aside grassland uphill from the forest, this rate was observed higher (4-57 and 5-41 µg N m^{-2} hr^{-1}, respectively). Atmospheric N$_2$ fixation in alder stand in intensively loaded test site was significantly lower than in less loaded test site (0.2-2.8 and 0.6-15 µg N m^{-2} hr^{-1}, respectively) being highest in July. The total N uptake is 140.2 and 204.8 kg N ha^{-1} yr^{-1}, and P uptake 10.8 and 15.1 kg P ha^{-1} yr^{-1}, accordingly in heavily polluted and less-polluted test sites.). According to high loading, the N and P retention value in heavily polluted site was relatively high: 0.48-42.8 kg N ha^{-1} yr^{-1} and 0.12-5.2 kg P ha^{-1} yr^{-1}. In less polluted stand, it was 0.24-4.6 kg N ha^{-1} yr^{-1} and 0.13-1.3 kg P ha^{-1} yr^{-1}. Riparian grey alder forests are effective buffers on stream banks and lake shores; they provide great interest as energy forest stands. From the point of view both productivity and nutrient retention, the optimal age to harvest them is 12-15 years.

INTRODUCTION

The Estonian Energy Forest Project was started in 1993 in order to promote the research and application of renewable energy resources in Estonia [28]. At the same time a joint Swedish-Estonian energy forestry research programme was established. Seven experimental willow plantations were established in 1993-1995, using the selected clones of *Salix viminalis* and *S. dasyclados*. The plant material originates from the Swedish Energy Forest Project. Two of these experimental plantations (Väike-Maarja, 0.2 ha and Aarike, 0.02) are used as vegetation filters for purification of municipal wastewater. The highly productive willow plantations (up to 35 t (d.w.) ha^{-1} yr^{-1}; [33]) have shown a satisfactory purification efficiency of organic matter, nitrogen and phosphorus (80-85, 20-50, and 45-85 %, respectively; see also [25]). There are altogether about 10 constructed wetlands of various type established in Estonia during the last five years. On the other hand, the number of natural wetlands used for wastewater purification or spontaneously formed seminatural wetland filters reaches several hundreds. Many of them can be converted into willow plantations to produce biomass for energy production. This paper gives an overview about the preliminary results of wastewater purification efficiency in three established willow plantations. Also, preliminary results of biomass production in the Aarike pilot plot are presented. For comparison, the above-ground biomass production figures of two older energy forest plantation are presented.

Beside willow plantations alder forests provide interest as potential biomass energy sources. They are typical riparian ecosystems in Europe which can retain and transform nutrient fluxes from adjacent intensively used territories. Therefore, riparian alder stands are commonly evaluated as buffer zones to protect water bodies against pollution. However, only few studies have been thoroughly carried out to investigate their buffering capacity and the results show contradictory findings ([27]; [7]; [38]). Due to fixation of atmospheric nitrogen by root nodules alders have been expected to act as an additional source of nitrogen pollution of water bodies. To clarify the influence of both internal and external loading in riparian alder forests, a study was carried out in two grey alder (*Alnus incana*) stands of different load in southern Estonia: one in the natural conditions without any significant nutrient input via groundwater/overland flow, and another downhill from the intensively fertilized arable land in the vicinity of a large pig farm. The main hypothesis was that from riparian grey alder stands with significant external nutrient load (manure application in upland field, ammonium deposition) significantly more nutrients will be leached than from the unloaded stands. The hypothesis was tested in 1994-1995. Some preliminary results of this comprehensive study will be presented in this paper.

MATERIAL AND METHODS

Study area and study sites

In May 1995 the Aarike pilot willow plantation for municipal wastewater purification was constructed. This three-basin serpentine pond system with rows of *Salix viminalis* clones on low beds in between the shallow (0.1 m) winding channel has an area 170 m^2 and an average hydraulic load of 41 mm d^{-1}. Planting density was 2 plants per m^2. A 20 cm basement of loamy till and clay layer avoids wastewater leakage into groundwater. The Aarike pilot plant is located within the Porijõgi River catchment [30]. Kambja and Saare energy forest plantations were established in 1993 with six clones of *Salix viminalis* and one clone of *Salix dasyclados*. The planting density was 2 plants per m^2. In both plantations all the seven clones were divided into four randomized subplots. Kambja plantation (area 0.3 ha) is situated on the well decomposed organic soil, pH in the upper 20 cm layer is 5.7, total N content 1.65 %. Saare plantation is situated on brown pseudopodzolic soil, pH - 5.3, nitrogen content - 0.08 %. After establishment of the plantations, in March 1994 the shoots were cut at the height of 5-10 cm to promote denser sprouting. In the establishment year and during the first half of growing season in 1994 weed control was performed mechanically, altogether six times.

Two different riparian grey alder stands have been chosen: one in the unpolluted Porijõgi River catchment (area description see [30]) and, second, in the vicinity of a Viiratsi pig farm (32000 pigs), Viljandi County. The Viiratsi study site has similar physio-geographical conditions like the Porijõgi River catchment. In both study areas, transects were established in autumn 1993 along topo-edaphic gradient. In the less polluted Porijõgi test site the following spectrum of communities, ordered from uphill to downhill, was analyzed: abandoned (former cultivated) grassland - wet meadow (dominating by *Filipendula ulmaria*, *Aegopodium podagraria*, *Cirsium oleraceum*, and *Urtica dioica*) - grey alder stand. In the heavily polluted Viiratsi test site the transect was established through the following communities: arable land (fertilized by pig slurry) - eutrophic grassland strip (*Elytrigia repens*, *Urtica dioica*) - young grey alder stand with wet meadow pattern (*Filipendula ulmaria*) - old grey alder forest. In landscape profiles rows of piezometers (3 rows in the Porijõgi transect and 5 rows in the Viiratsi study site, 3 replicates in each row) and study plots were established on the boundaries between the plant communities. Main N and P cycles and budgets have been assessed. In this paper some preliminary results of the soil water quality, tree biomass, production and nutrient uptake, nitrogen fixation and denitrification are presented.

Field investigations and laboratory analysis
In Aarike pilot plantation, four series of water samples (from May to September 1995) were taken from inflow and outflows of each bed. In the South Estonian Laboratory of Environment Protection water was analyzed for pH and BOD_5 values, and for NH_4-N, NO_2-N, NO_3-N, total-N, PO_4-P, total-P, SO_4, and Fe content. All water analyses were made following standard methods compatible with the international methods for examination of water and wastewater quality ([3]). In March 1993 the above-ground biomass production was estimated by weighing of all the shoots. The dry biomass was estimated after the subsamples were oven-dried (48h at 80 ^{o}C). In the end of growing season in 1995 the above-ground biomass was estimated by sample shoot method. First, the shoot diameter distribution was estimated for each subplot by measuring all the shoots of 10 randomly chosen stools. From each clone 25 shoots were sampled according to the diameter distribution and the oven-dry weight (48h at 20 ^{o}C) of shoots (bark and wood) and leaves was estimated. By using the linear regression of shoot cross sectional area at 55 cm above the ground and shoot dry weight and shoot cross-sectional area/leaf dry weight and by using shoot diameter distribution, the above-ground biomass of leaves and shoots was estimated. For Aarike site, the regression, established for the same clone in Kambja site was applied.
In grey alder study sites, water samples were collected and the groundwater depth was measured once to twice a month by piezometers. In the laboratory of the Estonian Agricultural University the filtered soil water samples were analyzed for NH_4-N, NO_2-N, NO_3-N, total Kjeldahl nitrogen (TKN), PO_4-P, total Kjeldahl phosphorus (TKP), SO_4, Fe, Ca. Soil bulk density, texture class, and field capacity was determined for each 20 cm of soil profile (up to 1.5 m depth). Hydraulic conductivity was estimated by using tracer (chloride) and pumping experiments [16]. Groundwater discharge was estimated on the basis of both Darcy's law and gauging with weirs installed in groundwater seeping sites.
Grey alder stands belong to the *Alnus incana - Aegopodium podagraria* community [14]. Average age of alders in the Porijõgi test site is 14 yr, in Viiratsi test site 40 yr. Above- and below-ground storages and production of alders have been estimated [8]; [5]; [32]. The atmospheric nitrogen fixation rate in soil was estimated by using the acetylene method [37]. To assess the denitrification rate acetylene as a nitrous oxide reduction inhibitor was used [41].

RESULTS AND DISCUSSION

Willow plantations
Biomass production
Biomass production values in the end of plantation establishment year were very low in natural soils, reaching 0.13 t ha^{-1} in Kambja and Saare in 1993. In Aarike, however, the plants displayed extremely vigorous growth, reaching the height 2.8 m after 3.5 month growth in September 95. The preliminary estimation of production in the first bed yields the above-ground production 2.4 and 0.7 t ha^{-1} in stems (bark and wood) and leaves, respectively. These high figures indicate the close to optimum growth conditions (lack of water stress , availability of nutrients on optimal proportions and unusually sunny and warm summer).
The above-ground biomass production figures for Kambja and Saare site are presented in Table 1.

Table 1. Above-ground production (t of oven-dry biomass ha^{-1}) of willow energy forest during the two first production years.

	Saare plantation	Kambja plantation
Stem production (two first production years)		
average for plantation	9.8	8.5
range for averages of 7 clones	4.6-16.7	7.2-10.8
range for all subplots	1.5-24.1	2.7-16.4
maximum production per clone	16.7**	10.8***
maximum production per subplot*	24.1**	16.4**
Stem maximum annual production per clone	13.8**	8.1***
Leaf production (second production year)		
average per plantation	2.9	2.5
maximum per subplot	6.18**	4.2**

*   - subplot area in Kambja- 108 m^2, in Saare 256m^2
**   - *Salix dasyclados*, clone 81090
***   - *Salix viminalis*, clone 78195

The high variability in production figures can be explained by two factors. First- by different success in establishment success. Some of the subplots may have suffered from weed competition in the year of establishment. One can expect the decrease of variability in the growth performance in different subplots in the coming years. Second factor of variability cam be the different susceptibility of clones to early and late frosts.

Nutrients removal
Nutrients removal efficiency (difference between inflow and outflow values; %) and retention capacity (g m^{-2} d^{-1}) are presented in Fig.1. Retention value R was calculated as follows:

$$R = [(Q_{in}*C_{in}) - (Q_{out}*C_{out})]/A \qquad (1)$$

where Q_{in} and Q_{out} - inflow and outflow values (m^3 d^{-1}), respectively; C_{in} and C_{out} - concentration values (mg l^{-1}), respectively; A - plantation area treated with wastewater (m^2).
Kirt (1994) has found a relatively high efficiency in the Väike-Maarja test site: 83, 43, 79, and 52 % for BOD5, total-P, NH4-N, and total-N, accordingly. The retention values in this area were 6.5, 0.15, 0.97, and 1.0 g m^{-2} d^{-1}, respectively. However, in the Väike-Maarja plantation the water flux is not exactly measurable because of exchange between the groundwater. Therefore, these relatively high retention values can be influenced by dilution effect.
In the newly established Aarike pilot plantation, no water losses via percolation could be found. Here, the average values of the purification efficiency of organic matter and total-N (73 and 25 %, respectively) are comparable with those for other plantations (Fig. 1). However, retention values for organic matter and total-P (2.1 and 0.06 g m^{-2} d^{-1}) are significantly lower. This is due to much lower load value relative to other plantation studied (see [31])..

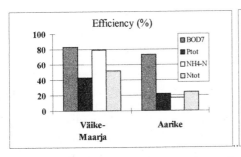
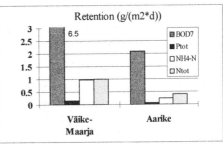

Fig. 1. Nutrient purification efficiency and retention in willow plantations (data on Väike-Maarja are calculated after [25]).

Riparian alder forests
Biomass production and nutrient uptake

Biomass, net production and nutrient uptake values in grey alder stands are presented in Table 2. Total biomass value in the less polluted Porijõgi stand was 115.5 t ha^{-1}. In the 40 yr old Viiratsi stand this was 158 t ha^{-1}, due to bigger diameter of stems (up to 110 cm at breath height) and bigger amount of branches. In opposite, because of intensive growth, the net production was higher in the Porijõgi test site (17.3 and 10.6 t ha^{-1} yr^{-1}, respectively). Nitrogen and phosphorus uptake by grey alders is relatively high in both study sites, however, this factor can explain only 8-10 % of total N removal and 10-12 % of total-P retention. Owing to higher productivity, the total N and P accumulation in both above-ground and below-ground biomass was significantly higher in the Porijõgi stand (204.8 kg N ha^{-1} yr^{-1} and 15.1 kg P ha^{-1} yr^{-1} in Porijõgi, and 140.2 kg N ha^{-1} yr^{-1} and 10.8 kg P ha^{-1} yr^{-1} in Viiratsi). Nitrogen content in wood and leaves was approximately the same in both test sites. However, the P concentration in total biomass of older and heavily loaded Viiratsi grey alder stand was 13.5 % higher than in younger and less polluted Porijõgi stand. In older stand, a half of assimilated P is allocated into leaves, in younger stand this share is only one third (Table 2).

Table 2. Biomass, production and nutrient uptake by grey alders in riparian alder forests, Porijõgi and Viiratsi test areas in Estonia.

		Porijõgi test site				Viiratsi test site			
		Plot area: 457.8 m2; number of trees per ha: 6110 (alders: 5240)				Plot area: 690.6 m2; number of trees per ha: 181 (alders: 1390)			
		Alder biomass t/ha	Alder production t/(ha*yr)	N kg/(ha*yr)	P kg/(ha*yr)	Alder biomass t/ha	Alder production t/(ha*yr)	N kg/(ha*yr)	P kg/(ha*yr)
Stems		81.7				104.7			
	woo		7	20.4	3.5		3.77	8.6	1.7
	bar		1.3	16.1	0.99		0.42	5	0.49
Branches									
secondary growth		7.9				23.7			
	woo		0.69	3.6	0.35		1.05	3.23	0.84
	bar		0.51	7	0.49		0.3	3.52	0.33
primary growt		0.89	0.89	21.9	1.74	0.59	0.59	11.6	1.2
Leaves		2.6	2.6	85,8	4.91	2.8	2.8	88.5	5.07
Roots		22.4	4.3 +	50	3.1	[26.2]	[1.7]	[19.8]	[1.2]
Litter									
	leave		2.5	70.9	2.93		[2.7]	[75.6]	[3.1]
	branche		2 +	11.2	1.1		[3]	n.e.	n.e.
Total ++		**115.5**	**17.3**	**204.8**	**15.1**	**158**	**10.6**	**140.2**	**10.8**

[3]	- preliminary estimated values	n.e.	- not estimated
t/ha +	- wood and bark together	Total ++	- without litter value

Nutrient retention

Retention of N and P was calculated according to Eq. 1. In Porijõgi, the groundwater filtration was calculated for 200 m^2 plots, in Viiratsi the calculation area was 800 m^2.

Nitrogen

Despite the significantly higher nitrogen load in the Viiratsi riparian buffer zone (eutrophic grassland - young alder stand - wet meadow - old alder stand) relative to the Porijõgi set-aside grassland - wet meadow - alder forest complex (0.4-4.3 and 2.0-62.1 mg N l^{-1}, respectively) the output concentrations were at the comparable level (0.4-2.1 and 0.5-3.0 mg N l^{-1}, correspondingly; Fig. 1). Atmospheric N deposition in Porijõgi and Viiratsi was estimated to be 10 and 15 kg N ha^{-1} yr^{-1}. Extremely high TKN content in Viiratsi soil water was caused by pig slurry application in the adjacent field in July and August 1994. Due to intensive fertilization during many years, soil in this field has been compacted and soil microfauna disturbed. This is why during the first part of the planned study period (from June 1994 to July 1995) the N content in upland soil water was always high. In the Porijõgi catchment, in opposite, the N input has decreased during the last three years since the agricultural activities had ceased in the upland field (see [29]). According to high loading, the retention value in Viiratsi was relatively high: 0.48-42.8 kg N ha^{-1} yr$^-$ (Fig. 2). In Porijõgi, it was 0.24-4.6 kg N ha^{-1} yr^{-1}. High buffering capacity in the Viiratsi study site can be explained by the following factors: (1) relatively high denitrification value, however, being less than in Porijõgi measurements; (2) relatively high plant uptake, being, again, less than in Porijõgi; (3) lower N$_2$ fixation.

Table 3. Measured atmospheric nitrogen fixation and denitrification values in less polluted Porijõgi and heavily polluted Viiratsi test sites.

| | Nitrogen fixation (µg N m^{-2} h^{-1}) | | | | | | Denitrification (µg N m^{-2} h^{-1}) | | | | | |
| | Porijõgi | | | Viiratsi | | | Porijõgi | | | Viiratsi | | |
	A	B	C	A	B	C	A	B	C	A	B	C
5-May-94	20.0	21.3	10.7	17.9	2.1	2.0	56.3	37.9	21.5	23.3	8.6	9.9
7-Jul-95	1.6	0.7	15.5	0.5	0.6	2.9	6.1	4.3	16.2	5.1	5.4	2.7
23-Sep-95	0.4	0.2	0.7	0.2	0.2	0.2	45.6	23.8	12.8	40.5	21.1	14.5

A - grassland; B - wet meadow; C - alder stand

Denitrification value varied between 12 and 21 µg N m^{-2} hr^{-1} in the Porijõgi test site. In Viiratsi , this value was 3-14 µg N m^{-2} hr^{-1} (Table 3). Nevertheless, in adjacent wet meadow and set-aside grassland uphill from the forest, this rate was higher (4-57 and 5-41 µg N m^{-2} hr^{-1}, in Porijõgi and Viiratsi, respectively). Despite a relatively small number of analyses, the main tendencies in denitrification intensity are comparable with the results of other investigations: (1) the main denitrifying activity was observed in spring and late summer [35];[40]; (2) a higher denitrification activity was found in the upper part of the slope (see also [13]; [34]; [24]; [40]). The latter tendency can be explained by significantly higher nitrate concentration in soil water.[36] have found up to 100 times higher denitrifying activity in slurry incubations than in black alder forest. Most probably, due to pig slurry application in Viiratsi the real values of denitrification could be higher than revealed by our observations, because our measurements were not carried out immediately after the slurry application.

Atmospheric N$_2$ fixation in soil of alder stand in Viiratsi was significantly lower than in the Porijõgi test site (0.2-2.8 and 0.6-15 µg N m^{-2} hr^{-1}, respectively). The highest values were recorded in July (Table 3). Surprisingly, the highest N$_2$ fixation values (up to 21.3 µg N m^{-2} hr^{-1} in Porijõgi and 17.9µg N m^{-2} hr^{-1} in Viiratsi) were observed in May in wet meadows and grassland communities without symbiotic N$_2$ fixers. The highest fixation was observed in May. Lower N$_2$ fixation in loaded Viiratsi test area can be explained by predominating of N assimilation over N$_2$ fixation when high concentrations of mineral N are present in the root medium of actinorhizal plants (see Troelstra *et al.*, 1992). However, our investigations show that the N_2 fixation in soil is a nonsignificant part of the whole N budget in both study plots.

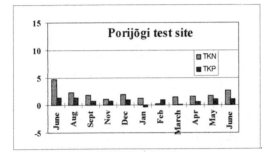 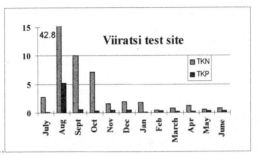

Fig. 2. Total Kjeldahl nitrogen (TKN) and total Kjeldahl phosphorus (TKN) retention in less polluted Porijõgi test site and heavily polluted Viiratsi test site (g m^{-2} d^{-1}).

Phosphorus
Similarly to N relations in ground water, the P concentration in output from intensively loaded Viiratsi test plot is not significantly higher than in Porijõgi, varying from 0.2 to 0.55 mg P l^{-1} and 0.08 to 0.65 mg P l^{-1}, respectively. Nevertheless, the input P values on the border of arable land and eutrophic grassland in Viiratsi are significantly higher than on the border of *Filipendula-Aegopodium* wetland and alder forest and in Porijõgi being 0.6-7.09 and 0.42-1.05 mg P l^{-1}, accordingly [32]. Again, the high P values in Viiratsi are caused by slurry application in the adjacent field. According to high load, the retention value in Viiratsi was high (0.12-5.2 kg P ha^{-1} yr^{-1}; Fig. 2). In Porijõgi, it was 0.13-1.3 kg P ha^{-1} yr^{-1}. There was found a slight P leaching in winter (0.13 kg P ha^{-1} yr^{-1}). High rate of P retention in the Viiratsi study site can be explained by the following processes: (1) uptake by alders: (2) accumulation in the soil.
Due to relatively low plant uptake and low leaching values, we suggest that the main portion of P retained is accumulated in the soil. In the long-term perspective, of course, this very high load can not be compensated by Fe, Al and Ca phosphate precipitation. This is, seemingly, the key process in P retention in Viiratsi. Likewise, some investigations suggest that permanently high N concentration in soil can cause the P leaching [2]. On the other hand, our earlier investigations demonstrate that riparian alder forests are effective buffers for phosphorus [30]. Even in riparian wetlands P can be retained due to micro-scale oxygenation variability within the wetland and, probably, owing to phosphorus inactivation by nitrate (see [1]).
Our results suggest that due to lower uptake in older *Alnus incana* stands (>20 yr), it is important to harvest these stands earlier.

CONCLUSIONS

(1) Willow plantations can be used as effective and cost-efficient biological purification systems, however, the leaching of wastewater into groundwater could be avoided.
(2) Growth and production of *Salix spp.* clones in plantations used for municipal wastewater purification can be several times higher than in plantations without wastewater addition. However, since our experiments with willow plantations are on the initial stage, these results are rather preliminary yet.
(3) From heavily loaded (pig slurry application on upland soils, atmospheric NH4 deposition) riparian grey alder stands no additional N and P leaching was found, thus, fast growing young grey

alder stands are no additional sources of nutrients and they can act as effective buffers on stream banks and lake shores.

(4) Grey alder stands provide perspectives as energy forest stands, from the point of view both productivity and nutrient retention, the optimal age to harvest is 12-15 years.

ACKNOWLEDGMENTS

This study was supported by the International Science Foundation grant No LLL100 and by the Estonian Science Foundation grants No 251 (1993-1996) and No 1766 (1995-1996).

REFERENCES

[1] J.M. Andersen, *Wat. Res.* 16. (1982).
[2] T. Andrusch *et al.,. Int. Revue ges. Hydrobiol.* 77. (1992).
[3] APHA, *Standard Method for the Examination of Water and Waste Water*, 15th edition, American Public Health Organization, Washington. (1981).
[4] H. Beaupied *et al., Can. J. For. Res.* 20. (1990).
[5] B. Berg, B. and G. Ekbohm, *Can. J. Botany* 7. (1991).
[6] J.G. Bockheim *et al., J. Environ. Qual.* 17. (1988).
[7] D. Binkley *et al., Ecology* 73. (1992).
[8] B.T. Bormann and J.C.Gordon, *Ecology* 2. (1984).
[9] F.S. Chapin III and R.A. Kedrowski, *Ecology* 64. (1983).
[10] S. Christensen *et al., Soil Sci. Soc. Am. J.* 54. (1990).
[11] B. Cote and C. Camire, *Can. J. For. Res.* 17. (1987).
[12] U. Doring-Mederake, *Scr. Geobot.* 19. (1991).
[13] J.H. Duff and F.J. Triska, *Can. J. Fish. Aquat. Sci.* 47. (1990).
[14] H. Ellenberg, *Vegetation Mitteleuropas mit den Alpen in ökologischer Sicht.* 3.Aufl. Ulmer, Stuttgart. (1982).
[15] B.A. Emmet, *Ambio* 22. (1993).
[16] R.A. Freeze and J.A. Cherry, *Groundwater*. Prentice Hall. (1979).
[17] H.L. Golterman, *Methods for Physical and Chemical Analysis of Fresh Waters*. Blackwell Scientific Publications. Oxford - Edinburgh - London - Melbourne. (1978).
[18] J.R. Gosz, *Wat. Res.* 12. (1978).
[19] P.M. Groffmann and J.M. Tiedje, *Soil Biol. Biochem.* 21. (1989).
[20] K. Huss-Danell, *Can. J. For. Res.* 16. (1986a).
[21] K. Huss-Danell, *Plant and Soil* 91. (1986b).
[22] T. Ingestad, *Physiol. Plant.* 50. (1980).
[23] G. Jacks *et al., Ambio* 23. (1994).
[24] T.E. Jordan et al, *J. Environ. Qual.* 22. (1992).
[25] E. Kirt, In P. Aronsson and K. Perttu (eds.) *Willow Vegetation Filters for Municipal Wastewater and Sludges. A Biological Purification System*. Swedish University of Agricultural Sciences. Report 50. Uppsala. (1994).
[26] K.M. Klingensmith and K. Van Cleve, *Can. J. For. Res.* 23. (1993).
[27] N. Knauer and Ü. Mander, *Z. f. Kulturtechnik und Landentwicklung* 30. (1989).
[28] A. Koppel, In P. Aronsson and K. Perttu (eds.) *Willow Vegetation Filters for Municipal Wastewater and Sludges. A Biological Purification System*. Swedish University of Agricultural Sciences. Report 50. Uppsala. (1994).
[29] V. Kuusemets *et al., Landscape and Urban Planning*. (1996).
[30] Ü. Mander *et al., Landscape and Urban Planning* 31. (1995).
[31] Ü. Mander and T. Mauring, In: L. Ryszkowski and S. Ba≈azy (eds.). *Functional Appraisal of Agricultural Landscape in Europe* (EUROMAB and INTECOL Seminar 1992). Research Centre for Agricultural and Forest Environment, Pol. Acad. Sci. Poznan, 1994. (1995).
[32] Ü. Mander *et al., Proc. Intern. Symp. on Ecological Engineering*, March 29-31, 1995, Östersund, Sweden. (1996).
[33] H. Obarska-Pempkowiak, In P. Aronsson and K. Perttu (eds.) *Willow Vegetation Filters for Municipal Wastewater and Sludges. A Biological Purification System*. Swedish University of Agricultural Sciences. Report 50. Uppsala. (1994).
[34] G. Pinay *et al.,.J Applied Ecology* 30. (1993).
[35] S. Struwe and A. Kjøller, *Plant and Soil* 128. (1990).
[36] S. Struwe and A. Kjøller, *For. Ecol. Manage.* 44. (1991).
[37] M.M. Umarov, *Pochvovedenije* 11. (1976).
[38] L.B.-M.Vought *et al., Ambio* 23. (1994).
[39] M.R. Walbridge and J.P. Struthers, *Wetlands* 13. (1993).
[40] D.E. Weller *et al.*, In W.Mitsch (ed.). *Global Wetlands:Old World and New*. Elsevier Science B.V. (1994).
[41] T. Yoshinari *et al., Soil Biol. Biochem.* 9. (1977).

Environmental Research Forum Vols. 5-6 (1996) pp. 197-202
© *1996 Transtec Publications, Switzerland*

Nutrient Reuse for Biomass Energy Production

T.S. Karlberg[1], I. Carlsson[2] and U. Geber[3]

[1] VBB Viak AB, P.O. Box 34044, S-100 26 Stockholm, Sweden

[2] Surahammar KommunalTeknik AB, P.O. Box 10, S-735 21, Surahammar, Sweden

[3] Swedish University of Agricultural Sciences, Department of Crop Production Science,
P.O. Box 7043, S-750 07 Uppsala, Sweden

Keywords: Biogas, Canary Reed-Grass, Co-Digestion, Energy Crops, Irrigation, Nutrient Reuse, *Salix*, Sludge, Wastewater Treatment, Willow Plantation

Abstract

Wastewater is treated in mechanical, biological and chemical steps at Haga central WWTP in the municipality of Surahammar, Sweden. Surahammar Kommunalteknik AB is a publicly owned company, responsible for water and wastewater services as well as for distant heating and waste management in the municipality. In co-operation with VBB Viak consultants and the Department of Crop Production Science at the Swedish University of Agricultural Sciences Surahammar Kommunal-teknik AB is looking at the possibilities of innovative solutions for nutrient reuse. The conditions for a systems approach on energy and materials are excellent.

- Four species of grass was irrigated with supernatant from the sludge digester blended with treated wastewater on a pilot scale 1994-1995. Test runs with co-digestion of ensiled grass and sludge was started during winter and spring 1995 but had to be interrupted due to operational problems. Test runs will be continued during autumn 1995. Starting in summer 1996, wastewater will be used to irrigate 20 hectares (200 000 m^2) of land close to the Wastewater Treatment Plant (WWTP). The grass will be ensiled at the treatment plant and co-digested with sludge to increase biogas production. Gas production is expected to increase by approximately 60 percent. Biogas is used for electricity and heat production for internal purposes.
- Biological phosphorus reduction will be tried in the WWTP to limit the use of chemicals for precipitation.
- Digested sludge will be used to fertilise willow plantations (*Salix*) on nearby fields. Harvested *Salix* will be used as fuel in the local heating plant. Depending on the quality of the sludge it may also be used for fertilising of land for food production. Further ashes from the heating plant should in the future be used for fertilising purposes.

Introduction

It is a matter of dispute whether or not modern wastewater treatment plants fulfil the criteria for a sustainable management of nutrients. Nitrogen is either passing through the plants converted from ammoniacal to nitrate nitrogen or converted to gaseous nitrogen in a biological nitrification/ denitrification process. Phosphorus is precipitated to sludge which is of varying quality. It is subject to an intensive debate whether wastewater sludge is or can in the future be suitable for agricultural purposes.

The challenge of nutrient reuse may be met through new techniques, such as urine separation and composting toilets, or through new approaches in the existing systems for wastewater management. Minimising energy consumption, optimising nutrient reuse and minimising the use and spread of hazardous substances are objectives to be met whichever system is chosen.

Household wastewater in Sweden of today contain phosphorus and nitrogen in almost relevant proportions for fertilising purposes. Phosphorus is however present in excess due to the phosphorus content in detergents. Generally speaking the wastewater receive 2 g of phosphorus and 13 g of nitrogen from each person each day. This is enough to fertilise a cultivated area of approximately 250 - 350 m^2. A community with 10 000 inhabitants thus would need access to between 250 and 350 hectares to reuse all nutrients in the wastewater.

Background
Surahammar municipality, with its 11 400 inhabitants, is situated some 50 km north of lake Mälaren and 150 km west of Stockholm. The area is dominated by forests and is only sparsely cultivated. There are roughly 2 800 hectares of agricultural land for food production in the municipality of which some 150 hectares is being converted for other types of production [1].

Surahammar Kommunalteknik AB is a publicly owned company, responsible for Haga wastewater treatment plant. The company is not only responsible for water and wastewater services in the municipality, but also for distant heating and waste management. The conditions for an integrated approach to water, waste and energy is therefore excellent.

Haga wastewater treatment plant receives wastewater from Surahammar and Ramnäs communities, the organic load amounting to 9500 p.e. It was designed to treat wastewater from 12000 p.e. and is hence somewhat oversized. The annual load of nitrogen and phosphorus is 48 tonnes and 8,9 tonnes respectively. The water is treated mechanically, biologically and chemically to a maximum discharge level of 15 mg BOD$_7$/l and 0,5 mg Phosphorus/l. [2]

The wastewater sludge is digested in a 400 m^3 continuos-flow stirred-tank high-rate digester. An additional 400 m^3 digester is today only used for sludge storage. After digestion the sludge is dried on sludge-drying beds. The biogas is utilised in a gas motor, producing electricity (40 kW) and heat for internal purposes. Sludge quality is good and the metal content in the sludge is below Swedish limits for land application. The wastewater treatment plant produces approximately 260 tonnes of sludge (dry matter) and 100 000 m^3 of biogas (60 000 m^3 of methane) annually.

Surahammar Kommunalteknik AB strives to adapt and change its activities to a sustainable water, wastewater and waste management and heat production. The company will contiously minimise the environmental impacts from its acvtivities and minimise energy consumption. In order to achieve this, wastewater nutrients should be made available for energy crop production and energy consumption in the wastewater treatment process should be minimised. For a long term sustainable system it is necessary that heavy metals and other hazardous compounds are not allowed to accumulate in soil or groundwater. The input of such compounds must therefore be limited. Switching from chemical precipitation to biological removal of phosphorus from wastewater will eliminate one heavy metal source. Further, limiting the environmental impact from wastewater must not be a question only about wastewater purification but also a question of the function of the sewer net and the behaviour of the citizens.

Sustainability is also a matter of an effective circulation of nutrients and other essential compunds used in production. Nutrients in wastewater must be brought back to the food producers with as small losses as possible. The Haga Project is aiming at a local system for nutrient reuse, as illustrated in Fig.1.

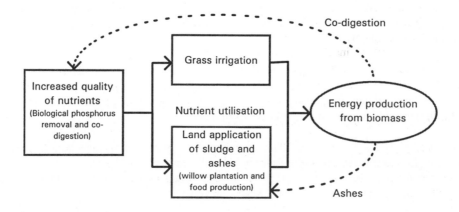

Figure 1   *The Haga Project - means and objectives*

The Haga Project
The Haga Project is focusing on energy crop production through wastewater nutrient reuse. The energy crop will be produced by means of (i) irrigation of grass with nitrogen-rich wastewater and co-digestion of grass and wastewater sludge and (ii) land application of sludge for willow production.

The sludge digesters are currently being modified to allow both parallel and sequenced operation so as to facilitate a high organic load. Further, the treatment plant is being modified for maximum flexibility and biological phosphorus removal will be tried on a full scale basis to find out whether this is a feasible way to limit the use of chemicals.

Irrigation of grass with nitrogen-rich wastewater
There are approximately 20 hectares of agricultural land available in the vicinity of the treatment plant. Surahammar Kommunalteknik AB has access to a further 150 hectares on a distance of no more than 5 km from the WWTP.

On the basis of preliminary results from a research project described further below, the 20 hectares close to the treatment plant will be irrigated with treated wastewater and supernatant from the sludge digester. The objective is to obtain a sufficient nutrient load (100-150 kg N/hectare) and a high hydraulic load (approximately 300-600 mm) so as to facilitate both nutrient uptake by vegetation and microbiological nitrification/denitrification. Since irrigation water will have a low phosphorus content, sludge may have to be applied regularly to guarantee the supply of phosphorus to the crop. The hygienic aspects have been considered important for the long-term sustainability of the project. For this reason untreated wastewater will not be used for irrigation.

The soil in the area has a high content of clay and has therefore large capacity of absorbing nutrients to soil particles. However, the impact of irrigation on soil and groundwater quality will be continuously monitored so that mitigating measures or change of irrigation regimes etc. can be undertaken should any adverse effects occur.

Irrigation will be done by means of a surface application system. It is of great importance, especially if in the future preliminary treated wastewater is to be used, to prevent the production of aerosols and also to prevent unwanted stripping of ammonia to the air.

Grass will be harvested two times a year at an estimated level of 8 tonnes of dry matter per hectare. It will be ensiled at the wastewater treatment plant and co-fermented with wastewater sludge in the digester to increase biogas production.

Digestion of grass gives a fairly high yield of biogas but operation is difficult due to the tendency of grass to form floating grass blankets in digesters and thickeners. Initial test runs described below indicate that the grass must be ground to a slurry with grass pieces of millimetre size before being fed into the digester and that the mixing will have to be intense.

Approximately 2 500 kg nitrogen and 100 kg phosphorus will be applied to the plots each year at a full scale basis. In return grass containing 160 tonnes of dry matter will be harvested, ensiled and co-fermented in the sludge digester. It is anticipated that the ensiled grass will contribute to the production of 40 000 m^3 of methane, representing 400 MWh annually [3].

Land application of sludge for willow production
The 150 hectares of available land on a greater distance from the wastewater treatment plant will be used for the willow plantations. *Salix* will be planted successively and harvested every four or five year, fertilised with sludge from the treatment plant. The impact of sludge application on soil and groundwater quality will be monitored. The chopped *Salix* can be incinerated in the heating plant of Surahammar Kommunalteknik AB.

Salix does not require high loads of fertilisers since the falling leaves are returning nutrients to the soil. A load of 10 kg phosphorus per hectare and year is estimated to be sufficient. Fertilisers will be added by application of sludge after each harvest. Fifty tonnes of sludge (dry matter) will be needed each year for the fertilising of 150 hectares of willow plantations. Ashes from the heating plant should also be used for fertilising purposes, supplying the plantation with additional phosphorus, potassium and other nutrients.

The average growth is estimated at 10-12 tonnes of dry weight per hectare and year representing an average energy production of 7 500 MWh heat.

Biological phosphorus removal
Biological phosphorus removal (Bio-P) is a method of activated sludge treatment which makes use of the ability of micro-organisms to "load" extra reserves of phosphorus under situations of stress. Since Haga wastewater treatment plant has more space than needed for the conventional process, it is possible to switch to biological phosphorus removal by means of minor changes within the existing volumes. It is anticipated that at least 50 percent of the precipitation agent used today could be excluded.

The biological treatment step of Haga WWTP is currently being modified to facilitate test runs on a full scale basis in one of two fully separate, biological treatment lines. Besides of the annual phosphorus removal capacity the test period will show:

-    The impact of low wastewater temperature on phosphorus removal efficiency
-    The impact of low-strength wastewater on phosphorus removal efficiency (during periods of snow melting)
-    Change in sludge and digester supernatant quality through biological phosphorus removal

Looking at the metal content in Haga sludge and in the precipitation agent used, it is clear that a switch to biological phosphorus removal does have a positive effect on the content of some metals, such as chromium and nickel but hardly no effect on lead and cadmium. Lead and cadmium being two of the most problematic substances we handle must thus be tackled in some other manner than by changing process in the wastewater treatment plant.

Early results

Grass irrigation

An area in the vicinity of Haga WWTP is currently used for a research co-operation project between Surahammar Kummunalteknik and the Department of Crop Production Science at the Swedish University of Agricultural Sciences with support from a research foundation associated with the Federation of Swedish Farmers. Four different grass species, reed canary grass (*Phalaris arundinacea*), Smooth brome grass (*Bromus inermis*), Meadow foxtail (*Alopecurus pratensis*) and Reedgrass (*Glyceria maxima*) were during summer 1994 and 1995 irrigated with a mixture of treated wastewater and supernatant from the sludge digester at different hydraulic (300 and 600 mm) and nutrient (150 and 300 kg N/hectares) loads. The grass was cut twice every season and the yield amounted in 1994 to between 7 000 and 10 000 kg dry matter per hectare. Reed canary grass gave the highest yield. Results from 1994 indicate that almost 100 percent of the nitrogen content in the wastewater is eliminated of which 70 percent can be removed with the harvested grass. [4]

The project will be reported after the 1996 irrigation season.

Co-digestion of sludge and ensiled grass

Before initiating co-fermentation on full scale basis (Sludge to grass in proportions 1,4:1) test runs will be performed. A programme was set up for the winter and spring 1994/95 to test the sludge treatment system for co-digestion of sludge and ensiled grass in the proportions 5:1.

Gas production will be followed and the general function of pre-mixing of sludge and ensiled grass, digester mixing and sludge preheating and the characteristics of ensiled grass, sludge, supernatant and gas will be monitored.

Operation troubles with the gas motor delayed the test runs but in early spring 1995 the tests could start. Sludge and ensiled timothy grass were co-digested in the proportions of 5:1. However the existing mixer showed to be unsuitable for the purpose and the tests had to be interrupted. The test runs will be continued when the mixing system has been modified.

Discussion

The Haga Project will illustrate the potential of nutrient reuse through two different types of cultivation projects. Eventually not only the level of energy utilisation but also environmental impacts, technical restrictions and economical realities will set the limits for the applicability of the project in Surahammar as well as in other municipalities in Sweden and elsewhere.

Between 250 and 350 hectares of arable land is needed to utilise all nutrients in the wastewater from a community by the same size as that of Surahammar. This far 170 hectares are available for energy crop production. Table 1 below indicates that treated wastewater may only be used in the vicinity of the treatment plant whilst sludge and urine from urine separation toilets, if at hand, is fairly economical to transport.

Table 1   *Nutrient content in different types of wastewater products (kg/m³)*

	Phosphorus	Nitrogen
Sludge, 20% TS	5	6
Urine and flush water from separation toilets	0,5	5
Digester supernatant	0,02	0,5
Treated wastewater	0,0005	0,02

Co-digestion of sludge with other easily degraded materials is an excellent way of stretching the capacity of an existing treatment facility. A large number of sludge digesters in Sweden are oversized due to the actual and future wastewater load and could be used for co-digestion of grass. On the countryside and in smaller communities and towns the combination of an oversized digester and available arable land should not be too unusual. Provided that the operational problems of co-digestion of sludge and grass can be solved, the technique is indeed promising.

Sludge quality is expected to improve with both co-digestion and biological phosphorus removal. Moreover the phosphorus emanating from grass can be expected to have a higher accessibility than the precipitated phosphorus. With an improved sludge quality, sludge could possibly be used for food production crops.

Willow plantations slowly gains popularity in Sweden. As long as there are any doubts about in which way sludge can influence the quality of crops for food production, fertilising of energy forest should be an excellent alternative for nutrient reuse. A close co-operation with local farmers is necessary.

The Haga Project will lead to increased costs for treatment work operation. It can and should be discussed whether the costs, which are indeed costs for an improved environment, should be born by the water customers, by the farmers or over governmental taxes e.g. on chemical fertilisers. The economical conditions for similar projects can however be improved by national and international (i.e. EU) decisions in the future.

The direct environmental impacts of the different parts of the project will be followed through monitoring and calculations. Except for the effects that can be easily measured and monitored there are others that are not so easy to measure but still important. The production of energy from biomass has generally no "greenhouse effect", cultivation of grass keeps the landscape open and irrigation and land application limits the need for chemical fertilising agents. On the other hand a long-term groundwater pollution or accumulation of heavy metals in soil can be difficult to measure but essential in a very long time perspective.

Through the above projects over 7 000 MWh from grass biogas and *Salix* may be produced annually. It is equivalent to 700 m³ of oil, enough to provide 280 Swedish private houses with heating and hot water.

References
[1] H. Engberg, Anaerobic digestion of sludge, solid waste and energy crops. 1994.
[2] Annual environment report 1994 for Haga wastewater treatment plant. Surahammar 1995.
[3] L. Brolin & L. Thyselius. Biogas ur energigrödor. JTI-rapport 97. Uppsala 1988.
[4] U. Geber & M. Tuvesson. Vallgräs för avloppsvattenrening och biogasproduktion. Konferensrapport 1995. Lantbrukskonferensen 23-24 januari 1995. SLU info. Rapport 187. Sveriges lantbruksuniversitet.

Environmental Research Forum Vols. 5-6 (1996) pp. 203-208
© *1996 Transtec Publications, Switzerland*

Ecological Engineering Treatment for Municipal Wastewater – Studies on a Pilot-Scale Slow-Rate Land Treatment System in Shenyang, China

Z.Q. Ou, T.H. Sun, P.J. Li and S.J. Chang

Institute of Applied Ecology, Academia Sinica, P.O. Box 417, Shenyang 110015, P.R. China

Keywords: Ecological Engineering, Land Treatment, Water Pollution Control, Wastewater Treatment

ABSTRACT

In this paper, the first pilot-scale ecological engineering slow-rate land treatment system (SR-LTS) in China was introduced with emphasis on its performance. The system occupying an area of 8.3 ha was built in 1986 for experimental purposes and had successfully operated for 6 years. Process parameters of the system are reported. Experimental data showed that the performance of the system was excellent. For the SR-LTS, concentrations of BOD_5, COD, total phosphorus and total nitrogen in the effluent were less than 5, 36, 0.2 and 5 mg/l and their removal rates were greater then 96, 84, 90 and 77%, respectively. For the rapid infiltration subsystem operating in winter time, effluent levels of BOD_5, COD, total phosphorus and total nitrogen were less than 10, 70, 0.3 and 20 mg/l with removal rates of greater than 92, 79, 86 and 50%, respectively. With the design process parameters and system configuration, the system could operate throughout the year with a constant wastewater treatment flow and without storage.

INTRODUCTION

Slow-rate land treatment is one kind of natural/ecological innovation/alternative techniques for the treatment and utilization of wastewater[1]. The slow-rate land treatment system (SR-LTS) reported in this paper was the first pilot-scale SR-LTS that employed paddy rice as its main vegetation. It was built in 1986 in the western suburb of Shenyang, the capital of the Liaoning province in Northeast China, where complicate wastewater containing various pollutants was available for experiments. It started to treat municipal wastewater from west Shenyang in 1987 and had successfully operated for 6 years. This experimental system occupying an area of 8.3 ha was required to treat a minimum daily flow of 600 tons of municipal wastewater around the year and effluent from the system was demanded to meet the effluent quality of conventional biological treatment (BOD ≤ 30mg/L, COD ≤ 60mg/L and suspended solids (SS) ≤ 30mg/L)[2,3].

DESCRIPTION OF THE SHENYANG SR-LTS

The climate in Shenyang is characterized by dry cold winter with a minimum monthly mean temperature in the year of -12.2°C in January and a minimum monthly mean precipitation in the year of 3 mm in December, and humid and hot summer (maximum monthly mean temperature 23.8°C in July, maximum monthly mean precipitation 178 mm in August)[2]. SR-LTSs can not operate in such cold winter. For the purposes of treatment of wastewater around the year, the Shenyang SR-LTS was, therefore, constructed into two subsystems as shown in Fig. 1: (1) a warm season SR-LTS (WS-SR-LTS) which covered an area of 7.3 ha and operated from late May to early November, and (2) a winter rapid-infiltration land treatment subsystem (W-RI-LTS) which occupied a total area of 1 ha, had no vegetation and operated from late November to next early May[2,3].

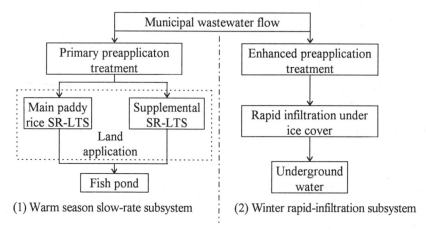

Fig. 1. Schematic of the Shenyang LTS

The WS-SR-LTS consisted of two parts: (1) a main SR-LTS with paddy rice as its vegetation, and (2) a supplemental SR-LTS with vegetation of willow (*Salix babylonica* L.), spiraea (*Spiraea salicifolia* L.), amorpha (*Amorpha frulicosa* L.) and sorghum for comparative studies[4]. This configuration was designed for the WS-SR-LTS to operate continuously in the period of late May to early November because there were about 25 non-operation days (10 days in July for the field to dry out to prevent tillering and 15 days in September for the rice to ripen) for the main SR-LTS during which rice can not be watered.

A systematic process design procedure was established for the design of the Shenyang LTS which can be found in References [2] and [4] together with detail design considerations. The key process parameters of the system are summarized below:

(1) The main SR-LTS:
    -Total area: 6.8 ha   -Field area: 5 ha
    -Pretreatment: primary treatment (a sedimentation pond followed by a stabilization pond)
           Sedimentation pond: 378 m³, Retention time (RT) of 1-4 hr
           Stabilization pond: 1200 m³, RT of *ca* 2 days
    -Operation days: 119 days (10 May to 15 Sept.) in rice growth season
                  56 days (10 Oct. to 25 Nov.) in autumn
    -Non-operation days: 25 days (11 to 20 July and 16 to 30 Sept.)
    -Hydraulic loading rates: 200.8 cm for rice growth season and 60.8 cm for autumn
    -Design depth of ground water in operation period: 0.3-0.5 m
    -Application cycle: 3 days (1 day application and 2 days of drying out)

(2) The supplemental SR-LTS:
    -Field area: 0.5 ha (10% of that of the main SR-LTS)
    -Pretreatment: as the same as the main SR-LTS
    -Operation days: 25 days in non-operation periods of the main SR-LTS and 56 days in autumn
    -Hydraulic loading rates: 200.8 cm in non-operation days of the main SR-LTS
                  60.8 cm in autumn
(3) Winter RI-LTS:
    -Total area: 1 ha   -Field area: 0.6 ha   -Initial soil permeability: 0.6-1.2 m/d
    -Pretreatment: Enhanced sedimentation with coagulants
    -Operation days: 165 days (26 Nov. to 9 May next year)
    -Hydraulic loading rate: 1650 cm
    -Design depth of ground water: ≥3 m
    -Application cycle: 3 days (1 day application and 2 days' rest)

RESULTS OF PERFORMANCE

Warm season SR-LTS

The characteristics of influent to the Shenyang SR-LTS are shown in Table 1. The performance of the preapplication treatment system is presented in Table 2. Average removal rates of BOD$_5$, COD, TOC and SS were 32, 16, 17 and 32%, respectively, which met the design values. Total nitrogen concentration did not change though alteration of its forms occurred[3].

Table 1. Characteristics of influent to the Shenyang SR-LTS (yearly means)

Year	pH (pH unit)	DO (mg/l)	EC (μmhos /cm)	BOD$_5$ (mg/l)	COD (mg/l)	TOC (mg/l)	TN (mg/l)	TP (mg/l)	SS (mg/l)	F. coli-form (no. /100 ml)	Cd (μg/l)
1987	6.9	0	913	110.0	281.6	80.4	19.0	2.3	127	NM	2.3
1988	6.1	NM	NM	96.8	181.3	54.2	18.9	1.8	54	3.48×10^9	5.4
1989	6.7	0.46	1024	95.6	234.2	69.5	24.0	1.6	67	NM	5.4
1990	6.7	NM	NM	96.7	216.8	NM	NM	2.5	121	NM	NM
1991	6.8	0	NM	92.0	226.2	57.6	16.6	3.1	70	NM	NM
1992	6.9	0	NM	166.2	289.2	77.0	18.1	1.2	NM	NM	NM
Means	6.7	0.12	969	109.6	238.2	67.7	19.3	2.1	88	3.48×10^9	4.4

NM = not measured; TN = total nitrogen; TP = total phosphorus.

Table 2. Characteristics of effluent from preapplication treatment (yearly means)

Year	pH (pH unit)	DO (mg/l)	EC (μmhos /cm)	BOD$_5$ (mg/l)	COD (mg/l)	TOC (mg/l)	TN (mg/l)	TP (mg/l)	SS (mg/l)	F. coli-form (no. /100 ml)	Cd (μg/l)
1987	7.0	0	916	59.1	167.3	55.7	18.9	1.8	91	NM	0.7
1988	6.4	NM	NM	67.5	141.3	42.2	17.2	1.6	45	3.68×10^8	4.7
1989	6.7	NM	1070	78.5	225.3	63.0	28.0	1.6	44	NM	4.5
1990	6.8	NM	NM	89.0	221.5	NM	18.4	2.3	95	NM	NM
1991	7.0	0	NM	77.8	203.3	54.8	19.3	2.8	23	NM	NM
1992	6.8	0	NM	67.4	243.0	63.4	15.1	1.8	NM	NM	NM
Means	6.8	0	993	73.2	200.3	55.8	19.5	2.0	60	3.68×10^8	3.3
Average removal rate (%)			-2.5	33.2	15.9	17.4	0	4.7	32	89.4	25.0

Experimental data (Table 3) showed the truly excellent performance of the main paddy rice SR-LTS in removal of conventional pollutants. BOD$_5$ was greatly reduced to 4.1 mg/l and removal rates were as high as 96%. COD removal rates were greater than 84% and concentrations declined from 200.3 to 35.7 mg/l in average. Both BOD$_5$ and COD concentrations in the effluent were much lower than those required. A characteristic of the SR-LTS effluent was low nutrient concentration. Total phosphorus (TP) contents were as low as 0.2 mg/l with removal rates up to 90%. Total nitrogen (TN) contents were 4.3 mg/l and much lower than the design value of C$_p$ = 10 mg/l, with removal rates of > 77%. Suspended solid concentrations were also lower than that required.

The ratio of BOD$_5$ to COD is an important parameter showing the biological treatability of wastewater and the purifying ability of a wastewater treatment process. Wastewater with a BOD$_5$/COD ratio of above 0.3 is usually considered to be readily biologically treatable while those of less than 0.2 are believed

Table 3. Characteristics of effluent from the main paddy rice SR-LTS (yearly means)

Year	pH (pH unit)	DO (mg/l)	EC (mmhos /cm)	BOD$_5$ (mg/l)	COD (mg/l)	TOC (mg/l)	TN (mg/l)	TP (mg/l)	SS (mg/l)	F. coliform (no. /100 ml)	Cd (mg/l)
1987	6.9	2.69	910	5.9	44.8	20.4	7.5	0.3	NM	NM	0.6
1988	6.6	5.68	-	2.0	18.5	7.4	3.6	0.1	26	4.41×10^7	2.0
1989	6.6	4.07	892	2.4	26.1	6.7	6.4	0.1	12	6.69×10^7	1.2
1990	6.9	NM	NM	6.8	53.3	-	4.8	0.3	19	NM	NM
1991	6.7	2.61	NM	3.2	36.7	8.9	1.2	0.1	16	NM	NM
1992	6.7	NM	NM	4.5	35.0	13.9	2.4	0.6	22	NM	NM
Means	6.7	4.00	901	4.1	35.7	11.5	4.3	0.2	19	5.70×10^7	1.4
ARR1 (%)			9.3	94.4	81.1	78.5	70.8	88.9	72	84.5	57.6
ARR2 (%)			7.0	96.3	84.4	83.1	77.8	90.5	79	98.4	68.2

ARR1 = average removal rates that only accounted for SR-LTS;
ARR2 = average removal rates that took the preapplication removal into account as well.

unsuitable for biological treatment[5,6]. The BOD$_5$/COD ratio dropped slightly from 0.46 to an average of 0.37 after pretreatment, which implied that only the organic pollutants most susceptible to biological degradation were removed by the preapplication treatment process. After treatment by the SR-LTS, BOD$_5$/COD ratio was significantly reduced to 0.11 which was lower than the level reckoned to be biologically untreatable. This ratio is also lower than the value (0.2) of well working biological treatment plants[7] showing that SR-LTS had a stronger purifying ability than conventional biological treatment process.

Fig. 2 presents two examples showing the high buffering capacities of the SR-LTS against the fluctuation of pollutant loading. Although the concentrations of BOD$_5$ and COD varied significantly in the influent from 13.9 to 183.9 mg/l and from 45.1 to 296.4 mg/l, the BOD$_5$ and COD levels in the effluent were very stable and below 4 (most in 1 - 2) mg/l and 30 (most in 9 - 20) mg/l, respectively.

The performance of supplemental subsystems was also excellent in removal of conventional pollutants (Table 4) with the same magnitude in respects of effluent levels and removal efficiency as the main system. Results showed that it was feasible for willow, amorpha and sorghum to operate in high hydraulic loading for short time as regulating measures.

Table 5 compares the effluent quality of the system with the *Environmental Quality Standard for Surface Water*[8]. 7 of the 18 indices (NO$_3^-$-N, NO$_2^-$-N, pH, cyanides, phenol, cadmium and arsenic) met the first class standard, 4 (chromium, lead, copper and zinc) met the second class standard, 4 (BOD$_5$, total phosphorus, dissolved oxygen and mercury) the fourth class, only 2 (COD$_{Cr}$ and K-N) were beyond the fifth class and 1 (B(a)P) was the third class. Except for mercury, all other trace elements met Class I or Class II standards as well as the organic chemicals compared[3].

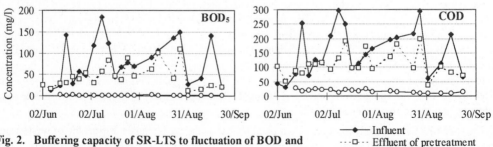

Fig. 2. Buffering capacity of SR-LTS to fluctuation of BOD and COD loadings (1988)

Table 4. Operation results of supplemental subsystems (yearly means of 1987 to 1989)[3]

Items	Influent (mg/l)	Willow SR-LTS Effluent (mg/l)	Willow SR-LTS Removal rate (%)	Amorpha SR-LTS Effluent (mg/l)	Amorpha SR-LTS Removal rate (%)	Sorghum SR-LTS Effluent (mg/l)	Sorghum SR-LTS Removal rate (%)
BOD₅	70.5	4.1	94.2	7.0	90.1	4.6	93.5
COD	178.0	35.0	80.6	36.5	79.7	30.9	82.6
TN	21.3	7.7	63.8	5.8	72.8	5.0	76.5
TP	1.7	0.2	88.2	0.3	82.4	0.4	76.5
SS	60	73	-21.7	36	40.0	85	-41.7

Table 5. Effluent quality of Shenyang SR-LTS

Items	Content (mg/l)	Class met	Environmental Quality Standard for Surface Water[8] Classification	Environmental Quality Standard for Surface Water[8] Suitable usage
pH	6.7	I	Class I-IV: 6.5-8.5	Class I: water catchment areas and
DO	4.0	IV	Class IV: ≥3%	state nature reserve areas.
BOD₅	4.1	IV	Class IV: ≤6mg/l	Class II: primary protection areas for
COD$_{Cr}$	35.7	>V	Class V: ≤25mg/l	centralized water sources of potable
NO₃⁻-N	1.3	I	Class I: ≤10mg/l	water supplies; protected areas for
NO₂⁻-N	0.06	I	Class I: ≤0.06mg/l	rare fish; fish/shrimp spawning sites.
K-N	2.6	>V	Class IV-V: ≤2mg/l	Class III: second class protection areas
TP	0.2	IV	Class IV-V: ≤0.2mg/l	for centralized water sources of po-
Phenol	ND	I	Class I: ≤0.002mg/l	table water supplies; protected areas
BaP	0.00004	>III	Class III: ≤0.0025μg/l	for common fish; protected areas for
Cyanide	0.005	I	Class I: ≤0.005mg/l	swimming.
Cd	0.001	I	Class I: ≤0.001mg/l	Class IV: areas for common industrial
Cr	0.020	II	Class II: ≤0.05mg/l	water supplies; and amenity water
Hg	0.0002	IV	Class IV: ≤0.001mg/l	supplies that the human body does
Pb	0.014	II	Class II-IV: ≤0.05mg/l	not directly come into contact with.
As	0.009	I	Class I-III: ≤0.05mg/l	Class V: water for agricultural irriga-
Cu	0.02	II	Class II-V: ≤1.0mg/l	tion and for general landscape re-
Zn	0.08	II	Class II-III: ≤1.0mg/l	quirements.

K-N = Nitrogen measured by the Kjeldahl method; TP = total phosphorus.

Winter RI-LTS

Depth of soil or underground water table, soil initial permeability (Ki) and hydraulic loading rates had significant effects on the effluent quality and removal efficiency of W-RI-LTS. A minimum depth of soil or underground water table of 2 m was found very necessary for effluent quality to meet requirements (Fig. 3(a)). The effects of Ki were not only on the effluent quality but also on the operation of the system. When Ki was higher than 1.2 m/d, BOD levels in effluent were over the requirement (30 mg/l) under design operation conditions (Fig. 3(b)). When Ki was lower than 0.6 m/d, ice cover could not build up and soil would be frozen after operation for 2 months. Feasible hydraulic loading rates under Shenyang conditions were in the range of 0.1 to 0.2 m/d. Below this range, the field area and costs of both construction and operation would increase while above the range, effluent quality would deteriorate (Fig. 3(c)). Application cycle was also important and an application cycle of 3 to 5 days was found appropriate depending on the temperate of air and influent. Under design process parameters, operation results (Table 6) showed that the effluent levels of BOD₅, nitrogen and total phosphorus were all blow the design values and met requirements though COD was a little bit higher.

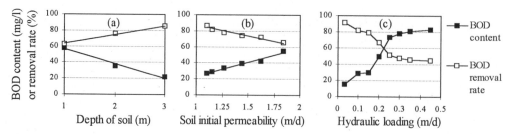

Fig. 3. Effects of soil depth and initial permeability and wastewater hydraulic loading on the performance of winter RI-LTS

Table 6. Operation results of winter RI-LTS (yearly means of 1989 to 1990)

Year	BOD$_5$ Infl. (mg/l)	Effl. (mg/l)	RR (%)	COD Infl. (mg/l)	Effl. (mg/l)	RR (%)	Total nitrogen Infl. (mg/l)	Effl. (mg/l)	RR (%)	Total phosphorus Infl. (mg/l)	Effl. (mg/l)	RR (%)
1988	118.3	8.4	92.9	277.3	47.7	82.8	27.8	7.5	73.0	1.66	0.13	92.2
1989	88.0	8.8	90.0	303.3	82.2	72.9	37.9	25.1	33.9	2.23	0.48	78.5
1990	155.4	10.3	93.4	441.9	71.6	83.8	27.8	15.2	45.3	1.96	0.20	89.8
Means	120.6	9.2	92.1	340.8	67.2	79.8	31.2	15.9	50.7	1.95	0.27	86.8

Infl. = influent, Effl. = effluent.

CONCLUSIONS

Slow-rate land treatment systems (SR-LTSs) were effective methods for wastewater treatment and utilization. The Shenyang system, with a main paddy rice SR-LTS and a supplemental SR-LTS with vegetation other than paddy rice and a field area of 10% of that of the main one, could operate through the whole growth period without storage. Under the design operation conditions, the system could produce high quality effluent. Plus a winter rapid infiltration land treatment system (RI-LTS), the whole system could operate continuously for a constant treatment flow of wastewater around the year with no storage. RI-LTS under ice cover was feasible in cold winter.

REFERENCES

[1] Z. M. Gao and X. F. Li (eds.), *Design Manual for Land Treatment and Utilization Systems for Municipal Wastewater*, Chinese Standards Press, Beijing, pp 1-458(1991).
[2] Z. Q. Ou, Z. M. Gao and T. H. Sun, Wat. Res., **26**(11), 1479-1486(1992).
[3] Z. Q. Ou, S. J. Chang, Z. M. Gao, T. H. Sun, E. S. Qi, X. J. Ma, P. J. Li, L. S. Zhang, G. F. Yang, S. H. Han, Y. F. Song, H. R. Zhang, L. P. Ren, and X. R. Qu, *Wat. Res.*, **26**(11), 1487-1494(1992).
[4] Z. Q. Ou, Z. M. Gao, T. H. Sun and S. J. Chang, *Wat. Treatment*, (6), 489-507(1991).
[5] Z. M. Gao, E. S. Qi, F. J. Zhang, P. J. Li, Z. S. Zhao, T. S. Dai, S. H. Han and X. R. Qu, *Studies on Pollution Ecology of Soil-plant Systems*, Z. M. Gao (ed.), China Science and Technology Press, Beijing, pp124-141(1986).
[6] E. S. Qi, Z. M. Gao, P. J. Li, Z. S. Zhao and L. P. Ren, *Studies on Pollution Ecology of Soil-plant Systems*, Z. M Gao (ed.), China Science and Technology Press, Beijing, pp142-153(1986).
[7] E. Hofssmann, 1995, Personal communication.
[8] NEPA of China, Environ. Qual. Stand. for Surface Water, GB3838-88, 1988-0405(1988).

Environmental Research Forum Vols. 5-6 (1996) pp. 209-214
© 1996 Transtec Publications, Switzerland

Project Autonomous Treatment of Sewages from Siberian Ecological House

B.D. Zhukov

Novosibirsk State Technical University, Marx prospect 20, 630092 Novosibirsk, Russia

Keywords: Autonomous Wastewater Processing, Individual House, Siberia, Soil-Plant Filter, Heat Accumulator, Physicochemical Sensors, Automatization, Bioreactor

ABSTRACT

The principles and practical realization of autonomous treatment and utilization of wastewaters from individual house in Siberia are discussed in the present report. The technology is based on the separate processing of the wastes of distinct origin. Automated purifying complex is proposed comprising soil-plant filter and anaerobic bioreactor placed nearby underground heat accumulator.

INTRODUCTION

Ecodoms (i.e., ecological houses) seem to be very promising for individual house building in Russia as they produce minimal negative effect on environment, possess enhanced public health standards and comfort level [1].

However, the lack of ecologically sound high performance autonomous systems for residential wastes treatment adapted to local conditions impedes the active building of ecodoms. The process of wastewater treatment at individual house should necessarily combine ecological and economic efficiency together with high reliability and productivity of facilities.

Decentralized wastewater purification, utilization of obtained water and useful recovered components provide substantial economic effect. It is profitable to process wastewaters on spot, on the one hand. But on the other hand, repeated processing of water in limited space can cause accumulation and formation of new toxic compounds in reprocessed water. That is, the sanitary-ecological requirements for the performance of water processing on spot should be stronger than usually.

Conventional technologies of municipal wastes treatment formally applied for the autonomous facilities of individual houses produce low effect as they intended for the processing of a huge amount of wastes at remote sites.

Operation conditions of autonomous processing are quite different. Uncontrolable mixing of the wastes of various origination is the most undesirable at operation of small local wastewater purifying systems. Such mixing violates wastes homogeneity and thus complicates biological treatment, content monitoring, automated systems operation, makes it difficult to recover and utilize useful admixtures from disposals.

Reference literature covers numerous approaches for the creation of autonomous water processing systems useful for individual houses. But only few authors consider the operation of purifying installations at cold climate like Siberian [2,3]. However, continental Siberian climate is responsible for not only long cold winter and short hot summer, but for sharp temperature gradient passing through zero point in spring and autumn as well.

Therefore, we try in this work to substantiate the main principles of the processing and utilization of wastewater from ecodom in the climate of Siberia and to consider the designs of ecologically sound, cheap (ca. 7-9% of the house total cost), highly efficient, adapted to the continental climate of Siberia technology of wastewaters treatment and utilization based on the separate processing of the wastes of different types.

The following principles form the basis of the technology:

1. Separate processing of the wastes of different origination.

2. Efficient water consumption provided by the high level of reuse of purified water and its useful components.

3. Combination of wastewater purification and utilization in the frameworks of unified technology.

4. Enhanced moisture capacity of the water purifying system as a whole.

5. Short period of passive retaining of water at ecodom allotment, composting of solid organic disposals at a biotoilet.

6. Processes automatization.

7. Close comformity of water purifying performance with the Public Health Standards.

WASTEWATERS SEPARATE PROCESSING

Basing on reference data, we suppose that purification systems comprising bioreactors and soil-plant filters are the most prespective for exploitation at Ecodom. However, admissible purification from sanitary and ecological viewpoint can be attained only at combination of these techniques with phisicochemical detoxication of water in a single compact design. Separate purification (i.e., without premixing of distinct wastes) of wastewaters followed by their utilization at a garden or greenhouse seems to provide the best possibilities for such combination.

According to this scheme, fecal wastes and hard food garbage are composted in a biotoilet whereas bathroom and kitchen waters are treated separately. In order to decrease the detergent content in processed water (and to save detergent at laundering) we propose to purify concentrated detergent solutions in a separate unit and to use them repeatedly.

At present the various modifications of this technology for practical use are under development. The function and operating principles of single units are well known and described in reference literature. Therefore we omit here their description but present the block-scheme of the whole technology together with our understanding how it can be adapted to cold Siberian winter.

Kitchen wastewaters passed through rough filter (1) or bathroom detergent solutions purified from dirt and diluted in block (2) enter filter-homogenizer (3). Coarse-dispersion fraction is retained, while filtrate with dissolved impurities is diluted by more pure water from anaerobic bioreactor (6), alkilized electrochemically and passed through heat exchanger (4) to soil-plant filter (5). Outgoing from it water is collected in anaerobic bioreactor. After anaerobic bioreactor water enters storage-conditioning unit (7) where it is prepared for crops watering. During conditioning water is detoxicated by UV-irradiation and, if necessary, diluted by pure water or that purified from block (9). A portion of water from anaerobic bioreactor is passed to aerobic bioreactor (8) for deep biological processing.

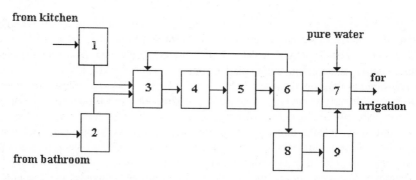

EFFICIENT WATER CONSUMPTION

Efficient water consumption is the basic component of the technology. Its main aims are to exclude the needless water discharge, to provide efficient utilization of wastewater and its admixtures with regard to the water balance of ecodom living.

Table 1 presents the supposed standards of water consumption at individual house.

Ecodom's water consumption is assumed to be ca. 160 L per inhabitant daily regarding the Sanitary Standards and Regulations (160-230 L/inhbt.daily). The consumption of pure water is decreased due to flushing out of a biotoilet with purified water and repeated use of purified detergent solutions. Substantial amount of purified wastewater will be used at greenhouse.

The summer in Siberia is hot and dry usually. Experience shows, that at least 150-200 L daily (per inhbt.) are required to wet 600-800 m^2 of open soil. This predicts the possible deficit of purified water for crops irrigation at Ecodom allotment during summer. Then pure or rain water can be used to compensate the deficit. So, saving of pure water will attain 150 L daily (per inhbt.) in summer.

Table 1. Supposed standards of water consumption at Ecodom

Water function	Water consumption, L/day.inhbt	
	pure	purified
Cooking	30	-
Washing	14	-
Bathing	80	-
Laundry	15-20	5-10
Compartment leaning	10	-
Rinsing:		
biotoilet	-	10
monitoring indicators	6	
Crops watering:		
open land and greenhouse	40	140
greenhouse (autumn-winter)	-	15-20
Evaporation via greenhouse	-	25-30
Evaporation on open air	-	25-30
Saving at the zone of soil-plant filter	-	60-70

*Calculation is made for a greenhouse of 25 m^2 and a family of 5 members.

In winter the purified water circulates mainly at greenhouse, evaporates and saves at reservoir comprising the zone of soil-plant filter.

UTILIZATION OF WATER AND ITS USEFUL COMPONENTS

P.O. Jenssen and A. Vain [4] adduce detailed arguments in favour of recovery and agricultural use of plant residues and organic matters.

We suppose, that the most suitable for ecodom method to utilize organics comprises wastewaters purification and using within the frameworks of unified technology. In this case fecal wastes and hard food garbage are composted in a biotoilet and are used as fertilizers, while dissolved organic substances after water filtering from suspended particles and additional processing are used for irrigation.

It is proposed to obtain and reuse water of two purity grades. Water of the first purity grade is intended for crops watering as it contains an optimum set of natural fertilizers. Water of the second purity grade is obtained from the water of the first type additionally processed in aerobic bioreactor and has quality indexes not exceeding the following presented in the Table 2:

Table 2. Quality indexes for purified water.

Index	COD	BOD	Suspended	NH_4^+	NO_3^-	PO_4^{3-}
below mg/l	35	10	2	0.2	1.2	0.4

The indexes agree with our data obtained at additional treatment of low concentrated waters in aeration tanks with immobilized microflora during 24 hours. □

Microbiological indexes are as follows: total bacteria number per 1 ml is below 100, E-coli index not exceeds 3 at a temperature of 20-25°C; pH is about 7 - 8.

We suppose to provide the strict ecological standards of wastewater processing by means of physicochemical automated control carried out with presetted frequency and automated water conditioning in a filter-homogenizer, water preparing unit for irrigation and at the inlet of bioreactor for additional processing.

ADAPTATION OF THE TECHNOLOGY TO WINTER CONDITIONS

It is impossible, of course, to clean water by plants on open ground in winter. However, our tests on ground temperature regimes have shown that purification and processing can be performed efficiently at underground anaerobic bioreactors compactly placed in the effective zone of ground heat accumulator.

According to the data granted us by Dr. A.V. Chernyshov, in the place of Novosibirsk (55 ° N.Latt.) under winter natural conditions and snow cover the ground at a depth has a temperature of ca. 3°C in the late winter. This temperature raises to 7-8°C by the autumn. Underground heat accumulator maintains substantially higher temperatures at the neighbouring areas (see Table 3) that provides perfect operation of soil filter and underground anaerobic bioreactor during temperatures drops and even at temporary switching of accumulator supply.

Table 3. Ground temperature at various depth and destination from the outer air line of ground heat accumulator in December-January, 1993-1994. Air line temperature is 20-23°C.

Depth, m	15 m, °C		1 m, °C	
	December	January	December	January*
1	5.3	3.3	9	-
2	4.9	4.2	18	6.7
3	6.8	6.7	20.3	9.1
4	-	-	20.6	9.9
5	7.1	6.2	21.4	10.3

* In 10 days after the heat accumulator supply switching of.

Additional heating of soil filter and anaerobic bioreactor can be provided by excess heat of wastewater, solar heaters and energy of small wind turbine supplying heat accumulator.

AUTOMATIZATION SYSTEM

Microprocessor-controlled units of water sorting according to composition are proposed to redistribute water for purification, utilization, conditioning. Software will be developed on the base of the results obtained at laboratory testing unit specially designed for technology adaptation to soil, climate and other local conditions.

It is proposed to monitor and adjust the following integral parameters of water quality: pH, Red-Ox potential, electroconductivity (total mineralisation), light dispersion (turbidity), fluorescence and temperature. The optical and potentiometric sensors of composition form the basis of sorting units. We have designed and constructed the prototypes of fluorescence indicators graduated against the content of surface-active substances and total organics, and light dispersion indicators to detect turbidity. The light indicators are flow cells equipped with optic systems forming light fluxes with preseted parameters at cells input and output. Filamentous light guide materials facilitate the using of sensors in any part of purification systems. Its principle design and description are given in [5].

Series-produced electrodes are proposed for using as potentiometric indicators for the detection of pH and Red-Ox potential. The set of ion-selective electrodes will be used for the content control of water for irrigation.

REFERENCES

1. I.A. Ogorodnikon, A.V. Avrorin, Surviving Together, 13, 59 1995).
2. P.O. Jenssen, T. Krgstad, T. Mohlem, Wastewater treatment by constructed wetland in the Norwegion climate: Pretreatment and optimization design, Proc. Int. Conf., Stedsund Falk Colledge, Sweeden (1991).
3. R. Crosby, A plan for a community greenhouse and waste treatment facility for cold climate regions, ibid.
4. P.O. Jenssen, A. Vain, Ecologically sound wastewater treatment. Concepts and implementations, ibid.
5. A.I. Lapshin, B.D. Zhukov, A simple flow fluorimeter-nephelometer, Proc. Conf. PGMI-92, The problems of metrology of hydrophysics measurements, Moscow, 189 (1992), in Russian

Environmental Research Forum Vols. 5-6 (1996) pp. 215-220
© *1996 Transtec Publications, Switzerland*

Graywater Treatment Options

G. Rasmussen[1], P.D. Jenssen[2] and L. Westlie[1]

[1] JORDFORSK - Centre for Soil and Environmental Research, N-1432 Aas, Norway

[2] Department of Engineering, The Agricultural University of Norway, N-1432 Aas, Norway

Keywords: Graywater, Literature Study, Graywater Composition, Column Study, Treatment Options

ABSTRACT

Graywater composition is examined and nature-based graywater treatment options reviewed. A literature study reveals that more research on graywater composition is necessary as it is influenced by people´s lifestyles. Content of heavy metals, tensides and organic micropollutants should also be investigated. Various treatment options are available, but performance is often poorly documented. Few fundamental data regarding greywater treatment processes and system sizing are available. Disinfection before reuse is usually recommended. Results from a column study using expanded clay biofilters (Leca) and sand, however, show good fecal coliform removal without additional disinfection. A Norwegian filter concept for cottages and single dwellings is presented. More testing is needed to find suitable treatment options and adequate design criteria for different areas and climatic zones.

INTRODUCTION

There are two main reasons why greywater (from baths, sinks and appliances) should be treated. One is to save water. Treated graywater can be used for purposes that do not require potable water, like irrigation and flushing toilets. The total water demand for a home could be reduced by 60% by instituting a graywater recycling system [1]. The second reason is to develop ecologically engineered wastewater treatment systems. Blackwater (from toilets) contains most of the nutrients in the wastewater and should thus be treated separately and used as a fertilizer. The greywater can be treated in a simple way, close to the home, then reused or disposed to a local stream or soil. The composition of graywater is different from ordinary sewage (black- and graywater),hence, technology for this type of wastewater needs to be developed.

Graywater treatment methods have been developed world-wide during the last 25 years. The graywater systems vary from simple filtration prior to subsurface irrigation and more advanced methods producing water of swimming water quality. The US Army has tested various biological, chemical, and mechanical methods to treat laundry and showerwater for the use in the field [2,3,4]. The US National Sanitation Foundation and the city of Los Angeles, CA, have tested several systems for treating greywater prior to irrigation [5]. In Berlin, development and testing of various greywater treatment systems is motivated by limited groundwater resources [1,6]. The Center for Applied Ecology Schattweid, Switzerland, and the University of Wisconsin, USA, have developed simple nature-based plant- and sand- filter systems for graywater handeling[7,8,9].

This paper presents results from a literature survey of greywater composition, as well as nature-based wastewater treatment technology. In addition some results from research conducted at the Centre for Soil and Environmental research (JORDFORSK) are presented.

GRAYWATER COMPOSITION

Most of the research on graywater characteristics was performed 15-30 years ago, whereas more recent studies tend to explain treatment methods. Table 1 shows graywater concentrations found in septic tank effluents.

Graywater composition varies greatly with lifestyle: family size, age of residents, eating habits, detergents used, etc. The concentrations will also vary with water use. Compared to combined sewage, graywater usually contains much less plant nutrients (nitrogen and phosphorus). Graywater can contain 1-33% of the total wastewater nitrogen [5]. Phosphorus content varies due to the composition of detergents. Kristiansen and Skaarer [12] found concentrations as low as 0,41 mg/l, or 15% of the total mass discharge from the household when using P-free detergents. Brandes [11] also found low P-values when omitting the laundry from his study. In Norway, detergents are phosphorus free, except for dishwashing and industrial detergents. If people use only P-free detergents, the nutrient content in the graywater can actually be lower than in treated wastewater from a sewage treatment plant with chemical precipitation [16].

Table 1: Graywater quality in septic tank effluents.

	Olsson et al [10] 1968	Brandes a [11] 1978	Kristiansen & Skaarer [12] 1979	Siegrist & Boyle [9] 1981	Bahlo & Wach [14] 1990	Schönborn & Züstb [8] 1994	Naturvårds-verket [15] 1995	This study 1995
BOD5(mg/l)	205	149	130c	178	289	142	187c	116c
COD (mg/l)	395	366	341	456	520	320		
SS (mg/l)		162	35	45			107	39
Tot N (mg/l)		11,5	19				6,7	42,2
NH4 (mg/l)		1,7	11,5		2,6	95,7		36,1
Kjeldal N	9,1	11,3		15,9				
Tot P (mg/l)	18,1	1,4d	1,3 (0,42d)	4.4	4,1	9,5	4 (1,0d)	3,97
Fecal colif. (log#/100ml)	6,15	6,15	5,08	6,2				

a Excluding laundry machine; bIncluding urine; cBOF7; dP-free detergents

Greywater contains more than 50% of the BOD in household sewage [17]. High numbers of fecal coliforms are usually found, while virus is usually not present unless someone in the home has an infection [18]. The content of heavy metals and organic micropollutants is generally low [15,19].

GRAYWATER TREATMENT OPTIONS

PRE-TREATMENT
Graywater contains hair, grease and soap that can cause problems in the following treatment process. Pre-treatment is therefore needed.

Septic tank
The most common pretreatmet is a septic tank. In a septic tank solids settle and materials like grease and hair float. In addition, anaerobic digestion reduces the organic matter content [20]. For treatment of greywater, a tank can be 50% smaller than for ordinary sewage [21].

Filters
After the septic tank, an anaerobic fixed medium upflow filter can further reduce soluble BOD and thereby reduce the clogging potential of the effluent in subsequent soil disposal systems [17]. Venhuizen [22] suggests that loading rates of > 400 l/m2 in a sandfilter can be supported for prolonged periods (10 years) using extensive anaerobic pretreatment.

Nylon or cloth filters are used for recovering washing machine water for irrigation purposes [5]. In Norway a cloth filter system is being developed by JORDFORSK, primarily for use in cabins.

Clivus has developed a stone roughing filter as an alternative to the septic tank. Particles are removed, and the graywater is being aerated [23].

The use of a home made rack filter is also reported [24]. A screen (rack) filled with chrushed stone, hay, wood shavings etc. is put into the upper part of a tank. The water is led through the filter and is collected in the tank below.

MAIN-TREATMENT

Sand- and soil- infiltration systems
Disposal of septic tank effluent to the soil via conventional subsurface drainfields is commonly employed in Norway. The graywater is applied to a leach field and treated while percolating through the soil. Treatment

efficiency is dependent on the soil type and depth to groundwater. In the Norwegian guidelines the loading rates are 1,3-4 cm/d (13-40 L/m^2d) dependent on site conditions [21].

A mound consists of sand layered on top of existing soil. The wastewater is treated through the mound before percolating through the natural soil. It is used where the ground is too close to the groundwater level, or where the soil is too dense to receive the raw graywater. JORDFORSK is monitoring the treatment efficiency of mounds. Results from samples collected from the groundwater, 2 meters downstream the mound, are listed in table 2. Loading rates depends on the sand and the local conditions.

Sand filters are used where the existing soil is unsuitable for infiltration (e.g. clay), or where the water is supposed to be reused after treatment. The water is collected after seeping through a sand filter of 60-100cm depth. Results from a study at the University of Wisconsin [9] are presented in table 2. Although fecal coliform concentrations were reduced by 99%, concentrations in the filter effluent remained relatively high (avg 6025/100ml). A coarse sand was used (d_{10}=1,37) and the loading rate was high (39 cm/d). Laboratory experiments at JORDFORSK using 80 cm sand with grain size of d_{10}=0,14 and loading rate of 10 cm/d show higher removal of fecal coliforms (0-125/100ml in effluent). Using a 1 meter sandfilter (d_{10}=0,24) and a hydraulic load of 0,7 cm/d, Brandes [25] observed less than 30/100 ml of fecal coliforms in the effluent. Increasing the load reduced the bacteria removal.

Soil-and plant- based methods
These systems utilize the soil for infiltration, while plants absorb nutriens and promote growth of microorganisms. The Center for Applied Ecology Schattweid in Switzerland developed a sand- and plant-filter system that has been used for more than nine years to treat a mixture of graywater and urine [7,8]. The liquid is led through a 12 m^2 sand filter and a 26m^2 planted soil filter in serial connection. The first step consists of sand and gravel (aerobic), and the latter contains a permanently clogged soil (anaerobic). The treated graywater is collected in a drain at the bottom of the latter filter and disposed of by underground infiltration. In Schattweid, the load is about 90l/person per day. Table 2 shows results from the experiment. There is still no detectable depletion of the phosphorus retention capacity. The low maintenance system is able to meet the rigorus Swiss effluent quality standards, even during the winter.

In Berlin, Sanitärsystemtechnik designed a graywater treatment system for apartment buildings using sandfilter with plants [6]. The graywater is pumped to a wetland consisting of two 4x5m^2 fields. After treatment it is collected, disinfected by UV, and pumped to a tank inside for toilet flushing. When air temperature drop below 15°C, only one wetland is used. This gives a larger heat input to a smaller area, thus freezing is avoided. BOD removal efficiency is shown in table 2.

In a German study [14] three filter beds in series were installed in a greenhouse. The first bed contains coarse gravel and reed (*Phragmites australis*); the second is a rush bed (*Schoenoplectus lacustris*) with fine-grained gravel; and the last is a sand bed planted with rush and iris (*Iris pseudacorus*). The graywater is applied to the surface of the first bed at 160 l/day. After a vertical passage down the first bed the water runs horizontally through the following two beds. For 2 persons the total area is 6 m^2 and the depth is 0,4 meters. Clogging of the filter beds did not occur within the 3,5 years of operation. Results are presented in table 2.

Clivus has developed a system where the graywater (excluding kitchen waste) is treated immediately after use[26]. The purpose is to avoid anaerobic conditions during storage. The graywater is pre-treated in a filter, then distributed on a planted soil bed. The water moves vertically through the bed and is collected at the bottom. The manufacturer claims that the water can be stored and used for surface irrigation after treatment.

Biofilters combined with naturebased methods
In Berlin, the graywater (excluding kitchen waste) from an apartment building is treated in the basement by a biorotor [1]. After treatment and settling, the water is disinfected by a UV filter. The water is used for toilet flushing. The water is further cleaned when led through containers with swamp-plants on a wall outside the building. The water meets EEC guidelines (<100/100ml) for content of fecal coliforms in bathing water. BOD removal is shown in table 2.

Future Eco-Systems Ltd has developed a black-and graywater treatment method called "Twins" [27]. Large particles are removed from the graywater by filtration and deposited in a composting toilet. The water flows below the toilet compost, so that heat is given to the compost. The water is then biologically treated using a tricling filter of the Vesimies type developed in Finland. Air drawn from the toilet chamber is constantly ventilating the filter. The treated water is released outside into a leach field. Results are listed in table 2.

Table 2: Removal rates (%) and effluent concentrations (mg/l) for various graywater treatment systems

Method	BOD5		COD		TSS		N		NH4		P		fecal colif.	
	%	mg/l	%	mg/l	%	mg/l	%	mg/l	%	mg/l	%	mg/l	%	/100ml
Mound	92	12,4a					95	0,42	92	<0,1	99	0,01		
Sandfilter[9]	97	6	78	112	83	10	43	3		0,3			99	6025
Sand-/plant filter [8]b	95	7,1	89	34					91	8,8	>89	1	>99	
Wetland [6]	96	5												
Reedbed[14]	97	<8	95	28					82	0.3	60	2		
Biorotor [1]	96	5												<100
Twins [27]	>90		>90			<25	>90							
Compact plant	93	7a	76	48	86	4,5					88	0,12	100	0

aBOF7 bincludes urine

Disinfection
Most authors recommend disinfection after treatment in order to kill the remaining bacteria. UV needs a short contact period, but the water must be free of particulate matter which otherwise will prevent the radiation from destroying bacteria [24]. Addition of chlorine and iodine to the water is efficient if sufficient amounts are used and adequate time (30 minutes) are given[24, 2]. Hypes tested the use of heat, and found that one would need 63°C for 30 minutes to kill all the fecal coliforms [2]. Ozone also has the ability to destroy bacteria and is used in graywater treatment systems [5].

RESEARCH AT JORDFORSK
JORDFORSK develops simple and low-maintenance treatment methods. The goal is to provide treatment options for cabins, single dwellings and apartment buildings. Sand and Leca are tested as filtermedia for graywater treatment in cold climate. Leca is clay that has been heated until it expands (like popcorn). Previous experiments show good P removal. Leca is light in weight, easy to obtain, permeable, and it comes in various size ranges. The disadvantage is that it requires energy for production.

Compact treatment plant for cabins
Most vacation homes today have running water. Therefore a permit is required for sewage disposal [21]. This permit is only given when soil infiltration is possible, or where a collecting sewer exists. These requirements cannot always be met. People ignore the law and dispose of sewage untreated, thus polluting streams and wells. JORDFORSK is therefore developing a compact treatment plant for use in cabins. A filter first mechanically removes larger particles. This is followed by a tank containing a filtermedium (Leca and sand). After treatment the water can be disinfected by UV treatment. To date, two samples have been analyzed, and the results are presented in table 2.

Column study
JORDFORSK is currently conducting a laboratory experiment treating graywater in columns (cylinders) filled with sand and Leca. Four different filter media are used in duplicate: sand, crushed Leca (< 4mm diameter), 2-4 mm Leca and 4-10 mm Leca. Each column is 14,4 cm in diameter and 80 cm in height. Daily loading rates of 10- and 50 cm (cm/d) are being tested (Table 3). Temperature is regulated at 10-12°C.

Graywater is collected from a septic tank outlet. Input to the tank comes from a family consisting of 4 adults. Effluent value of BOD, SS, P, N and NH4 are shown in Table 3. Results in Table 3 represent mean removal rates for four samples, using duplicates, collected during a 12-week period.

Not surprisingly, the fine grained filtermedia show the best treatment results for all parameters. Both crushed leca and sand remove fecal bacteria very well. However, receiving loading rates of 10 cm/day the sand started ponding after 22 days. After 112 days the Leca has not ponded.

Table 3: STE-Septic tank effluent concentrations (mg/l) and removal efficiency (%) using different filter media and two different loading rates. Preliminary results

| | STEa | 10 cm/day - 8 doses/day | | | | | | 50cm/day - 39 doses/day | | | |
| | | sand: d10=0,14 | | 0-4mm leca | | 2-4mm leca | | 2-4mm leca | | 4-10mm leca | |
		%	sdev	%	sdev	%	sdev	%	sdev	%	sdev
BOD7	90 - 163	87,2	9	97,4	2,4	63	22,2	65,2	27	64,5	16,8
SS	24 - 59	84,4	5,5	85,6	5,1	63,6	13,3	74,5	14,2	64,1	20,4
Tot P	2,6 - 6	98,6	0,9	92,9	9,1	20,7	24,3	14,1	15,3	11,5	16,6
Tot N	23,8 - 72,3	12,7	11	4,2	3,7	1,5	4,4	4	5,5	4,1	7
NH4	23,8 - 58,3	31,2	35,6	42,6	39,8	19,9	33	29,6	42,9	25	37,3
pHb	7,7 - 7,9	8-8,4		8,5-9,3		8,1-8,9		7,4-8,4		7,5-8,5	
Fecal colif.b	>1*104	0-125		1-45		860-3600		120-13800		600-15000	

a All STE values are in mg/l except fecal colif.(#/100ml) and pH; beffluent values; sdev=standard deviation

The 2-4mm Leca is not able to treat the graywater as well as the finer filter medium. However, it can handle a higher load. Table 3 shows that increasing the load from 10- to 50- cm/d does not change the treatment efficiency. Also increasing the diameter of the Leca to the range of 4-10mm does not change the results substantially. The reason for this is not clear, but frequent "wetting" may be advantagous for the biofilm. The microorganisms might have a large capacity to break down BOD; but where they do not receive enough "food", they work less efficiently. Yet another reason could be the pH. The pH of the Leca is high because it contains dolomite on the surface. As the Leca is being flushed with graywater, the pH goes down. Because the pH drops faster where the Leca is frequently washed, the high-load columns might have a better environment for the microorganisms. Bacteria removal in the columns filled with 2-4mm Leca and 4-10mm Leca has varied.

Nitrification processes have been slow, especially in columns with Leca, but seem to improve. This might also be caused by high initial pH values. The pH in the Leca is now decreasing.

DISCUSSION AND CONCLUSIONS

The composition of graywater varies. Most of the research on graywater composition was conducted years ago. Detergents have changed, and phosphorus has been substituted by other compounds. It is important to gain more information on the content of heavy metals, organic pollutants and tensides in graywater. For safe reuse of treated graywater it is important to know what to remove during treatment, or what to avoid putting into the sink. More knowledge on how various lifestyles impact the graywater characterization would also be useful concerning greywater treatment system design.

This literature search has revealed a variety of greywater treatment methods. Systems based on mechanical, biological or chemical methods have been tested and are available on the marked. Several nature-based systems are also beeing tested or are in use. Common for many of these systems are that their performance is not well documented. According to the National Association of Plumbing-Heating-Cooling Contractors [5], several graywater treatment systems are being sold in the USA without first being tested. This is also occuring in Norway, where untested technical and nature-based graywater treatment systems are on the market. These systems could be damaging for the reputation of graywater treatment. Guidelines and approval routines for treatment options should thus be developed.

Table 2 shows that the systems achieve high removal of nutrients and organic components. That is for a given load and system size. Few of these publications or reports show how they found the optimum size for BOD removal. Our studies show how important it is to test various filtermedia and loading rates to optimize treatment and loading capacity. Because greywater treatment systems will be needed where space is often limited, it is important that they are compact. Further work is necessary in order to develop adequate design criteria for greywater treatment systems for cabins, single dwellings and apartment buildings.

Fecal coliforms represent a health hazard. Removal efficiency of bacteria is lacking in many studies. Disinfection of the water before reuse is, however, often recommended. Excluding disinfection and removing the bacteria using natural processes would be a better ecological solution. The sand and plantfilter [7] and sandfilter [9] achieved 99% removal of bacteria. Still, in the latter study, over 6000 fecal coliforms per 100 ml water was left in the effluent. Bacterial counts are dependent on grain size and loading rate [25], and can thus be manipulated. Fine sand and crushed 0-4mm Leca show very good results for fecal coliform removal. It should be possible to achieve effluent concentrations of <100/100 ml of fecal coliforms if the combination of filtermedia and loading rate is right.

Effluent quality requirements depend on what the water will be used for, where it will be disposed of and what regulations exist. Removal of fecal bacteria is very important if the water will be in contact with people, while for the use of subsurface irrigation it might be unneccesary. Nutrients and BOD must be removed before disposing the water into a lake or river, while this is of less importance when used for irrigation. If the water will be disposed of in the ground or to waterbodies, governmental requirements often need to be met. Because of varying requirements it is important that many graywater treatment options are available.

Studies in Switzerland [8] and Berlin [6] show that plantbased systems work during winter time. Except from these studies, little is mentioned about low temperature treatment efficiency. Testing of nature based greywater treatment systems in cold climate is therefore needed. JORDFORSK will use columns, the compact treatment plant and mound systems to gain experience using low temperatures.

The cost of the various systems is seldom mentioned. By reducing the wateruse at the same time the sewage production, resources and money can be saved. However, low operational cost is common for nature-based systems.

REFERENCES

[1] Sanitärsystemtechnik, Graywater Recycling in Urban Areas, Berlin (1994)
[2] W. D. Hypes, Proc. of the 5th Nat. Confr. on Individ. On-Site Wastewater Syst., NSF, p 209 (1979)
[3] M. Tleimat, B. Tleimat, M. A. Friedman, T. E. Stycznski and S. Schwartzkopf, Desalination: The Int. Journ. on the Science and Techn. of Desalting and Water Purif., vol 87, no 1-3, p97 (1992)
[4] Proc. Water Reuse Symp. III: Future of Water Reuse San Diego, CA, AWWA, Aug 26-31 (1984)
[5] Assment of On-Site Graywater and Combined Wastewater Treatment and Recycling Systems, National Association of Plumbing-Heating-Cooling Contractors, Falls Church, VA, USA (1992)
[6] Sanitärsystemtechnik, Biologische Grauwasserreinigungsanlage, Berlin (1991).
[7] J. Heeb & B. Züst, Proc. of the Int. Conf. in Ecol. Eng. for Wastewater Treatment, March 24-28, Stensund, Sweden p199 (1991)
[8] A. Schönborn & B. Züst, gwa, 8/84, 74. Jahrgang (1994)
[9] R. L. Siegrist & W. C. Boyle, Proc. of the 3rd National Symp. on Individual and Small Community Sewage Treatment, Dec. 14-15, Chicago, Il, ASAE, p176 (1981)
[10] E. Olsson, L. Karlgren & V. Tullander, The National Swedish Institute for Building Research, Stockholm, report 24 (1968)
[11] M. Brandes, Journal Water Pollution Cont. Fed., Nov, p 2547 (1978)
[12] R. Kristiansen & N. Skaarer, Vann. 2 (1979)
[14] K. E. Bahlo & F. G. Wach, Proc. of the Int. Conf. on the Use of Constructed Wetlands in Water Pollution Control, Cambridge, UK, 24-28 September, p215 (1990)
[15] Naturvårdsverket, Vad innehåller avlopp från hushåll, report 4425 (1995)
[16] A. Schönbeck, Tanums county, Sweden
[17] R. Laak, Compost Science, Nov/Dec, p 29 (1977)
[18] R. L. Siegrist, J. E. H., vol 40, no 1, July/aug, p5 (1977)
[19] W. D. Hypes, in J.T. Winneberger (ed.), Manual of Grey Water Treatment Practice, Ann Arbor Science p79 (1974)
[20] R. L. Siegrist, Proc. of the 2nd Northwest On-Site Wastewater Disposal Short Course, Univ of Washington, Seattle (1978)
[21] Miljøverndepartementet, Forskrifter om utslipp fra separate avløpsanlegg (1992)
[22] D. Venhuizen, Proc.of Conserv 90, August 12-16, Phoenix, Az, 1289 (1989)
[23] A. Rockefeller & C. Lindstrom, Compost Science, Sept/Oct, p22 (1977)
[24] P. Lombardo, Onsite Management, May/June p45 (1982)
[25] M. Brandes, Ontario Ministry of the Environment, Report no. W 68 (1977)
[26] Clivus Compost AB, Stockholm
[27] F. W. Smith, NSF 7th National Confr. on Individual Onsite Wastewater Systems, p99 (1980)

Environmental Research Forum Vols. 5-6 (1996) pp. 221-226

Nitrogen and Phosphorus in Fresh and Stored Urine

D. Hellström[1] and E. Kärrman[2]

[1] Division of Sanitary Engineering, Luleå University of Technology, S-971 87 Luleå, Sweden

[2] Department of Sanitary Engineering, Chalmers University of Technology,
S-412 96 Göteborg, Sweden

Keywords: Nitrogen, Phosphorus, Urine, Storage, Separation, Wastewater, Sewage, Recycling

Abstract

One strategy for nutrients recycling from wastewater is to handle urine separately and use it as a fertilizer for agricultural applications. At present, there are only a few applications of urine separation in Sweden and, in most of the applications, the urine is separated in the toilet and stored in a tank. This study was divided into an investigation of the composition of fresh urine and a study of nutrient retention and transformation during storage of stored urine. The aim of the investigation of fresh urine was to determine the content of phosphorus and nitrogen in urine per person and day, and to study the daily variation of the same substances. Average values were 1.0 gP/day and 13 gN/day. The study of daily variation showed that the largest part of the quantities of phosphorus and nitrogen will appear in urine, produced at home, between 17.00-08.30. The aim of the storage studies was to estimate nitrogen losses and changes of pH and NH_3-N concentration during different storage conditions. Laboratory experiments with undiluted urine were carried out during a period of three months. The results indicate that the nitrogen concentration decreases in sampled urine through some separation process in the urine, but that losses through evaporation of ammonia were small. The increase of NH_3-N and pH is a slow process in undiluted urine and is inhibited by acidification or a low storage temperature.

Background

Today, there is an ongoing discussion concerning nutrient recycling from human urine and faeces. Phosphorus is a non-renewable substance and is therefore important to recycle. By using sewage sludge from wastewater treatment plants as a soil improver in agriculture, the greater proportion of phosphorus from wastewater can be recycled [1]. In spite of this, there is a concern over the quality of, and health risks related to, the use of sewage sludge. In the sewer system, urine and faeces are mixed with household wastewater, stormwater and industrial discharges, and will inevitably be polluted by heavy metals and toxic organic substances. These substances will be found in sewage sludge. As a result of this, sewage sludge is often disposed to a landfill. A strategy for recycling of nutrients is to handle urine separately and use it as a fertilizer [2]. A literature review concerning the content of several substances in the fractions; greywater, urine and faeces has been carried out by Sundberg [3]. One part of Sundbergs combination of data shows that urine is the source of around 70% of phosphorus and around 90% of nitrogen in blackwater (wastewater from water closets). The fact that urine contains almost all nitrogen from food increases the advantages of urine separation. Examples of advantages of recycled sewage sludge are nitrogen recycling to agriculture and minimisation of discharges of nitrogen to receiving waters. Sundbergs [3] study is based on data from studies of very specific groups of persons. The intake of food varies between groups of persons [4], which motivates the analysis of urine from a wide population of people.

There are only a few applications of urine separation in Sweden as yet. In most of the applications, the urine is separated in the toilet and stored in a tank. The tank is emptied by a farmer who collects the urine to use it as a fertilizer for agricultural applications. It is important that the system for collection, storage and handling of human urine is constructed for minimizing losses. The experience from storage and handling of animal urine is that nitrogen losses can be large. Thus, it is also desireable to study how different storage conditions influence the risk of losses of nitrogen, during storage and distribution of human urine. It is also known that urea in contact with wastewater relatively quickly decomposes into ammonia. Thus, it is interesting to study how fast the speciation of nitrogen in urine changes. The main losses are expected to occur through ammonia evaporation and thus the pH, concentration of total nitrogen and ammonia were investigated.

Objectives

The aims of the experiments in this study were:
- To estimate the content of phosphorus and nitrogen in whole day samples of urine, and to study the daily variation of the same substances.
- To study the losses of nitrogen as well as changes in pH and ammonia concentration under different storage conditions.
- To study the effect of urine handling on ammonia and nitrogen concentration.
- To compare the urine quality after storage of undiluted urine with the quality of stored urine from houses with systems for urine separation.

Experiments with fresh urine

Experimental

The aim of this study was to determine the content of phosphorus and nitrogen from one person during one day (24 hours), and to describe the daily variation of the same substances. One sample containing the total urine volume from one day was collected from each person, which means that the study can only describe urine in general, and not show any differences between various groups of people. The studied group of people is office working adults (not manual labour workers). The study includes analysis of daily samples of urine from 30 persons, with ages between 23 and 62 years. The sample of people included 15 females and 15 males. The daily variation was studied for 10 of the 30 persons. For these 10 persons each urination was collected separately, which means that 4-10 samples were collected per day for each person. Phosphorus analysis was made by a HACH method. The methods for analysing total phosphorus and phosphate were compared. No significant difference between the two methods in terms of concentration was found, so the less complicated phosphate method was chosen. The nitrogen analysis was carried out by means of Dr. Langes method for total nitrogen.

Results

The volume of urine and the content of phosphorus and nitrogen from the 30 persons in the sample are presented in table 1. The results of phosphorus and nitrogen for the whole group correspond with results from analysis of urine from persons with a normal diet [5]. There is a notable difference in quantity of phosphorus between females and males (see table 1). This is probably a result of the lower body weight of females. In a biological handbook, phosphorus in urine is related to body weight [6]. Urine from a person with a body weight of 70 kg is usually in the range of 130-300 mg N/day,kg and 10-15 mg P/day, kg. Recalculated to a value per person with a body weight of 70 kg gives 9-21 gN/pd and 0,7-1,1 gP/pd, which are values close to the results presented in table 1.

Table 1   Volume of urine and content of phosphorus and nitrogen.

	Volume (l/person,day)	P (g/person,day)	N (g/person,day)
The whole group	1.5 ± 0.5	1.0 ± 0.4	13 ± 3
Females	1.5 ± 0.5	0.8 ± 0.3	12 ± 3
Males	1.5 ± 0.5	1.1 ± 0.3	14 ± 3

The results from the studies of the daily variation of phosphorus and nitrogen in 10 persons urine are presented in figure 1.

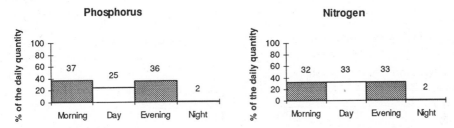

Figure 1   *Daily variation of phosphorus and nitrogen in urine from 10 persons.*

The day has been divided into four periods, in figure 1. The morning period is between 05.00 and 08.30 (15% of the day), the day period is between 08.30 and 17.00 (35% of the day), the evening period is between 17.00 and 24.00 (30% of the day), and the night is between 00.00 and 05.00 (20% of the day). It can be noticed from figure 1 that the morning urine includes around a third of the total quantity of phosphorus and nitrogen from one day. One reason for this is that the test persons had not urinated for several hours before the morning urination. All the ten test persons were urinate at least one time during the morning period. Only two of the ten test persons urinated during the night period. Test persons are usually at home during the time periods, represented with filled staples in figure 1, and at work during the time period, represented with an unfilled staple, and consequently 75% of the daily quantity of phosphorus and 67% of nitrogen can be found in the household urine. Concerning nutrient recycling, it would therefore be advantageous to give priority to the construction of urine separation systems in residential houses, while urine separation at work may be less beneficial.

Experiments with stored urine

Experimental
Two laboratory experiments were performed to investigate how the nitrogen and ammonia concentration changed during storage. A third experiment was performed to investigate the loss of nitrogen by poor handling after storage. In addition to these experiments, samples of stored urine were taken from an "ecological village". In water closet systems for separation of urine and faeces there will always be some amount of flushwater and probably some contamination of faeces into the urine storage tanks. Therefore, it was desireable to compare the results from the laboratory experiments with the composition of stored urine from an "ecological" village with a separated toilet system. The village, Björsbyn, is located about 5 km from Luleå. Urine was sampled from a urine tank with a volume of 10 m³. The urine has been continously collected since september 1994 and the samples were taken in May 1995. The tank is placed underground which means that the temperature was below 5°C during the storage period.

In the laboratory experiments, bottles containing 80-85 ml of urine were used. In experiment 1, was the urine sampled by decanting 60 ml from each bottle, after storage. The aim of the decanting procedure was to investigate if there are any separation processes *in* the urine during storage. Instead of decanting, the whole samples were used for analysis in experiment 2. All samples in the three experiments have been frozen until analysis for NH_3-N, total nitrogen and phophorus. The pH was determined immediately after sampling (before freezing). The preparation of samples for measurement was done according to Swedish standard (SIS 02 81 02, SIS 02 81 31 and SIS 02 81 34). The concentrations of total nitrogen, NH_3-N and total phosphorus were then measured by using an automated procedure for analysis (an autoanalyser - TRAACS 800 - Bran+Lubbe). It should be noted that both dissolved ammonia and ammonium are measured in the NH_3-N analysis.

The laboratory experiments were performed with undiluted urine from five males and five females, ages between 25 and 50 (see table 2).

Table 2   Quality of the urine, used in experiments 1-3.

	pH	Tot-N, g/l	Tot-P, g/l	NH_3-N, g/l	N/P
Morning urine	5.88	8.5	0.94	0.50	9.09
Day urine	6.08	7.25	0.98	0.41	7.44

Experimental design
Experiment 1 was run as a two level factorial design experiment, where effects of several storage conditions were investigated. A 2^5 factorial design allows a preliminary screening of factors that might effect urine quality without unnecessarily detailed experiments. A 2^5 factorial design requires 32 samples. Two levels of exposure were tested for five parameters:

- *Storage time*. The urine was stored for 9 and 64 d, respectively.
- *Temperature*. The samples were either stored at room temperature (about 23 °C) or in a refrigerator (about 6 °C).
- *Acidified/unacidified*. A dosage of 0.60 ml 4 M H_2SO_4 /100 ml urine was used to achieve an initial storage pH 3 in 16 of the 32 storage bottles.
- *Type of urine*. The urine was separated into two categories, morning- and dayurine, see table 2. The time-periods are defined in figure 1.
- *Sealed/unsealed storage*. Sixteen (16) samples were stored sealed and the remaining were stored unsealed. The samples were further placed in a sealed plastic box with a volume of three litres.

Only two different storage times were investigated in experiment 1. Thus, to follow the changes in concentrations as a function of storage time, experiment 2 was performed. In experiment 2, sixteen samples of mixed urine (50 % of each urinetype) stored for 1, 2, 5, 9, 15, 30, 64 and 87 days. Eight of the samples were stored in 4-6°C and the remaining were stored at room temperature. The urine was mixed after storage, and thus the effect of precipitation/sedimentation was eliminated.

In experiment 3 urine samples, stored for 9 and 64 days respectively, were stirred in an open 400 ml beaker for 24 hours. Before mixing, the urine had been stored in sealed bottles at room temperature. The pH, ammonia and nitrogen concentration were measured at selected intervals during the stirring period.

Results
In experiment 1, it was found from the factorial design, that the changes in nitrogen concentration only depend on length of storage time. No other significant effects were observed. The average concentration of nitrogen stored for 64 d was about 7.4 g tot-N/l and urine stored for only 9 d had

an average concentration of 10.1 g tot-N/l. The 95 % confidence interval for the difference in nitrogen concentration between urine stored for 9 and 64 days respectively, was (1.71, 3.57) g tot-N/l. The 95% confidence interval for the difference in nitrogen concentration between urine stored during 9 and 64 days respectively, was (1.71, 3.57) g tot-N/l.

Of the studied variables, time, addition of acid and storage temperature and interactions of these, had a significant effect on the NH_3-N concentration. It is noteworthy that the increase in NH_3-N concentration is detectable for short storage times only if the urine is stored in room temperature *and* stored without addition of acid. For long storage times it seems to be no increase in NH_3-N concentration, if the temperature is kept low and acidification has been done. If one of these variables is on the high level, the increase is marginally (from 0.45 to 0.52-0.56 g amm-N/l). It is only during long storage time, high temperature and no acidification that the increase is "dramatic" (from 0.45 to 1.25 g amm-N/l).

Considering pH it is noteworthy that a combination of high temperature, long time, no acidification and unsealed storage gives the highest increases. The final pH with this combination was about 8.1-8.6. As a comparison it could be mentioned that storage in unsealed bottles placed in a refrigerator gives neglible increase in pH even if the urine was stored for 64 days.

From experiment 2, no decrease in nitrogen concentration could be detected. The concentration of NH_3-N and the pH increase during storage in room temperature but not when the samples were stored in a refrigerator (see figure 2). It should be noted that the NH_3-N concentration increases during the whole period of analysis.

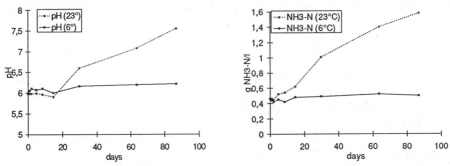

Figure 2   *Changes in pH and NH_3-N during storage in 6°C and 23 °C, respectively.*

In experiment 3 an increase in pH for both 9 days and 64 days storage time was found. A small increase of NH_3-N and nitrogen concentration due to evaporation could be noted during both runs. No losses of nitrogen were detected.

The samples for analysis in the storage tank at Björsbyn were taken at three different depths. There was no large difference in concentration between the samples. The urine contained 0.90 g tot-N/l, 0.77 g NH_3-N/l, 0.037 tot-P g/l and 0.036 PO_4-P g/l. The N/P ratio was 24.7. The pH in the tank was 8.86.

Discussions and conclusions

The study of stored urine indicates that losses of nitrogen are small during storage of human urine. The decrease of nitrogen concentration during storage in experiment 1 is probably explained by a separation process within the urine, because the differences between different storage conditions were insignificant. The statement about small nitrogen losses during storage is also confirmed by the fact that it was impossible to detect any losses of nitrogen in experiments 2 and 3, and that the N/P-ratio was high in the stored urine from the "ecological" village.

It has been shown that factors such as storage pH and temperature are important for changes in urine quality. The conversion of organic nitrogen into NH_3-N is a slow process, favoured by high temperature and inhibited by addition of acid. It is interesting that about 85 % of the nitrogen in the long stored urine from the "ecological" village occurs as NH_3-N. This is probably explained by the relative long storage time and the dilution due to flush water and probably some fecal contaminants. It is interesting to compare the results from Björsbyn with the ratio between NH_3-N and total nitrogen in domestic wastewater and urine, which normally is 0.8. The relatively high concentration of NH_3-N combined with a high pH indicates that there is a risk for nitrogen losses due to ammonia evaporation during handling of stored urine. Thus, the problem with nitrogen losses from stored human urine needs to be further investigated.

Wastewater systems with urine separation have advantages compared to systems where nutrients are recycled from sewage sludge. Urine contains less quantities of heavy metals and toxic organic substances, but includes almost the whole quantity of nitrogen from food intake. There is still one aspect of urine separation to consider; the whole quantity of phosphorus from food intake will not be found in the urine. From a study of food habits in Sweden [7], the daily intake of phosphorus was determined as 1,3 g/p,d for adult females and 1,5-1,8 g/p,d for adult males. A comparison to the present study of fresh urine in table 1, shows that 30-40% of phosphorus from intake will not be found in urine. There is also a content of phosphorus in faeces [8]. An example of a study of metabolic balance in adult males [8], shows that for two different diets the faeces include 40% and 50% of phosphorus from intake, respectively. Recycling of phosphorus from faeces and maybe even greywater should be taken into consideration, while constructing a wastewater system with urine separation. Phosphorus is a non-renewable resource and the phosphorus from wastewater should therefore be recycled.

The results of the studies of daily variations of phosphorus and nitrogen in urine show that the urine separation system should on the first hand be applied to houses where people live. The morning, evening, and night urine contained 75% of phosphorus and 68% of nitrogen from the test groups one-day sample of urine.

Acknowledgement

This study was supported by the Swedish Council for Building Research.

References

[1] E. Kärrman, Proceedings of the Conference Ecotechnics'95, held at Östersund, Sweden, March 29-31 (1995).
[2] H. Kirchmann and S. Pettersson, Fertilizer Research (in press) (1995).
[3] K. Sundberg, Swedish Environmental Protection Agency, Report 4425 (1995) (in Swedish).
[4] W. Becker, H. Enghardt and A. Robertson, National Food Administration of Sweden, Report (1994) (in Swedish).
[5] E.S. West, W.R. Todd and J.T. van Bruggen, Macmillan, New York, USA (1966).
[6] P.L. Altman and D.S. Dittmer, Federation of American Science for Experimental Biology, Maryland, USA (1974).
[7] W. Becker, Vår Föda, 8/92, 349-362 (1992), (in Swedish with an English summary).
[8] H. Spencer, C. Norris, J. Derler and D. Osis, J. Nutr., 121, 1976-1983 (1991).

Environmental Research Forum Vols. 5-6 (1996) pp. 227-232
© *1996 Transtec Publications, Switzerland*

Integration of Sanitation into Natural Cycles: A New Concept for Cities

R. Otterpohl[1], A. Albold[1] and M. Grottker[2]

[1] Otterpohl Wasserkonzepte GbR, c/o Dr.-Ing. R. Otterpohl,
Kanalstrasse 52, D-23552 Lübeck, Germany

[2] Fachhochschule Lübeck, FG Siedlungswasserwirtschaft, c/o Prof. Dr.-Ing. M. Grottker,
Stepensonstr. 3, D-23562 Lübeck, Germany

Keywords: Sustainable Sanitation, Vacuum Toilets, Vacuum Sewerage, Biogas Plants, Anaerobic Treatment, Digestor, Greywater, Constructed Wetlands, Domestic Waste, Biowaste, Nutrient Cycle, Agriculture

ABSTRACT

Water closets and flushing sewerage systems have first been installed in very humid countries with small population densities. Even with the best economically feasible wastewater treatment the problem created by mixing faeces with a lot of water can not be compensated. New technologies applied in future should no longer mix the human food cycle with the water cycle. This requires toilet-systems with no or very low water consumption. Highly concentrated organic material may be composted or treated in anaerobic reactors. Composted or digested sludge from households only will be fairly unpolluted and can be used as the best possible fertiliser for agriculture. By these technologies the inorganic substances taken from the soil by the farmers can be returned in about the appropriate quality and quantity. There are also positive effects on the energy balance with savings by avoided aerobic treatment and avoided production of artificial fertilisers. Another advantage is the option to treat a good portion of organic solid waste from households in the same systems.

A pilot project for a new settlement of about 300 inhabitants in Lübeck, Germany shall demonstrate the feasibility of a new integrated system with vacuum toilets and pipes for semi-centralised anaerobic treatment. Greywater will be treated in decentralised biofilm systems. Stormwater is collected, retained and infiltrated in a through and drain trench system. This way no centralised sewerage system will be necessary for this settlement.

REVIEW OF TODAY'S TRADITIONAL SANITATION CONCEPT OF INDUSTRIALISED COUNTRIES

Wastewater can really mean "waste of water". The use of 6 up to 30 litres of water for flushing the toilet once may not be a problem in humid countries with a very thin population density. But this can not be justified in densely populated regions where clean water resources are exceedingly difficult to find. In arid regions it does not make sense anyway. Water consumption is not only a question of personal habits but also of technologies installed in the homes. With growing population densities and a decline in water quality all societies should care about the importance and the outstanding role of water and to use it wisely. It should become a principle to have water supply from the nearest appropriate resources. Here people of the communities can care for a sustainable management for their own water. This would also stop unlimited usage of water as the limits become obvious in the regions management.

Quality and quantity of wastewater are a consequence of water usage. Lower water consumption and the application of easily degradable detergents make the state-of-the-art wastewater treatment easier. But even with the best possible end of the pipe technology about 20 to 30 % of nitrogen and about 5 to 10 % of phosphorus remains in the effluent. This is not acceptable for hundreds of years with a high population density. Mineable reserves of phosphorus will only last for about 100 - 150 years and should also for this reason not be wasted to the receiving waters. The nutrient levels in the seas are causing serious algae blooms already now. For these reasons the systems applied should get all nutrients back to agriculture. They can replace artificial fertilisers which have a high energy demand in their production, too. Also energy that would be needed for aerobic treatment of the diluted wastewater can be replaced by a small win on energy. All in all modern wastewater treatment plants seem to be an end of the pipe cure for an ill-defined overall system.

POSSIBLY SUSTAINABLE SANITATION CONCEPTS

The word 'sustainable' is often misused by naming completely conventional technology with some minor improvements. The most popular definition was made by the World Commission on Environment and Development (Brundtland Report, [1]) and defines sustainable development as 'development that meets the needs of the present without compromising the ability of future generations to meet their own needs.' This concept is helpful for judging about future water management, too. Some simple assumptions can change destructive discussions to constructive work. Numbers of future world population are not important for decisions. All probable numbers will give the same hard restrictions for acting in water management.

Instead of diffuse discussion about all the problems of the world the concept 'sustainability' should be operationalised. In the context of water management a simple analogy is helpful: When rivers became heavily polluted a concept 'cleaning rivers' was operationalised to create standards for wastewater treatment. This was done without understanding biology and chemistry of rivers in detail. Rivers could be cleaned with analysing the overall situation. Concern about polluted rivers is basically the same as concern about the fast increasing damage to the ecosphere on a global level. From the well known global problems we can derive needs for standards and rules for acting on the local or regional level. The block in thinking comes from keeping it on a global level for the action to be taken, too. This would be adequate for somebody who rules the world - otherwise it is a brain game. From analysing the global situation it is fairly simple to derive general rules for acting on the local level. The same procedure as contributing to saving a river is done with the concept 'sustainability' to contribute to save the ecosphere.

Some qualitative guidelines for possibly sustainable sanitation concepts are:

◎ Mass balances must integrate into natural cycles forever ➙ recycling of nutrients to the soil instead of accumulation in receiving waters (even oceans are not endless)
◎ No reduction or degradation of resources (water and soil) in the long run
◎ Low energy demand and supply from renewable resources
◎ Economical and maintainable solutions (construction and operation) for long-term application

Possibly sustainable systems in water management can be found by comparing their mass- and energy balances. Due to the complex global implications and the unknown future development there is no way in finding the ultimate sustainable system. It is not necessary either, as enough information for all basic decisions is available. This way possibly sustainable systems can show adequate solutions for serving future needs.

The same care that is taken today for a river ecosystem by cleaning wastewater can be taken for the future ecosphere including coming generations by locally applying rules derived from analysing the global situation. This way 'sustainability' is operationalised...

The comparison of different feasible systems in water management by their mass- and energy-balances makes decisions possible. The ability of a system to recycle nutrients, keep water clean and its water- and energy demands can lead to a judgement as 'possibly sustainable'. Our traditional sanitation system is not sustainable at all. New approaches have to be defined and researched for future application even if they can not be implemented everywhere right away. We have to show a ways for subsequent replacing of the traditional systems when they have to be renewed.

OBJECTIVES IN FORWARD PLANNING FOR SANITATION CONCEPTS

Objectives in forward planning for the future should be defined free of existing structures. If the systems with most advantages are not compatible with today's structures in the respectively regions or countries a solution for a replacement with the time may be found. None of our sewerage and treatment systems will last forever thus making changes in the long run possible. In many countries of the world sanitation in not yet in a satisfying state. New solutions can turn out to solve problems better or to avoid problems. Main objectives in forward planning for sanitation systems are:

1. Systems in sanitation must strictly avoid spreading of diseases
2. Nutrient contents of human waste must be completely recycled to food production (Food- and water cycle should be kept separated)
3. Systems must have a low water/energy consumption and not accumulate nutrients in ecosystems
4. Use and operation of systems must comply with social and religious rules
5. Costs of installation and operation should be reasonable (environmental damage has to be calculated, too)
6. Operation should be easy and not only possible for highly specialised experts
7. Systems should rather avoid problems than solving self-made ones

Participation of random selected non-expert groups of citizens for more objectives is essential for successful projects. The groups can be organised according to Dienel [2] with up to 25 citizens who are willing to participate. They receive a small fee for participating and are informed by the professionals and other groups concerned of the project with their different views. Work can be organised in small groups of up to 5 people and define objectives or solutions in the whole group. The procedure must be finished within a relatively short period to limit possibilities of influencing participants. Application of this method in different planning procedures gave very good results. Experts tend to stick to their special knowledge too much and get a chance this way to consider different aspects.

A PILOT PROJECT FOR A POSSIBLY SUSTAINABLE SANITATION CONCEPT FOR URBAN AREAS

Classification for adequate treatment processes of liquid and solid waste from households leads to the 3 groups represented in table 1.

Table 1:   Classification of domestic waste and wastewater for adequate treatment processes

Classification for treatment	appropriate treatment	
Group 1: **Biodegradable solid waste and low diluted faeces with urine**	anaerobic or composting	related to food cycle
Group 2: **Grey wastewater (greywater) from bathrooms, washing machines and kitchen with little nutrients**	aerobic	related to water cycle
Group 3: **Non-biodegradable solid waste**	reuse as raw material	

Examples for feasible possibly sustainable sanitation concepts based on the above table are:

1.   Vacuum closets (VC), anaerobic treatment with co-treatment of organic household waste (biogas-plant), anaerobic sludge reuse in agriculture in growth periods (decentralised/semi-

centralised biogas plant with vacuum pipes, larger plants on farms with storage, gas usage)

2.   Composter for faeces and organic household waste, application of compost to agriculture (care of users necessary, with more than 2 storey buildings rooms below toilet needed, leachate used for garden, compost removal in appropriate season)

3.   Flushing toilets with treatment in aquacultures with algae and fish in decentralised or semi-centralised greenhouses.

For 1. and 2.: Treatment of grey wastewater e.g. in aeration tanks, aerated trickling- or sandfilters (urban areas), in constructed wetlands or ponds (rural areas) with effluent to reuse (irrigation, toilet flushing), to infiltration or else to receiving waters. (see [3] for details on greywater treatment) Further concepts and some implications for developing countries can be found in [4] and [5].

After discussion of the above concepts for a special project it turned out, that only the vacuum system would be accepted. The implementation of composters for a settlement of 300 people is difficult because of the missing acceptance. Concept 3 was not possible because of space limitations. This way for a planned settlement in the city of Lübeck near the Baltic Sea in Germany, a concept according to no 1 in the list above is planned by the authors. It is meant as a pilot project to research the concept in practice and over several years. The implementation will be made without public subsidies to demonstrate economical feasibility, too. Funds are acquired only for the research part. All components of the project are used in different structures since many years and are therefore well developed. The new part is the following combination of systems which was published first by Otterpohl and Naumann [6]. The system consists mainly of:

◎   vacuum closets (VC) with collection to anaerobic treatment with co-treatment of organic household waste in decentralised/semicentralised biogas-plants, recycling of digested anaerobic sludge to agriculture in growth periods and use of excess biogas

◎   decentralised treatment of grey wastewater in aerated sandfilters or constructed wetlands and infiltration within the stormwater storage- and infiltration system

◎   stormwater collection for reuse, collection of excess-stormwater in through and drain trench for retention and infiltration [7]

A sketch of the principle of such a system is presented in figure 1. The figure is not meant for showing all the details but shall give an idea of the concept with collection and treatment of faeces. At the digestor a vacuum pumping station will be installed. The pumps are redundant and have an extra unit for the case of failure. Pressure in the system is 0.5 bar operating both the vacuum toilets and the vacuum pipes. Pipes in houses are dimensioned 40 mm, those in the ground 50 mm. They have to lie deep enough to be protected against freezing and must have down-bows about every 30 meters to create plugs of the transported matter. Vacuum toilets technology is well developed for ships, aeroplanes and fast trains. There are already some implementations in flat buildings for saving water. Noise is a concern with vacuum toilets but modern units meet the German norms.

Mass- and energy-balances indicate some major advantages of this system. It seems to be able to close the nutrient cycle, which is mixed up with the water cycle in the traditional sanitation system with flushing toilets. The proposed system requires about 50% higher investment costs for the water management part of the settlement compared to the connection to nearby sewers. It needs only about a half to 2/3rd of operation costs compared to the municipal fees for water and waste. Operation costs including personnel are about half to 1/3rd of those of the wastewater fees of the town. To cover the higher investment a company open for financial and organisational participation of inhabitants can build the structures and charge people for the service. Fees for waste are considered half with treatment of biowaste. The fees for water reduce to the price of drinking water, wich will be needed in smaller quantities. All in all calculations indicate economical feasibility even for this pilot project. For economical reasons such a system should serve 500 - 1000 inhabitants. Digestors are much cheaper when they are larger and loose less energy. If the proposed system shall be implemented for smaller units there should be a collection unit instead of a digestor. In this case a digestor dimensioned for the demand can be installed on the farm utilising the sludge. This digestor could be of an adequate size to make seasonal storage. However for the pilot project digestion will be done on the site to demonstrate the feasibility even for this sub-optimal size. Faeces will be hygienised by heating the feed to 70 °C. This energy is not lost when feeding starts before the matter cools down.

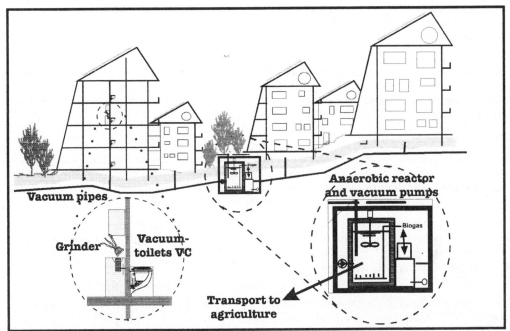

Figure 1:   **Principle of the vacuum system for collection and transport of faeces with biowaste and semicentralised anaerobic treatment**

One concern for treatment of faeces mixed with urine in digestors is the ammonia concentration. If it is higher than about 2500 to 3000 mgN/l a self poisoning of the system can occur. Vacuum toilets use about 1 litre of water per usage resulting in NH_4-N concentrations of about 2000 mgN/l. There-fore special care has to be taken especially with the start-up phase of the digestor.

Another concern about the proposed system is the energy demand. The traditional system requires about 10 watts/person for wastewater transport and treatment and additional 10 watts for water supply [8]. Compared to this the planned system saves about 25% of the water supply lowering the demand to 7,5 W/P. The vacuum system requires approximately 3 W/P while the digestor produces about 7 W/P from biogas. From the biogas about 1/3 have to be used for heating of the reactor and hygienisation of the feed leaving 4-5 W/P for use. The transport of the sludge to agriculture can be in a wide range. With a medium one way transport distance of 25 km and two truck transports a month the energy demand is about 3-4 W/P with a wide variation. All in all the overall energy demand of the system is about 10 W/P compared to the 20 W/P of the traditional concept. The sludge from the digestor can avoid production of artificial fertiliser. The production of 1 kg nitrogen for fertilisers from the air costs about 1000 Wh. Each person produces about 12 g of nitrogen per day allowing energy savings of 12 Wh/d or 0.5 W/P.

Transport of the digested sludge to agriculture is not very difficult in smaller towns and medium size cities. For large cities with millions of inhabitants served by the vacuum system transport distances grow because of the large quantity of sludge and the long way through the city. However adequate solutions could be found by making use of the infrastructure for rail-transport of persons. These systems are unused from the late night-time to the early morning hours. For example there could be transport wagons with standardised containers for subways being loaded underground and moved to the railway at adequate places.

Decentralised treatment of grey wastewater should be done by biofilm processes. Appropriate technologies are aerated sandfilters, constructed wetlands, rotating disk plants and trickling filters. Treated greywater can be feeded to the stormwater storage- and infiltration system, making double use of this structure. Greywater is relatively easy to treat because it has low contents of nutrients. There may even be a lack of nutrients for incorporation during the start-up of the greywater treatment system. As soon as there is a sufficient biofilm the micro-organisms can reuse nutrients released by lysis.

The proposed system is one possibly sustainable system for cities of the future and shall be evaluated in this pilot project now. From the mass balances it seems to fit into natural cycles, save a lot of water and can be operated with little energy-input. From first calculations it seems to be economically feasible, too.

CONCLUSIONS

The traditional sanitation concept has several severe disadvantages: it needs too much water, dilutes faeces and rises nutrient levels in the seas even with very advanced treatment plants. 'Sustainability' is a concept like cleaning rivers and can be operationalised by defining rules for local action. Comparing systems by their respective mass- and energy-balances show possibly sustainable ones.

A promising possibly sustainable sanitation concept for humid urban areas is:
- ◎ source control for faeces and urine with vacuum toilets and co-collection of biowaste to anaerobic digestors
- ◎ Decentralised aerobic treatment of greywater
- ◎ Infiltration of stormwater with trough and drain trench system to avoid sewerage system

This system shall be evaluated in a new settlement for 300 inhabitants in the city of Lübeck, Germany. It is a system that may be one adequate solution for sanitation in urban areas of humid countries. Conditions for application have to be evaluated for the future to implement new systems gradually. Collection of faeces by a vacuum system with feeding the sewerage system or transport to the digestors of an existing treatment plant can be an intermediate step. Time for a finally complete change would be when existing sanitation systems are written off technically.

REFERENCES

[1] n.n., World Commission on Environment and Development
        Our Common Future, p 43, Oxford University Press, Oxford, England (1987)
[2] Peter C. Dienel, Die Planungszellen - eine Alternative zur "Establishment-Demokratie"
        Westdeutscher Verlag, Opladen, Germany (in German) (1995)
[3] Erwin Nolde, Betriebswassernutzung im Haushalt durch die Aufbereitung von Grauwasser,
        Wasserwirtschaft und Wassertechnik 1/95, S. 17ff (in German) (1995)
[4] Ralf Otterpohl, Matthias Grottker, Appropriate sustainable sanitation concepts. UNESCO / IHP
        Symposium 'Integrated water management in urban areas', 26-29 Sept. in Lund, Sweden
        (1995)
[5] Grottker, Matthias, Khelil, Amar. (1989) Urban drainage strategies for small communities in
        developing countries, Proceedings of the Conference on Design and Optimization of Small
        Wastewater Treatment Plants, Trondheim, Norway
[6] Ralf Otterpohl, Jörg Naumann, Kritische Betrachtung der Wassersituation in Deutschland in:
        Umweltschutz, Wie? Symposium: Wiviel Umweltschutz braucht das Trinkwasser? Hrsg.:
        Kirsten Gutke Verlag, Köln, ISBN 3-540-56410-1 (in German) (1993)
[7] F. Sieker, Plädoyer für die allgemeine Anwendbarkeit des Versickerungsprinzips, SuG Nr. 23, S.
        3 ff (in German) (1993)
[8] Willi Gujer, EAWAG/ETH Zürich, Personal information (1995)

Environmental Research Forum Vols. 5-6 (1996) pp. 233-240
© *1996 Transtec Publications, Switzerland*

Comparison of the Performance of Alternative Urban Sanitation Concepts

W. Gujer

Swiss Federal Institute for Water Science and Technology (EAWAG) and
Swiss Federal Institute of Technology (ETH),
CH-8600 Dübendorf, Switzerland

Keywords: Compost Toilet, Energy Requirement, Nutrient Recycling, Storm Events, Treatment Performance, Urine Separation

Abstract: Four alternative concepts for urban sanitation (traditional technology, compost toilets, urine separation and a combined strategy) are compared with regard to different aspects of system performance: Nutrient removal from wastewater, requirements for energy and chemicals, recycling of nutrients, water consumption, combined sewage overflow during storm events, etc. Urine separation proves to have merit relative to today's technology of central treatment of combined wastewater. Compost toilets for feces separation are only an interesting alternative if they are combined with urine separation.

Introduction

Centralized, large wastewater transport and treatment systems are typically based on conventional technology which has evolved from decades of engineering experience. These systems are designed to provide their service reliably and for low cost. Energy consumption and labor are important cost factors for these systems, they are therefore minimized. Nutrients are not yet considered to be a valuable resource, agricultural use of the sludge produced is therefore in competition with alternative ways of sludge handling (incineration, landfill, etc.).

Ecologically engineered systems are typically developed for small, decentralized treatment in low population density areas. They give priority to recycling of nutrients and reuse of resources, frequently at increased requirements for labor and possibly energy.

In the future sustainable technology will have to be utilized in high population density urban areas. Its goals will possibly be to minimize energy consumption and maximize resource recycling.

In this report I would like to compare the performance of traditional technical solutions for urban sanitation with that of alternatives, which have been developed under the notion of ecological engineering or more sustainable technology.

Characterization of the systems

In order to compare the performance of different sanitation systems, four alternatives are introduced below. Clearly none of these four systems will ever be applied in its pure form; we will always observe combinations of these four and additional possibilities for sanitation.

Conventional System

The conventional system considered is designed in analogy to the existing urban water management system in the city of Zurich: Water supply uses a mix of spring-, ground and surface water. 0.4 m^3 d^{-1} cap^{-1} are distributed for domestic as well as industrial and trade use. 0.65 m^3 d^{-1} cap^{-1} of wastewater (from water supplied, infiltration and storm water) are treated in modern wastewater treatment plants which include nutrient removal and tertiary filtration (according to European standards).

All information used to describe the performance of this system stems from the yearly reports of the water supply as well as the urban drainage authorities of the city of Zurich [1][2]. It relates therefore to full scale and practical experience with these systems.

Compost Toilets

For this alternative system it is assumed that the conventional system is changed only with regard to the fact that compost toilets are used to retain feces, but that all other aspects of urban sanitation remain unchanged. Urine is assumed to be completely separated, directed to the public sewer and treated in the centralized facilities.

One compost toilet is assumed serve 3 persons. It includes continuous forced ventilation with a ventilator of 6 W (2 W cap^{-1}). In order to maintain an aerobic precomposting process in the toilet 80 g d^{-1}cap^{-1} of dry structural material (saw dust, etc.) must be added to the toilet.

It is clear, that such an alternative system is not very appealing, however in order to evaluate whether a development into the direction of sanitation alternatives is attractive or not, it is good practice to isolate and analyze single steps rather than to postulate an integrated change of the system which may not be realized in full extent based on the background of the existing system.

Urine Separation

An alternative to compost toilets with feces separation would be consistent separation of urine. Larsen and Gujer [3] propose a system, where urine is collected in individual houses and is then released into the existing sewer such that all urine arrives simultaneously at the treatment plant. If this release is during the night, the urine would only slightly be diluted by nearly unpolluted infiltration water. At the entrance of the treatment plant the concentrated wastewater is collected from the sewer and is directed towards separate handling facilities for urine, where nutrients could be extracted.

This proposed system has many advantages even if only a fraction of the urine is separated. An interesting aspect of this system is that the transition period from an existing to a new system yields advantages in all intermediary phases [3]. Further it appears that this strategy would be acceptable for the general public even under today's urban conditions.

Several possibilities exist to handle urine. P.e. decentralized partial nitrification of urea to NH_4NO_3 could stabilize nitrogen in the urine. A summary of further rather conventional technologies is given by Siegrist [4], they include precipitation as $MgNH_4PO_4$, air stripping to NH_4SO_4 or biological nitrification / denitrification. Within season a significant fraction of separated urine could be used as a fertilizer in decentralized urban agriculture and gardening. Here no specific solution is suggested since only little full scale experience with these technologies exists.

Combined Strategy

Finally compost toilets and urine separation could be combined. This would lead towards complete separation of sanitary wastewater (blackwater) and would leave the treatment of graywater and industrial as well as trade effluents to the public treatment plants.

For simplicity it is assumed here that urine separation and compost toilets would just be combined similar to the individual strategies. In reality decentralized alternatives for urine handling would probably dominate.

Table 1: Comparison of material and energy flux in the four alternative urban sanitation systems. All information is given per day and per capita. The absolute values are derived from the present situation in the city of Zurich. Each inhabitant includes a proportional share of industrial and trade effluents.

System design:	Traditional strategy	Compost toilets	Urine separation	Combined strategy	Units
Water consumption	0.40	0.39	0.35	0.34	$m^3d^{-1}cap^{-1}$
Wastewater production	0.72	0.71	0.67	0.66	$m^3d^{-1}cap^{-1}$
Pollutant loading of treatment plant					
Organic C	58	41	53	36	$gCd^{-1}cap^{-1}$
Total N	10	8.5	2.5	1.0	$gNd^{-1}cap^{-1}$
Total P	2.8	2.2	2.0	1.4	$gPd^{-1}cap^{-1}$
Electricity consumption: Total	17.2	18.2	>12.1	>14.3	$W\,cap^{-1}$
Water supply	11.0	10.7	9.6	9.4	$W\,cap^{-1}$
Wastewater treatment	8.2	6.5	5.0	4.4	$W\,cap^{-1}$
Compost toilet (forced Ventilation)	-	2.0	-	2.0	$W\,cap^{-1}$
Production from Biogas	-2.0	-1.0	-2.5	-1.5	$W\,cap^{-1}$
Urine handling	-	-	unknown	unknown	
Material and chemicals for operation:					
Saw dust (structure)	-	80	-	80	$gTSd^{-1}cap^{-1}$
'Methanol' for denitrification	-	10	-	-	$gd^{-1}cap^{-1}$
Ferrous sulfate for P precipitation	25.5	18.0	13.0	10.2	$gd^{-1}cap^{-1}$
Urine handling	-	-	unknown	unknown	
Sludge and compost production					
Digested sludge (dry)	80	58	70	40	$gTSd^{-1}cap^{-1}$
Compost (dry)	-	80	-	80	$gTSd^{-1}cap^{-1}$
Products from urine handling	-	-	unknown	unknown	
Area for P use (22 kg P $ha^{-1}a^{-1}$)					
Digested sludge	460	360	330	230	m^2cap^{-1}
Compost	-	100	-	100	m^2cap^{-1}
Urine handling	-	-	130	130	m^2cap^{-1}
Loss of heat due to ventilation *	-	13	-	13	$W\,cap^{-1}$
Transport of compost and sludge	0.5	1.5	0.5	1.5	$W\,cap^{-1}$
Handling of urine	-	-	unknown	unknown	
Total primary energy requirement**	52.1	82.2	>36.8	>68.4	$W\,cap^{-1}$

*   Assumption: An additional 60 kg dry air $d^{-1}cap^{-1}$ (= 50 $m^3d^{-1}cap^{-1}$) is heated from 10°C to 20°C with a relative humidity of 50%. This amount of air is also required in order to evaporate approx. 200 g $d^{-1}cap^{-1}$ of water.

** Sum of all forms of energy accounted for in this table. Electricity is multiplied by 3 in order to account for its increased value. Organic material added (saw dust, methanol) is converted to energy. Gray energy and energy for urine handling is not considered.

Comparison of the Performance of the Four Systems

The four systems are designed in order to yield equal performance in view of water pollution control. Differences will result with regard to the materials and energy used and the amount and the location of products generated. Table 1 gives a summary of different results:

- Water consumption and wastewater production will be affected by the frequency and the amount of water used to flush conventional toilets (60 l $d^{-1}cap^{-1}$ including public toilets).

- The pollutant loading of the treatment plant depends on the composition of the materials treated by non conventional technologies. Feces is rich in organic material and phosphorus, urine contains primarily soluble nutrients, such as nitrogen and phosphorus [3].

- The electricity consumption of the water supply is proportional to the water consumption (pumping). In wastewater treatment plants electricity is primarily used to drive the aeration system. The degradation of organics and the requirement for nitrification determine the amount of oxygen used. Urine separation results in a situation where nitrification will not be required any more since the wastewater will nearly be nitrogen limited. Compost toilets require ventilation, it is possibly easier to control loss of heated air under winter conditions if forced ventilation is used. Biogas production depends on the amount of biodegradable organics that reach the digester. Primary sludge (which includes a substantial fraction of feces) yields large amounts of biogas whereas secondary sludge from biological systems does not. If nitrification of the wastewater is required (no urine separation) biological solids are stabilized in the treatment plant due to the high solids retention time required - this results in a fractional loss of biogas and requires additional oxygen for treatment.

- The operation of compost toilets requires the addition of carbon rich organic materials in order to provide the structure to support aerobic conditions and a high C : N ratio in the precomposting phase of the feces. A lack of organic materials will make it difficult to obtain denitrification in wastewater treatment plants. Feces separation will result in the requirement of the addition of organic substrates (p.e. methanol) in order to obtain sufficient nitrogen removal. Ferrous sulfate is a chemical (frequently a waste product) for simultaneous precipitation of phosphorus. The amount of ferrous added depends on the amount of phosphorus to be precipitated after the requirements of the biomass have been satisfied.

- The production of digested sludge (here expressed as dry solids) depends on the amount of organics and the precipitates from phosphorus removal that reach the digester. The production of compost depends on the amount of feces and the organic material added to the toilets.

- The total requirement of agricultural area for the recycling of the nutrients is governed by the amount of phosphorus contained in the products. These products are generated either decentralized at the source (compost) or in a centralized facility (treatment plant) and may require storage and transport to the site of utilization.

- Ventilation of compost toilets requires the use of preheated air under winter conditions. The moisture and odor content of this air would make it difficult to use its heat content in heat exchangers. In most cases this air would have to be heated in addition of today's heat requirement. A minimum of air is required in order to evaporate some of the moisture of the materials introduced into the toilet.

- Compost and digested sludge will have to be collected and transported to the user. Decentralized technology requires more energy (and labor) for product collection.

- The sum of primary energy used in the different sanitation alternatives is composed of electricity, extra heat requirements, chemical energy in organic materials (which otherwise could be utilized as fuel) and energy used for transport. Electricity is multiplied by three in order to convert this energy to primary energy.

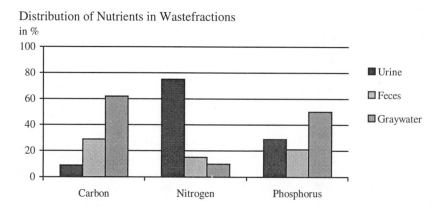

Figure 1: Distribution of the nutrient content of the different contributions to today's wastewater. Basis is the present situation in the city of Zurich. Graywater includes trade and industrial effluents.

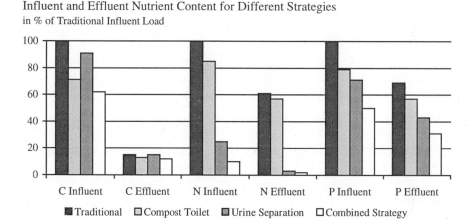

Figure 2: Comparison of the nutrient content in the influent and the effluent of the treatment plant with different sanitation strategies, without nutrient removal in the treatment plant. The influent in the traditional strategy is chosen as 100%.

Discussion

Nutrient Distribution in Waste Fractions

Figure 1 is a summary of the fractions of nutrients in the different waste streams. Whereas the largest fractions of carbon and phosphorus are contained in graywater (at dilute concentrations however) nitrogen is dominantly contained in urine. The phosphorus content of graywater may in the future be reduced if a ban for phosphorus in all consumer products and not only in washing powders would become active (The relevant technology is mostly available). In Figure 1 it is particularly interesting to compare the widely differing composition of feces and urine.

Figure 2 summarizes the carbon, nitrogen and phosphorus content of the treatment plant influent and effluent for the four different sanitation strategies. It is assumed that nutrient removal (P precipitation, denitrification) is not operative. For all nutrients 100% is equal to the nutrient load of the influent in the traditional strategy. The residual organic carbon in the effluent strongly depends on the characteristics of the graywater and is only marginally affected by the sanitation strategy. Nitrogen may become the limiting nutrient if urine is separately handled, this explains why nitrification / denitrification may not be necessary if consistent urine separation is achieved. As long as phosphorus is dominantly contained in graywater, phosphorus removal to a different extent will be required in the central treatment facility if discharge requirements typical for northern Europe are applicable. It is well known that treatment of graywater is no easy task - partially due to nutrient limitations.

Conclusion: The nutrients (N, P) in today's wastewater originate primarily from urine and graywater. In order to simplify the required centralized treatment technology (omit P precipitation and nitrification / denitrification) it is essential that phosphorus and nitrogen are controlled at the source. Urine separation and a phosphorus ban in consumer products are possible strategies.

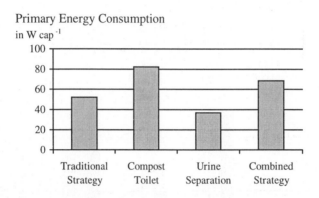

Figure 3: Consumption of primary energy for urban water management (including water supply and waste water treatment) in the four sanitation strategies. Primary energy includes electricity, heat, organic chemicals and transport. Gray energy is not included, electricity is multiplied by three in order to account for its value. Energy for separate urine handling is not included.

Energy Consumption

Figure 3 summarizes the consumption of primary energy for the four sanitation strategies for the entire task of water supply and wastewater treatment. Due to the fact that compost toilets require extra heating energy and organic materials for toilet operation and wastewater treatment, this strategy results in extra energy consumption. Urine separation has the advantage that less electrical energy is used for wastewater treatment, since nitrification is not necessary. An additional energy requirement for urine handling is however not included in these figures, it may be substantial, depending on the technology used.

Conclusion: The traditional sanitation strategy is based on large scale, industrial technology. It is difficult to compete with its optimized energy requirements. Compost toilets are probably not advantageous from the point of view of energy consumption. Depending on the handling technology used, urine separation may be competitive.

Chemical Resources

Figure 4 is a summary of the operating materials (chemicals) used in the four sanitation alternatives. With decreasing amounts of phosphorus in the influent to the treatment plant, decreasing amounts of precipitants ($FeSO_4$) are necessary. Methanol is only required if compost toilets are used to retain carbon rich material but not the dissolved nitrogen. Saw dust supports the operation of compost toilets. Depending on the technology applied, urine handling might result in very high requirements of chemicals if the nutrients are to be stored in dry form for seasonal application. These chemicals are not included in Figure 4.

Conclusion: The traditional strategy results in the lowest requirement for chemicals. The use of alternative strategies for urban sanitation can only achieve equal performance of water pollution control if additional materials are used. For phosphorus the recycling into gardening and agriculture is equally achieved in all four alternatives. For nitrogen, recycling is best obtained with urine separation, relevant technologies are however not economically interesting at this moment (See also [4]).

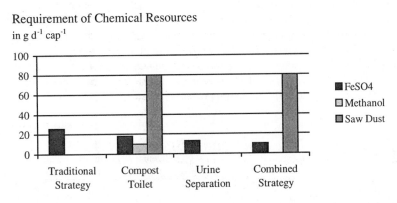

Figure 4: Requirement for chemical resources in the four sanitation strategies. $FeSO_4$ (Ferrous Sulfate) is required for P precipitation, Methanol serves as an additional substrate to support denitrification, 'Saw Dust' is required as a structural additive in compost toilets. Methanol and 'Saw Dust' are also included in the primary energy requirement.

Land Requirement for Nutrient Recycling

A typical population density in a residential urban area is in the order of 60 persons ha^{-1} (or 167 m^2 cap^{-1}). This leaves less than 50 m^2cap^{-1} of area which could potentially be supplied with nutrient rich compost or urine. Since 100 to 230 m^2cap^{-1} of area could be supplied with the nutrients contained in this waste fraction (See Table 1) collection and transport of these products towards agricultural areas is mandatory. Together with the nutrients accumulated in the digested sludge from the treatment plant an area of 460 m^2cap^{-1} is required for the recycling of all the phosphorus in the waste streams. This amount of land may typically be found in the environment of urban areas, at least if agriculture does not depend on the import of nutrients in the form of fodder, a procedure which does not appear to be sustainable on the long run.

Conclusion: All strategies with decentralized handling of waste or nutrient streams will depend on a collection system for the products of the waste handling procedures. Centralized waste handling will minimize transport requirements, here collection is achieved via the transport in the sewers with a minimum of external energy requirements.

Performance During Storm Events

During storm events combined wastewater must be discharged with only little treatment to the receiving waters. Today the loss of ammonium in combined sewage overflow (CSO) is of major concern because of possible toxicity of NH_3 to fish. Urine separation has the advantage that ammonium will not be contained in the CSO. Feces might cause a hygienic and aesthetic threat to receiving waters which is reduced with compost toilets.

Conclusion: During storm events the traditional sanitation strategy is inferior to the alternatives evaluated.

Conclusions

Four alternative sanitation strategies for urban areas are compared with regard to performance, energy and chemical resource requirements etc. Table 2 is a summary of the ranks of these four alternatives with regard to a variety of evaluation criteria. Before a specific strategy is chosen, it would of course be necessary to include more criteria such as investment and operating cost, reliability, error tolerance, etc. From Table 2 it is evident however, that compost toilets are not a suitable alternative to the traditional system under today's urban conditions. Urine separation may however prove to have potential if a technology is found which allows for cheap and reliable recycling of the soluble nutrients in urine. The combined strategy of compost toilets and urine separation combines the advantages of urine separation with the disadvantages of compost toilets and can probably not compete with the traditional strategy.

Table 2: Summary of the ranks of the four alternative sanitation strategies with regard to a variety of evaluation criteria (Rank 1 = best, 4 = worst scenario).

System design:	Traditional Strategy	Compost toilets	Urine separation	Combined strategy
Water consumption	4	3	2	1
Pollutant loading of treatment plant	4	3	2	1
Wastewater composition for efficient treatment	2	3	1	4
Problems arising during storm events	4	3	2	1
Total Electricity consumption	3	4	1*	2*
Material and chemicals required for operation:	2	4	1*	3*
Sludge and compost production	2	4	1*	3*
Area requirement for phosphorus recycling**	1	2	3	4
Possibilities for nitrogen recycling	3	4	1	2
Total primary energy requirement	2	4	1	3
Public acceptance	1	3	2	4
Total Ranking Points:	28	37	17*	28*

*   These ranks do not include the (negative) effects of separate urine handling

**   Priority is given to the area which is required for the utilization of products generated in central facilities - these do not require a separate collection system.

Literature

[1]   Wasserversorgung Zürich, Geschäfts- und Untersuchungsbericht 1992

[2]   Tiefbauamt der Stadt Zürich, Stadtentwässerung, Jahresbericht 1992.

[3]   Larsen T. and W. Gujer (1996) 'Separate Management of Anthropogenic Nutrient Solutions', Submitted for review, Wat. Sci. Tech.

[4]   Siegrist H.R. (1996) 'Nitrogen removal from Digester Supernatant - Comparison of Chemical and Biological Methods', Submitted for review, Wat. Sci. Tech.

Environmental Research Forum Vols. 5-6 (1996) pp. 241-246
© *1996 Transtec Publications, Switzerland*

Production of High Quality Soil from Pottery Waste and Sewage Sludge

H. Obarska-Pempkowiak

Faculty of Hydrotechnics, Technical University of Gdansk,
ul. G. Narutowicza 11/12, PL-80-952 Gdansk, Poland

Keywords: Sewage Sludge, Solid Waste, Reed Beds, *Phragmites communis*, Composting, Soil, Pottery Waste

Abstract

In a pottery factory at Łubiana (Gdańsk district, Poland) technological wastewater together with surface run-off is directed to a sedimentation pond. A total of 312 m^3 per year (45 t dry matter per year) of sediments settlers there within one year and has to be utilized together with 31.0 m^3 per year (53 t dry matter per year) of material from a sand filter. Analysis of composition revealed that the gatherd sediments comprised some 20% of organic matter as well as 6.2 mg N/g and 0.6 mg P/g. Apart from this, in the factory some 70 m^3 per year (16 t dry matter per year) of sewage sludge is separated from domestic wastewater in an Imhoff tank. This is removed to a domestic wastewater treatment plant localized 15 km away from Łubiana. In 1994 new technology of utilization of generated sediments (both from the sedimentation pond and the Imhoff tank) was introduced. It is based on dewatering of sewage sludge in four reed beds (21x10 m each) and composting material rich in organic matter obtained there with sediments recovered from the sedimentation pond. After 6-9 monthes of composting a soil like material rich in organic matter (15%), organic nitrogen (0.9%) and phosphorus (0.08%) is produced absent of phatogens and excessive heavy metals. The material supports growth of trees and grass due to suitable structure and fertilizer properties. The total amount of material obtained within one year will be some 100 t dry weight. This will be applied in recultivation of clay excavation areas during the next few years.

Introduction

One of major problems connected to the operation wastewater treatment plants is the excess of sewage sludge. Most often purification of wastewater of organic matter and biogenic elements is of primary concern due to the need of protection of surface waters. Utilization of sewage sludge ecompassing dewatering, chemical and biological stabilization and actual utilization is, in Poland, quite often left to rather inefficient processes. Sewage sludge is rich of nutrients (nitrogen and phosphorus) and organic matter. This suggests utilization of sewage sludge as organic fertilizer, especially in sandy soils (in Poland some 30% of cultivated land are sandy soils). Unfortunately microbiological and parasitological properties of sewage sludge prevent its direct utilization as fertilizer. This is frequently accompanied by excessive content of heavy metals [1, 2].

In several industries utilization of refuse is of primary concern due to toxicity or/and volume. Elaborated neutralization procedures are employed in case of toxic refuse. Chemically neutral refuse causes problems due to the volume and as a source of dust. Most often no sensible way of utilization but storing is employed.

In this paper a method of simultaneous utilization of sewage sludge and industrial refuse composed of pottery clay in the Table Pottery Plant in Łubiana (northern Poland) is presented. The method is based on effective dewatering and stabilization (chemical and biological) of sewage sludge in reed beds followed by composting with industrial refuse. The product of composting, a soil - like material, is utilized in recultivation of clay extraction areas.

Sewage and solid refuse management in the plant
The production of pottery is based on quartz and koalinite clay which after milling are mixed with plastomers and water. After initial dewatering by means of filtration moulds are filled with the mixture and pottery is produced in ovens.

Liquid waste generated in the course of production comprises very fine suspension. This is removed in primary settlers and by flocculation (flocculants are utilized) in two reaction tanks.

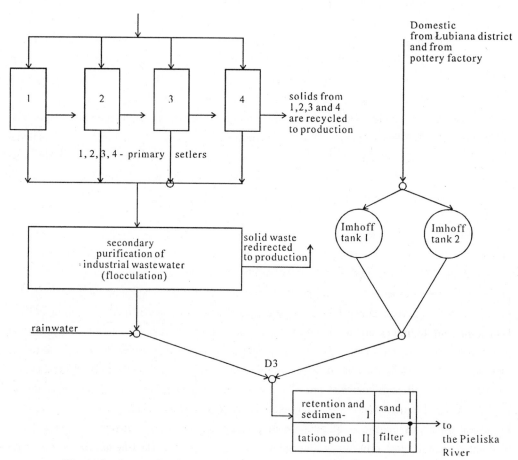

Fig. 1. A scheme of technology wastewater treatment applied in pottery factory

Floculated sediments are dewatered by means of filtration, and reutilized in production. Purified water of the total volume of 467 m³/day is directed to the sedimentation pond. To this pond rain water and water after sewage treatment are also directed. Rain water is collected by means of rain sewer from the area of 12 ha. Concentration of suspended matter in water draining to the retention and sedimentation pond from rain sewer is about 1200 mg/l. In the rainy weather the volume of water from this source is about 500 m³/day. From sedimentation pond some 102 t/year of inorganic sediments originate.

Sanitation sewerage drains sewage from houses inhabited by employees of the plant and from offices to two Imhoff tanks. Sedimentation volumes of the tanks are equal to 14 m³ and 39 m³. Total volume of domestic sewage directed to the sedimentation pond is 239 m³/day.

Industrial sewage (after purification), rain water and domestic sewage (after purification) meet in the sewage well D3 (Fig. 1) and are directed through a common sewerage to the sedimentation pond. In the pond the final purification of wastewater takes place. The pond (13m x 36m) consists of two chambers. Water purified of suspended matter is drained from the pond through a sandy filter and a drainage system. Apart from collecting suspended matter, the pond due to its large volume (650 m³), retains rain water of heavy rains, together with excessive suspension it carries.

Quality and quantity of the solid waste generated in the plant
Solid waste originate from three sources synonymous to three sorts of wastewater (Table 1). Composition and quantity of solids in the domestic wastewater are stable. It was found [4], that in this particular instant composition of solids collected in the Imhoff tanks differ depending on the season. Thus there are „winter" and „summer" sediments comprising 7% and 10% dry matter respectively in winter (November, December, January, February) or summer (March, April, May, June, July, August, September, October).

Table 1   Volumes and amount of sediments generated in the Table Pottery Plant in Łubiana

Type of sediment	Volume m³/year	Amount t/year	Comments
Domestic from Imhoff tank - „winter" - „summer"	70.0 109.0	5.2 10.9	not stabilized (fermented) partly stabilized
Mixed, from sedimentation pond	312.0	45.0	high organic content due to improper operation of Imhoff
Sand from sand filters	31.0	53.0	almost „pure" sand
Total	491.0	141.1	

Composition of sediments in the retention tank was established experimentally (Table 2). Samples of sediments were collected at the inlet to the tanks in the middle part, and close to the sand filter (outlet). The measurements revealed that sediments differed substantially from both suspended matter of industrial wastewater, and sediments of the Imhoff tanks. Organic matter consisted 32% of sediments at the inlet, while at the outlet it consisted just 19% of sediments [3]. Inorganic portion of sediments consisted predominately (from 86% to 98%) of fine particles (0.75 > d > 0.102 mm). Properties of sediments from the retention pond are presented in Table 2. The fine suspension causes kolmatation of the sand filter. Therefore sand has to be replaced twice a year. The volume of sand is 15.6 m³, which additionally makes some 53 ton of inorganic sediments per year. The balance of sediments generated at various sources in the plant is presented in Table 1.

Table 2   Quality sediments collected from different parts of the retention and sedimentation pond

Parameter	Sampling place		
	At the begining	In the middle	At the outlet
moisture, %	88.02	87.53	87.33
dry matter content, %	11.98	12.47	12.67
mineral substances, %	67.50	72.63	81.48
organic matter, %	32.50	27.37	19.52
total nitrogen, mg/g dry matter	6.07	6.25	6.76
total phosphorus, mg/g dry matter	1.21	0.44	0.66
density, g/cm³	1.18	1.20	1.22
0.75 > d > 0.102 mm	86.3	-	98.3

Utilization of organic and inorganic solid waste

Up to 1993 solid wastes generated in the plant were either transported to a nearby sewage treatment plant (15 km away from Łubiana) or stored in piles. Fine suspension was discharged to the Pieliska River. The reconstruction of the waste management system in 1993 brought about changes in the manner the solid wastes were utilized. The new approach was based on the following principles (Fig. 2):

- construction of one additional Imhoff tank, to accomodate increased stabilization time of sediments in winter,
- introduction of activated sludge treatment of domestic wastewater after the Imhoff tanks,
- separation of suspension from industrial wastes and its reutilization in the production,
- final purification of liquid wastes in the retention pond,
- local utilization of solid wastes (sediments from the Imhoff tanks, sediments from the retention and sedimentation pond, sand from filters).

Up till now the first two presumptions have not been implemented. Nevertheless the sludge originating in the existing Imhoff tanks is stabilized locally in 4 beds grown with reeds. Each bed (21m x 10m) is covered once in two monthes with a layer of 10 cm of sediments from an Imhoff tank. This is followed by sediments from the retention and sedimentation pond. Sediments from sand filters are stored separately. Materials originating from reed beds and sand are mixed in the proportion 1:1. Up till now one such event took place. About 100 t of soil like material was produced. The material, dark in colour, lacking of smell, and not attracting insects was characterized by the following properties: organic nitrogen: 0.82%, phosphorus: 0.11%, organic matter: 10.7%.

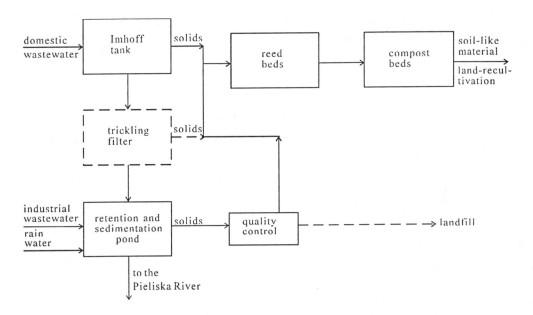

Fig. 2. The scheme of utilization of solid wastes in the Table Pottery in Łubiana

The first portion of the material was utilized to recultivate pottery clay excavation areas filled with used pottery forms (40 t/year) and broken pottery (90 t/year). Of 1.5 ha, of areas allocated for recultivation, 0.05 ha was actually covered with the new material and planted with grass and willow. The second event is planned to take place in spring 1996.

Conclusions

Sediments from Imhoff tanks and solid industrial wastes can be jointly neutralized in reed beds. Dewatering and biological stabilization take place in the beds. Next the sediments are composted to produce fully stable, soil-like material. The soil-like material can be locally utilized in recultivation of areas devastated with clay excavition and storing of broken pottery.

References

[1] K. Hoffmann, Proceedings of the Conference: Use Constructed Wetlands in Water Pollution Control, Cambridge, 269-277 (1990).

[2] E.H. Nicoll, Small water pollution control works. New York, John Wiley & Sons, 394 (1988).

[3] H. Obarska-Pempkowiak, Technical University of Gdańsk, Faculty of Hydrotechnics, manuscript, 7 pp (1993).

[4] H. Obarska-Pempkowiak, K. Banach, Technical University of Gdańsk, Faculty of Hydrotechnics, manuscript, 28 pp (1994).

Environmental Research Forum Vols. 5-6 (1996) pp. 247-248
© *1996 Transtec Publications, Switzerland*

Workshop 3:

The Social Side of Wastewater Treatment: Politics and Markets

Summary and Conclusion by the Chairperson

S. Fassbind

Seemattstrasse 17a, CH-6330 Cham, Switzerland

The speakers Maciej Dzikiewicz, Eduardas S. Davidavicius, Anders Finnson and Anna Peters, François Gigon, Joachim Niklas, Thomas L. Crisman, Janusz Niemczynowicz presented mostly technical details of their work and their projects and/or general ideas about the present and the future in a financial and political context.

The big question was: How can we convince government and the people to build natural wastewater treatment plants even if these plants could be costlier than conventional systems (of which all the costs never were fully accounted for). Eco-colonialism is outdated, the construction industry is a powerful lobbyist, engineering people usually do not know how to sell, new techniques are "uncertain, not proofed". Thinking in marketing terms, we have to find a USP (unique selling proposition), which means that product, price, distribution and communication have to be thought over and perfected all the time and in every case. And there have to be found incentives, bonus systems, attractions instead of punishment in order to induce the people to follow the ecological path.

Possibilities and ways how to convince:
- the people by deeds and actions (if possible with pilot projects), by talks person to person, which may be a big effort, by education in schools and during everybody's whole life time
- the industry that it can save money and improve its image
- the politicains by the good quality of the plant, by efficient work providing jobs for local people, by the low costs of planning, of construction, of maintenance (accounting in middle and long terms).

Success often beginns on a local scale with one or a few persons and the solution is cheaper and faster achieved in the end. We can help local people to find local solutions so that they can do what they really want to have and can earn a living by building and maintaining it. We can help create an autonomous development.

Decision:
Our workshop group wants to keep in touch and start cooperation.

Environmental Research Forum Vols. 5-6 (1996) pp. 249-254
© *1996 Transtec Publications, Switzerland*

The Use of Macrophytes for Wastewater Treatment in Rural Areas of Poland – Social Aspects

M. Dzikiewicz

Water Supply Foundation Skwer Kard. S., Wyszynskiego 6, PL-01-015 Warszawa, Poland

Keywords: Reed Bed, Aquatic Plant, Duckweed, Costs, Acceptability, Social Aspects

Abstract

The number of wastewater treatment plants in rural Poland has increased in recent years . But there are not many natural systems among them . This paper presents a brief overview of the reed bed systems and aquatic plant systems currently in operation and underconstruction in Poland , especially regarding capital and operating costs , performance , reliability and acceptability by local communities , regulatory and funding agencies . This should give the answer to why there are not more natural systems in rural areas. This paper discusses prospects for and limitations to such systems under Polish conditions , where there are the same high standards of nutrient removal for big municipal plants and small units (> 5 m^3/ d).

Introduction

Restoration of the democratic system in Poland in 1989 brought changes in economic policy . Freely elected local governments became responsible for each public experditure made in their communities. Because the budget of the community is usually too small for the needs . There is always the problem of what should be done first and how much it will cost . In such competition wastewater treatment projects are at the end of the list of needs in many communities. In fact the costs of sewage treatment systems are highest among all needs and without outside support are beyond the reach of most if not all rural communities .

According to statistical data , 38.2% of the population of Poland live on 2,599,101 farms in rural areas. The number of farms connected to water supply systems have increased rapidly in the last few years . In 1994 it reached 45.7 % of all farms . The rest of households are still using shallow wells for water supply . But the number of farms connected to wastewater treatment systems (4% of all farms) show that the process of construction of such facilities in rural area has just started . 49% of households are not equipped with flush toilets . There are privies still in use . Most of the farms have cesspools or on-site treatment poorly designed and constructed which cause ground water pollution . This situation is due to the effect of long lasting neglect of private agriculture by communist authorities. The neglect of rural areas in the line of civilisation caused , and still causes many social problems . Lack of infrastructure is also an obstacle in the development of small industry in these area. Local governments which understand this situation and have public support and willingness to solve the problem try to build at least one wastewater treatment system in their community . Therefore, local governments have been and still are, searching for more cost-effective solutions .

Macrophyte - based wastewater treatment systems in Poland

In the early nineties the concept of treating wastewater with aquatic macrophytes-based systems was introduced to Poland by West European countries and USA . The experiences of these countries were successful and a number of systems have been built there. Design manuals have been published [3,5,6]. These systems attracted a great deal of attention in Poland for the following reasons: low construction and maintenance costs; simple construction no mechanical equipment and, possibly, no need for electrical equipment; low energy requirements; more flexible and less

susceptible to shock loading; environmentally acceptable with potential for wildlife conservation .
Aquatic macrophyte-based wastewater treatment systems may be classified into [2] :
1. Floating macrophyte treatment systems ,
2. Submerged macrophyte treatment systems,
3. Emergent macrophyte treatment systems divided into:
    - constructed wetland with sub-surface horizontal flow of wastewater
    - constructed wetland with pulse- loaded vertical flow of wastewater
    - constructed wetland with surface flow of wastewater
    - integrated systems.
In Poland only floating and emergent macrophyte treatment systems have been built .

Aquatic plant systems

Floating macrophyte treatment systems, sometimes called aquatic plant systems, which have been
built in Poland are designed and constructed as duckweed systems by only one commercial firm
representative of American corporation . Duckweed (*Lemna sp.*) is a small , green , freshwater plant
with a leaflike frond a few millimetres wide and a short root usually less than 1 cm long. It has one
of the fastest reproduction rates and , therefore , doubles the area covered , every 4 days . Duckweed
systems are used as tertiary treatment after aerated ponds . Growing plants form a mat covering the
water surface of a specially designed pond . Because this mat is susceptible to movement by wind ,
patented hexagonal floating barrier grids are used to minimise disturbance. The duckweed plant
mat prevents light penetration into the water column , suppressing the growth of algae. Biological
reactions in the water column , from aerobic to anoxic to anaerobic are stabilised , resulting in
efficient treatment within this system. Beside the floating barrier grids the pond is divided by
hydraulic baffles which provide channelled ,serpentine flow to maximise detention time within the
system and under the duckweed plant mat. Growing duckweed plants assimilate nutrients [8].
Periodic harvesting to thin the duckweed plant population permanently removes these nutrients
from the water column . A patented floating harvester (only one in Poland) or a simple boat with
an operator that ride over the top of the floating grid is used for excess plant removal . Harvested
duckweed biomass may be used as compost or animal fodder. The Zootechnic Institute in Kraków is
currently researching the use of duckweed from wastewater treatment as a feed for poultry and
livestock.

Table 1. Duckweed systems in Poland

Site	Flow	p.e.	Depth	Surface area	BOD mg/l in	out	TSS mg/l in	out	N mg/l in	out	P mg/l in	out	Date of const.
	m³/d		m	ha									
Kochcice	650	5850	2.5 - 3.0	1.05	450	30	386	50	40	30	23	5	1992
Rossocha	110	770	2.8	0.37	350	25	300	30	60	30	25	5	1992
Cieszanów	200	1360	3.0 - 3.1	0.58	340	30	370	50	65	30	12	5	1993
Mniów	150	1330	2.5 - 3.0	0.46	442	15	552	30	78	30	15	5	1993
Raków	330	1980	3.85- 4.0	0.60	300	15	290	20	60	15	23	1.5	1993
Wręczyca	643	3860	2.95	1.11	300	30	300	50	50	30	10	5	1994
Przyrów	500	2500	2.5- 2.55	1.09	250	30	286	50	35	30	7.6	5	1994
Stopnica	300	2250	2.5	1.0	375	15	400	30	79	30	13	5	1994
Kargowa	1008	10765	2.9-3.35	2.47	534	15	567	30	56	10	19.7	1.5	1994
Łąka	663	4975	3.0	1.48	375	30	400	50	68	30	12.6	5	1994
Barwice	700	2800	2.5 - 3.0	1.34	200	20	300	20	60	25	11	4	1994
Przykona	200	1600	2.5 - 3.1	0.51	429	30	498	50	58	30	16	5	1994

Because of high nutrient removal requirements all duckweed systems in Poland are supplemented with chemical dosing equipment for seasonal application of iron precipitates . Twelve duckweed systems have been built in Poland . A list of these plants and their design details is shown in Table 1. Next four systems are expected to be complete this year.

Reed bed systems

Most emergent macrophyte treatment systems or constructed wetlands which have been built in Poland use common reeds (*Phragmites australis*) . So the term reed bed systems is used here to describe all systems including those systems that have also been called otherwise in Poland . It is more difficult to identify the design criteria used in existing systems than the term . There is no consensus regarding the design of reed bed systems in Poland . Design approaches appeared to range from "trial and error " to semi-rational using design models based on other European experiences and data . Reed bed systems are divided into three groups according to the following basis of design :

- European design and operations guidelines for reed bed treatment systems prepared by the EC/EWPCA Emergent Hydrophyte Treatment Systems Expert Contact Group [3], mostly derived from Danish experience [1] ; characterised by the use of sand or gravel with horizontal or vertical flow or integrated systems with both flows in few stages ,
- Kickuth reed bed systems or Root-Zone systems design and build by Kickuth licensees ; characterised by the use site soil "improved " with the additions to be stipulated in each case ,
- Independent designs where each system has a different configuration ,or other details such as depth of water or media ,type or size of media ,slope of bed , length to width ratio ,level of pretreatment.

The key features of all these systems are : rhizomes of the reeds increase and stabilise the hydraulic conductivity of the media and the reeds supply oxygen to the rhizosphere where wastewater is treated by bacterial activity in aerobic and anaerobic zones .

There is not an inventory of reed bed systems in Poland . The available data and information were gathered in Table 2. This Table shows that the total number of reed beds systems is not very impressive.

Capital and operating costs

The capital costs of small wastewater treatment plants in Poland vary widely, depending on : contractor prices , region of the country , inflation rate , chosen equipment (Polish producers are much more cheaper) contracts ,technical standards, cleaning efficiencies and supervisors . Therefore, the general opinion that the capital costs of natural treatment should be lower than conventional systems has not been prooved . Fig.1 shows the capital costs of duckweed systems on a unit flow and a unit population equivalent basis respectively to their increases and years of construction . These costs are rather higher than conventional systems due to the high cost of imported know-how and equipment from the USA .

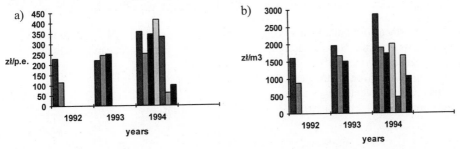

Fig.1. Capital costs of duckweed systems : a) per unit of population equivalent , b) per unit of flow.

Table 2. Reed bed systems in Poland

No.	Site	System type [a]	Purpose	Flow m³/d	p.e.	Surface area m²	Cost new zloty	Date of const.
1.	Wawrów	H/ I	II	208	1300	3500	117 800	1991
2.	Gralewo	H/I	II	200	1300	3325	435 100	1993
3.	Golczew	H/I	II	27	180	1000		1991
4.	Rokitno	H/I	II	42	125	1200		1991
5.	Myślibórz	H/I	II	25/ 40	140	980	213 000	1994
6.	I łowo Osada [b]	H/I	II	747		9550		1993
7.	Margonin [c]	H/I	III	250		141		1992
8.	Babimost	H/I	III	1000		1000		1994
9.	Ostrów n.Gopłem	H/I	III	40		150		1995
10.	Łubowo	H/I	III	332	2200	4900	236 400 [d]	const.
11.	Gronowo Elb.	H/EU	III	350	1600	7950	387 100	1993
12.	Bogaczewo	H/EU	III	50		1200		1992
13.	Krzelów	H/EU	III	100	500	660	196 000	1995
14.	Leszczyny	H/EU	II	53	350	4500	222 300	1994
15.	Kościelec	H/EU	II	110	500	4325	190 000	const.
16.	Nowa Słupia	H/EU	II	325		6480	692 900	const.
17.	Darżlubie	H,V/EU	II i III	242	750	3600	315 000	1995
18.	Sobiechy	H,V/EU	II i III	7.6	48	492	20 000	1994
19.	Dębowa Kłoda	INT	II	150	1500/ 2500	6600	180 000	1994
20.	Przywidz	H/INT	II i III	22	150	870		1992
21.	Jarosławiec	FW	III	45/ 1200	300/ 6000	7000		1993
22.	Wieżyca	FW	III	120	855	18370		1992
23.	Bujny	H/K	II	50	300	1000	115 000	1993
24.	Dąbrowa	H/K	II	20	150	430	55 000	1994
25.	Zarzęcin	H/K	II	20	150	430		const.
26.	Lesznowola	H/K	II	15	100	500	30 000	1993
27.	Kletno	H/K	II	7,6		250	25 000	1993
28.	Bolimów	H/K	II	80/250	1200	1800	554 800	1994
29.	Goworów	H/K	II	8		200	17 000	1994
30.	Szklarka	H/K	II	14		400	43 000	const.
31.	Złochowice	H/K	II	10		256	34 000	1995

[a] Types : H - subsurface horizontal flow system ; V - vertical flow system ; INT - integrated system; FW - surface flow system ; Basis of design : I - independent ; EU - European guidelines K - Kickuth licensees.
[b] natural wetland adopted for treatment
[c] only cattails (*Typha*) vegetation
[d] only costs of reed beds

Construction costs were available from 20 reed bed systems . The values in Table 2. include preliminaries , civil and mechanical and electrical engineering costs. They do not include design and supervision or land acquisition . There are still too few data to report the range of capital costs of reed beds on a unit flows or population equivalent basis. For smaller populations (flow < 50 m^3/d) reed beds are clearly cheaper to install than conventional systems . The capital costs of bigger units can range from much higher to astonishingly lower than conventional systems .

Relatively few data have been accumulated on operating costs of duckweed and reed bed systems . Operating costs of reed bed systems which rely upon gravity without energy requirements and feed by sewer systems may be very low . However , operating costs are usually higher in duckweed systems and the larger reed bed systems especially with intake of sewage brought by trucks where shift operators or guards are employed .

Performance

In Polish wastewater treatment plants with a larger flow than 5 m^3 /d , discharge standards are for BOD (30 mg/l) , suspended solids (50 mg/l) , total nitrogen (30 mg/l) total phosphorus (5 mg/l) . Since 1991, all plants have been designed and constructed to meet these official standards . In some cases the environmental authorities can increase the demands . These high standards are the main limitation for natural systems in Poland . Performance levels in these systems decline somewhat in the winter, but it is still believed that effluent quality can meet these high standards by designing a system which recognises seasonal variations . Thus far, this has not been proven due to the lack of performance data . Monitoring requirements in Poland for small wastewater treatment plants are minimal . The regional regulatory authorities may require collection of only one sample of small plant influent and effluent every two or three months . In some regions they do not require any collection at all . However , all plants have standard checks twice a year by the State Inspectorate for Environmental Protection agencies . As a result , these plants usually do not generate data more frequently than required . Sometimes they have only the control data . A few locations had collected data more frequently during start-up or from periodical research study . Therefore , there is no scientifically valid documentation of all duckweed and reed bed systems performance.

Reliability and acceptability

The need for water pollution control in rural Poland is a good occasion for propagation and application of macrophyte based wastewater treatment systems . Seminars and workshops for environmental engineers designed for exchange of experiences of reed bed systems with foreign experts from UK , Germany and Denmark [1,4,7] have been organised . However , there is not much evidence that reed beds systems built in Poland are effective and reliable . Some of the systems, especially independent designs had maintenance problems during the start up period due to faulty construction ,clogging and short circuiting of flow . Sometimes they had to be reconstructed .

Quite a lot of environmental courses, trainings , and meetings for local governments have been organised (e.g. 148 courses) all over the country by the Water Supply Foundation and were attended by 3752 participants . These courses promote natural systems as low cost , low maintenance solutions . Acceptability of reed bed systems by local governments and regulatory agencies is still rather poor . Reasons for this are as follows :

- the consulting companies which promote these systems are new , small with young enthusiastic designers, without references ; even if they have them, they do not always meet the clients' expectations;
- the capital cost of these systems are difficult to estimate without study of local soil ,groundwater level or sources of needed media materials ;
- the construction of these systems are too simple to work well ; there is some uncertainty regarding operations , maintenance , repair and improvement of these systems ;

- there are not many performance data available especially during the winter season ; usually in rural areas of Poland, wastewater treatment plants are built together with sewage systems so in the beginning the flow is too small and it is difficult to proove later performance ,
- all reed beds require more space than conventional systems this is not always available, and if it is , neighbouring inhabitant may complain of nuisances like odours ,flies etc.
- most consulting engineers and agency staff prefer "proven technology" to new ones .

The last four points also apply to discussion of reliability and acceptability of duckweed systems .

Conclusions

Macrophyte-based wastewater treatment systems such as reed bed and duckweed systems have been used in rural areas of Poland . The number of these systems is not very impressive . However, some of these systems have not always been reliable and cost-effective method for wastewater treatment . Further work is necessary to optimise the design approach for the reed bed system option. There is a lack of performance data at the moment making it difficult to evaluate these systems . This will gradually change as more data become available. As the uncertainties now associated with the use of macrophytes are resolved ,these methods of treatment will gain better acceptability by local communities , regulatory and founding agencies, assuming their rightful place alongside conventional systems .

References

[1] Birkedal K., Brix H., Johansen H., Wastewater treatment in constructed wetland - designers manual , Technical University of Gdansk (1993).

[2] Brix H., Schierup H.H., The Use of Aquatic Macrophytes in Water Pollution Control. Ambio **18,** 100 (1989).

[3] Cooper P.F.,(ed.) European Design and Operations Guidelines For Reed Bed Treatment Systems. WRc Swindon UK(1990).

[4] Dzikiewicz M., Geller G., Hofmann K., Schori U., Oczyszczanie ścieków na filtrach gruntowo-roślinnych . Water Supply Foundation - Centre for applied ecology Schattweid. Cedzyna (1994)

[5] Constructed Wetlands and Aquatic Plant Systems for Municipal Wastewater Treatment Design Manual. EPA/625/1-88/022 ,U.S. EPA, Cincinnati ,Ohio (1988)

[6] Natural Systems for Wastewater Treatment - Manual of Practice FD-16.Water Pollut. Control Fed.,(1990).

[7] Kickuth R., Karsch M., Schroll K., Dzikiewicz M., Sakowski M., Malarski R., Gąsiorowski M., Oczyszczalnie korzeniowe System Kickuth . Water Supply Foundation . Łódź (1994).

[8] Zirschky J., Reed S.C., The use of duckweed for wastewater treatment . Journal WPCF **60,** 1253 (1988)

Environmental Research Forum Vols. 5-6 (1996) pp. 255-260
© *1996 Transtec Publications, Switzerland*

Better Marketing Chances for Advanced Soil Filter Technology in Combination with Publicly Financed Labour

J. Niklas, M. Korak and S. Hagenbuch

ÖKOTEC GmbH, Rosa-Luxemburg-Strasse 39, D-14806 Belzig, Germany

Keywords: Soil Filter, Water Treatment, Natural Systems, Public Labour, Sludge Composter, Biomass Production, Cost-Reduction, Employment Effects

Abstract

As an example for many European rural areas the former Eastern German county of Branden-burg was evaluated for better acceptance of natural waste water treatment systems. Low population densities and a majority of the villages with fewer than 500 inhabitants require small decentralized treatment plants. Scarce finances and a high unemployment rate of mostly former farmworkers dictate economic solutions. This paper describes an attempt to improve the cost effectiveness of soil filters by combination of technical innovations, production strategies and a socio-economic approach to construction and maintenance. Technical improvements were:

    - replacement of septic tank by a composter as mechanical pretreatment
    - circulation system for reduction of the performance/area ratio
    - reuse of nutrients by biomass production.

Additionally the cost reduction effects of public employment support used in construction of waste water treatment plants was evaluated. The cost reduction potential is 25 - 30 %.

Introduction

The former Eastern German so called "new counties" are in average less populated than the old counties of the Federal Republic. This results in a connection rate to municipal sewer systems of about only 50% [1] compared with 86% [2] in the old counties. Soon after the reunion of east and west the official strategy was a copy of the centralized systems in the old counties. In the meantime the failure of this strategy is obvious: The planned waste water treatment plants (WWTP) were very large, to match the existing philosophy in the west. But the expected financial resources were not available for such oversized installations as well as for the huge sewer systems. Now some of the municipal water authorities, which already have built WWTP's are facing a financial collapse.

The economic progress, so urgently expected and needed by the people has been delayed. One reason is the lack of waste water treatment (WWT) concepts, that prevents the establish-ment of industries. People and the small communities are poor in these counties. The unemployment rate sometimes exceed 30%. The proficiency of these people in mainly connected with agriculture.

The official policy prevents water treatment in some areas almost totally by fixing the effluent standards for undergound infiltration on drinking water levels. Due to the lack of water courses in Brandenburg in many cases there is no watercourse near the villages.

The situation is adverse in every respect. Successful implementation of WWTP's requires highly efficient systems, that are inexpensive in construction and maintenance and might even help to improve the social infrastructure.

Natural systems, especially soil filters seemed to have advantages in this respect. This was the challenge to further improve the technology and the way of application. The improvement included mechanical pretreatment, biological treatment and using a filter bed for production of biomass. Fig. 1 shows the system layout of a WWTP for 300 population equivalents (p.e.).

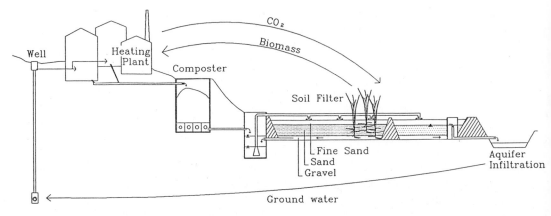

Fig. 1: System sketch of the treatment plant

Composter as mechanical pretreatment
The composter (Fig. 2), developed by Hermann WURSTER, is one of the essential parts of the system. Here the solids settle on a filter layer, that is installed on top of aeration stones or polyethylene pallets. The walls have permeable covers to enhance air flow. For sufficient aerobic decomposition it is necessary to add straw or wood choppings in biweekly to monthly intervals, depending on the size of the composter.

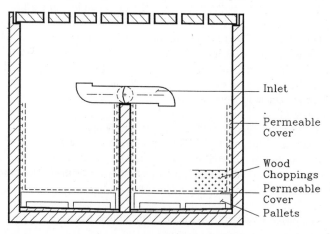

Fig. 2: Cross section of a composter for small treatment plants

Small treatment plants need only two compartments. Precast concrete parts can be used. If the composter is dimensioned with 300 l/p.e. the first compartment will be filled in two years. During the filling of the second compartment the decomposition process in the first compartment proceeds, causing a remarkable loss of volume. After two years of decomposition the material can be removed and used in agriculture and landscaping. A layer of about 20 cm is left in the compartment as microbial starter.

Bigger treatment plants (>100 p.e.) should have more than only two compartments to make sure the volume of each compartment is not too big for sufficient aeration (Fig.3). The upper limit for economic use of the composter seems to be between 500 and 1000 p.e.. Above this size an EMSCHER tank in combination with sludge dewatering in reed beds seems to be advantageous.

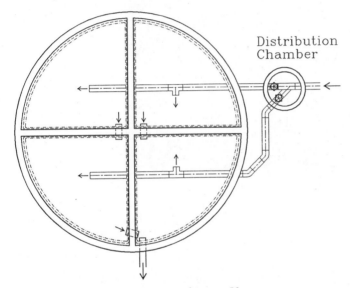

Distribution Chamber

Fig. 3: Composter for treatment plants > 50 p.e.

In comparison with a septic tank the composter has following advantages. It:
- reduces costs by avoiding transport of sludge
- reduces the amount of material to be handled
- produces fertilizer instead of waste
- makes the handling less uncomfortable (odour-free)
- keeps the waste water in an oxygen enriched state.

Sand filter bed as biological treatment
Fig. 1 shows a longitudinal section of the total system, Fig. 4 a photograph of the installation soon after planting. The filter is loaded on the surface and has a vertical infiltration. To fully utilize the whole surface area of the filter bed a minimum waste water level of 30 mm is required with each loading cycle. For storage of the necessary amount the drainage layer of part one is used. This layer is connected with the pump chamber. At a certain water level in the pump chamber the pump is put into operation by a level switch. The water is distributed

on the surface of part one and two. The backflow ratio can be regulated by valves. In part one the infiltrated water mixes with the influent in the drainage layer. A backflow ratio of 2Q was used during the first two years of operation.

Part two

Part one

Fig. 4: Filter bed in spring of the first vegetation period

The water tables in part one can be adjusted by the level switch, in part two by adjusting the elbow in the outlet man hole.

The total area of 900 m² is parted in the longitudinal direction to allow for alternate loading. During the regular operation one side was used for at least 4 weeks continuously, even with overloading by 400 persons. There is some evidence that a specific area of only one m²/p.e. would guarantee sufficient nitrification.

The vegetation consists of different grass and tree species that seemed to be promising for high biomass productivity. The trial consists of 24 test plots, 7,5 x 2 m² each in part one and 16 test plots, 6 x 2 m² each in part two. For statistical accuracy 4 test plots were planted for each species.

The following species were used: *Salix viminalis* "Mötzow", *Populus canadensis*, *Populus sinensis*, *Phragmites australis* var. *pseudodonax*, *Miscanthus sinensis* "Giganteus", *Miscanthus sacchariflorus*, *Andropogon gerardii*, *Sorghastrum nutans*. The goal of this trial is to assess the productivity under heavy irrigation and high nutrient load in a good aerated light soil and in the same soil with a high water table.

As a final treatment step the soil filter part two could be replaced by an irrigation area, where the nutrients could be used by plants, for example the same species like in the soil filter. The minimum area to allow profitable harvesting and processing of the biomass seems to be about

20 ha. The growth of the willows exceeded two meters/year. The stem circumferences amounted up to 26 cm. The first harvest will be after the third vegetation period. After cutting the stems will resprout.

Treatment efficiency
The performance of the treatment plant is summarized in Table 1. Several series of water samples were taken at various intervals and at different water table levels. The data show a very good decomposition of the organic matter and a good nitrification. Phosphate elimination is high because of the iron content in the sand and added iron granulate. Concerning the composter effluent, the high oxygen content is remarkable. The other values are within the range of septic tanks.

The soil filter effluent was analyzed in the first year (1993) under different subsoil water table levels, "Free outflow" and "Water table 10 cm below surface". The higher water table caused slightly more anaerobic conditions with slightly higher values of P, NH_4-N, COD and lower values of O_2 and NO_3-N. Overload situations (400 p.e./450 m²) did not have any remarkable influence on the effluent quality. Analyses of the water control authorities between 1993 and 1995 showed an improvement of the nitrogen elimination.

Table 1: Performance of the soil filter, values in mg/l

	Inflow n = 11	Outflow free n = 3	Outflow high n = 8	Contr. Authority Data (5)
Oxygen	4,9	8,0	4,7	-----
COD	315	<18	24	26
BOD_5	-----	-----	-----	<3
NH_4-N	>45 - 66	<2	3,6	1,8
NO_3-N	3,1	40,7	29,9	8,4
PO_4-P	>8 - 10,7	0,5	0,7	0,3

Publicly financed labour
The official policy in Brandenburg is to only support large centralized WWT authorities and not single municipalities directly. These large organizations tend to favor central solutions with oversized WWTP's. So in many cases the only chance for independent villages to realize WWTP's is publicly financed labour. The cost savings using publicly financed labour amount to 30 - 40 %. This amount is greater than the usual governmental financial support for such projects, and so it is more cost effective than asking for governmental assistence.

Two feasibility studies showed, that the main bottleneck connected with pulicly financed labour is the performance guarantee. The employment support companies (ESC) are not allowed to act in the same legal manner as construction companies. In fact a competition in the market has to be avoided. Publicly financed labour is restricted to projects that could not be realized by other means and should be combined with training. WWT projects therefore need the cooperation of ESC with a construction company. A special section in the specifications is necessary, to specify the tasks for the construction company especially in terms of machinery work and works where performance guarantees are critical. In the specifications the training of the ESC workers is included as special tasks. Ideally the construction company can employ some of those workers after the project. Maintenance of the WWTP will also be included in the training.

Conclusions

The construction costs for a WWTP can be reduced to about 80 % by reduction of the area requirements. An additional reduction of 25 - 30 % is possible by using publicly financed labour. The cost reduction of the sewer system by publicly financed labour amounts to 20 - 30 %.

The described approach seems worthwhile to be transferred to other European areas with similar structure. Still more research work is needed for the ultimate area reduction and the additional benefits of using the effluent for irrigation.

Literature

[1] DOHMANN, M. and B. BÖHNKE: Studie zu verfahrenstechnischen Konzepten beim stufenweisen Aufbau von kommunalen Abwasserreinigungsanlagen und betrieblichen Optimierungsmöglichkeiten. Studie Ministerium für Umwelt, Naturschutz und Raumordnung des Landes Brandenburg, 1992

[2] FEHR, G. and H. SCHÜTTE: Leistungsfähigkeit und Wirtschaftlichkeit der Abwasserentsorgung im ländlichen Raum. Korrespondenz Abwasser, **39**, 818 - 830 (1992)

Environmental Research Forum Vols. 5-6 (1996) pp. 261-266
© *1996 Transtec Publications, Switzerland*

Wastewater Management Outside Urban Areas:
A Conceptual Approach to Integrate Economical,
Practical and Ecological Constraints with Legal and Social Aspects

F. Gigon and Y. Leuzinger

Bureau Natura, Le Saucy 17, CH-2722 Les Reussilles, Switzerland

Keywords: Method, Wastewater, Small Water Circuit, Quality Assessment, Swiss Regulations

ABSTRACT

According to Swiss regulations, disposal of wastewater into streams and other surface waters must obey strict requirements in terms of chemicals (quantitative) or at least global criteria such as colour, taste, etc. (qualitative). Meeting these criteria for small streams would almost invariably lead to the need of connecting to a main treatment plant through a sewage network. This is already the case for the great majority of urban and suburban areas (over 90% in canton of Bern). However, such policy rapidly becomes financially unsustainable as soon as extra-urban context is concerned, especially in very diffuse habitat or very remote contexts. Private plants or other means of treatment must then be envisaged, treated water being released into nearby streams (even though legal requirements cannot always be met). From there on, it becomes necessary to establish new criteria aiming at :

1. Finding financial limits beyond which full legal compliance cannot reasonably be asked for. Beyond these limits, connection to a sewage system has to be abandoned. Locally treated wastewater can then exceptionally be released into surface waters, partly disregarding the relevant regulation.
2. Finding ecological criteria to evaluate the minimum level of purification to be achieved before treated water can be released. Is the pollution burden acceptable? Can the stream be "sacrificed" or does it really deserve to be kept virgin?

From there on, any decision becomes essentially of political nature. But any political decision has to be supported by a wide acceptance to avoid being labelled "arbitrary". Public acceptance therefore must be based on an easy, cheap and transparent decision process. This process has to include ways to guarantee that the quality of surface waters (possibly soil and/or ground water in case of infiltration) will be preserved. It should also lead to a better respect for the new trends in surface waters ecosystems protection.

INTRODUCTION

According to Swiss regulations, disposal of wastewater into streams and other surface waters must obey strict requirements in terms of chemicals (quantitative) or at least global criteria such as colour, taste, etc. (qualitative). Meeting these criteria for small streams would almost invariably lead to the need of connecting to a main treatment plant through a sewage network. However, such policy rapidly becomes financially unsustainable as soon as extra-urban context is concerned, especially in very diffuse habitat or very remote contexts. Private plants or other means of treatment must then be envisaged, treated water being released into nearby streams (even though legal requirements cannot always be met). From there on, it becomes necessary to establish how exceptions to the legal frame can be envisaged.

THE LEGAL CONTEXT

The Swiss legal context consists of at least two levels: the Confederation level and the canton level. In any case, the national requirements must be fulfilled but there may be some additional requirements set up at the canton level. In this case, the general *goals* to be achieved regarding water

protection are given by a federal law, which lays the basis to[1]:
preserve natural biotopes which harbour indigenous fauna and flora;
preserve fish water;
preserve water systems as a landscape element.

And concerning wastewater:
It is forbidden to lead, directly or indirectly, substances into water when there is a risk to pollute the latter. Infiltration of such substances is equally forbidden.
Polluted water must be treated. Discharge into water or infiltration are subject to cantonal authorization.

The general *means* to apply to achieve these goals are also given[2]:
Cantons must make sure that communal authorities establish their own general planning for water disposal. This planning secures protection and adequate discharge of water originating in inhabited areas.

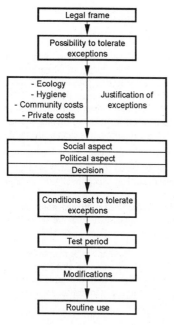

Fig. 1: problem situation and constraints

More precisely[3], qualitative criterion (such as sludge, turbidity, colouring, animal and vegetal communities, etc.) as well as quantitative criterion (target values for NH_4, NO_3, NO_2, P, PO_4, DOC, pH, etc.) must be reached in surface waters. However, exceptions can be tolerated:

For building hosting few inhabitants or outside constructed areas, cantonal Authorities can recommend less strict specifications.

It follows that outside of urban areas, public communities, sometimes very small ones, will be responsible for wastewater management. They will have to make choices as to the applicable techniques, given the consequences on surface waters quality and financial implication. If such cases happen to be numerous, it will be necessary to have a solid basis to make sound and consistent decisions. Financial, practical and social issues are then of equal importance (fig. 1).

THE SITUATION IN CANTON OF BERN

The extra urban situation in canton of Bern can be considered an important and meaningful one: even though more than 90% of wastewater is treated in proper plants, about 3000 individual cases presently need water treatment planning. This situation is likely to become more preoccupying in the next future, since agriculture is changing fast and farmland is likely to be more and more grouped in bigger farms. Farmhouses will then be converted into dwellings for non agricultural population and their wastewater will not be mixed and recycled with other organic farm waste any more. For the next 20 years or so, about 10'000 such cases are likely to be handled in the communes of canton of Bern. It means that on top of the good planning basis it will need, not all the cases can possibly be handled in full legal compliance. Therefore, the choices are:

1. Ignore (or postpone) legal compliance and water protection: this actually is not a *choice* but the long or mid term situation if no method can be proposed to alleviate financial problems in small

---

[1] Loi fédérale sur la protection des eaux, art. 1, 6.1 and 7.
[2] Ordonnance générale sur la protection des eaux, art. 11.
[3] Ordonnance sur le déversement des eaux usées, art. 1 and 11

communities. Those communities will often not be in a position to face their legal financial responsabilities and cannot really be forced to. The proposed concept precisely aims at avoiding this "choice".

2. Design an acceptable procedure in order to process cases where feasible and/or affordable, which means, sooner or later, on a financial basis only. Moreover, since owners will have to participate, the question to consider them on an *egal footing* still acutely arise.

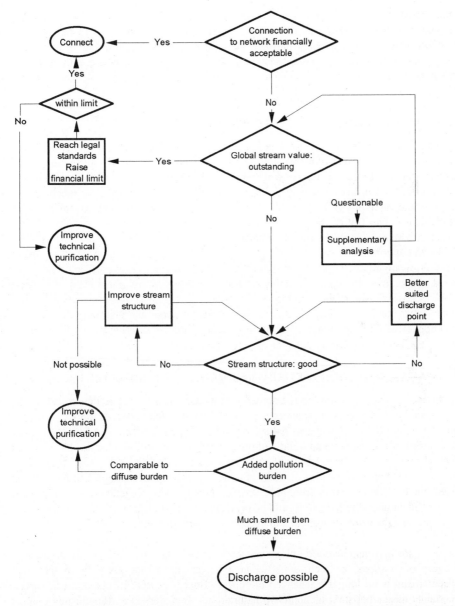

Fig. 2: Conceptual flowchart (still under development). Infiltration remains a spare possibility at every stage, at least theoretically (but at a low level of priority).

3. Process all cases but allow exceptions to some of the regulations, i.e. the quality of released water. A fair amount of money can so certainly be spared in this way, but to make this an efficient process, it is necessary to base it on rational and objective criteria: some studies will be necessary. Part of the spared money *must* be devoted to conduct those necessary supplementary studies. The trade off is: spare little money on supplementary studies and potentially spare substantial amounts on reduced equipment. It will also often be a trade-off between *status quo* and water quality improvement *on the whole* (as opposed to *in full*).

The demonstrated concept aims at finding sound criteria. Based on these criteria, a way to render decision making easy is being developed. Such a "method of exceptions" has to be calibrated and verified, which can be achieved only partially on selected sample situation. It is part of such a concept to be used, followed up and subsequently altered. The concept is presently in the making and has not been tested under real conditions yet, but sample situations have been isolated and investigated.

PRIMARY CRITERION: THE FINANCIAL THRESHOLD

Basically, connecting wastewater to sewage networks is a legal requirement *everywhere provided that it is considered opportune and can reasonably be required.* Practically, technology and costs must be comparable to those in usual constructed areas. The problem lies then in a good estimate of what the usual costs are, because costs are not proportional to the number of inhabitants in any building, given the variability of building size. They are not proportional to the building commercial (or insured) value either. But considering that:
1. outside urban areas, building are rather small,
2. if they do not connect to the sewage network they should build a small private treatment plant,
3. costs of private treatment plants are roughly equivalent to connection costs in urban areas,
it seems reasonable to set the limit as roughly equivalent to a private treatment plant. If a special risk exists (such as possible water table pollution or damage to particular ecosystems/biotopes, etc.) then this financial limits can be raised by 10, 30, 50 %, etc. At this stage, the first basic decision is taken: within the applicable financial limits, connection to an existing sewage network will be applicable. Outside of these limits, different possibilities have to be examined (fig. 2).

SURFACE WATERS PROTECTION AND RUNNING WATERS ECOSYSTEMS PROTECTION

Legally, surface waters protection is guaranteed by the "Ordonnance sur le déversement des eaux usées", provided qualitative and quantitative aspect can be fulfilled (see above). These legal requirements aim at protecting water as natural resource and vital environment for fish. However, it must be considered that the legal (and cultural) context has changed since 1975 (when this *Ordonnance* originated); some other legal texts have been elaborated since then (see *Loi sur la protection des eaux* above). The idea of water is nowadays much more comprehensive: not only the water itself but the whole stream (riverbed, shores, accompanying vegetation, actually the whole structure depending upon the stream) should be protected. So "water protection" includes the *stream ecosystem*, not only the *water ecosystem* and even less the *water content*.

Stream ecosystems may also be good linking elements between valuable biotopes or even considered as a category of biotopes[4] as such; a particular management is thus justified. In this sense, an artificialised stream is no longer a balanced biotope and needs *repair*. In our context, *repairing* a stream certainly means boosting its ability to naturally purify water. Even though this optimisation may be *qualitatively* well known, it is *quantitatively* not so well known and experts are very

---

[4] This connecting function is highlighted by the *Loi cantonale de la protection de la nature* in canton of Bern.

reluctant to speculate over calculating it. However, it is sure that a well structured and varied stream will perform self purification more efficiently than an artificialised or ill structured stream. So revitalisation and enhancement of self purification, i.e. acceptable pollution burden, fit together in the global concept of running waters ecosystems management. It is therefore not unreasonable to propose some *exceptional thresholds*, provided these new limits depend on the given examined streams and, even more important, that these particular streams can be examined in a standard, simple and economical way.

SOME PARTICULAR CONSTRAINTS AND OPENINGS

There are anyway conditions which need special care, in order to define new thresholds adequately:

- The stream features fish fauna. Fish being very sensitive to some chemical factors (NH4, NO2, etc.), known tolerances for those factors must be respected. In small streams anyway, it has been shown [1] that these tolerances may vary according to the specifications of discharged waters: for example, fish can be more tolerant to dissolved organic carbon (DOC) in a well purified water (complete nitrification) than in a less purified water.
- Fish can be absent from streams which harbour an otherwise very interesting fauna (macroarthropods in particular). One could argue that all the streams are at least potentially able to harbour fish, which would restrict the possibilities of adapting some legal limits. However, sample studies in two contrasting regions show that very small streams are often likely to be the recipients of wastewater where flow rates (though permanent) and slopes will never be adequate for fish development. Such waters' quality has already proven in some previous studies to be *biologically* very good[5] and some very interesting fauna was then identified [2]. It is thus not unfounded to take streams without fish into particular account and propose again exceptional thresholds.
- The considered stream belongs to a special *natural objects inventory*: the procedure of including biotopes or other natural units into one of the established inventories also participate in the *Loi sur la protection de la nature* and must be integrated to this conceptual approach. Here also, exceptions can be defined regarding the release of wastewater, but thresholds will have to be stricter.
- Infiltration of wastewater is foreseen and ground water protection has to be taken into account. So far, only the drinking water side of the problem has been addressed. Practically, it has been handled by means of "protection zones", classified according to their proximity and relationships to known fresh water springs. The biological aspect has never been addressed and very little is known actually about the possible self purification underground where pathogens are of prime concern. Infiltration has therefore to be considered a spare possibility when no streams will be able to be the recipient of wastewater. The financial threshold for compulsory connection to an existing sewage network has then to be raised accordingly.

TWO BASIC PRINCIPLES OF THE PROPOSED METHOD

1. A functional stream unit must be defined: data will be collected for each unit as a whole. Basically, a unit consists of the chunk of stream comprised between two intersections. If this unit is too long, it can be divided or conversely several units may be grouped (according to predefined procedures).
2. The pollution burden must be acceptable for each stream unit. Successive units cannot be grouped

---

[5] Biological value is determined by standardised methods such as the sprobiontic index, the *Indices biologiques globaux normalisés (IBGN)*, or other methods using fauna and/or flora as bioindicators[3, 4].

together to achieve the water quality requirements. This is an axiomatic basis to legitimate the foreseen exceptions and to guarantee that these exceptions will not deteriorate the general streams system.

DATA COLLECTION AND HANDLING

1. Collection of data must be easy and standard: data characterising any stream unit will be collected using forms (typical situations are illustrated and can be identified in the field).
2. Filled forms will result in attributing values to stream units. A subset of these values will typify the present value of the units, that is as a natural object. A second subset will represent the estimated ability of the unit to accept wastewater. This second set does not rely on a self purification calculation but on a classification in wastewater tolerance categories; attributing the desired "exception" level to the stream unit will be based on this tolerance.
3. Finally, according to collected field data and defined limits and thresholds, decision must be made as to connect, release, find alternatives, etc. This will be made easier by using a standard evaluation document (decision making aid in the form of a dual choice chart). Adjusting the treatment level (*technical purification*, fig. 2) will be decided upon using a catalogue of the available methods and their related output.

CONCLUSION

It is hoped that the basis for wastewater management, purification and release have been built in this conceptual approach. Adjusting some of the proposed thresholds certainly needs more investigation in the form of testing, follow up and improvement. It is actually obvious that a certain amount of arbitrary is unavoidable in choosing the thresholds, given the present state of knowledge and the practical context. However, this is an easier goal to achieve when a solid basis exists, in the form of a standard and easy to follow protocol in the centre of an open process. This open process, together with the integrated procedure seems to be the best guarantee for obtaining the required public recognition: technical, financial, ecological and a widened legal scope (different laws taken into account) are so to speak *embedded* in the process right from the beginning.

AKNOWLEDGEMENTS

We would like to thank Mssrs. E. Baer, R. Hostettler, R. Wahli, Drs. U. Ochsenbein and A. Von Känel for their support and assistance.

BIBLIOGRAPHY

[1] W. Gujer, V. Krejci & E. Eichenberger, Gas-Wasser-Abwasser **62**(11), 546 (1982).
[2] Natura, Valeur biologique et paramètres chimiques des différents ruisseaux, complément à l'étude d'impact de la N16 La Roche-St-Jean - Tavannes.
[3] V. Sladecek, Arch. Hydrobiol. Beih. **7**, (1973).
[4] Norme française NFT 90-350, Détermination de l'indice biologique global normalisé (selon J. Verneaux), afnor, (1992).

Environmental Research Forum Vols. 5-6 (1996) pp. 267-272
© *1996 Transtec Publications, Switzerland*

Water and Wastewater: 13 Crucial Issues for Future Development

A. Finnson and A. Peters

Swedish Environmental Protection Agency, S-106 48 Stockholm, Sweden

Keywords: Water Supply System, Wastewater System, Systems Analysis, Resource Management, Phosphorus, LCA

ABSTRACT

The Swedish Environment Protection Agency (EPA) has started a project called "Systems Analysis - water supply and sewerage". By means of cooperation with municipalities and agriculture, the aim is to produce a checklist and collate practical examples, including thorough environmental and resource evaluations of various types of water supply and sewerage systems.

The paper presents the ideas of the project together with a recently compiled list of 13 crucial environmental and resource issues regarding water supply and sewerage.
The list is a first rough guideline for municipalities which are intending to enlarge or make alterations in water supply and wastewater systems.

INTRODUCTION TO THE PROJECT "SYSTEMS ANALYSIS - WATER SUPPLY AND SEWERAGE

Today in Sweden, there is an increasing interest in investigating alternatives to conventional wastewater treatment. Several small municipalities have started discussions on replacing parts of their existing systems for wastewater treatment with solutions with for example separation of urine or where treatment processes are operated outside ordinary basins and tanks. Accordingly, the discussion on advantages and disadvantages of new and old systems from an environmental point of view is quite intense in Sweden.

Encouraged by the work with Agenda 21 in municipalities and the debate on our future water and wastewater systems in Sweden the Swedish EPA initiated the project "Systems Analysis - water and wastewater systems". To better manage the project it has been divided into two phases, a pre-study and a main study. This paper mainly presents the pre-study of the project, the leaflet containing an overview presentation of the criteria - "*13 Crucial Issues*" and the strategy for the main study.

Aim of the project
The aim of the project is to answer three crucial questions. The aims of the pre-study have been to analyse the question "What is an environmentally compatible water and wastewater system?" and to identify a strategy for the main study. The aim of the main study is to find answers to the following questions "How can environmentally compatible systems be designed and how can they be implemented?

The pre-study
The pre-study started in September 1994 and ended in June 1995. To make it possible to judge and

compare different kinds of wastewater treatment systems some kind of criteria have to be set up [1]. The aim of these criteria should be to make it possible to carry out relative appraisals of suitability from, for exemple, hygienic, environmental and resource management aspects. The aim of such criteria should not be to find "the one and only" wastewater treatment system but to enable relative comparisons between different systems to be made. The result of appraisals of different systems will vary depending on the local conditions such as urbanization, geology, topography, climate etc.

The following main criteria have been chosen in the pre-study for the environmental assessment of water and wastewater systems:
- *Hygiene*
- *Conservation of resources and environmental impact*
- *Technical and socio-economic criteria*

The technical and socio-economic criteria include areas which have not previously been regarded as having any impact on the environmental compatibility of water and wastewater systems. When carrying out an overall assessment of a water and wastewater system many of the sub-criteria of this criteria are prerequisite for a development towards environmental compatible systems. Such sub-criteria can be economy, reliability, user-friendliness, impact on society, and responsibilities.

Seminars with researchers, the environmental movement, agriculture, municipalities and water and wastewater business have been an important task in the phase of designing reliable criteria for the systems analysis of water and wastewater systems.

The pre-study has resulted in the following publications:

- [2] A leaflet with the criteria presented in an easy readable overview "*13 Crucial Issues Regarding Water Supply and Sewerage*" (published in Swedish but presented in English in this paper). Target group: municipality politicians and the general public

- [3] The report "*Environmentally Compatible Water and Wastewater Systems - a Proposal for Criteria*". (In Swedish). Target group: public works engineers in the municipalities, consultants

- [4] The report *"Comparison of Different Water and Wastewater Systems - a Literature Review"* Target group: consultants and researchers (In Swedish).

The main study
The main study will start in autumn 1995 and will end in 1997. The aim of the main study is for the Swedish EPA together with the agriculture sector and the municipalities, to find answers to the two questions "How can environmentally compatible systems be designed and how can they be implemented?
The strategy to answer the questions is to:
- provide a tool for the municipalities for environmental assessment of their existing and future water and wastewater systems.
- carry out a comprehensive overview of existing and possible future instruments for a development towards more environmentally compatible water and wastewater systems.

The environmental assessment tool will be developed from case-studies at three to four municipalities and be based on a methodology related to Life Cycle Assessments (LCA). In each

municipality a comparison will be made between the existing water and waste water system and:
- a proposal for a water and wastewater system that can be implemented within ten years.
- a vision of a water and wastewater system that can be implemented within fifty years.

The comparisons will be made according to criteria and sub-criteria presented in the pre-study. Most effort will be put in to the comparison of the sub-criteria that we have identified as the most important: hygiene, energy flow, phosphorus flow and nitrogen removal.

The different case-studies will form the basis of the guideline for how to compare different water and wastewater systems. The guideline will be published together with the case-studies in spring 1997.

CRUCIAL POINTS FOR FUTURE WORK

The discussions in the pre-study resulted in the proposals in [2] and [3]. These proposals will form the basis for the work in the main study to develop a method for assessment of different kinds of water and wastewater systems.

The crucial points for the work in the main study are:
- to convert the ordinary LCA methodology into a methodology which can be used for systems and not just for products.
- that the municipalities can carry out their visions for water and wastewater systems with sufficient amount of accurate data for the development of the assessment method.
- that the participating municipalities feel deeply committed to the project - from the municipal politician to the public works engineer and the local farmers.

When the main project is ended and the method is published in 1997 the project will be only half completed. The experiences and the publications of the main study have to be spread by an information project. The information project will consist of seminars, lectures and visits to the municipalities which have been working in the project.

And that part, to inform new municipalities about the main study and its assessment methodology, will quite possible be as challenging as carrying out the main study itself.

THE LIST "13 CRUCIAL ISSUES REGARDING WATER SUPPLY AND SEWERAGE"

This part of the paper presents an English translation of the leaflet *"13 Crucial Issues Regarding Water Supply and Sewerage"* with criteria presented for municipality politicians and the general public. The list is a first rough guideline for municipalities which are going to enlarge or make alterations in water and wastewater systems.

These issues are addressed to the municipalities and must be considered before making alterations to existing water and wastewater systems or building new ones, or carrying out work under local Agenda 21 projects etc.

Introduction
Have you, like many other people, ideas about the design of future water and wastewater systems? The answer probably varies depending on local conditions, although the basic problems are in fact

common to all areas. Here at the Swedish EPA we believe that the following 13 points may be of assistance as you try to find the answers. We also think that the 13 points should provide a stimulus for new ideas and improvements in the field of water and wastewater systems.

The Swedish EPA is now pursuing two important issues in a project called `Systemanalys-VA'` (`Systems Analysis - water supply and sewerage'). Firstly, by means of cooperation with municipalities and agriculture and others, we aim to produce a checklist and collate practical examples, including an evaluation of various types of water and wastewater systems in terms of environmental objectives. We will also be initiating research in the fields in which we are able to identify significant gaps in our knowledge.

Greater environmental compatibility of water and wastewater systems **and** agriculture will mean that a greater proportion of nutrients will have to be returned to agricultural land. The ecocycles will then be closed and environmental pollution will be reduced. Increased use of the resources contained in sewage often goes hand in hand with a reduction in pollutant emissions. And such a reduction is important. One in six Swedish lakes is now seriously eutrophied. Discharges of sewage are one of the main causes of this.

We need daily access to water. One of our natural basic needs is to have water in our taps at home. This produces waste water and sewage which must be dealt with. How is this to be achieved in a way which is sustainable in the long term?

Hygiene
The water and wastewater systems of today came into being to reduce the risk of epidemics in the growing cities and towns. It is easy to forget this. Naturally, it is not possible to choose new types of water and wastewater systems without being aware of the hygienic risks and how to overcome them. One of the most serious risks is the spread of infectious diseases to man and animals. Salmonella, viruses and intestinal parasites are examples of sources of infection which may be spread via sewage. Diseases may also be spread to man via crops fertilised using sewage or indirectly via flies, rats or other animals. Diseases may also reach man via meat or milk from infected livestock or poultry. Faeces is the most potentially dangerous constituent in sewage. But bath water, washing-up water and storm water may also contain sources of infection. Urine is the least troublesome constituent, but this also depends on how it is handled. If the following rules of thumb are used, the risk of spreading infection is significantly reduced. Sewage products (apart from urine) which are not treated in any way should be stored for at least six months. The time between fertilising with sewage products and harvest should be at least ten months for fodder-plants, root vegetables and vegetables which are normally consumed raw.

1. It is essential that all forms of treatment and disposal of sewage or sewage sludge break the routes of transmission. When ecocycles between society and agriculture are closed it is even more important to be aware of how diseases can be spread. Who possesses competence in this area in your municipality?

Conservation of resources and environmental impact
An environmentally compatible water and wastewater system must not have an impact on the environment, even in the long term. Preserving the environment is largely a matter of conserving resources. Some natural resources are renewable, some are not. Sewage contains both kinds of resources. Phosphorus is the most problematic of the finite resources. What happens to the following `water and wastewater resources'` in your municipality?

2. Phosphorus is essential and cannot be replaced by any other substance. In its natural form it is finite; at present rates of extraction it is estimated that there is only enough extractable mineral phosphorus to last another 200 years or so. It is therefore important to take various steps to minimise our losses of phosphorus. One way of doing this is to recycle phosphorus in sewage for food production. The discharge of phosphorus into watercourses in your municipality is not only a waste of resources, it is also a cause of contaminated aquatic systems and ecological impoverishment. What possibilities are there in your municipality to increase the amount of phosphorus recycled to the agricultural sector?

3. Nitrogen is just as essential as phosphorus, but there is an almost unlimited supply of this element. Air comprises 80 per cent nitrogen. However, energy is needed to turn the nitrogen in air into fertiliser. The main reason for recycling nitrogen to the agricultural sector is if emissions can thereby be reduced more efficiently. But recycling is also worthwhile if less energy is required to extract nitrogen from sewage than from air. Emissions of nitrogen cause increased algal growth, particularly in marine and coastal areas.

4. Potassium, micronutrients and **humic substances** are all significant to agriculture, but their supply is limited in the same way as that of phosphorus. It has not yet been possible to establish a causal link between emissions of such substances to water and any pronounced adverse environmental effects. Nevertheless, it is naturally advantageous if they can be recycled for use on agricultural land.

5. Agricultural land is a resource which must be protected in order to assure long-term food production capacity. The sewage or sewage sludge used on agricultural land must not contain heavy metals or slow-degrading substances having an adverse effect on the soil or the food produced, not even in the long term. This is the requirement stipulated by the Swedish EPA for sludge used on agricultural land. What steps are you taking to reduce the harmful substances contained in sewage and sewage sludge?

6. Water is the basis of our existence and is our most important `food'. There is generally a plentiful supply of water in Sweden, although not everywhere. What steps is your municipality taking to secure sufficient availability and good quality of water in the long term?

7. Energy in sewage is present in the form of heat in the water and chemical energy, stored in organic matter. This energy should be utilised where the result is a net gain of energy. What is the potential for this in your municipality?

8. Raw materials are used when water and wastewater systems are built and operated. Construction normally involves the use of cement, gravel, steel and energy, for example. Operation consumes energy as well as precipitation chemicals and organic carbon. How can use of the existing system be optimised? Could reduced use of raw materials result in a more environmentally compatible water and wastewater system?

Technical and socio-economic criteria
No chain is stronger than its weakest link. Water and wastewater systems must be in working order and reliable at all stages, from water supply source, via dwellings, pipes, to treatment, transport and use in agriculture, all year round. It is therefore necessary to attempt to evaluate technical solutions as well as laws and contracts, the competence of personnel, transport requirements, userfriendliness.

9. Financial aspects must naturally also be considered. How can costs for a future water and wastewater system be estimated if all process stages are to be included? Do investment or operating costs predominate? Who pays? Is what is best in terms of private or municipal finance also best in socio-economic terms?

10. Reliability in operation. How are operational reliability and maintenance requirements assessed for different kinds of water and wastewater systems? Could minor technical problems or disruption in transport or organisation cause major problems? If new systems are introduced, how will it be possible to ensure that present and future systems operate in parallel during a transitional period?

11. Competence requirements. What competence is needed to run the facilities properly? Are qualified technical personnel needed, or should the users themselves be able to cope with all stages? How should the system be designed in order for different kinds of user to be able to operate parts of the system themselves?

12. Impact on society. Water and wastewater system issues affect overall social planning. There is an interdependency between water and wastewater systems and infrastructure. How will overall social planning be affected by future water and wastewater systems? Will flexibility and coordination potential increase or decrease?

13. Responsibility. Ultimately it is the municipality that is responsible for water supply and sewerage. Who is to be responsible for a future water supply and sewerage system? Will parties outside the municipality be responsible for different components? How can the municipality guide developments in the right direction? What degree of preparedness is necessary to assure a good quality environment for future generations and to be able to intervene if something goes wrong?

Summary of the list
Many pieces of a complex jigsaw will have to fall into place in order to design an environmentally compatible water and wastewater system. The important pieces are encapsulated in the 13 issues outlined above. You will not be able to by-pass any of these issues when you plan an environmentally compatible water and wastewater system. They are all needed to make the jigsaw complete.

REFERENCES

[1] J. van der Graaf, F. Fastenau and A. van Bergen, Practical Performance of Various Systems for Smallscale Wastewater Treatment During a Two-Year Field Test. Water Science and Technology vol **21**, 1-12, (1989).

[2] Swedish EPA, 13 Crucial Issues Regarding Water Supply and Sewerage, Swedish EPA (Naturvårdsverket Förlag), ISBN 91-620-9624-9, (In Swedish), Stockholm, Sweden (1995).

[3] Swedish EPA, Environmentally Compatible Water and Wastewater Systems - a Proposal for Criteria, Swedish EPA (Naturvårdsverket Förlag) Report 4429, ISBN 91-620-4429-X, (In Swedish), Stockholm, Sweden (1995).

[4] E. Kärrman, Comparison of Different Water and Wastewater Systems - a literature review, Swedish EPA (Naturvårdsverket Förlag) Report 4432, ISBN 91-620-4432-X, (In Swedish), Stockholm, Sweden (1995).

Environmental Research Forum Vols. 5-6 (1996) pp. 273-278

Eco-Policy and Market

E.-S. Davidavicius and E. Ramoškiene

Institute of Chemistry, Goštauto 9, LT-2600 Vilnius, Lithuania

Keywords: Product, Production, Eco-Cycle, Waste Treatment, Environmental Protection, Quality, Management, Environmental Output, Environmental Input, Utilization, Standardization, Certification, Environmental Label, Market

Abstract Products must be designed from the start so that they contain as little material and consume as few resources as possible in their manufacture and use. The Lithuanian experience with eco-cycle design, which has a long tradition especially in industrial wastewater treatment, is presented.
There are many social, legal, economic, and technical problems between the cradle and grave of manufactured goods. Those discussed include: environmental protection, consumer protection, voluntary and mandatory eco-certification and eco-labeling, export and import laws, and technologies.

Environmental quality management systems are now becoming more and more important not only in West Countries but also in a European State with transition economy. It might also serve as a basis for sustainable development, which considers the needs of future generations.

If society is to function in a sustainable long term fashion, products must be designed from the start to they contain as little material and consume few resources as possible in their manufacture and use. And the products which are produced must be recycled to maximize the re-use of raw material. And finally we should minimize the wastes and their harmfulness. Theoretically it can be showed in traditional "from cradle to grave" scheme (Figure 1).

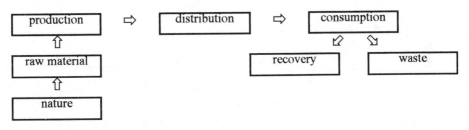

Figure 1. Traditional "from cradle to grave" scheme

We have to follow the products from raw material to manufacturing, then to distribution as well as the use of the products. If we can not close the cycle we get wastes which we have to get rid off in a safe way. Therefore we have to choose raw material that can be recycled, so we can make the out-take of raw material from the nature as small as possible. The products must be designed to facilitate separation of materials, re-use and recycling, and must contain a minimum of harmful substances. To reduce the risk to man and nature, we need a good waste management, with a good solution for waste, either it is land filling or incineration or anything else.

The directives of Action Programme on the Environment presented by the European Community are based on the following:
- preventive action shall be taken where possible,
- reduction shall take place at the source,

- the polluter shall pay (Polluter Pay Principal - PPP).

The Commission of the European Communities made a communication "Community strategy for waste management". This strategy ended up with a directive on waste 91/156/EEC. These are three main principles in EC's management policy:

- prevention or reduction of the waste production - especially the content of hazardous substances,

- recovery of the waste production,

- safe disposal of any waste which cannot be recycled or re-use.

The return of products must be the policy to the environmentally sound products. In reality now we have one-way economy (Figure 2), where environmental consumption (material and energy) and discharges (waste and emission) are increasing heaving (Figure 3). Our goal is a closed-cycle economy (Figure 4). In the typical pattern of a closed-cycle economy environmental consumption are stabilized on a certain equilibrium level (Figure 5) [1].

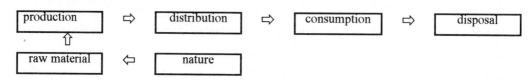

Figure 2. One-way economy scheme

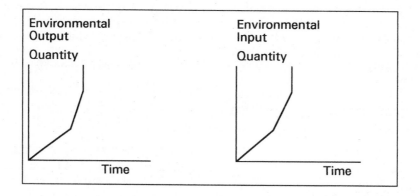

Figure 3. Environmental output and input in a one-way economy

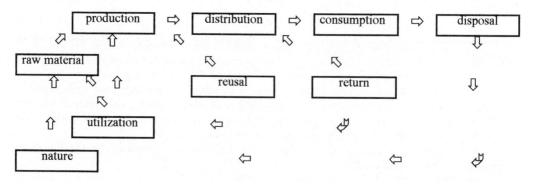

Figure 4. Closed-cycle economy scheme

The environmental effects of product's are largely determined by the inputs that are used and the outputs that are generated at all stages of the goods and product's life. Changing any single input, either to alter the materials and energy used or to influence a single output, likely will affect other inputs and outputs.

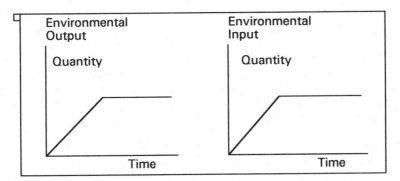

Figure 5. Environmental output and input in a closed-cycle economy

Inputs fall into two broad categories: materials and energy. Material inputs to the development, manufacture, distribution, and disposal of products can produce several environmental effects. These effects can include accelerated depletion of renewable and non-renewable recourses, impairment of land use, environmental or human exposure to hazardous materials, excessive waste, and emissions to land, air or water.

Outputs generally fall into four categories: emission to air, effluents to water, waste and noise and vibration.

Measures to prevent industrial pollution of the environment are efficient when realized being as a system blocking the impact of pollution sources. The most interesting experience in Lithuania have Joint-stock company INECO. This company was a leader in scientific researches and introduction of a new method in ecology in the former USSR.

One is mistaken if one is to think that every-thing is done after a waste water treatment plant is constructed. Waste water treatment is only one link of the chain. Already "at the entrance" into the treatment plant the question arises: What must be the amount and composition of basic production waste to provide a reliable operation of a waste water treatment plant and the necessary quality of treatment? A more complicated question arise "at the exit": What is to be done with the waste obtained after treatment? Toxic substances extracted from industrial waste water must be decontaminated!

In the most widely-used "one-way" economy scheme waste is disposed of in special land fills or incineration. Other countries experience dificulties because of enormous expenses for the construction and maintanenance of such land fills.

The so-called closed-cycle economy scheme with regeneration and secondary use of toxic materials in production require great capital investment and appear to be expedient only in specific cases. Often the manufacturing of reagents for waste treatment exceeds the volume of basic production processes.

Experience has proved that in many situations treatment of waste and subsequent utilization in other production is the most advantageous way. It is expedient from economic and reliable from ecological points of view. The production generating great amounts of waste but followed by utilization are more preferable than those with less waste but without utilization.

Lithuanian experience showed the utilization to be the best alternative. On one hand, it almost eliminates the necessity of land fills and on the other hand, it makes it possible to avoid great expenses and complicated problem related to the introduction of closed-cycle systems in individual plants.

Test and experiments proved that waste water treatment sludge of machine-tools plants can be decontaminated and used as a raw material in the manufacturing of building materials. After a certain modernization the Plant of structural ceramics in Lithuania has been fitted to the utilization of sludge of other kinds of industrial waste. Waste water treatment sludge is utilized in the manufacturing of tiles in the first centre of utilization. In the second one, lubricants-coolants waste with abrasive and metal dust impurities, surfactants, non-regenerable oil products, paper production waste, waste water with oil and road dust premises, are utilized in clay aggregate concrete.

Experimental work has been carried out to examine the possibilities of the utilization of metals-containing sludge in glaziers manufacturing. A centralized system for production with sludge additives of ferrite's with magnetic properties is under design. Experiments are carried out of metal-containing etching waste to be used as micro elements in fertilizers.

A regional programme of industrial waste-water decontamination covering basic production processes, waste water decontamination and utilization of waste, was initially developed for machine-building and metal-working enterprises. Later the experience was expanded to other industries: foundries, fur and tannage industry, textile and paper manufacturing, etc [2].

Products has some effects on the environment during its production, distribution, use or disposal. Each of these effects may range from slight to significant: they may be short-term or long-term; and, they may occur at the global, regional or local level. Requirements in products standards may influence significantly the extent of these environmental effects. Relationship between products standards and environmental aspects is showed in figure 6 [3].

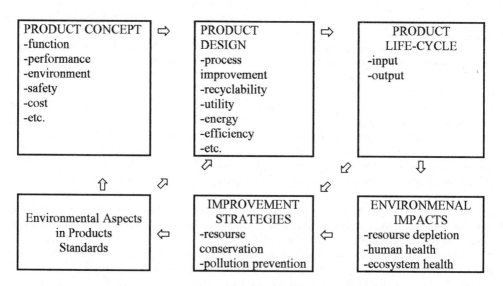

Figure 6. Relationship between products standards and environmental aspects

Privatization and market create new conditions for environmental protection. These changes will have a beneficial effect if coupled with a consequent environmentally friendly legislations, standards and should be harmonized with those of the EC to the greatest possible extent. Consumers, producers and government have to play an active role. The last step of productions and goods life-cycle is characterised by technical methods. The well-tried method must be "finalised", the processes, parameters are drafted in standards and other regulations. Lithuania like other Eastern European countries must rapidly adopt the accreditation and certification systems to the requirements of EC, EFTA and Eastern Europe countries. We must prove that products and production are "environmentally friendly" and correspond with requirements of guality.

In carrying out the tasks we use approach of managing environmental protection and quality in products and production (Figure 7).

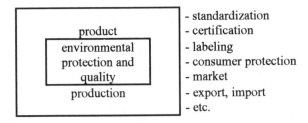

Figure 7. Environmental protection and quality management aproach in products and production

ISO Technical Committee 207 was given the task to standardize in the field of environmental management tools and systems. It is hoped that a major benefit of international environmental management standards will be the arrest of proliferating national and regional standards. These latter types of standards create unnecessary cost for organizations while raising potentially serious barriers to international trade.

In Lithuania the technical work of environmental protection will be carried out through Technical Committee "Environmental Protection". Committee structure include the following subjects:
- environmental management,
- engineering of environmental protection,
- quality of environment,
- natural resourses.

In 1995 the Environmental Label was introduced in Lithuania. This sign may be awarded to products which, in comparison with other products serving the same purpose, are particularly acceptable in terms of environmental protection.

References

[1] J.Mayr. Austrian experiences with the segregate collection of waste and the return of products. Extended producer responsibility as a strategy to promote cleaner products. Invitational expert seminar Trolleholm castle, Sweden, May 4-5 (1992).
[2] J.Budilovsky. Non-waste production in the region. The economist of Lithuania, 3 (1991).
[3] ISO CD 14060. Guide for the inclusion of environmental aspects in products standards.

Environmental Research Forum Vols. 5-6 (1996) pp. 279-284
© *1996 Transtec Publications, Switzerland*

Necessity of Integration of Urban Water Management with Management of Societies

J. Niemczynowicz

Department of Water Resources Engineering, University of Lund,
Box 118, S-221 00 Lund, Sweden

Keywords: Global Pollution, Large Cities, Urban-Rural Interaction, Water Treatment

ABSTRACT

Cities constitute the most important source of global environmental pollution. Water management in urban areas is governed by an old technical paradigm. Uncritical application of this paradigm in developing countries brought disastrous results manifested in land degradation, pollution and misery. Old methods of utilization and recycling resources have been abandoned and substituted with alien expensive wasteful technology. A vicious circle of loosing nutrients and carbon in agriculture, causing land degradation, hunger and urban migration, cannot be broken by traditional technology. In order to brake this circle development of novel approach and new technology of urban water management and change to sustainable agriculture is necessary. Technological change must be accompanied by basic changes in all sectors, social and central structures, educational and research programs and in lifestyle.

INTRODUCTION

Water management in urban areas is governed by an old technical paradigm developed mainly in England at the beginning of this century. Existence of a paradigm means that a consensus reached at that time, defining what is a "normal" way of solving water and sanitation problems, is still valid. This fact has enormous consequences, because large number of traditionalists, i.e technicians, experts, professors and decision makers operate within this frame and are unwilling to admit that societal goals has changed so much that further application of the old paradigm will be devastating for human societies and the world. Technologies and approaches developed and applied within the domain of *ecological engineering* came from people standing outside of the old paradigm. They have potential to brake down old paradigm and create conditions for progress.

Traditionalists, or "Continuators" say: developed world showed the way how sanitation and drainage should be organized in a non-polluting way. Look at Stockholm: the large city where salmon can be cached in a middle of the town, there surrounding waters are so clean that you may bath and enjoy watersports. This is a result of wise policy of construction of water treatment facilities that efficiently purify all waters from the city. If there is some more problems, they may be solved by construction of additional treatment steps, sludge treatment plants large-scale detention of stormwater, real-time operation of sewage systems in order to optimize the pollution load at the outlet.

"Rebels" say: Traditional high-technological methods of sewage treatment are very expensive in terms of money, resources and energy. They deplete natural resources of the world and cause loss of nutrients necessary for food production. They emerged from a single-disciplinary approach and old philosophy of dilution resulting in open material flows and global pollution. These technologies cannot possibly be considered as viable alternatives for developing world, because introduction of this technology for all people lacking sewerage and treatment facilities, i.e. about 90 % of world's urban population, would engage too large part of existing natural resources, exhort economies and, in finally, it would only move environmental problems to most remote areas of the world.

Of coarse, none of these opinions is entirely true. But bearing in mind the future need of feeding exponentially growing urban population, there is no doubt that urban water management and sanitation issues are central for all philosophy of future development not only of cities, but also of urban-rural interacting areas, especially in developing countries. There is no doubts that *sanitation and drainage can no longer be considered as a separated, single-disciplinary task*. There is a vital connection between technology used and sustainability, between deteriorating water quality and water scarcity, between water use and energy production/consumption. There is also a vital connection between urban water management and possibility to increase food production to feed growing population worldwide. Thus, management of water resources and water quality issues become closely related to almost all human activities.

INVESTMENTS IN OLD TECHNOLOGY

Industrialized world have invested enormous sums and resources in traditional technology of water disposal and treatment, and it continues to do so by construction of additional facilities. Meanwhile, problems with environmental impacts increase steadily. It was finally realized that sludge cannot be used in agriculture without reservations. This resulted in large investments on construction of new facilities for dewatering, treatment, composting and incineration of sludge. For example, New York City has spent 700 millions USD on construction of eight large sludge dewatering plants and plans for four new treatment works. On top of this come running costs of about 100-300 USD per dry tone of sludge. One can wander if all resources, energy and money invested in improvements added to existing traditional technology really contribute to improvement of global environment, or it will rather create enormous problems for future generations.

POPULATION GROWTH

Average population increase between 1960 and 1990 was 75%, but in Asia where growth is fastest, population increased by 158 % and in Africa by about 135. Year 1994, the world population was 5479 millions. Urban growth in developed counties is linear, but exponential in less developed countries. In Africa where urban population increase fastest, 70% of people will be living in cities year 2020. This indicates that during next 40 years production of food must increase at least three times. Thus, it is important that nutrients that are lost with wastewater will be used in agriculture for food production. Traditional wastewater treatment technology cannot provide that.

Urban population of the world has increased 2.5 times during last 30 years. Year 2000 number of megacities (above 10 millions) will increase to 23, and cities above 5 millions to 60, 80% of which most situated in developing countries. Cities in Africa grow fastest, more than 10% per year, and the growth is going on without any control, mostly in slum and squatter areas, about 80% of residents of Addis Ababa, and 70% of residents in Calcutta live in squatter areas without access to safe potable water and any sanitation facilities. Totally in the world, 2 billion people lack safe drinking water and 3 billions lack sanitation. Almost all of them live in developing countries, and most of them - in urban areas. Problems of megacities can be put on top of all well known environmental problems of the world to-day. Water shortage is a rapidly growing, municipal and industrial wastes are disposed to the environment, infrastructure and wastewater treatment facilities are usually by far insufficient. Present situation with respect to pollution of air, land, and water, as well as lack of basic water and sanitation facilities in these cities creates living conditions that are nothing less than derision of human dignity. Women and children suffer most.

Lack of sanitation and treatment facilities results in enormous global environmental impact. This means also gigantic loss of life supporting resources such as Phosphorous, Nitrogen and Carbon that, instead to be used in agriculture for food production, fill up rivers with polluted water.
In many places raw wastewater is used for irrigation. This, in turn, if done without any consideration of risks, brings health hazards.

RESULTS OF (OLD) TECHNOLOGY TRANSFER

Growth of cities can, ironically, be accounted to good will of the industrialized world. During 50-ties massive international efforts were directed at effectivization of agriculture in developing countries by introduction of new crops, new agricultural routines and massive increase of irrigation. These methods not always took into account local soil- and climate conditions as well as social structure, religion and traditions. After initial growth of productivity, long-term results of these help programs were not satisfactory. On the contrary, according UNDP [1] from total area of 4.7 billion ha 1988, 1.235 billion ha, i.e. 25% of total agricultural area in the world have experienced decrease of productivity due to human impacts. Erosion, salinization, water logging and desertification resulted in loss of arable land. On top of this, "effective" agriculture, directly copied from wester countries, in many cases impoverished soils through removal of carbon and nutrients resulting in increased need of fertilizing. For example, in arid areas of West and Central Africa carbon content in soils has decreased by more than 50% that reduced productivity by about 50%. The need of using fertilizers increased from year to year and is now about 5-6 times higher that, f.ex. in Sweden. This was one of initial reasons for poverty of people dependent on agriculture and resulting mass migration to cities. Figures 1 summarizes the results of application of traditional technologies in the South, figure 2 suggests some necessary actions to brake existing vicious circles. Cities were growing without any control mostly as slum and squatter areas. In order to change this situation, international aid organizations begun in 70-ties to invest large sums in water supply, traditional sanitation and wastewater treatment in developing countries. Resource and funding intensive technology could not keep pace with increasing population. For example, total necessary capital cost of water and sanitation was estimated to 60 billion USD year 1986 [2]. Such level of investments was not realistic. Later estimation of costs only for Latin America amounted to 20 billion USD for a period 1986-1990. Only 856 millions were invested by the World Bank. From the beginning of 1990, good will of international organizations has, in a way, enfeebled. One have given up the race with exponentially increasing population of big cities in developing countries.

COASTAL POLLUTION

About 60 % of world's population and most of all large cities is situated by the see coasts. Year 2000 this figure will increase to about 70 %, i.e. six billions people will live in these areas [3]. Most of people in this area are dependent on fishing industry that supplies majority of their protein intake. Coastal waters constitute complex mostly productive ecosystems that are nurseries of all life in the world's sees. As a result of pollution by urban effluents and other activities such as damage to coral reefs, removal of mangrove, biological productivity of coastal waters, particularly in tropic and subtropic areas have been strongly reduced. In South East Asia and around Caribbean islands about 70-80 % of mangrove disappeared during latest 40 years. About 90% of corral reefs in this area is damaged by eutrophication and sediment eroded from land areas [4]. In many cities untreated wastewater is released to the coastal water via see outfalls. For example, Ipamena outfall that releases 6 m^3 wastewater from Rio de Janeiro per second via 2.4 m pipe and will increase in near future. Many new sea outfalls will be put in operation. The origin of this global disaster can be also accounted to failure of technological water and sewerage paradigm that has shown too costly and not adjusted to socio-economical conditions of developing countries.

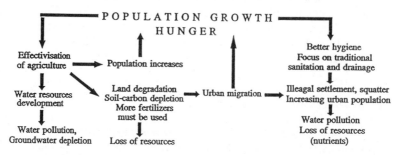

Figure 1. Vicious circles, i.e. feed-back mechanisms leading to present disastrous situation in developing countries.

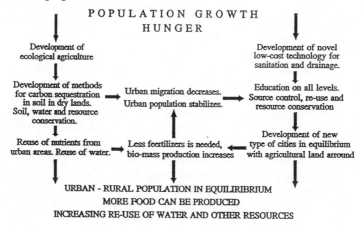

Figure 2. Possible ways out of vicious circles.

FUTURE PERSPECTIVES

Developed countries still put enormous resources in money in traditional technologies. These countries can economically afford additional steps of treatment, nitrogen removal, sludge treatment, real-time operation etc. Question is if the world can afford loss of resources that will be a result of these improvements. Rapidly developing countries, with Malaysia and Thailand and South Korea as examples, copy western traditional solutions. Rivers in urban areas are put in concrete; pollution of rivers, groundwater and coastal waters is still increasing at accelerated pace. One may wander if all errors of developing countries must be really repeated in developing countries before new approach can be adopted? One can wander if content in "transfer of knowledge and technology" is really responding to needs of sustainability?.

For majority of cities in less developed countries in Africa and Asia, traditional treatment technology is far beyond the economical potential. According [5], countries with GNP per capita less than 500 USD annually not only lack resources to construct treatment plants but also cannot maintain them. Paradoxically enough, positive changes can be expected in these countries in the future. People try to find new less expensive and more environment friendly solutions. Facing the lack of will and capacity of solving the problems on a central level, people begin to organize themselves and, in some cases, have achieved success on a local level by initialization of bottom-up initiatives and using local traditional knowledge and solutions. One of well known examples of such development is the so called Orangi Pilot Project in Karachi region with 800 000 inhabitants. One dynamic person, Akhter Khan, home latrines for 70 000 private houses were constructed with 15 % of normal costs. In Amman, Jordan stabilization ponds are successfully used for treatment of 57 000m^3/day, and the effluent is used to irrigate citrus and olive trees [6]. Calcutta wetlands are another example. First i 1993 has this system been recognized by central authorities as an ecological treatment and bio-mass production plant, i.e. object worth protection and further development. After that new wetlands are being developed in Calcutta for the same purpose.

POSSIBLE SOLUTIONS

Without bringing down population increase in the world and particularly in cities there is no hope that complex problems of large cities can be successfully solved. Urban migration must be slowed down by creation conditions in which living in rural areas will be more attractive. Thus, major changes in agricultural development are necessary. Such development must be based in local conditions, traditional technologies and crops. In dry regions devastation of land must be stopped by introducing carbon sequestrating crops. Western technology used for sanitation and wastewater treatment is not available for majority of developing countries due to economical reasons. The only hope for these countries is that novel, ecologically sound and economically efficient technology of urban water management will be developed and implemented in a near future. Ancient technologies developed during centuries in developing countries have been abandoned in the name of "civilization" and "progress". Many of problems in these countries originates in erroneous assumption that western solutions were superior those previously existing. Large development programs executed both on rural areas and in cities have failed because they were based in alien and unfamiliar life style, culture and resource-extensive, costly technology. Problems of large cities of today can be seen as a consequence of this misunderstanding.

Thus, new approach and novel technical solutions must be based on resource conservation principle as an opposite to treatment at the end-of-pipe. Efforts must be taken to introduce small-scale, low-cost technology based in local traditions, decentralized and ecologically sound, i.e. such that can close material cycles between agricultural land and a city. Recycling of nutrients must be in focus in order to decrease water pollution and reduce need of fertilizers. Examples of such technologies are: dry sanitation, wetland treatment, soil and soil-aquifer filtration, oxidation ponds, integrated solar and aquaculture. In arid regions it is not very wise to use treated drinking water for toilet flushing. Dry sanitation, i.e. composting or separation toilets constitutes a viable alternative that should be further developed, adjusted to local traditions and implemented (or re-established) in many cities. It has to be noted that dry sanitation for disposal and re-use of human manure is an ancient system known and used for centuries in many dry countries of Africa and Asia. For example, in Yemenite towns dry toilets separating urine and faeces are used for centuries even in multi-story houses. The traditional system of excreta disposal in the town of Ourgala in the Algerian desert consists of composting latrines. Still, such systems was abandoned in the name of "progress" and substituted by water closets. Thus, ancient, well functioning, and resource recycling systems developed in agreement with the environment using experience of generations, were named "old-fashioned" and substituted by systems that are called "modern", but in reality constitute a step back in terms of deep ecological context.

In humid tropics the use of biological systems, integrated aquaculture and farming systems for wastewater treatment have in many places been abandoned in the name of development process. New physical and social planning should cross sectorial borders. Water-related problems in large cities touch upon all elements of water cycle: water, land, air and energy, i.e. they are connected to all human activities within river basins. Resource conservation principle is a guideline for future actions, It can show what actions and what technologies are heading in desirable direction.

CONCLUSIONS

Urban water management and sanitation issues are central for all philosophy of future development not only of cities, but also of urban-rural interacting areas, especially in developing countries. If growing demand of clean water are to be met, and if sanitation in large cities is to be improved, a holistic approach to resource management must be applied. This requires basic changes not only in applied technologies but also in the education systems, aid programs, social habits, policies, structure and management of the societies.

A vicious circle of loosing nutrients and carbon in agriculture, causing land degradation, hunger and urban migration, cannot be broken by traditional technology. Massive investments in traditional urban infrastructure and sanitation cannot catch-up with accelerated urban papulation growth. In order to brake this circle three elements are necessary: 1) Development of novel approach and technology of urban water management and wastewater treatment. 2) Revision of rural food production routines and change to ecological agriculture in order to slow down growth of cities. 3) Basic change in approach to physical city planning, and in all sectors. Instead of ever growing cities, new agro-industrial areas in a balance with new small urban units should be developed. Transportation within these units will be by walking or cycling, connection between them - by light rail, buss. Grate change in approach and technology will not occur in developing countries. It must first be developed, proven and implemented in countries from the North that possess necessary research potential. All this depends on how fast the old technical paradigm will be broken.

Centralized, sectorial efforts not always function satisfactorily. There is however many examples of successful cases when bottom-up approach was applied. Cross sectorial actions with force of local initiatives should meet all support. The goal is to empower strong groups that are capable to enforce their own solution on a local level. The role of governments is then to make these initiatives possible by increase of general educational level and mobilize necessary financial means.

Two conclusions may be drown from this story. Firstly, that's us i.e. industrialized world, that has contributed to present misery in agricultural areas and in large cities of developing countries. It is also here, where previous mistakes can be repaired by development of new approach and new technology economically affordable for developing countries and more sound for the environment. Secondly, technological change that is taking place right now must be accompanied by other basic changes. Several societal and central structures as well as educational and research programs must change. Present economy, that does not considers environmental damage and throughput of resources as costs, must also change. Our lifestyle will inevitably undergo deep changes.

References

1. Megalli, N., (1992): "Hunger versus the environment: a recipe for global suicide". Our Planet, UNEP, vol.4 No. 6, pp 4-11.
2. Rooy, C., Doyle, B.,A., (1992): "Focus on Africa". Waterfront No 2, June.
3. Ambio Vol.22 No.7 (1993) Special Edition on Royal Colloquium: "Tropical and Subtropical Coastal Management: A question of Carbon Flow in a Sectorized Society". Royal Swedish Academy of Sciences, Stockholm, Sweden.
4. Ngoile, M.,A.,K, Horrill, C., J. (1993): "Coastal Ecosystems, Productivity and Ecosystem Protection: Coastal Ecosystem Management". Ambio, Vol. XXII, No. 7, pp 461-467.
5. Grau. P., (1994): "What next?" Water Quality International No.4, 1994, pp 29-32.
6. Pescod, M.,B. (1994): "Agricultural use of treated urban wastewater in developing countries". New World Water, Sterling Publication, Editor Roy Harris. pp 104-106.

Environmental Research Forum Vols. 5-6 (1996) pp. 285-288
© *1996 Transtec Publications, Switzerland*

Workshop 4:

Floating and Submersed Plants in Wastewater Treatment

Summary and Conclusion by the Chairperson

D. Porath

Department of Life Sciences, Ben-Gurion University of the Negev,
P.O. Box 653, Beer-Sheva 84105, Israel

The recycling potential of waterplants is, to a certain extent, misunderstood and even ignored in our intensive, high-tech modern trials to recycle wastewater. We should remind ourselves that a single real waterplant which is naturally well-adapted to complete its life cycle on wetlands has been used for thousands of years in the economical recycling of wastewater on a commercial-industrial scale. Most, or even all, technologies that we attempt to establish nowadays should, in my humble opinion, be examined and compared with those methods used in rice (*Oryza sativa*) aquaculture.

The studies in this workshop, as listed in the table, concisely represent the major scientific groups active in this field. Progress in these areas should encourage adoption in our modern world, an efficient recycling agent – the waterplant.

Workshop 4: A comprehensive display of the R&D projects in wastewater management represented

Investigators & Location	Waterplant Species	Major Studies on Wastewater Management
1. Austin & Tchobanoglous California USA	*Eichornia crassipes*	Characterization of treated waste water stability in warm climate
2. Furukawa & Fujita Japan	*Nasturtium Officinale* *Impomoea aquatica*	Natural purification capability in a polluted river of the temperate zone
3. Jamil K. India	*Eichornia crassipes* *Pistia stratiotes*	Removal of toxic metal ions
4. Jana et al. India	*Eichornia crassipes*	Nutrient removal from eutrified water in warm climate
5. Josephson et al. Massachusetts USA	not indicated	Nutrient removal, mamely NH_4^+, iron and phosphorus-containg waste products
6. Valtere et al. Latvia	*Lemna minor* *Elodea canadensis* *Pistia stratiotes*	Nutrient removal from dairy wastewater in temperate and cold conditions
7. Landolt et al. Switzerland	Lemnaceae	Physiological diversity among duck-weed species and clones
8. Ozimek, T. Poland	*Lemna minor*	Nutrient removal from wastewater in temperate climate
9. Poole, W.D. Minnesota, USA	*Lemna sp.*	Patented floating barriers and other engineering devices
10. Porath et al. Israel	*Lemna gibba* cv. Galilee	Selection of clones for specific novel industrial products

Note: The work of Greenway et al from Australia was also discussed in this workshop although *Melaleuca* is considered as a typical wetland plant, like rice which was mentioned in the text above.

In general, the success of any R&D project, especially the biotechnology of wastewater management, needs a constructive mutual effort across different disciplines, as illustrated by the following scheme:

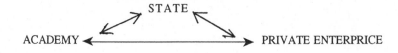

Acknowledgement
Finally, I wish to thank the Conference Board, especially Andreas Schönborn and Juerg Staudenmann for their faithful and kind efforts to enable me to participate and chair this workshop.

THE R & D
TOWER OF BABEL

Environmental Research Forum Vols. 5-6 (1996) pp. 289-296
© *1996 Transtec Publications, Switzerland*

Duckweeds (*Lemnaceae*): Morphological and Ecological Characteristics and their Potential for Recycling of Nutrients

E. Landolt

Geobotanisches Institut ETHZ, Zürichbergstr. 38, CH-8044 Zürich, Switzerland

Keywords: Duckweed, *Lemnaceae*, Recycling of Nutrients

Abstract

Duckweed species belong to the smallest flowering plants and grow on or below the water surface. They are highly productive (up to 130 kg dry weight per ha and day) and have a remarkably high protein content (up to 45 % of the dry weight). They can also tolerate and fix high amounts of nitrogen, phosphorus, and heavy metals. Therefore, they are used as test organisms in plant physiology and as phytoindicators for herbicides, heavy metals and other toxicants, for waste water treatment and as an energy and protein source. Cultivation and harvest of the plants is easy. Different species and clones may have quite different responses to temperatur, light, nutrient concentration, and toxicants. Lemnaceae are quite useful in recycling projects if these differences are understood and the plants are used in an appropriate combination of different species or with other aquatic species.

1. Introduction

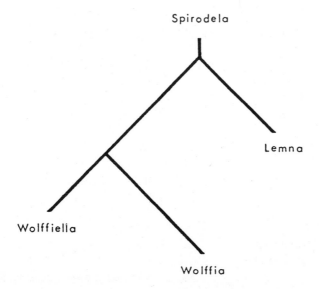

Fig. 1. Supposed taxonomical relationship between the genera of the Lemnaceae (from [1]). *Spirodela* and *Lemna* belong to the subfamily Lemnoideae, *Wolffiella* and *Wolffia* to the Wolffoideae.

The duckweed family (Lemnaceae) includes the smallest and most reduced species of flowering plants. A monograph of the family exists [1, 2] in which all literature on Lemnaceae (more than 3000) until 1987 are surveyed. In the most reduced genus (*Wolffia*), the plant consists of a not much differentiated embryo-like group of cells, the so-called "frond", which is able to perform all necessary physiological functions. This frond is, in the extreme (*Wolffia globosa, W. angusta*), globular to ellipsoid and not more than 0.5 mm long. Propagation is mostly vegetative by budding. Sexual reproduction is very rare in most species. Only a few species regularly flower and fruit and are thus able to colonize seasonal waters and survive the dry season with fruits. Fronds are very sensitive to dryness. Exposed to dry air they die within a few hours. However, if they are tightly enclosed in mud they stay alive for many days or, in the case of *Wolffia cylindracea,* for many months. The family consists of 4 genera and 35 species. It belongs to the monocotyledones and is probably nearest related to the Araceae. However, the often assumed relationship to *Pistia* is rather superficial and phylogenetically unproven. The relationship within the family is demonstrated in Fig. 1.

2. Morphological and ecological characteristics

Spirodela and *Lemna* have a flat or gibbous frond and several roots (*Spirodela*) or only one root (*Lemna*). The new fronds develop in two pockets at the proximal side of the frond, forming groups of two to several fronds. The new ones separate from the mother frond after several days. If flowers are formed, they originate from the same pockets, one on a single side or sometimes one on each side. The flower consists of a small scale, two 4-loculed stamens and one pistil. *Wolffiella* and *Wolffia* have no roots and one pocket at the proximal end where the fronds are formed. Flowers develop in one cavity or two cavities (some species of *Wolffiella*) on the back of the frond. No scale and only one 2-loculed stamen is present. They are thin in *Wolffiella* and either orbicular to ovate and float on the surface of the water or ribbon- to tongue-like and below the water surface. In *Wolffia* they are orbicular to cylindrical or boat-shaped and float on the water's surface. (Fig.2).

Fig. 2. Groups of fronds of various Lemnaceae seen from above or from the side(f) (after [3]

a.	*Spirodela polyrrhiza* (x7)		d.	*Wolffiella hyalina* (flowering, x7)
b.	*Lemna aequinoctialis* (flowering, x7)		e.	*Wolffiella oblonga* (x7)
c.	*Lemna valdiviana* (fruiting, x12)		f.	*Wolffia brasiliensis* (x7)
g.	*Wolffia brasiliensis* (flowering, x20)			

A	anther		Ov	ovary
Ap	apex		Pa	papule
AS	air spaces		Pi	pigment cell
B	base		Po	pocket
BP	basal part of the frond Pr		Pro	prolongation
DP	distal part of the frond R		R	root
F_0	mother frond		RC	root cap
F_1	daughter frond of the first generation		Se	seed
F_2	daughter frond of the second generation		Sta	stamen
Fl	flower		Sti	stipe connecting daughter frond and mother frond
Fr	fruit		Stig	stigma
Ne	nerve		Tr	tract of elongated cells connecting stipe and node
No	node			

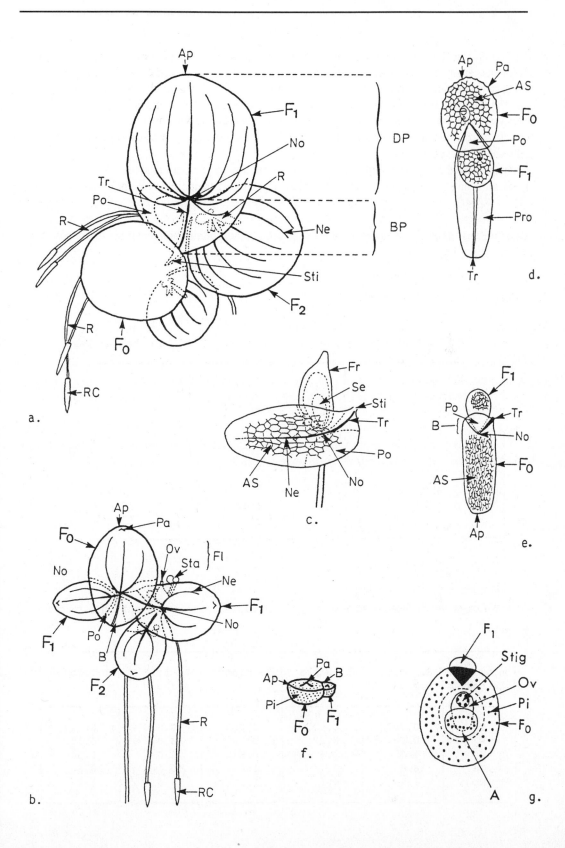

The Lemnaceae float free in the water. Roots or appendages may stabilize the frond in the water. However, the plants get transported by water currents, waves, or wind to the shore and dry out. They are restricted to small stagnant waters or attached to other water vegetation like *Phragmites, Scirpus, Typha, Eichhornia* etc. Natural habitats are small lakes, lagoons, ponds, ditches, flooded areas, slowly flowing brooks, and even wet rocks. On ponds or ditches they can cover the water surface with many layers forming dense mats of up to several cm thickness. The lower layers are forced to live heterotrophically from organic substances released by the upper layers.

Lemnaceae need a relatively high concentration of nutrients to grow fast (Tables 1, 2) depending on the species and on environmental conditions (Fig. 3). The limiting factor in waters for Lemnaceae is mostly

Table 1. Range of nutrient content in waters with Lemnaceae (chemical elements in mg/l, conductivity in μS/cm) [from 1].

characteristics of the water mineral content	absolute range	range of 95% of the samples
pH	3.5 - 10.4	5.0 - 9.5
conductivity	10 - 10900	50 - 2000
Ca	0.1 - 365	1.0 - 80
Mg	0.1 - 230	0.5 - 50
Na	1.3 - >1000	2.5 - 300
K	0.5 - 100	1.0 - 30
N	0.003 - 43	0.02 - 10
P	0.000 - 56	0.003 - 2
HCO$_3$	8 - 500	10 - 200
Cl	0.1 - 4650	1 - 2000
S	0.03 - 350	1 - 200

Table 2. Lowest yearly mean of some nutrients in waters of Central Europe with Lemnaceae [after 4].

nutrients in mg/l	S. poly-rrhiza	L. gibba	L. minor	L. tri-sulca	L. minuta
P	0.007	0.027	0.006	0.010	0.012
N	0.04	0.22	0.04	0.04	0.18
K	1.3	2.6	0.9	2.0	1.5
Ca	18.8	23.7	11.6	40.0	18.6
Mg	1.1	4.5	1.1	4.6	1.1

phosphorus. Long term growth is only possible with a minimum phosphorus concentration of 0.02 mg/l in *Lemna minuta* and 0.4 mg/l in *S.pirodela polyrrhiza* [4]. The reason for the need of a high nutrient level

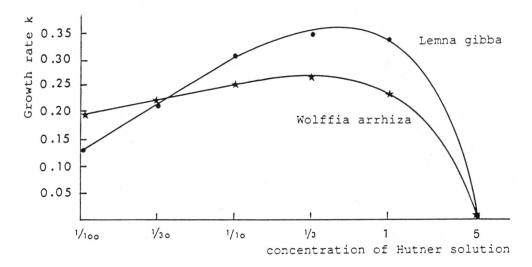

Fig. 3. Growth rate of *Lemna gibba* and *Wolffia arrhiza* in relation to the concentration of Hutner solution [after 5].

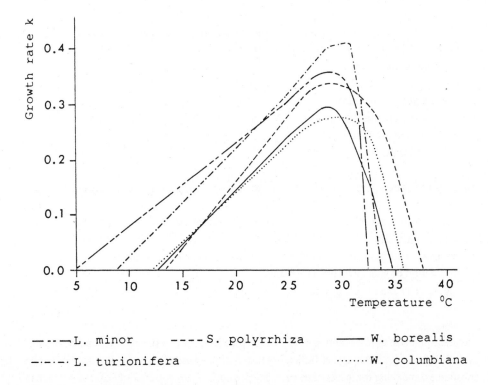

Fig. 4. Growth rate of different Lemnaceae species in relation to temperature [after 7].

in Lemnaceae is that they take up the nutrients by the lower surface of the frond which is rather small in area compared with the area of the root hairs of other plants. Growth conditions for each species and even for different clones are different. Several species (e.g. *L. gibba, L. minuta, L. turionifera, W. hyalina*) absorb ammonium preferably when both ammonium and nitrate are available, others prefer nitrate (e.g. *W. angusta, W. microscopica, W. australiana, L. obscura*) [2]. The need of other nutrients and the tolerance towards certain heavy metals and toxins is also very different in various Lemnaceae clones [6]. Fig. 4 shows the growth rates of 5 different species in relation to temperature. Some species still grow at temperatures near zero whereas others need temperatures above 15° C. There are species (*L. trisulca*) dying at temperatures around 30°, others (*S. polyrrhiza, L. aequinoctialis*) still grow at temperatures above 35°.

Lemnaceae are able to grow quite rapidly. Some species double their biomass under optimal conditions within 24 hours. Theoretically, the dry weight increase of Lemnaceae cover of a pond could reach $20 \, g/cm^2$ or $200 \, kg/ha$ within a day or more than $52 \, t/ha$ and year. Actually, best long-term results under natural conditions and in warm climate do not exceed $25 \, t/ha$ and year [8].

A high protein content is remarkable for Lemnaceae. The protein level may reach up to 45% of the dry weight under optimal conditions. All essential amino acids are present in sufficient amounts except methionine which reaches only half of the percentage of the reference pattern of FAO, and tryptophane from which only traces can be detected [9, 10].

3. Economic importance

Lemnaceae have some characteristics which make a versatile application promising:

- simple culture conditions (aseptic conditions possible, low space demands)
- vegetative propagation (genetically uniform)
- high multiplication rate
- high protein content
- easy harvest possibilities
- high nutrient demand
- ability to accumulate heavy metals
- low desease rate

Accordingly, utilization is possible in the following directions (for a comprehensive survey see [11])

- test plant for phytophysiological experiments and for detection of toxicants
- source of protein
- removal of surplus nutrients and heavy metals from (waste) water
- production of energy and phytochemicals
- regulator in aquatic ecosystems

The use of Lemnaceae as test plants are very wide-spread. Many physiological processes could be demonstrated and cleared with Lemnaceae. Indicatory tests for heavy metals, growth factors, herbicides, allelopathic substances, and other toxicants have been developed. Lemnaceae are rather sensitive to many substances but not very specific in their reaction. Mostly the effect on growth rate, frond size, dry weight, root length, and anthocyanin production has been studied.

The possibility of Lemnaceae as a source of protein is economically very important. The protein yield of duckweed per area is higher than that of most other crop plants and, in waste water systems, may reach up to 12 t per ha and year [12]. Table 3 shows the yield of duckweed protein and other substances in Thailand compared with other crops . Duckweed is used as food for dairy heifers, pigs, ducks, poultry, fish and prawns. Many more animals feed on duckweed in the wild. Duckweed as human nutrition is known in eastern Asia where *Wolffia globosa* is cultivated in small ponds. Recently , Porath and co-workers were able to cultivate *Lemna gibba* as a vegetable plant and to sell it sucessfully in supermarkets of Israel. The use of Lemnaceae protein via extraction seems to be promising but is not yet being carried to any large extent.

Table 3. Content of protein, carbohydrate and fat in different crop plants from northern Thailand [from 10].

	% dry weight	$kg\ ha^{-1}\ yr^{-1}$				
	Wolffia	Wolffia	soya	nuts	rice	corn
protein	19.8	2080	303	229	71	179
carbohydrate	43.6	4589	255	164	849	1451
fat	5.0	533	158	397	4	87
fibres	13.3	1398	44	21	3	40
ash	18.3	1928	41	20	5	24

The removal of certain substances by Lemnaceae is widely utilized. Lemnaceae have a great capacity in assimilating N, P, K, Ca, Na, and Mg, especially in heavyly loaded waters. Daily, it is possible to remove up to 4.7 kg /ha N, 1.6 kg /ha P and 2.1 kg /ha K[8, 14].
In waters with a N and P content below 4 mg/l [13], however, Lemnaceae are not able to reduce this level to a great rate. A reduction of the phosphorus level down to 0.1 mg /l, as it is proposed for strict water quality standards in the USA cannot be achieved with Lemnaceae alone [14]. A mixture of different water plants might do better. Also within Lemnaceae, not all taxa behave exactly the same. It could be possible to use a variety of species in a mixed culture to optimally remove the surplus nutrients at different concentrations. A series of two to three different species with similar climatic but different nutrient level demands could improve the purifying capability. For instance, a mixture of *L. gibba* and *L. minuta* could work the same as in nature, where waters with high concentrations of nutrients are covered with *L. gibba*; *L. minuta* being very scarce. As soon as the nutrient content becomes lower during summer, *L. minuta* begins to spread and to finally dominate. *Salvinia* and *Azolla* have an even better capability of utilizing low concentrations of nutrients.
The disadvantage of Lemnaceae for water purification is the demand of a large water surface area. [15] calculated a minimal area of 300 m^2 per 100 inhabitants. The problem is the small absorbing surface of Lemnaceae which corresponds to not much more than the surface area of the water. There are some submerged living taxa of Lemnaceae (*Wolffiella* species, *L. valdiviana, L. trisulca, L. tenera*) which could improve the situation. However, these species are rather sensitive to pollution and could be used at best in the last stage of purification. The advantage of Lemnaceae is the versatile utilization potential of the Lemnaceae

yield. The harvest is easy, the Lemnaceae cover can just be skimmed off. The plants can easily be transported, dry faster than most other water plants, are very digestible for most herbivores, and are quickly composted if used as fertilizer.

The possibility to accumulate heavy metals and other toxicants such as polychlorinated biphenyls and pesticides is another economic potential for Lemnaceae. The purifying capability is best at lower concentrations of heavy metals up to 1 mM. An accumulation factor of up to more than 100,000 is possible at such low concentrations in Al, Cu, Mn, Sb, Ti, Fe, Zn. Heavy metals and other toxicants must be extracted from the dry mass of Lemnaceae.

4. Conclusions

Lemnaceae are economically very valuable plants if applied according to their qualities and environmental demands. Since they grow throughout all climatic regions except arctic and desert areas, there will be available best adapted species and clones for each climate, environmental conditions and economic purpose. In many cases a mixture of species or clones might be preferable to be adapted best to seasonally varying climates and environmental conditions. As Lemnaceae do not colonize all niches within a water body it is advisable in some cases to combine Lemnaceae with water plants rooting in the soil like *Phragmites, Scirpus, Typha, Nasturtium, Ludwigia* etc. However , some of these plants (e.g. *Myriophyllum, Nymphaea, Nuphar, Cabomba, Brasenia*), have allelopathic effects on Lemnaceae and should not be grown together with duckweed.

References

[1]   E. Landolt, Veröff. Geobot. Inst. ETH, Stiftung Rübel 71, 566 pp. (1985)

[2]   E. Landolt, R. Kandeler, Veröff. Geobot. Inst. ETH, Stiftung Rübel, 95, 638 pp. (1987)

[3]   F. Hegelmaier, Engelmann, Leipzig, 169 pp. (1868)

[4]   A. Lüönd, Veröff. Geobot. Inst. ETH, Stiftung Rübel 80, 116 pp. (1982)

[5]   E. Landolt, Ber. Schweiz. Bot. Ges. 67, 271-410 (1957)

[6]   J. Sarosiek, H. Wozakowska-Natkaniec, C. Niemiec. Acta Univ. Wratisl. 530, Prace Bot. 25, 11-39 (1982)

[7]   D.M. Docauer, Ph.D. Thesis Univ. Mich. 223 pp. (1983)

[8]   D.D. Culley Jr., R.W. Myers. Baton Rouge (polycopy), 6 pp. (1980).

[9]   L.L. Rusoff, E.W. Blakeney, Culley D.D., J.Agr.Food Chem. 28, 848-850 (1980)

[10]   K. Bhantumnavin, M.G. McGarry, Nature 232, 495 (1971)

[11]   F.R. Ruskin , D.W. Shipley (eds.). Natl. Acad. Sci. Washington DC (1976), 148-154.

[12]   G. Oron, D. Porath, H. Jansen, Biotechn. Bioeng. 29, 258-268 (1987)

[13]   E. Rejmankova, Proc. 1st Intern. Wetlands Conf. (New Delhi), 397-403 (1982)

[14]   D.D. Culley Jr., J.H. Gholson, T.S. Chisholm, L.C. Standifer, E.A. Epps, US Environ. Prot. Agency, Ada, Oklahoma, 166 pp. (1978)

[15]   J.F. Martin, G. Caudron, E. Jan, E. Kofman, Tech.Eau Assain. 376, 23-34, 377, 21-37, 378/79, 19-25 (1978)

Environmental Research Forum Vols. 5-6 (1996) pp. 297-302
© *1996 Transtec Publications, Switzerland*

Usefulness of *Lemna minor* in Wastewater Treatment in Temperate Climate – Myth or Fact?

T. Ozimek

University of Warsaw, Institute of Zoology, Department of Hydrobiology,
ul. Stefana Banacha 2, PL-02-097 Warszawa, Poland

Keywords: Pleustonic Macrophytes, Wastewater, *Lemna* System

Abstract. In Poland 12 "Lemna system" have worked for wastewater treatment. The role of *Lemna minor* in stripping of nitrogen and phosphorus in these systems is presented. The limitations of using *Lemna minor* in such system is disscussed.

Introduction

Among different ecotechnological systems for treating wastewater - the system based on *Lemna* was commenced by Lemna Corporation. During the last years it is promoted in Poland [1]. In Poland three species of *Lemna* occur, *Lemna minor* is the most common. A lot of data on biology and ecology of *Lemna minor* existing in literature were obtained in controlled conditions in laboratory [2,3]. Investigations on utilisation of *Lemna* in wastewater treatment were carried out mainly in warm climatic conditions [4,5,6].

The aims of my paper is answer to the question - Is the *Lemna minor* efficient in sewage treatment in temperate climate?

Assumption of Lemna system

"Lemna system" is recommended in Poland as inexpensive, easy to operate and efficient in treatment of wastewaters [1]. Such system consists two ponds, primary aerated pond (depth 3-4 m) and secondary with *Lemna* (depth 2-4 m). Ponds are provided in additional equipment like nitrifying chambers and coagulants to precipitate phosphorus. The area of ponds depends on quality and quantity of flowing wastewater. Usually 4 m^2 per inhabitant is needed. To overcome wind movement secondary pond have a barrier system that forms a floating gird on water surface.

The costs of such system is estimated on 25 to 50% on costs of conventional treatment plant for the same amount of sewage.

According to the authors [1], "Lemna system" *Lemna* effect on wastewater treatment on two ways:

- directly by stripping and accumulation large amounts of nutrients which can be removed from system with duckweed crop.

- indirectly by changing of physical and chemical conditions in the water column what cause inhibition of algae growth and development of mosquitoe larvaes.

Limitation of Lemna minor

The role of *Lemna minor* in reduction of nutrients in wastewater is proportional to its biomass per unit area and nutrient concentration in its tissue. The biomass is a result of clonal reproduction and growth. It rarely reproduces sexually [7]. Among the abiotic factors esential for reproduction, growth and suvivorship of *Lemna minor* - light intensity and temperature are the most limiting and the most variable in temperate climate. *Lemna minor* has high optimum of temperature and light [8]. Generation times in duckweed clones under aseptic, optimal temperature and continuous light conditions, is usually 2.0-2.5 days [9,10]. In field conditions (in ponds supplied with wastewater) in 10-15°C water temperature doubling time range from 12 to 7 days. Due to temperature and light intensity growing season of *Lemna minor* lasts about 7 month in temperate climate. The rest time of a year *Lemna minor* lays at the bottom in bed light conditions. My results from field and laboratory experiments shows that in temperature 4°C and light intensity (PhAR) 5 W m^{-2} *Lemna* servive but it does not produce new fronds (Fig.1.1).

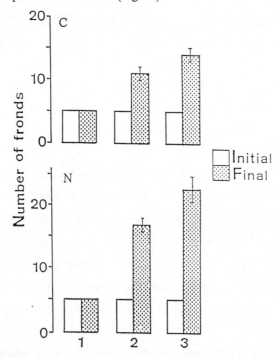

Fig.1. Reproduction of two clones *Lemna minor* (C- from Central, N- from North Poland) during two weeks in continuosly winter conditions (1), after transplantation from winter to summer conditions (2) and in continuosly in summer conditions (3).

After transplantation of plants from winter conditions to the same summer conditions (temperature 20oC, light intensity (PhAR) 30 W m^{-2}) to ponds supplied with sewage the reproduction of clone from North Poland was significantly higher (Anova, P=0.01) than reproduction rate of clone from Central Poland (Fig.1.2). The significant differences (Anova, P=0.01) between rate of clonal reproduction of these two clones were observed too when they grew continuosly in summer conditions (Fig. 1.3). The clone from North Poland produced higher biomass than the clone from Central Poland (Fig. 2). Significant differences (Anova, P=0.01) between production of dry weight by two clones were detected in regard to all concentration of nitrogen.

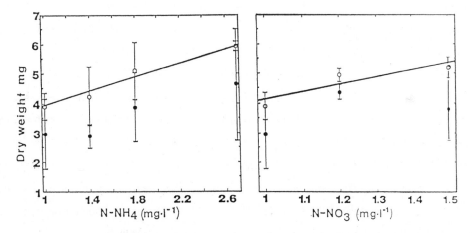

Fig.2 The final dry weight of two clones (white spots - the clone from North Poland, black spots - the clone from Central Poland) *Lemna minor* after one months growth in enclosures with different concentration of nitrogen (5 replicates for each concentration of nitrogen, initial dry weight - 1mg).

As reported [11] clones from the same geographical region behave more similarly than those from different region. My results indicate that differences between clones from the same region may be siginificant too. So, it is important which clone of *Lemna* we use to wastewater treatment. It would be good to organize clone banks in countries where "Lemna system" is so strongly promoted.

Biotic factors such as intra specific relations and inter specific relations with other plants, effect of herbivores; bacteria and fungal contamination play a great role in functioning of *Lemna* in nature [12]. There are not such problems in aseptic laboratory cultures.

When area is limited the crowding of fronds occurs and intra specific relation can be strong due to competition for light and contamination by different metabolitic substances. So, when *Lemna minor* covers whole water surface its multiplication of fronds decreases [13].

Authors of "Lemna system" [1] claim that *Lemna* inhibit growth of phytoplankton by light deterioration. Is it possible oposite situation? The phytoplankton grows much earlier in the season than the vascular aquatic plants in temperate waters [14]. In Rosocha "Lemna system" even in June 1994 *Lemna* did not occur on water surface, but phytoplankton was very abundant. Continued high nitrogen loadings may favour filamentous algae [15]. Some of them can inhibit by allelopathy growth of vascular plants [16]. Several studies have shown that duckweed are very sensitive to allelochemical phytotoxycity which can effect on decreasing of biomass production [17].

A large variety of invertebrates is commonly associated with macrophyte community. Various invertebrates consume the fresh macrophytes or in different stages of decay [18,19,20]. This is well known that *Lemna minor* has high protein content (30-40% of dry weight) and it is acceptable fresh and dry as animal food [21]. So, ponds with *Lemna* are a very good sources of food for animals, invertebrates (snails, larvae of insects) and vertebrates (e.g. waterfowl). Why they can not utilise *Lemna* crop? Ponds with *Lemna* is not isolated system. It is open and possible to occupy by many animals, e.g. *Cataclysta lemnata* or *Lymnea stagnalis* which can live in polluted waters and can use *Lemna* as source of food [18, 22]. A great number *Lymnea stagnalis* was noticed on *Lemna* in ponds supplied with wastewater in Pruszków near Warsaw. Animals effect on realise of nutrients from plants and accelerate of theirs cycling.

In climatic conditions of Poland 1.5 t dry weight ha-1 year-1 of *Lemna minor* can be yielded using three harvests [21]. For comparison, grown in warm water enriched with domestic sewage *Eihchornia crassipes* produces 4.0 t dry weight ha^{-1} day^{-1} [24]. Maximum amount of nitrogen and phosphorus possible to remove with yield of plants is presented in Table 1.

Table 1. Annually biomass and accumulation values of nitrogen and phosphorus in *Lemna minor* in optimal conditions of temperate climate

Dry weight [*] $t·h^{-1}·yr^{-1}$	Concentration [**] % dry weight		Accumulation $t·ha^{-1}·yr^{-1}$	
	N	P	N	P
1.5	6.5	2.8	0.09	0.04

[*]   Acc.[23]
[**] Acc.[25]

At mean this is only about 0.5% of nitrogen and 1.0% of phosphorus which inflow during a year to pond of Lemna system in Poland. Presented in Table 1 values are only potential ones , obtained in optimal conditions. Usually they are markedly lower.

Lemna minor populations are not uniform physiological age [26]. At any time during the growing season populations are comprised of plants at various physiological ages because of continual vegetative reproduction. Senescence and died fronds do not uptake but realise nitrogen and phosphorus to water. The balance of uptake and realise of nutrients by plants depends on proportion of alive to died fronds (Fig.3).

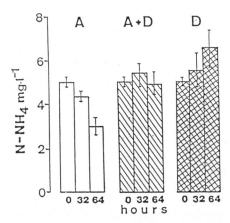

Fig.3. Effect of cultivation of *Lemna minor* populations at various physiological ages (A - 100% of alive fronds, A+D - 50% of alive fronds and 50% of died fronds, D - 100% of died fronds) on concentration N-NH$_4$ (mean + SD, n=5) in water. *Lemna minor* was cultivated in enclosures in ponds supplied with wastewater (Pruszków near Warsaw, July 1990).

Biomass utilisation

Authors do not precise how many crops of *Lemna* should be harvested from Lemna system during a year in temperate climate. Should crops be harvested regulary or occasionaly? Who should decided about that? How area with plants should be leaving to regrow a new crop so that the wastewater treatment becomes a continual process? According to literature [23,27] and my own results only three harvests are possible in temperate zone.

How changes content of protein in *Lemna* during growing season? How crops of *Lemna* can be use? The crops of fresh duckweed can be acceptable as food for animals only after washing. Fresh biomass of duckweed must be quickly used due to quick rate of decopmosition, and occupying by saprophytic organisms. Sun-drying of crops would probably be the most economical method, but is subject to variable weather conditions. The authors of "Lemna system" in Poland [1] do not precise how the *Lemna* crops should be used and what cost will be for that operation.

Conclusions

1. *Lemna minor* play a small role in wastewater treatment in temperate climate.

2. It seems that cost of cultivation, harvesting and utilisation of *Lemna* crop is higher than *Lemna* effects on wastewater treatment.

3. There are a many gaps in our konwlege on reasons of limitation *Lemna* role in wastewater treatment.

4. *Lemna minor* has a great plasticity in environmental equirements. The investigations towards to select the clones the most fit to grow in cold temperature and pollution environment should be begin.

Acknowledgements

The papers was prepared in cooperation with Division of Land and Water Resources, Royal Institute of Technology from Sweden. Special thanks are due to student Olga Sakwińska, who helped me in some laboratory experiments.

References

[1] A. Kowal, Hydro, Kielce, (1994).
[2] A. Luond, Ber. Geobot. Inst. ETH, 70, 118-141, (1980).
[3] D. Porath, Y. Efrat & T. Arzee, Aquat. Bot., 9, 159-169, (1980).
[4] D. Porath, B. Hepher & A. Koffen, Aquat. Bot. 7, 273-278, (1979).
[5] N. Boniardi, G. Vatta, R. Rota, G. Nano & S. Carra, Biochem. Eng. J., 54, 41-48, (1994).
[6] M.S. Hassan & P. Edwards, Aquaculture, 104, 315-320, (1992).
[7] W.S. Hillman, Bot. Rev., 27, 1-221, (1961).
[8] G.J. Filbin & R.A. Hough, Limnol. Oceanogr., 30, 322-334, (1985).
[9] E. Landolt, Aquat. Bot., 1, 327-332, (1975).
[10] E. Rejmankova, Aquat. Bot.,1, 423-428, (1975).
[11] E. Landolt & W. Dann, Ber. Geobot. Inst. ETH, 50, 86-96, (1983).
[12] L. Vasseur & L.W. Aarssen, Oikos, 65, 233-241, (1992).
[13] T.A., DeBusk, M.D. Ryther, M. Hansiak & L.D. Williams, Aquat. Bot., 10, 133-142, (1991).
[14] B. Moss, Hydrobiologia, 200/201, 367-378, (1990).
[15] G.P. Fitzgerald, J. Phycol., 5, 351-359, (1969).
[16] T. Ozimek, Wiad. ekol.,38, 13-34, (1992).
[17] C.F. Cleland & O. Tanaka, Aquat. Bot., 13, 3-20, (1982).
[18] A. Kołodziejczyk & A. Martynuska, Ekol. pol., 28, 201-217, (1980).
[19] D.M. Lodge, Aquat. Bot., 41, 391-404, (1991).
[20] R.M. Newman, J. of The North American Benthological Society, 110, 89-114, (1991).
[21] D.D. Culley & E.A.Epps, J. Water Pollut. Control. Fed., 45, 337-348, (1973).
[22] A. Piechocki, PWN, Warszawa, (1979).
[23] S. Bernatowicz & S. Wolny, PWN, Warszawa, (1969).
[24] B. Wolverton & R.C. McDonald, New Scientist, 12, 318-320, (1976).
[25] D. Dykyjova, Folia Geobot. Phytotax., 14, 267-325, (1979).
[26] J.I. Harper, Academic Press, London, (1977).
[27] D.I. Sutton & W.H. Ornes, J. Environ.Qual., 4, 367-370, (1975).

Environmental Research Forum Vols. 5-6 (1996) pp. 303-306
© *1996 Transtec Publications, Switzerland*

Natural Wastewater Treatment with Duckweed Aquaculture

W. Poole

Lemna Corporation International Inc.,
1408 Northland Drive, Suite 310, MN 55120-1013 St. Paul, USA

Keywords: *Lemna*, Duckweed, Biomass, Nitrification, Wastewater, Nitrogen, Phosphorus

ABSTRACT

The Lemna®System is a duckweed based aquatic system. A surface mat of these tiny floating plants is stabilized over the surface of the lagoon. This cover produces a quiescent, dark water column which promotes settling and inhibits photosynthesis. Algas and other materials that settle to the bottom of the lagoon are subjected to anaerobic digestion. Biochemical Oxygen Demand (BOD) reduction is accomplished by both anaerobic and aerobic decomposition.

Periodic harvest of a portion of the plant population is necessary to maintaining a healthy, vibrant crop of duckweed. Harvesting promotes the growth of young growing plants that assimilate nutrients more rapidly than mature plants.

Nitrogen removal can occur through plant growth and microbial transformations. Applications with strict effluent ammonia limits may require supplemental nitrification methods. Lemna Corporation has developed and patented a submerged nitrification system consisting of a floating modules containing aerated media.

Phosphorus (P) removal can occur through plant growth but natural precipitation is an important mechanism especially during the non-growth season. Additional P removal can be incorporated into the design by employing alum addition in the pond.

Strategies which balance land intensive pond processes with power and chemical intensive processes will lead to optimized designs which will fit the economics of the location.

INTRODUCTION

Duckweeds (family Lemnaceae) are small free floating plants that grow in a wide variety of climates throughout the world. Duckweed aquaculture systems rely upon the existence of a mat of duckweed plants to provide the conditions to treat wastewater.

The Lemna®System, utilizes a matrix of barriers on the water surface to protect the plants from the wind and to maintain an even distribution of plants upon the water surface. A grid of these floating barriers effectively partitions the water surface into 9.3 m^2 (100 ft^2) square cells providing ideal growth conditions for the duckweed plant mat. Flexibility in the positioning of these barriers allows the use of this system on virtually any pond.

LEMNA PLANTS

There are approximately 40 different species of Lemna plants found world wide. These and other floating plants are often found in wastewater environments where they compete with algae and other plants for nutrients.

THE LEMNA POND ENVIRONMENT

An even plant mat is formed when the plant is protected from the wind with floating barriers. The plant mat can grow to a thickness of up to a few cm. Usually more than one species of duckweed is found intermingled with roots and debris from older duckweed plants. Young plants have greater buoyancy than the older plants. Thus, the younger plants tend to rise to the surface of the mat while the older plants move to the bottom of the mat. The plants tend to stratify in the mat when physically disturbed by rainfall and harvesting.

Aerobic conditions exist in the mat and a short distance below the surface. The organic loading will dictate the depth to which this aerobic zone extends. Generally, the oxygen levels just a few cm under the mat are lower than within the mat.

The duckweed mat acts as a thermal barrier to the water column below. As a result the temperature gradient within the pond is fairly stable throughout much of the year and inversions are short lived if they occur at all. During daylight hours, the duckweed mat can be as much as 10 _C warmer than the water below. These higher temperatures in the mat result in faster reaction rates where BOD decomposition, nitrification, and reproduction of the plant will occur more rapidly than normal.

The physical layer formed by the duckweed mat intercepts light as well as heat energy while inhibiting gas exchange. A duckweed mat of 1 cm in thickness can intercept more than 90 percent of the light. Thus, light availability for algae in the water column is drastically reduced.

TREATMENT MECHANISMS
BOD Removal
BOD removal is the result of anaerobic, anoxic, and aerobic conditions within a duckweed pond. Anaerobic decomposition of organic materials occurs in the sediments and in or throughout the anoxic water column. The presence of the duckweed mat maintains the anaerobic conditions even under low organic loadings. Thus, bacterial populations are not subject to changing oxygen regimes.

The physical barrier of the mat also inhibits the gaseous by-product of anaerobic digestion (CH_4/H_2S) from escaping to the atmosphere. The aerobic conditions in the mat and in the water just below the mat support populations of bacteria which oxidize some of these anaerobic by-products.

TSS Reduction
TSS is reduced by a number of mechanisms in a duckweed system. The most important is the reduction of sunlight. Light interception by the duckweed mat is proportional to the mat density. A mat thickness of about 1 cm is sufficient to remove more than 90 percent of the sunlight. Thus, the algal population has limited light for growth.

Additionally, the calm environment created by a floating barrier grid and the duckweed mat is conducive to settling. Algae, bacteria, mineral matter and other debris settle under the slow moving conditions in the pond. Even sediments disturbed by the activity of harvesters normally will settle within a few hours.

Phosphorus Removal
Removal of P in a duckweed controlled environment is achieved through a combination of plant uptake and chemical precipitation. Phosphorus is utilized by the plant tissue in its growth process. Generally, P can represent as much as 0.5 to 1.5 percent of the plant tissue dry mass. The level of P uptake by the duckweed will depend on the concentration of P in the water and the balance of other nutrients.

The removal of P from the system by the mat is dependent upon the total growth and subsequent harvest of the mat. A system can be designed and operated to remove the majority of P through this process. However, the land requirement can be substantial.

The density of the plant mat (measured in kg/m^2 or lbs/ft^2), represents the mass of fresh biomass per unit area. The density of the biomass is maintained within a range which will provide the maximum growth and a relatively young population of duckweed. A young plant population tends to have a higher P uptake rate than an older plant population due to its accelerated growth rate.

Specially designed aquatic harvesters are used to remove a portion of the mat at intervals of several days to several weeks. The frequency of harvesting depends upon the growth rate of the mat, the temperature, and the treatment objective.

Precipitation of P compounds may also occur in Lemna®Systems. Since algal growth is inhibited by the shaded environment under the duckweed mat, P tends to be available in solution where precipitation reactions occur. Some of the P, for example, will precipitate and settle as calcium phosphate. The extent of this natural precipitation will depend on the inorganic chemical makeup of the water.

In order to meet stringent limits or to deal with colder periods of the year when biological reactions slow down, the natural phosphorus removal capabilities of a Lemna Pond can be supplemented by alum injection. This process is carried out in the last half of the Lemna Pond, where the concentrations of BOD and TSS are lowest in order to minimize the amount of alum required.

The aluminum phosphate solids formed are minimal and will not create a significant deposit at the bottom of the Lemna Pond. The deposited material is chemically stable and will not dissolve. The Lemna Pond is designed so that this small amount of deposit does not require removal during the design life of the facility.

Nitrogen Removal

One method of Nitrogen (N) removal occurs as a result of plant uptake and microbial transformations. The plants contain 3 to 6 percent N on a dry weight basis. Nitrogen removal by the growth of the plant mat can be optimized by conducting frequent harvests during the growth season.

Nitrification and denitrification may also occur in the duckweed mat. Nitrifying bacteria have been isolated from duckweed mats grown on wastewater. Denitrification of nitrate to gaseous forms can occur in anaerobic micro-sites within the mat or in the water column and sediments. The resulting loss of nitrates will reduce the total N in the wastewater.

These nitrogen removal mechanisms are expected to be most significant in the summer and less significant during the other parts of the year.

SUBMERGED NITRIFICATION REACTORS

Lemna Corporation developed a submerged nitrification system to meet more stringent ammonia limits. The concept of using submerged nitrification reactors is a relatively mature concept in the United States. Such reactors require a separate building or other such construction, and include a system of blowers, pumps, air diffusers and a media for attached growth reactions.

The use of submerged reactors in a Lemna®System pond requires special design considerations. The reactor must be isolated from the rest of the pond. Baffling or a separate tank structure must be constructed to maintain hydraulic separation from the rest of the pond.

Partitioning of the reactor volume into two or more units increases the effectiveness of the reactor by maintaining the concentration of ammonia at the highest concentration possible without dilution into the total volume of the reactor.

Ideally, baffles are positioned so that flow enters the first reactor unit where it is retained an average of 1 or more hours before the water moves into the second reactor through portholes. After additional time, the water moves into another chamber for further treatment or discharge. The actual retention time in the reactor is governed by the flow through the pond system.

Since the duckweed pond is an anaerobic environment, the water has a very low DO and potentially can contain compounds which may interfere with the nitrification process. Therefore, a preaeration diffuser is usually installed in the first zone of the LNR.

Vigorous aeration of the water is important to move water in close contact with the media surface and supply the necessary oxygen for the reaction. The rate limiting step in a microbial process such as nitrification is the diffusion of reactants into the microbial biomass.

Insufficient alkalinity may also limit the reaction. In some cases additional alkalinity may be necessary to allow reaction to go to completion. However, this has not been done as yet in a LNR, even though alkalinity levels as low as 30 mg/l have been measured.

EXAMPLE OF A FULL SCALE LEMNA®SYSTEM WITH A NITRIFICATION REACTOR INSTALLATION

Description of Plant

The town of Mamou, Louisiana, USA has a wastewater treatment system consisting of a two cell aerated pond 1.2 ha (2.9 A) with a depth of 3 meters. It has a 12 day detention time at design flow. The aeration pond is followed by a 3.6 ha (8.9 A) duckweed pond with a depth of 2 meters and a 30 day DT. The air supply for the WWTP is provided by two 20 h.p. blowers with a third blower available as standby.

The daily design flow is 2,300 m^3 (0.60 MGD). The actual flow averages slightly above design flow (0.62 MGD) with peak monthly flows as high as 157% of design (0.94 MGD). Daily flows have been measured above 200% of design flow. These large flows are due to extremely high infiltration into the collection system.

The nitrification reactor, located near the effluent of the Lemna®System pond consists of two parallel channels in formed by hydraulic baffles. A fine bubble diffuser in the first part of the first channel provide pre-aeration as the water enters. The remaining channel space contains 30 nitrification modules. Each module consists of PVC media, 1.83 m x 1.83 m x 1.52 m deep (6 ft. x 6 ft. x 5 ft. deep), suspended with floats. The top of the media is 15 cm (6 in.) below the water surface and the bottom of the module is about 30 cm (1 ft)   above the pond bottom.

PVC aeration racks, suspended below each module provide a uniform distribution of air across the media. Aeration is supplied to each aeration rack at a rate of 4.3 l/s (9 cfm).

Several parameters including flow, temperature, dissolved oxygen, ammonia, nitrate and alkalinity have been monitored through a combination of analyses on-site and at a local laboratory.

System Performance

This system was designed to meet the limits of 10/15/2 mg/l of BOD/TSS/NH$_4$

Results for a period between July, 1994 and March, 1995 are summarized. The aeration pond removed BOD from an average of 102 mg/l to an average of 33.3 mg/l.

The Lemna Pond reduced BOD to an average of 23.3 mg/l. This is lower than anticipated in contrast to systems following facultative ponds in which high concentrations of algas can be removed leading to a reduction of BOD.

The LNR system removed much of the remaining BOD from an average of 23.3 mg/l to an average of 4.0 mg/l at the effluent.

Although duckweed pond reduction of BOD has been hindered by hydraulic overloads, the LNR has dramatically reduced effluent BOD concentrations to less than 5 mg/l consistently year around.

There has been consistently excellent ammonia removal performance of the system.

Significantly greater than design flows during the rainy period of February 1995, which hydraulically overloaded the LNR, allowed ammonia excursions up to 4 mg/l for a 3 week period.

SUMMARY

A duckweed plant mat controlled by a matrix of floating barriers provides a specialized environment for wastewater treatment. The mechanism by which duckweed ponds remove pollutants from wastewater is related to the duckweed plant mat at the surface of the water. The complete cover of plants on the pond prevents light transmission into the water column. Algae is primarily controlled by shading, settling, and inhibition.

BOD is removed from the water column by oxidation processes anaerobic processes in the sediments and water column.

Nutrient removal is accomplished through plant uptake, microbial processes, and precipitation. The harvesting of the plant mat can be used to increase the removal of nutrients by the duckweed biomass. Additional measures to enhance the nitrification rate such as submerged reactors are feasible in a pond.

The Lemna®System, complete with a nitrification reactor can provide effluent ammonia concentrations consistently below 2 mg/l. Design is based on reducing ammonia levels by providing aeration and mixing to a submerged media with a high surface area and maintaining a healthy attached-growth culture of nitrifying bacteria.

Environmental Research Forum Vols. 5-6 (1996) pp. 307-312
© *1996 Transtec Publications, Switzerland*

The Duckweed *Lemna gibba* cv. Galilee: A Biotechnological Agent for Recycling Ammonium Ions into Available Proteins

D. Porath

Department of Life Sciences, Ben-Gurion University of the Negev,
P.O. Box 653, Beer-Sheva 84105, Israel

Keywords: Biotechnology, Duckweed, *Lemna*, Ammonium Ion Assimilation, Nitrogen-Metabolism, Available Proteins

Abstract

The collegiate efforts of scientists, farmers, industry experts and others occupied more than two decades before the first novel and comestible product of the duckweed family appeared on modern supermarket shelves. The domesticated duckweed *Lemna gibba* cv. Galilee has unique palatable and crisp characteristics as compared with other closely related clones. While initial economic success is resulting from marketing the duckweed product as a natural health food, this was not the major goal of the original biotechnological R&D. The main goal had been to use duckweed for recycling the liquid fraction of fermented cattle manure by generating a "ready-to-serve" leaf protein biomass. Although this remains a dire need in modern intensive farming, and even though the recycled duckweed biomass was proven to be a useful available protein source, commercial acceptance of this method was rather limited.

At present there are three basic challenging biotechnological research lines in duckweed aquaculture. (1) The biomass production of ca. 35 known duckweed species results from cell division rather than cell expansion. The activity of protein synthesis is predominant in duckweed biomass production. (The known industrial crops such as maize and potato accumulate carbohydrate as the result of intensive photosynthesis.) (2) The preferential ammonium ion uptake, followed by its rapid assimilation through the GS-GOGAT pathway, indicates that such "built-in" recycling of amino acids is important in the overall process of protein synthesis. (3) Recently, during our upscaling study, we observed that the doubling of chelated iron concentration in the growing solution resulted in significant accumulation of available iron which seems to be linked to a specific protein. A further upscaling study underway will enhance commercialization of this novel protein product.

Introduction

Floating and submerged plants, also defined by botanists as hydrophytes, are affected directly by the quality and chemical composition of the water milieu surrounding them. This fact has motivated environmental scientists and engineers to prefer water plants over terrestrial plants in wastewater management [1]. The biotechnological perspective represented in the present study summarizes the major generic research carried out over more than two decades to domesticate the globally-distributed duckweed (Lemnaceae) family and to develop it for several industrial applications.

The trials to domesticate clones from the duckweed family were encouraged also by the study of Bhanthumnavin and McGarry [10] who suggested adopting the duckweed *Wolffia arrhiza* as an aquatic crop and nonconventional source of cheap protein. This minute duckweed species is grown traditionally by farmers in Thailand and Burma. In parallel, another group of scientists, Pirie and his collaborators [11], suggested to concentrate protein from leaves as a by-product to be obtained from forest trees felled for use in the local wood industry. Such a biotechnological approach to recycle industrial waste was proposed to help overcome malnutrition problems common in developing countries. Other technologies were developed independently to recycle energy from cattle manure. After fermentation and utilization of the resulting methane as an energy source, the solids remaining are used as an improved processed organic manure. The liquid fraction still remains as dairy farm waste and a severe ecological nuisance. In Israel today, only two dairy farms are recycling cattle manure on a semi-industrial scale.

Living plant collection and plant tissue techniques are prerequisite biotechnological methods in the modern breeding of higher plants. These methods were adopted for use in several research studies on duckweed, as shown in Table 1.

Table 1. A comprehensive display of subjects and goals, reviewed by Landolt and his collaborators [2,3,4,5] as opposed to the biotechnological* approach of the present overview.

Biosystemic investigations	The biotechnological approach	References
Defined high multiplication rates found in many duckweed species and clones grown aseptically.	High biomass production of a selected clone (*Lemna gibba* cv. Galilee).	[6,7]
High content of nutritive protein which contains essential amino acids such as lysine and methionine.	Preferential ammonium ion uptake followed by intensive assimilation through the GS/GOGAT cycle. A "built-in" recycling system.	[8]
The natural floating habit of the duckweed family.	Separable biomass production of duckweed in open ponds, enabling the design of harvesting and engineering devices for wastewater treatment.	[9]

* Note: The widely divergent use of the term "biotechnology" has led the author to suggest an *ad hoc* definition: The technology needed to convert and upscale a biological process or system for application in industrial products.

Aseptic collections of duckweed clones; the clue in domestication and breeding research.

The aseptic collections of duckweed clones provide infinite possibilities in biotechnological domestication and breeding research. Selecting an appropriate clone suitable for biotechnological purposes out of more than one thousand already defined is a tedious task. However, a simultaneous recording study under laboratory and field conditions, followed by comprehensive study of clones from different research groups, proved more challenging and yielded satisfying results [6,7], as shown in Table 2 and Fig. 1.

Table 2. Some morphological criteria which were tested in *L. gibba* clones potentially suitable for human consumption. Note the stability of cultivar Galilee in the outdoor milieu after transfer from the aseptic laboratory conditions. Clustered long fronds, with a gibbous structure and short roots, are considered palatable characteristics.

Clone	Attached fronds (no)		Frond length (mm)		Gibbsity thickness (mm)		Root length (mm)		Selection index*
	lab	outdoor	lab	outdoor	lab	outdoor	lab	outdoor	
cv. Galilee	4.9	3.5	5.3	5.6	-	1.4	7.0	8.0	3.43
ETH 7021	3.7	2.2	5.0	2.8	-	0.5	13.0	21.0	0.15
ETH 7784	4.6	2.6	5.0	4.2	-	0.8	5.0	10.0	0.87
ETH 8703	4.9	3.5	4.5	5.3	-	1.4	6.0	30.0	0.87
ETH 8381	4.0	1.5	5.0	3.6	-	0.4	8.0	13.0	0.17

* The ratio of clone size to root length.

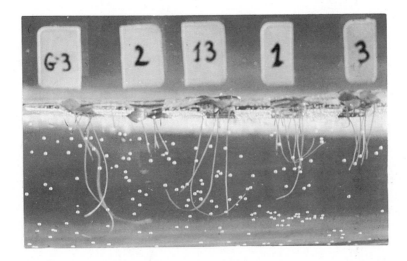

Fig. 1. The significance of aseptic collection of duckweed clones. Note clone #2 with its short roots, defined today as *L. gibba* cv. Galilee. This cultivar is used as a source for comestible products by Israeli industry [5]. This duckweed is used also as a model for intensive N-metabolism [8]. The clone #G3 on the left, was preferred by Culley et al. [1] for waste management and animal feed. The apparent long roots of this clone exclude it as a ready-to-serve palatable plant.

A combined multi-looped recycling aquaculture of fish and duckweed.

In the years 1980-1985, an R&D project for combined fish and duckweed aquaculture was set up in order to create an efficient water and nutrient recycling aquaculture (see Scheme A). Although the scientific and engineering problems involved were solved [12,13,14], the upscaling and industrialization stages were not implemented. The R&D project's successful biotechnological results, including its novel pilot plant, did not sufficiently convince the scientific decision makers to obtain appropriate funding support from industry and national government sources.

The duckweed *L. gibba* cv. Galilee as a comestible industrial product

During the intensive biotechnological studies on duckweed described in Scheme A, a rather amateurish trial to test the use of duckweed as "water sprouts" in home cooking was carried out, and yielded surprisingly successful results. The palatable appearance of the domesticated duckweed from the laboratory studies encouraged nonprofessional home economics colleagues to prepare several dishes for staff meals and parties. Although their "beginners' luck" was gratifying, and although a number of venture investors showed interest in pursuing this avenue further, no commercial research and development project resulted. It took nearly five years to receive certification from the authorities on each of the two Israeli patents currently pending. The economic and socio-legal difficulties are discussed by Tsur [15] in a specific case study report of the Second International Conference on Ecological Engineering.Some initial information in German was published recently [5]. Fig. 2 shows comestible duckweed with short roots compared to uncultivated duckweed with rather long roots.

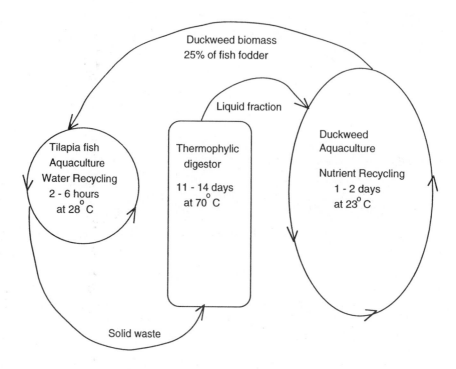

Scheme A. A multi-looped recycling model of combined fish (*Tilapia* sp.) and duckweed (*Lemna gibba* cv. Galilee) aquaculture.

The water recycling system lasts 2-6 hours, discarding the solid fish wastes and avoiding ammonification as well as enriching the water with oxygen. The daily nutrient recycling system via duckweed aquaculture takes place after 10-14 days of thermophilic digestion of the solid wastes accumulated in the intensive fish culture. The fermented dissolved organic and mineral compounds are used as nutrients for duckweed growth. Fresh duckweed biomass regularly substitutes up to 30% of the total fish fodder [13]. The energy gained from the methane fermentation can be used to heat the system in cold seasons.

Fig. 2. Domesticated (*L. gibba* cv. Galilee)*versus* uncultivated duckweed on the right side. Note the elongated roots which are common in the uncultivated duckweed biomass.

A biotechnological perspective toward novel industrial products using duckweed

In 1991, a two-year R&D project supported by the Israel Ministry of Trade and Industry finally led to successful industrialization of the duckweed aquaculture system and opened the way to our search for promising new uses for duckweed.

In connection with R&D level marketing research and product quality studies carried out for duckweed as a potential comestible product, it was observed experimentally that the ash content of cultivated domesticated duckweed contains considerable amounts of minerals, among them potassium and iron. Further study indicated that, in comparison with the traditionally iron-rich spinach, the iron content in duckweed forage is quite high. Enrichment of the growing solution with chelated iron source doubled the iron content and boosted other beneficial constituents such as ascorbic acid (Vitamin C). The preliminary findings are summarized in Table 3.

Table 3. Experiment in enrichment of comestible duckweed (*L. gibba* cv. Galilee) nutrient solution with chelated iron solution. Numbers indicate mg/g dried biomass.

	Iron	Manganese	Ascorbic Acid	Oxalic Acid
Duckweed regular nutrient solution	46	75.5	1.5	1605
Duckweed enriched with chelated iron	108	8.5	16.0	1735
Spinach (control)	22	5.0	0.9	3780

The results show a clear increase in iron and ascorbic acid content and a parallel decrease in manganese. Frozen spinach from the supermarket shelves was used as a control. It is rather clear that the duckweed can be used as a source for mineralic iron nutrition.

It was also found that a considerable amount of the iron binds to a high molecular weight protein, which enables isolation of an iron-rich proteinous fraction. Oxalic acid, which interferes with iron uptake by humans, is rather low in duckweed plants as compared to spinach plants. Fractionation of the mashed duckweed allows separation of the different nutritive components and their use as nutritive additives to become industrial raw materials. Further development of such novel products will be a challenge for research scientists as well as for the biotech industry.

References

[1]   D. D. Culley Jr., E. Rejmankova, J. Kvet, J. B. Frye, J. World Maric. Soc. **12**, 27 (1981).
[2]   E. Landolt, Veröff. Geobot. Inst. ETH, Stiftung Rübel, Zurich **71**, 566 (1986).
[3]   E. Landolt, R. Kandeler, Veröff. Geobot. Inst. ETH, Stiftung Rübel, Zurich **95**, 638 (1987).
[4]   A. Lueoend, Veröff. Geobot. Inst. ETH, Stiftung Rübel, Zurich **80**, 116 (1983).
[5]   E. Landolt, Ber. Geobot. Inst. ETH, Stiftung Rübel, Zurich **59**, 132 (1993).
[6]   D. Porath, B. Hepher, A. Koton, Aquat. Bot. **7**, 273 (1979).
[7]   D. Porath, Y. Efrat, T. Arzee, Aquat. Bot. **9**, 159 (1980).
[8]   E. B. Monselise, D. Kost, Planta **189**, 167 (1993).
[9]   W. D. Poole, Proceedings, 2nd Internat. Conf. on Ecological Engineering, Waedenswil, Switzerland (in press).
[10]   K. Bhanthumnavin, M. G. McGarry, Nature **232**, 495 (1971).
[11]   N. W. Pirie, Nature **253**, 239 (1975).
[12]   I. G. Gaigher, D. Porath, G. Granoth, Aquaculture **41**, 235 (1984).
[13]   D. Porath, G. Oron, G. Granoth, Proceedings, 5th Internat. Symp. on Agricultural Wastes, ASAE, Chicago, 680 (1985).
[14]   B. Hepher, Y. Pruginin, Commercial Fish Farming, Wiley & Sons, New York, 261 (1981).
[15]   Y. Tsur, Proceedings, 2nd Internat. Conf. on Ecological Engineering for Wastewater Treatment, Waedenswil, Switzerland (in press).

Environmental Research Forum Vols. 5-6 (1996) pp. 313-318
© *1996 Transtec Publications, Switzerland*

On the Stability of Water Hyacinth Wastewater Treatment Systems

D.C. Austin and G. Tchobanoglous

Department of Civil and Environmental Engineering, University of California-Davis,
206 Walker Hall, CA 95616 Davis, USA

Keywords: Water Hyacinth, Ecosystem Stability, Ecotechnology, Activated Sludge, Wastewater, Natural Systems, Ecological Engineering

ABSTRACT

Differences in the stability of water hyacinth and activated sludge treatment systems used to treat domestic wastewater have been observed. An explanation for these differences is presented in terms of the physical and biological properties of the respective treatment ecosystems. Suggestions are put forth for improved characterization of treatment ecosystems and their stability.

INTRODUCTION

Consistent attainment of desired effluent parameters is an essential quality in any wastewater treatment system. As more ecologically based technologies are used to treat wastewater, questions of system performance stability invite comparison with traditional treatment technologies such as activated sludge. Questions concerning the stability of the treatment ecosystem also may clarify certain elements of ecological theory that seek to explain why one biological system is inherently more stable than another. The purpose of this paper is to consider the stability of two different types of ecosystems used for the treatment of domestic wastewater. Differences have been observed in the stability of water hyacinth (*Eichornia crassipes*) and conventional activated sludge treatment systems. The differences between the two systems are discussed in light of relevant ecological theory.

PARAMETERS USED TO QUANTIFY TREATMENT PERFORMANCE AND STABILITY

Comparison of wastewater treatment systems requires use of standard parameters. Standard parameters used to quantify treatment performance are typically set by historical precedence, regulation and accepted engineering practice [7].

Treatment Parameters

Common treatment parameters expressed in mg/L are: biochemical oxygen demand (BOD), carbonaceous biochemical oxygen demand (CBOD) in which oxygen demand of nitrifiers is suppressed, total organic carbon (TOC), chemical oxygen demand (COD), and suspended solids (SS). Other parameters include turbidity (NTU's) and microbiological measures such as number of total fecal coliforms per 100 mL of effluent. It is important to note that lumped or aggregate parameters such as suspended solids, BOD, TOC, COD and turbidity are of limited value in answering the basic removal mechanisms involved [14].

Process Performance Stability

System performance and stability were analyzed using probability plots of standard performance criteria (SS, TOC and CBOD). The probability plot is an analysis of variance in which each point represents the percentage of time an effluent parameter is recorded at or below a certain value. In a perfectly stable system effluent values will be constant, resulting in a plot of zero slope. A system with significant variability will result in a plot with a steep or discontinuous slope.

COMPARISON OF THE TWO TREATMENT SYSTEMS

The treatment ecosystems compared in this paper are the San Diego water hyacinth wastewater treatment pilot plant [15] and a number of conventional activated sludge systems. The San Diego water hyacinth treatment system is comprised of a rotary drum and a disk filter for primary treatment followed by a step-feed, shallow plug-flow, water hyacinth covered reactor with recycle (see Fig. 1a). In-channel aeration is also used. The flow diagram for a conventional activated sludge secondary treatment system is given in Fig. 1b.

The performance and stability of the water hyacinth system and the activated sludge systems is presented in Figures 2a and 2b, respectively. The water hyacinth treatment system exhibited significantly more consistent effluent values as evidenced by the slope of the probability curve and the uniformity coefficient defined as the ratio of the P_{60} to P_{10} values. The markedly steeper slopes observed with activated sludge systems are indicative of a less stable system.

Differences in the stability of the two treatment systems may be inherent properties of the type of treatment ecosystem they contain. Accounting for these differences requires that clear definitions of ecosystem and stability be available suggesting why a water hyacinth treatment system might be inherently more stable than an activated sludge system.

ELEMENTS OF ECOSYSTEM STABILITY

The concept of ecosystem is commonly used, but ambiguous. A precise definition of ecosystem has been elusive because varying criteria for observation have been used to define it [1]. Odum [9] considers it the basic ecological unit, but other concepts are common. Most definitions that do not select a particular set of observational criteria are statistically confounded, amounting to little more than an ecosystem being defined as a spot where organisms interact with each other and their environment over some period of time. Stability of an undefined system is meaningless.

Definition of Ecosystem

Ecosystems are typically defined in terms of the methods used to observe and quantify them. A large part of the ecological sciences is concerned with population dynamics or community structures, such as predator-prey relationships or forest succession. Another branch of the ecological sciences bases its observations on Odum's classic concept of the ecosystem involving transfers of energy and matter as organisms interact with each other and their environment. It is possible for the same territory observed using these different criteria to yield substantially different, yet scientifically valid, concepts of the important elements comprising the ecosystem [1]. Because the biological treatment of wastewater involves flows of energy and matter between organisms and their treatment volume, Odum's concept of the ecosystem appears to be the more useful for engineering applications.

For the purposes of this paper, treatment ecosystems can be defined functionally in terms of transfers of energy and material as measured by effluent parameters on a mass balance basis. These criteria are consistent with both current ecological theory and standard engineering practice [1, 5]. Defining the treatment ecosystem in terms of transfers of energy and matter between organisms and their treatment volume links stability of effluent parameters functionally to the stability of the treatment ecosystem. The consistency of effluent parameters defines ecosystem stability, allowing for empirical investigation of the underlying mechanisms operative within the treatment system.

Stability and Diversity

There is an extensive body of literature on the relationship between ecological stability and diversity. It seems intuitively clear that a more biologically diverse treatment system is likely to be inherently

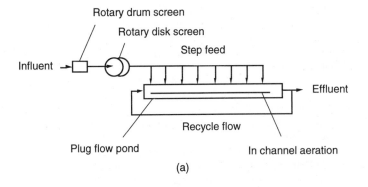

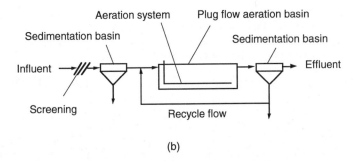

Figure 1 Wastewater treatment systems: (a) water hyacinth wastewater treatment system at San Diego, CA and (b) conventiontional activated sludge treatment system

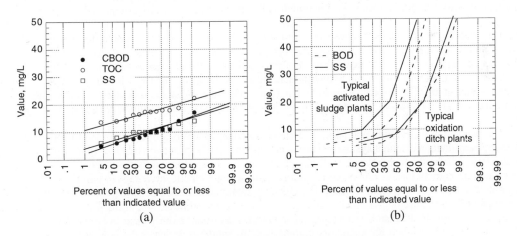

Figure 2 Process stability based on effluent quality parameters:
(a) water hyacinth wastewater treatment system at San Diego, CA [12, 15] and
(b) conventional activated sludge treatment system [16]

more stable than one of lessor diversity. Unfortunately, a simple relationship between diversity and stability is not only equivocal, but contrary to both mathematical models and empirical evidence. Gardner and Ashby demonstrated, using the results obtained with mathematical models, that increasing the number of organisms with highly connected interrelationships can quickly destabilize a system [4]. Indeed, the notion that stability can be related to species diversity measured as a tally of total species is widely discredited in theoretical ecology [2, 6]. Comparison of activated sludge and water hyacinth systems offers support for this position. Activated sludge is comprised of a complex ecosystem characterized by rich species diversity of bacteria, protozoa, rotifers and other invertebrates [5]. If stability can be correlated with species numbers, why should adding just one more species, such as water hyacinth, among hundreds create a more stable system? It is more likely that physiological and structural alterations of the treatment volume by the water hyacinth provide the mechanisms for greater stability. If stability is to be linked in some convincing fashion to diversity within the treatment ecosystem, then diversity must be a quality entailing some measure other than mere number of species.

WHY THE DIFFERENCE IN THE TWO ECOSYSTEMS?

The stability performance data presented above were gathered in other studies unrelated to this paper. Therefore, no hypothesis concerning treatment ecosystems can be directly supported by the data. However, in light of what is known about wastewater treatment systems and ecosystem dynamics, plausible reasons for the differences in stability between the two ecosystems can be suggested and testable hypotheses proposed.

Activated Sludge

Activated sludge treatment systems are large-scale bacterial enrichment cultures. A portion of the heterotrophic bacteria which form flocs and settle in the secondary clarifier is returned to the aeration chamber. Return of activated sludge causes floc forming bacteria to dominate the aeration chamber ecosystem. A fundamental reason why activated sludge works so well with respect to the removal of organic matter is that a great many species of bacteria are heterotrophic floc formers. Bacterial species composition may change every few days, but as long as they rapidly consume soluble carbon and form flocs the treatment ecosystem is functionally stable as measured by effluent BOD and SS.

Apart from acute toxic shocks that destroy microbial populations, performance instability in activated sludge chiefly occurs when a non floc-forming populations of bacteria become dominant in the treatment ecosystem. Causes of instability are poorly understood from an ecosystem perspective [10], but are sufficiently well correlated with factors such as sludge age or BOD loading to guide effective management strategies. One known activated sludge ecosystem process is that protozoa, particularly ciliated protozoa, reduce turbidity by heavily grazing suspended single celled bacteria [3].

It is inevitable that undesirable populations of bacteria will at some time outcompete floc forming bacteria in an activated sludge system. There is nothing to stop the process within the treatment ecosystem. The aeration chamber is well mixed, allowing any perturbation in the influent to propagate throughout the system. A perturbation giving a slight kinetic advantage to non floc forming bacteria over floc forming bacteria can lead to domination of undesirable populations and consequent performance degradation. Only careful management strategies requiring a skilled and relatively large operational staff can keep these tendencies in check. Even so, thirty day average effluent values are standard for regulatory compliance. At shorter time scales the inherent instabilities of activated sludge would make compliance unnecessarily difficult.

Water Hyacinth

Water hyacinth roots provide a large surface area, relative to the treatment volume, colonized by attached bacteria. Nobel [8] notes that the root surface area of plants is usually 200-400 times that of the leaf surface area. Root biofilms and their extent are undoubtedly critical to the performance stability of water hyacinth systems. Hawkes [5] notes that activated sludge treatment systems have long been recognized as less robust than biofilm filters. What advantage would root biofilms have over biofilms on an nonliving surface such as spun polyester?

Answers to this question await empirical comparison, but some biological facts should be considered. Aquatic plants such as water hyacinths translocate air to their roots. Mass flux of oxygen to the roots is insufficient for aeration of the treatment volume, but is sufficient to support aerobic attached bacteria. Another important fact is that vascular plants passively exude or actively secrete photosynthetically fixed carbon in a variety of forms susceptible to microbial degradation [11]. The plant's physiological response to its environment determines the type and quantity of compounds released to it's root zones. An inert filter surface cannot physiologically regulate the metabolic activities of attached bacteria.

Leakiness of roots may have important implications for physiological conditions within the treatment volume. When roots are passing oxygen and carbon substrates to attached bacteria, it is likely that physiologically significant gradients develop in the biofilm from the root surface to the wastewater interface. Biofilms tend to self-organize by bacterial species along a variety of gradients. Along water hyacinth roots these gradients are likely to be based on redox potential and specific carbon substrates regulated by plant physiology. Whereas bacteria life cycles range from 20 minutes to a few hours, water hyacinths live for weeks to months. Thus it appears that the physiological conditions induced by the roots tend to be a stabilizing factor in comparison to the more transient physiological conditions created by bacteria alone in a well mixed volume such as activated sludge. Additionally, the species composition of the attached bacterial biofilms will be significantly influenced by root physiology.

NEED FOR BETTER CHARACTERIZATION OF ECOSYSTEM PROCESSES

Differences in performance stability between the two treatment systems have plausible explanations as outlined above. Mere plausibility, however, is not a sufficient explanation of treatment ecosystem stability. Research employing an ecosystem approach is essential to gain an understanding the underlying mechanisms of stability in water hyacinth wastewater treatment systems.

Variability of effluent parameters beyond a restricted range is indicative of performance instability as indicated previously. Within biological systems variability is a product of internal dynamics of the treatment ecosystem and its reaction to variability of influent parameters. Standard wastewater measurements, unfortunately, may be too crude to distinguish important ecosystem processes. Suspended solids cannot be used to distinguish between a raw fecal particle and a root hair of the same mass. TOC and COD have no direct relation to physiological processes. The most commonly used parameter, BOD, has no stoichiometric validity [13]. Turbidity cannot distinguish between suspended bacteria and humic acids blocking light. Unless the shortcomings of these parameters are taken into account little insight into treatment ecosystem processes will be gained. More sensitive parameters, such as use of specific organic compounds, are needed in addition to those mandated by regulatory agencies.

It is also important that standard analytical procedures be adopted to document stability. Probability analysis based on standard non lumped parameters is an excellent choice. Stability as measured by a specific parameter can be referred to unambiguously in terms of it's probability distribution.

Standardized analysis of stability allows for meaningful comparison between various treatment systems. For a given parameter, alteration of ecosystem structure or inputs can help determine what structural components mediate the energy or material flow in question. Questions such as how the type and degree of biological diversity affect the robustness of the system can be investigated more rationally.

SUMMARY
Water hyacinth treatment systems appear to be more stable than activated sludge systems. Reasons for the greater stability are probably related to physiological and structural diversity provided to the treatment volume by roots. Defining stability in terms of the consistency of effluent parameters on a mass balance basis will allow for more meaningful comparisons to be made. Combining a probabilistic analysis with an investigation of transfers of energy and matter within the treatment volume may lead to useful insights into the biotic mechanisms underlying stability in plant based wastewater treatment ecosystems.

BIBLIOGRAPHY
[1]   T. F. H. Allen and T. W. Hoekstra, *Toward a Unified Ecology*. Columbia University Press, New York. 1992.
[2]   T. F. H. Allen and T. B. Starr. *Hierarchy: Perspectives for Ecological Complexity*. University of Chicago Press, Chicago. 1982.
[3]   C. R. Curds. Protozoa, in *Ecological Aspects of Used-Water Treatment*, Vol. 1 Biological Activities and Treatment Processes, Chapter 5. 1975.
[4]   M. R. Gardner and W. R. Ashby. Connectance of Large Dynamic (Cybernetic) systems: Critical Values for Stability. Nature 228:784.
[5]   H. A. Hawkes. Activated Sludge, *in Ecological Aspects of Used-Water Treatment*, Vol. 2 Biological Activities and Treatment Processes, Chapter 3. C.R. Curds and H.A. Hawkes, Eds. Academic Press. New York. 1983.
[6]   J. J. Kay. On the Nature of Ecological Integrity: Some Closing Comments, in *Ecological Integrity and the Management of Ecosystem*. S. Woodley, J. Kay, G. Francis, Eds. St. Lucie Press. 1993.
[7]   Metcalf & Eddy, Inc. *Wastewater Engineering, Treatment, Disposal and Reuse*, Third edition. McGraw-Hill. 1991.
[8]   P. S. Nobel. *An Introduction to Biophysical Plant Physiology*. W. H. Freeman and Company, San Francisco. 1974.
[9]   H. T. Odum, *Systems Ecology: An Introduction*. John Whiley and Sons, New York (1983).
[10] E. B. Pike. Aerobic Bacteria, in *Ecological Aspects of Used-Water Treatment*, Vol. 1 Biological Activities and Treatment Processes, Chapter 1. 1975
[11] A. D. Rovira and C. B. Davey. Biology of the rhizosphere, p. 153-240, in E. W. Carson (ed.) *The Plant Root and its Environment*. 1974.
[12] City of San Diego. 1995. Current performance data. January 1995 through July 1995.
[13] E. D. Schroeder. The importance of the BOD plateau. Water Research. Pergamon Press. Vol. 2, pp. 803-809. 1968.
[14] G. Tchobanoglous. Constructed Wetlands And Aquatic Plant Systems: Research, Design Operational, And Monitoring Issues, In *Constructed Wetlands for Water Quality Improvement*, Edited by G. A. Moshiri, Lewis Publishers, Boca Raton, FL, 1993.
[15] G. Tchobanoglous, F. Maitski, K. Thompson, and T. H. Chadwick. Evolution and Performance of the City of San Diego Pilot-Scale Aquatic Wastewater Treatment System Using Water Hyacinths. Research Journal WPCF, Vol. 561, Nos. 11/12, pp 1625-1635, 1989.
[16] Water Environment Federation. *Design of Municipal Wastewater Treatment Plants*. WEF Manual of Practice No. 8 ASCE Manual and Report on Engineering Practice No. 76. Water Environment Federation. Alexandria. 1991.

Environmental Research Forum Vols. 5-6 (1996) pp. 319-324
© 1996 Transtec Publications, Switzerland

Advanced Treatment of Secondary Effluent and Polluted Municipal River Water through the Utilization of Natural Purification Capability

K. Furukawa and M. Fujita

Department of Environmetal Engineering, Faculty of Engineering, Osaka University,
Yamadaoka 2-1, Suita, Osaka 565, Japan

Keywords: Pak-Bung, Water Cress, Water Lettuce, Living Filter, Advanced Treatment, Purification of River Water, Contact Oxidation, Porous Concrete

ABSTRACT

Actual treatment plant composing of hydroponic type wastewater treatment(HWT) process, which utilizes the natural purification capability, was designed for the treatment of domestic wastewater of 7,100 inhabitants. HWT plant(width 3 m, total length 120 m, depth 1.2 m) was equipped with normal advanced wastewater treatment facility. Aquatic plants(pak-bung:*Ipomoea aquatica*, water cress:*Nasturtium officinale*) were successfully cultivated in HWT plant as nutrient absorber. Based on the harvested weight, nutrient removal(N and P) capacities of these aquatic plants were determined to be as follows;*I. aquatica*: 1.49 t-N/ha/y, 0.239 t-P/ha/y, *N.officinale*:2.27 t-N/ha/y, 0.401 t-P/ha/y. Harvested pak-bung was shown to be effectively used as the raw extracting material of peroxidase, because of its high peroxidase content(400 U/g-DW). This study demonstrates that implantation of pak bung and water cress is an appropriate option for nutrient removal of wastewater.

Pilot plant(width 0.8 m, total length 24 m, depth 1.4 m) for the direct purification of urban river water polluted with grey water by the combination of contact oxidation and HWT was constructed. Influent BOD, SS and anionic surfactants were effectively reduced under hydraulic retention time of 4 hours in summer season. Cultivation of aquatic plants(water lettuce:*Pistia stratiotes* and water cress) was succeeded by using the effluent from contact oxidation basin. Direct purification of polluted river water using contact oxidation and HWT was proved to have a high application potential.

INTRODUCTION

There remains high concentration of nutrients(N and P) in the secondary effluent of domestic wastewater. These nutrients cause the eutrophication of receiving waterbodies. Nutrient removal of wastewater through the usage of floating and unattached macrophytes is receiving increased attention and has been the subject of a number of investigations[1,2]. But this process is not commonly applied owing to the following reasons; (i)wide plant area is required, (ii)stable treatment is difficult due to the changes in purification capability under different climate conditions, (iii)effective harvesting and disposal method of grown aquatic plants are left unsolved.

Small urban rivers in Japan have been polluted with grey water from the area of no sewage works service. Direct purification of these urban river waters is also attracted much attention as a transitional mean until sewage works is provided and several actual treatment plants, which mostly use the contact gravel bed treatment method, were constructed for the reduction of SS and BOD. Hence the water quality of polluted urban river water is quite similar to that of secondary effluent, simultaneous BOD, SS and nutrient removal may be possible by the proper combination of contact gravel bed treatment and the cultivation of aquatic plants.

In this study, treatment capability of HWT(hereafter this process is termed a living filter), which consists of a contact oxidation tank and hydroponic type cultivation tank, was evaluated for the economically and energy saving advanced wastewater treatment and the direct purification of polluted urban river water.

MATERIALS AND METHODS

Living filter

Figure 1 depicts the configuration of advanced wastewater treatment plant which incorporates living filter. This actual treatment plant was constructed to receive the domestic wastewater of 7,100

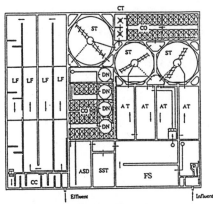

Fig. 1. Living filter configuration.

FS:Flow stabilization tank, AT:Aeration tank,
ST:Settling tank,CO:Contact oxidation tank,
CT:Coagulation tank, DN:Denitirification tank,
ASD:Aerobic sludge digester,
SST:Sludge storage tank, LF:Living filter,
CC:Chlorine contact chamber

Fig. 2. Longitudinal section of living filter.

inhabitants. This treatment plant is located at the upper stream of Yono River, whose surface water is an important source of water supply of Ikeda city (Osaka, Japan) located at the downstream of this river. Hence, the stringent effluent standards(BOD < 5 mg/L, T-N < 5 mg/L, T-P < 1 mg/L) were imposed to this treatment plant by the Osaka prefectural government. In order to achieve this strict effluent discharge limits, living filter(width 3 m, total length 120 m, depth 1.2 m) was equipped with normal advanced wastewater treatment facility which consists of activated sludge process, chemical phosphorus precipitation and biological denitrification.

Figure 2 presents the longitudinal cross section of living filter designed. 70.7 m^3 of biofilm contactors(Hishipakkin FK-40, specific surface area 71 m^3/m^2, Mitsubishi Plastic Co. Ltd., Tokyo, Japan) were set at both wall sides of water channels 1 and 2. 237 pieces of plastic baskets(each of width 42cm, length 58 cm, height 14.5cm) packed with gravel were set on the biofilm contactors. Half of these baskets were dipped in wastewater. Corridors were provided in the central of water channel for the maintenance of aquatic pants. Pak-bung(*Ipomoea aquatica*) was cultivated during summer to autumn season, and water cress(*Nasturtium officinale*) was cultivated during winter to spring season as the nutrient absorber. About 20 strains of pak-bung were implanted on the gravel beds at the end of May. Harvesting was carried out every 2 weeks after complete cover of gravel bed with pak-bung. Germinated stems were cut for its harvesting. On the other hands, 20 strains of water cress were implanted on gravel beds at the end of December. Grown water cress over 30cm height were cut for its harvesting after complete cover of gravel bed with water cress. This living filter was constructed in a greenhouse without heating function.

Direct river water purification treatment plant

Figure 3 shows the flow diagram of pilot plant for direct purification of polluted river water which was constructed at the drains of Tatuta River in Nara Prefecture. This plant consists of a primary settling tank(2.8 m^3), a contact oxidation tank(13.4 m^3) with biofilm contactors(Hishipakkin FK-40, specific surface area 71 m^2/m^3), a contact oxidation tank(3.4 m^3) with 100 pieces of porous concrete(30 % void space, 10cm diameter and 20cm length), cultivation tank of an aquatic plant (3.4 m^3) and final fish

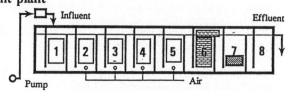

Fig. 3. Flow diagram of treatment plant for direct
purification of polluted river water.
1:settling tank, 2-5:contact oxidation tank,
6:porous concrete tank, 8:fish tank
7:aquatic plant cultivation tank

cultivation tank(3.4 m^3) in series. Polluted river water was directly pumped up and was used as the influent. The influent flow rate was controlled by triangular weirs. Influent flow rate was set at

80m^3/d initially owing to secure the stable biofilm attachment on contact materials. After recognizing the well growth of biofilm, the plants was operated under various loading rates. Water lettuce(*Pistia stratiotes L.*) was implanted at the middle of June, covering about 5 % of the surface area of aquatic plant tank. After 2 months, water lettuce completely covered the surface area of this tank. Harvestings were conducted every 2 weeks after full cover of entire surface area with water lettuce canopy. Water lettuces were cultivated until end of December. Water cress were implanted on the gravel bed(2.4 m^2)constructed on the upper portion of contact oxidation tank with porous concrete at January. Harvesting of water cress were started after its full cover of gravel bed by cutting the grown stems. Sludge accumulated on the bottom and sloughed sludge were withdrawn every 6 months.

Analytical methods

Unless otherwise stated, water analyses were performed in accordance with the Standards Methods[3]. Anionic surfactants were measured as ethyl violet active substances(EVAS)[4].

RESULTS AND DISCUSSION

«Living Filter»
Treatment Capabilities

Owing to the small number of residents at the start of operation of wastewater treatment facility and the unexpected big amount of ground water leakage, both amount of inflow wastewater and wastewater strength were low compared with those of design value. Therefore, influent from flow stabilization tank were directly introduced to the living filter as influent and the wastewater treatment was conducted only by passing the influent through living filter.

Figure 4 shows the treatment capability of living filter. Seventy nine and 72.0 % of TOC were removed at the treatment temperature of 20-28 °C and 14-20 °C, respectively. Effluent TOC concentration below 5 mg/L could be consistently obtained. Moreover, excellent clear effluents with transparency of more than 1.2 m were also obtained. The average T-N removal of 51.3 % and 37.2 % were achieved at the treatment temperature of 20-28 °C and 14-20 °C respectively within the test volumetric T-N loading rates. Figure 5 presents the relationship between the volumetric T-N loading rates and nitrification efficiencies during the treatment. Here, nitrification efficiency was defined the concentration ratio of effluent NO$_3$-N to effluent T-N. Volumetric T-N loading rates more 3.2 g-N/m^3/day and 2.5 g-N/m^3/day were required to achieve 50 % of nitrification efficiency at 20-28 °C and 15-20 °C, respectively. Within test loading regions, 37-51 % of T-N were removed only by living filter treatment. But these removal efficiencies were still significantly higher than the nitrogen uptake capabilities of test aquatic plants. By combining these results with the progress of nitrification during the treatment, amount of nitrogen removed by denitrification is regarded to be relatively high in this living filter treatment. T-P removal of 56.8 % and 41.4 % were achieved at the treatment temperature of 20-28 °C and 14-20 °C respectively only by this living filter treatment.

By summarizing the treatment results of this living filter, wastewater treatment capability of living filter could be estimated as shown in Table 1. Presently, number of residents increased to more than 1,000 and effluent standard could not attain only by living filter treatment. Therefore, secondary and tertiary treatment facilities were operating before living filter treatment. Living filter is now functioning as the final polishing filter as designed.

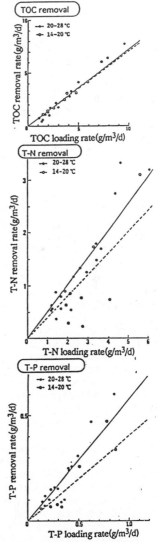

Fig.4. Treatment capability of living filter

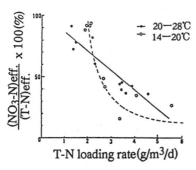

Fig. 5. Nitrification in living filter.

Table 1. Treatment capability of living filter

Temp.	Treatment capability (t/ha/year)		
(°C)	BOD	T-N	T-P
20-28	85	38	4.4
14-28	50	25	3.1

Plant growth

Cultivation of pak-bung and water cress on the gravel beds were succeeded using nutrients contained in influent domestic wastewater. Adequate harvesting and maintenance of suitable plants density were required for significant nutrient removal by these aquatic plants. After full coverage of gravel beds with these aquatic plants, biweekly harvesting was appropriate. 750 kg (fresh weight) of pak-bung was harvested during 125 days of living filter treatment. Whereas 2,395 kg (fresh weight) of water cress could be harvested during 7 months of cultivation. By using these crop yields and N and P contents of aquatic plants, N and P removal capabilities of these two aquatic plants were calculated as shown in Table 2.

Table 2. Crop yield, T-N and T-P removal rate of aquatic plants
cultivated on domestic wastewater

Aquatic plant	Wastewater conc. (mg/l)			Temp. (°C)	Crop yield (t-DW/ha/y)	Nutrient removal rate	
	TOC	T-N	T-P			(tN/ha/y)	(tP/ha/y)
Pak-bung	2.1-41.9	3.6-18.8	0.4-2.4	20-32	25.8	1.49	0.239
Water cress	10-50	10-30	1-4.3	10-20	45.5	2.27	0.401

Crop yields, N and P removal rates for water cress were unexpectedly high compared with that for pak-bung. The causes of this difference are as follows; (i) water cress can be cultivated at high cultivation density because of its vertically growth, (ii) long term harvesting (March to July) is possible for water cress. Harvesting period for pak-bung is short under climatic condition like Japan, because it takes long time (about 2 months) to harvest pak-bung after implanting. The nutrient removal capabilities of these aquatic plants were proved to be inferior to that for water hyacinth, but these two plants have the advantages of being used for food stuff[5].

Effective reuse of these harvested aquatic plants as food stuff has the following drawbacks; (i) mental barrier for taking plants cultivated on wastewater as food, (ii) higher expenses for harvesting. Alternative reuse of these harvested aquatic plants was investigated and found that pak-bung contained higher amounts of peroxidase (EC. 1.11.1.7)[6], which is one of the industrially important enzyme. Table 3 shows the peroxidase content of pak-bung. Based on the crop yield of pak-bung and its peroxidase content, 5.54×10^7 U/ha/month of peroxidase was possible to extract from pak-bung cultivated on wastewater effluent. Owing to the difficulties of cost estimation required for extraction, the judgment of market price of harvested pak-bung as the raw material for peroxidase production is impossible. But harvested pak-bung should have some market value by considering the market price of peroxidase (2,500 yen/5,000U). Hence, the living filter process, which can remove nutrients and produce valuable by product, might be a promising advanced wastewater treatment alternatives in future.

Table 3. Peroxidase activity of pak-bung

Portion	Peroxidase activity
Leaf	26.9 U/g-wet weight
Stem	28.1
Root	41.5

«Direct River Water purification»
Treatment results

Fig. 6 presents the daily changes in TOC, SS, T-N and EVAS concentrations during 250 days of operation. Fig. 7 shows the changes in TOC loading rates. Plant was operated under TOC loading of 80-130 g-TOC/m³/d during summer to autumn seasons. The loading rates were decreased to 50-70 g-TOC/m³/d during winter season as the decrease in treatment temperature. After 100 days of operation, TOC removals were improved and effluent TOC concentrations below 10 mg/L were consistently obtained. After full coverage of aquatic plants cultivation basin with water lettuce canopy, SS were effectively removed within only short hydraulic retention time, indicating high SS trapping capability of this plant. This high SS removal function may attributed to the high SS trapping function of root hairs of water lettuce. Effluent transparencies more than 50 cm were obtained. Owing to the area limit for aquatic plants cultivation basin, only 20-30 % of T-N were removed during the treatment. Unexpectedly high concentrations of EVAS(> 3 mg/L) were detected in influent polluted river water, but EVAS was efficiently removed below 1 mg/L, which is the limiting concentration of foaming in river stream.

Plant growth

Water lettuce was cultivated for about 200 days. Water lettuce was covered with the surface of aquatic cultivation tank after one month of cultivation on polluted urban river water. Plant density could be satisfactorily controlled by harvesting about one third of waster lettuce existed in the aquatic plants cultivation tank every 2 weeks. Water lettuce showed the best growth on polluted urban river water in September to October. About 30 kg(wet) of water lettuce could be harvested during this treatment period. Figure 8 depicts the changes in cumulative harvested weight and standing weight of water lettuce during the treatment. By taking consideration of the increase in standing weight, total amounts of water lettuce grown in plants cultivation tank reached to 35.7

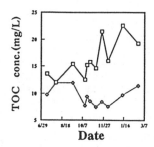
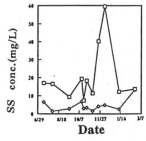
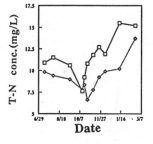
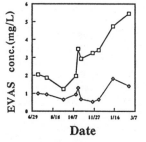

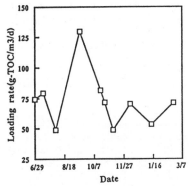

Fig.7. Changes in loading rates.

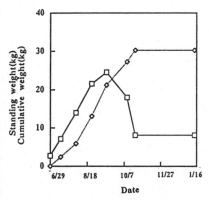

Fig. 8. Growth of water lettuce.

Fig.6. Diary changes in TOC, SS, T-N and EVAS

kg(wet) during 200 days of operation. Table 4 shows the nutrient removal capability of water lettuce, which was determined experimentally in this study. As shown in this Table, nutrient removal capability of water lettuce was proved to equals to those of water cress and pak-bung.

Table 4. Nutrient removal capability of water lettuce

maximum crop yield	939	t(wet)/ha/y
	51.6	t(dry)/ha/y
maximum N removal rate	1.51	t-N/ha/y
maximum P removal rate	0.342	t-P/ha/y

Different from pak-bung and water cress, water lettuce can not be utilized as food stuff. But water lettuce was proved to possess outstanding characteristics for decorating this kind of treatment plant compared with another aquatic plants. And also harvested water lettuce has the market value as aquatic plant for decorative aquarium.

Water cress was successfully cultivated in winter season by replace of water lettuce. Water cress was reported not to survive in hot climate. We are continuing experiments in order to make clear the effective cultivation term and purification capability of water cress in this direct river water purification plant.

CONCLUSIONS

Treatment capabilities of advanced wastewater treatment plant and direct purification plant for polluted urban river water, both of which compose living filter, were experimentally evaluated and the following conclusions were obtained.
(i)Pak-bung and water cress could be successfully cultivated on domestic wastewater and the implantation of these aquatic plants is an appropriate option for nutrient removal of domestic wastewater.
(ii)Direct purification of urban river water polluted with grey water using living filter was proved to have a high application potential.

REFERENCES

(1)R.Dings, Journal WP CF, **50**, 833(1978).
(2)C.M.Rebecca and B.C.Wolverton, Economic Botany,34, (1989).
(3)APHA, Standard Methods, American Public Health Association, Washington, DC.(1990).
(4)E.Nakamura, N.Tuyuzaki and H.Namiki ,Industrial Wat.,No.421,21(1993).(in japanese)
(5)S.Hashimoto, K.Furukawa and Y.Ozaki, Journal Ferment. Technol.,63,343(1985).
(6)K.Furukawa and S. Hashimoto, Jap. J. Water Treat. Biology.,27,47(1991).

Environmental Research Forum Vols. 5-6 (1996) pp. 325-330
© *1996 Transtec Publications, Switzerland*

Aquatic Plants for Wastewater Treatment and Recycling Resources

K. Jamil

Indian Institute of Chemical Technology, Hyderabad 500 007, A.P., India

Keywords: *Eichhornia*, *Pistia*, Heavy Metal, Effluents, Biological Processes, Biosystem

Abstract - This investigation details the experiments and designs with acquatic weeds like Eichhornia and Pistia which have the tremendous biological capacity to remove dissolved toxic organic and heavy metals from acquatic environments, and also detoxify the acquatic environment by physiological mechanisms. Based on these principles stepwise lagoon systems were designed. The details of this unique biosystem useful for industrial application is discussed.

Introduction

Water is an important commodity which is recently becoming highly polluted. As a result of this, several living forms which are better suited to the polluted waters have outbursted in our acquatic environments causing ecological imbalance.

Every water body has its own capacity to maintain its natural state through self purification process, however when the discharge of pollutants is heavy, the process of self purification of water bodies is adversely affected and the water remains polluted. The EPA_US has listed 129 pollutants occuring in industrial waste waters [1] of which 13 are metals. Attention has been focussed on heavy metals as environmental pollutants since the occurence of "Itai-Itai" and "Minimata" diseases caused by Cd and Hg poisoning. Chromium and Nickel are also toxic metals entering into the streams, from various industrial effluents [2-4].

Acquatic plants are important components of lakes, ponds, rivers and streams ecosystems. Acquatic weeds like Eichhornia crassipies, Pistia stratiotes, Lemna minor etc., have drawn much attention for removal of heavy metals and organic pollutants [5-8].

Pollution abatement through efficient and cost-effective alternatives to conventional processes had been carried out by utilizing nature's own technology for instance, acquatic plants like Pistia, Eichhornia and Ceratophyllum which help in the removal of toxic metals from acquatic environments. Methods so far developed for removing heavy metals from acquatic environments appear to be highly expensive hence available weeds which have the physio logical capacity to absorb these toxic inorganics as well as organic pollutants were tried with some success. This biological system is a dynamic system, hence a study was undertaken to utilize these living systems to sponge the toxic metal contaminants from various effluents containing high concentrations of the soluble metals.

Principle involved in utilizing acquatic plants for metal pollution removal

Acquatic plants are important components of lakes, ponds, rivers and streams ecosystems. Inspite of their nuisance characteristic (i.e., when they outgrow and become a threat to the water bodies), the ecological and environmental significance of these plants, especially their capability to improve water quality, has created substantial interest in utilizing them for beneficial purposes. Plants contain a complete harmony of enzymes which ensure the oxidation of essential intermediates of molecular oxygen. It is fortunate that in nature the qualitatively significant routes are existing for biological reduction of oxygen produced by different pathways which yield H_2O directly.

Fig 1 : Oxygen Toxicity

1 = Metalloprotein oxidase   2 = Flavoprotein oxidase   3 = Various enzyme oxidase
4 = Superoxidedismutase   5 = Peroxidase   6 = Catalase
7 = Photoreduction by illuminated chloroplasts.

In plants, chloroplasts are organelles with high rates of oxygen turnover during the day. The H_2O_2 and O_2 act as regulators of photosynthetic CO_2 fixation and of dioxygen as a Hill acceptor also maintains and regulates photophosphorylation. Industrial wastes containing various metals disturbs the vegetational pattern of acquatic habitats. As early as 1948, Dymond [9] had suggested the possibility of using water hyacinth for removal of nutrients from waste water effluents. Since then, these plants have found their use in abating acquatic pollution, especially in removing the toxic and non-destructive heavy metals. We have also expanded this study and suggested its use as a Biofilter [10-15].

Materials and Methods

Acquatic plants used in this study include Eichhornia crassipies and Pistia stratiotes. A stock culture of these plants was maintained in the laboratory pond at ambient temperature and 60-70% relative humidity.

Chemicals used : Soluble salt solutions of the metals   were prepared as 1000 ppm stock solutions of Zn, Cd, Cr, Cu and Ni, and various dilutions were used in the experiments.

Experimental Designs

A)   Glasshouse/Labscale experiments : In order to derive more precise and definite conclusions the plants wer exposed to known concentrations of metal solutions (10.5 and 100ppm) (Fig 2) and the bioaccumulation of the metals was determined by Atomic Absorption Spectroscopy(AAS) or Inductive Coupled Plasma Atomic Emission Spectrophotometry (ICP-AES). This gave a quantitative estimation of the metal uptake by the plants.

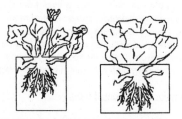

Fig 2 : The rootmass of both the acquatic weeds is almost similar.
Glass House Experiments.

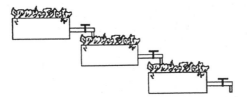

Fig 3: Diagrammatic design of stepwise lagoon system.

<u>Design of stepwise lagoon system</u> : About one meter cube asbestos containers were fixed in stepwise system (Fig 3) and various dilutions of the effluents from industries was poured in the containers. After a retention time of 48 hours, the effluents were flushed out into the next container and again after 48 hours into the third container. Finally, the water which was collected from the last and fourth container was found to be free of toxicants (Figs. 3 and 4).

Fig 4 : Stepwise lagoon system

Results and Discussions

Approaches and methodologies are the two most important components in designing a study. The approach to a problem determines the technique and methods to be used. In the present investigation, we have designed a stepwise lagoon system after careful laboratory and glasshouse studies. In the laboratory experiments, it was observed that Eichhornia could remove 75% of Zn, 62% of Cd, 62% of Cr, 72% of Cu and 59% of Ni in 48 hours, at 100ppm concentration (Table 1). Pistia could remove much higher concentrations of heavy metals from acquatic medium i.e., 68% of Zn, 58% of Cd, 58.8% of Cr, 66% of Cu and 53% of Ni in 48 hours. A rapid absorption of all these metals was observed in the first 12 hours of exposures, followed by a slow bioaccumulation (results not shown). A linear increase in the uptake was found wiht increasing concentration of metal ions in both the macrophytes (Eichhornia and Pistia) (Table 1)

Industrial wastes/effluents are complex, comprising of innumerable chemical compounds. Even the most sophisticated chemical analysis are just inadequate to identify all the biologically active compounds present in the waste. Another difficulty is that the chemical composition keeps changing from batch to batch and from time to time. Hence, Bioassays provide a more direct measure of

Table 1: Uptake of Heavy metals by aquatic macrophytes

Sl. No.	Metal conc.in ppm	Uptake of metal by water-hyacinth in 48 hrs		Uptake of metal by water-hyacinth in 48 hrs	
		ppm	%	ppm	%
1.	Zinc				
	10	9.20 + 0.34	92.0	8.82 + 0.31	88.2
	25	21.10 + 1.15	84.4	20.23 + 1.12	80.9
	50	39.75 + 3.45	79.5	36.80 + 1.44	73.6
	100	75.50 + 1.78	75.0	68.00 + 2.96	68.0
2.	Cadmium				
	10	7.30 + 0.27	73.4	7.30 + 0.51	73.0
	25	17.5 + 1.40	70.2	16.20 + 1.20	64.0
	50	31.6 + 2.70	63.0	30.02 + 3.50	60.0
	100	62.30 + 3.22	62.1	58.30 + 3.01	58.3
3.	Chromium				
	10	8.02 + 0.43	80.2	7.84 + 0.51	78.4
	25	16.80 + 1.27	67.4	16.50 + 1.00	66.0
	50	32.10 + 1.78	64.2	31.15 + 2.30	62.3
	100	62.30 + 2.89	62.3	53.80 + 3.20	58.8
4.	Copper				
	10	9.00 + 0.32	90.0	8.40 + 0.33	84.0
	25	20.00 + 0.88	80.2	19.23 + 1.00	76.9
	50	38.40 + 3.52	76.8	34.15 + 1.81	68.3
	100	72.00 + 2.33	72.0	66.00 +2.51	66.0
5.	Nickel				
	10	6.98 + 0.46	69.8	6.36 + 0.42	63.6
	25	15.50 + 0.29	62.0	14.58 + 0.75	58.3
	50	29.50 + 0.86	59.0	28.00 + 2.18	56.0
	100	57.00 + 2.08	57.0	53.00 + 3.20	53.0

Fig. 5 : Uptake of heavy metals by water hyacinth

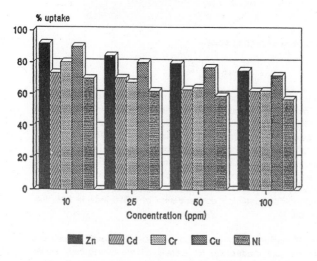

Fig. 6 : Uptake of heavy metals by
water lettuce

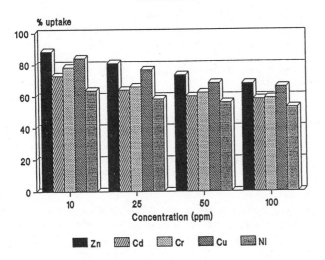

environmental toxicity as the results reflect the combined effect of environmental variables and chemicals [19]. The effluents obtained from Leather Industry (containing large quantities of chromium), Pesticide Industry (containing copper) and Drug Industry (containing Ni) were processed through the lagoons after careful chemical analysis. It was found that the effluents had to be diluted (1:1 or sometimes more) and then transferred to lagoons or containers. The retention time in the lagoons was generally 72 hours after which the effluents were collected in the second container, then third and lastly, fourth container contained good water free of the toxic metals and which could be recycled or used for various purposes. The concept of utilizing acquatic weed for pollution control is not new [7, 10, 11, 16 and 17] and these plants, besides removing the heavy metals are known to remove other organic contaminants as well. Another useful observation on chemical speciation and bioavailability of soluble metals was reported by Nor and Chang (1986). Biological remedies for pollution reduction have received increasing attention since the 1980's and have attained wide spread recognition in the wake of certain environmental catastrophies such as MIC leakage in the Bhopal Tragedy or oil spills on water bodies, etc. The increase in bioremediation engineering applications has been fostered in party by our expanding knowledge of how the dynamic biosystems work, or how these chemicals are metabolized by existing microbes, plants and our ability to rationally design novel metabolic capabilities using genetic engineering, or by using cell-free enzymes, cell components, or cell products, and more recently macrophytes.

References

1.   Keith L.H. and Telliard (1977). Priority Pollutants - I. A prospective view. ES&T Special Report. Environ. Sci. Tech.. 13(a) 416-420.

2.   Royle H (1975). Toxicity of Chromic acid in the Chromium Plating Industry. Int. Environ. Res., 10, 39-53.

3.   Baetjein A.M., A.M. Berningham, P.E. Enterlino, W. Mertz and J. Pierce II (1974). Effects of Chromium Compounds of human health. In: Chromium, Natt. Acad. Sci., Washington, 42-73.

4.   Fornstner G, and G.T.W. Wittman (1981). Metal Pollution in the acquatic environment. Publisher : Springer-Verlag, Berlin, Heitderbug & NY.

5.   Wolverton B.C., and R.C. McDonald (1975). Water Hyacinth and alligator weed for removal of Pub. and Hg from polluted waters. NASA Tech Memorandum TM-X-72723.

6.   Wolverton B.C. and M.M. McKnown (1976) Water Hyacinth for removal of phenols from polluted waters. Aquat. Bot., 2, 1991-201.

7.   Wolverton B.C. R.M. Barlow and R.C. McDonald (1976). Application of vascular acquativ plants for pollution removal, energy and food production in a biological system. In: Biological Control of Water Pollution. Publishers : University of Pensylvania Press, 141-149.

8.   Kaiser Jamil (1991). Environmental Biology of Water Hyacinth. Monograph Publ. by Avichal Sciences Foundation, Gujarat, India.

9.   Dymond G.C. (1948). The Water Hyacinth, a Cinderella of Plant world. In: J.P.J. Van Vuren (Ed) "Soil Fertility and Sewage". Doves Publishers, NY., pp 221-227.

10.   Kaiser Jamil, P.V.R. Rao and M. Zafar Jamil (1987). Studies on the efficacy of Eichhornia crassiper (Mart) Solns. in removing heavy metals from aquatic medium and their effects on the plant tissues. In: "Aquatic plants for water treatment and resource recovery". Eds K.R. Reddy and W.H. Smith, Publ.Magnolia Publishing Inc.,Florida, USA, pp 479-486.

11.   Kaiser Jamil, S.S. Madhavendra, M.Z. Jamil, and P.V.R. Rao (1987). Studies on Water Hyacinth as a biological filter for treating contaminants from agricultural wastes and industrial effluents. J. Environ. Sci. & Health (Part B), 22, 103-112.

12.   Saber Hussain and Kaiser Jamil (1989). Bioaccumulation of heavy metal ions and their effect on certain biochemical parameters of water hyacinth weevil, J. Environ. Sci. Health (Part B), 24, 251-264.

13.   Kaiser Jamil (1989) Aquatic eco-environments: The concern of every living being. In: "Management of Aquatic Ecosystems", Eds.Ved Prakash Agarwal, pp 183-188.

14.   Kaiser Jamil, Saber Hussain, Ishrath Anees, and U.S.N. Murthy (1995). Toxic effects of Lead on the Biochemical parameters of the hyacinth weevils Neochetina eichhornia through trophic levels of food chain", J. Environs. Sci. & Health (Part A) In press.

15.   Fayed, Sami E. and Hussein I, Abd El Shafy (1985). Accumulation of Cu, Zn, Cd and Pb by aquatic macrophytes. Environment International 11, 77-87.

16.   Rogers, H.H. and D.E. Davis (1972). Nutrient removal by Water hyacinth. Weed Sci., 20,423.

17.   Sutton D.L., R.D. Blackburn and W.C. Barlow (1971). Response of aquatic plants to combination of endothal and copper. Weed Sci., 19, 643-651.

18.   Nor H.M. and H.H. Cheng (1986). Chemical Speciation and bioavailability of copper. Uptake and accumulation by Eichhornia Environ. Tox. & Chem. 5,941-947.

19.   Kaiser Jamil and K.N. Jyothi "Enhanced reproductive potential of Neochetina bruchi (hustache) fed on water hyacinth plants from polluted water bodies". Current Sci., 57, 195-197 (1988).

Environmental Research Forum Vols. 5-6 (1996) pp. 331-336
© 1996 Transtec Publications, Switzerland

Reclamation of Eutrophic Waters: Nutrient Removal by Waterhyacinth in Response to Qualitative and Quantitative Variations of Fertilizers

B.B. Jana, S. Datta and S. Saha

Fisheries & Limnology Research Unit, Department of Zoology, University of Kalyani,
Kalyani-741235, West Bengal, India

Keywords: Reclamation, Wastewater, Waterhyacinth, Nutrient Removal, N/P Ratio

Abstract

Nutrient removal capacity of waterhyacinth was examined in simulated models of wastewater in response to qualitative and quantitative variations of fertilizers. Removal of nitrogen and phosphorus was significantly higher in the mixed enrichment than that of N and P seperately. Maximal growth in the N-enriched treatment was related to the highest tissue N/P ratio of waterhyacinth, though the nutrient removal was maximum in mixed treatment. The amount of total N and total P in the tissue of waterhyacinth tended to increase as a direct function of the dose of fertilizers.

INTRODUCTION

Water, an integral component of the biosphere, is severely threatened by the heavy demands of modern civilizations. Unchecked consumption patterns and industrial policies, unsustainable agricultural practice, unlimited population growth and ill conceived development policies led to mass scale deteriorations of ecologically important aquatic ecosystems in India. As a result, a vast number of tropical pond wetlands are converted into eutrophic or polluted systems which are no longer suitable for fish culture. There has been a notable shift during the past decade in emphasizing towards environmental restoration. The recovery potential of a damaged ecosystem depends on the extent of damage caused, type of perturbation and mode of operation of the stressor [1]. Several direct options such as, diversion of sewage input and its treatment, inactivation of nutrients in the input water, hypolimnetic aeration, artificial bubbling, liming of acidified lakes, alum treatment, biomanipulation, etc. are now in use, and are practised with success in many waters. Several sophisticated ecotechnological models adopted to control eutrophication mostly in large and deep lakes in advanced countries are not suitable for tropical small and shallow ponds utilized for fish farming in developing nations.

Use of biolgical agents in the reclamation of tropical pond wetlands for fish culture is of great ecological significance because of their ease of handling, and as component mermbers, they may cause less perturbation in the pond ecosytem. Several preliminary studies [2,3] have indicated the possibilities of their use as biofilter in the treatment of wastewater. It is hypothesized that a vast number of tropical eutrophic pond wetlands prevailing in India might be profitably used for fish farming through restoration of biological agents. The present paper attempts to examine the nutrient removal capacity of floating waterhyacinth (*Eichhornia crassipes*) in simulated models of wastewater in response to qualitative and quantitative variations of fertilizers.

MATERIALS AND METHODS

Two experiments with simulated models of wastewater were performed during October 1994 through June 1995 to examine the nutrient removal capacity of waterhyacinth (*Eichhornia crassipes*); the first trial has been dealt with qualitative variations of nutrients, whereas, the second in terms of quantitative differences. Twenty four plastic pools (300 l) were filled with dechlorinated groundwater (pH-7.2) and were subjected to initial enrichment with P and N separately and with P and N together. Single superphosphate and potassium nitrate were used as sources of P and N respectively. The dosage for each of the fertilizers was 7.5 mg/l, and the ratio between the single superphosphate and potassium nitrate used in mixed treatment was 1:1. Freshly collected young stages of waterhyacinth was introduced in equal wet weight (1.75 kg) into each of the three treatments, while a subset without waterhyacinth served as control. Replication was four fold in all the six sets tested.

In order to examine the responses of waterhyacinth to nutrient doses, thirty two plastic pools (300 l) were used. They were filled with dechlorinated ground water (pH-7.2) and were subjected to initial enrichment with P and N at four different dosages. Single superphosphate and potassium nitrate (1:1) were applied at the rate of 7.5 mg/l, 12.5 mg/l, 17.5 mg/l and 22.5 mg/l into each of the thirty two plastic pools grouped into eight in quadruplicate. Equal wet weight (1.25 kg) of waterhyacinth was introduced into each of the four dose treatments, while the remaining batches without introduction of waterhyacinth served as respective control.

Water samples were collected from plastic pools every week in the first trial, whereas every other day in the second experiment until the nutrient level declined to marginally low values. The concentrations of nutrient parameters (orthophosphate, soluble reactive phosphate, total phosphate, nitrite-N and nitrate-N) of water were determined according to the standard methods [4,5]. Samples of water were also examined for other water quality variables. At the time of termination, all the waterhyacinth were harvested from each pool and their growth increments were recorded in terms of dry and wet weight. The materials were subjected to assay for total-N and total-P following the methods described in Jackson [6]. One way analysis of variance was applied to find the significant differences of different treatments. Duncan's multiple range test was used to compare the treatment means ($P < 0.05$).

RESULTS
Effect of qualitatively different fertilizers on nutrient removal
Removal of OP was found to be as high as 90% from the mixed (P+N) treatment, and 35% from the P-enriched treatment within seven days of waterhyacinth introduction (Table-1). Removal of SRP was recorded to be 31% higher in the former when compared with the latter. Cosequently, OP or SRP were removed completely from the water within two weeks in the mixed treatments, and three weeks in the P-enriched treatment. This suggests that removal of P from the P-enriched water body might be enhanced by the addition of N fertilizer. The loss of nitrite-N was maximum during the first week, which then tended to decline as a function of time in all the treatments except in the N-enrichment where the maximal loss occurred during the second week. There was complete disappearance of nitrate from the mixed treatment within seven days of waterhyacinth introduction, whereas, the removal efficiency of 78% was observed in the N-enriched treatment. This again suggests that addition of P is necessary in the N-enriched water body in order to accelerate the nitrate removal by the waterhacinth.
Effect of qualitatively different fertilizers on water quality
Introduction of waterhyacinth in any of the fertilizer treatments resulted in decline of pH (7.69-8.42) when compared with their control counterparts (8.17-9.05). There was no marked differences of bicarbonate alkalinity and total hardness of water between treated and control groups (Table-2). The decline of DO was 0.69-2.29 mg/l in the former treatments. Differences between treated and control groups were maximum in mixed treatment than in either P or N treatments.

TABLE-1. Nutrient removal (%) by waterhyacinth in response to qualitative variation of fertilizers in different weeks of observation.

Parameters	P			N			P + N		
	1st week	2nd week	3rd week	1st week	2nd week	3rd week	1st week	2nd week	3rd week
Ortho-P	34.68	22.75	13.74	-	-	-	90.51	-	-
SRP	17.00	23.09	9.05	-	-	-	47.97	0.4	-
Nitrite-N	14.82	0.82	2.12	136.00	185.00	176.28	130.00	94.00	70.25
Nitrate-N	-	-	-	78.43	-	-	-	-	-

TABLE-2. Range and mean (±SE) of major water quality parameters in qualitatively different dose treatments with (+W) and without (-W) waterhyacinth introduction.

Parameters	P		N		P + N	
	+W	-W	+W	-W	+W	-W
pH	8.03-8.23	8.63-8.77	7.69-7.97	8.17-8.52	7.96-8.42	8.76-9.05
Alkalinity (mg/l)	245.66+ 14.30	270.00+ 12.93	193.17+ 17.81	191.08+ 18.37	230.16+ 22.61	216.00+ 25.68
Hardness (mg/l)	185.16+ 14.60	190.25+ 11.04	143.00+ 14.22	151.16+ 12.88	156.74+ 21.90	170.50+ 22.53
D. O. (mg/l)	9.76+ 0.54	10.45+ 0.36	9.60+ 0.51	10.96+ 0.45	10.81+ 0.46	13.10+ 0.56

Effect of qualitatively different fertilizers on waterhyacinth growth

The growth of waterhyacinth was maximum in the N-treatment (2.69 kg) followed by P (2.37 kg) and mixed treatments (2.08 kg) (Fig.1). The growth reduction of waterhyacinth in the mixed treatment was primarily due to the effect of N rather than the effect of P.

Effect of qualitatively different fertilizers on P and N contents of waterhyacinth

Of all the treatments, the amount of total-N was lowest (0.904%) in the P-enriched treatment ($P < 0.01$). The amount of total-N in the N-enriched treatment (0.976%) was maximum. This was not significantly different ($P > 0.05$) from the mixed treatment (0.972%). The amount of total-P, on the other hand, was highest (0.152%) in the P-enriched treatment not differing significantly ($P > 0.05$) from the mixed treatment (15%). The amount of P in the N-enriched treatment was the lowest (0.142%) of all ($P < 0.01$).

Dose-induced nutrient removal

The rate of removal of nutrent parameters in different days was variable depending upon the fertilizer dose, suggesting a strong degree of interaction between the dose and nutrient removal efficiency. Differences of OP removal among the four doses of fertilizers tested were not significant ($P > 0.05$) until three weeks of waterhyacinth introduction. After three weeks, OP removal was maximal in the second dose, which was significantly higher (P < 0.01) than the lowest dose (Fig.2).

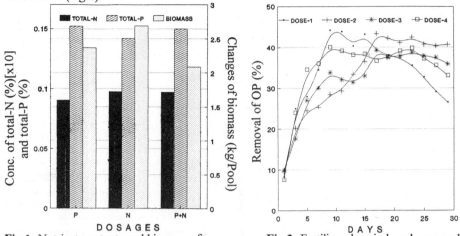

Fig.1: Nutrient contents and biomass of waterhyacinth in response to different fertilizers

Fig.2: Fertilizer dose induced removal per cent of orthophosphate

Highest dose of fertilizers induced the removal of SRP at its highest degree duiring most of the period of investigation (Fig.3) In general, treatment differences became pronounced after three weeks of waterhyacinth introduction. The trend of variations in the loss of SRP in these four treatments during the fourth week was in the order: highest dose > third dose > second dose > lowest dose.

During most of the period, the maximal loss of total P occurred in the lowest dose which was not sigificantly different (P > 0.05) from the highest dose. However, the rate of loss in the second dose was relatively higher (P < 0.01) than other treatments during the fourth week of waterhyacinth introduction. The rate of nitrite loss was maximal in the second dose treatment which was distinctly higher (P < 0.05) than the remaining treatments for most of the period. The rate of nitrate loss in the lowest dose was significantly higher (P < 0.01) than some other treatments (Fig.4). As a result, there was complete absence of nitrate from the water within seven days of waterhyacinth introduction in the lowest and second dose treatments, and within ten days in the third and fourth dose treatments.

Dose induced changes of water quality

Introduction of waterhyacinth resulted in decline of water pH (7.78-8.65),carbonate alkalinity (0-28.67mg/l) and rise in free carbon dioxide (1.0-9.66 mg/l). The values of DO ranged from 6.96 to 8.86 mg/l in the treated groups, and from 7.43 to 9.90 mg/l in the controls. The concentrations of bicarbonate alkalinity and total hardness did not differ much among these treatments, nor between the treated and control groups (Table-3).

Dose induced changes of waterhyacinth growth

The net weight of waterhyacinth tended to increase gradually with increasing dose of fertilizer till the third dose, but a decline in biomass was registered in the highest dose. The net gain in biomass was 2.22 kg, 1.88 kg, 2.16 kg and 2.13 kg in the third, first, second and fourth dose, respectively.

Dose induced changes of P and N contents of waterhyacinth

Examination of body composition of waterhyacinth at the time of harvest revealed that the amount of total-P (0.206%) and total-N (1.77%) were maximum in response to highest dose of fertilizer (P < 0.01). The amount of total-P and total N in the waterhyacinth tended to decline in amounts when grown with reduced load of fertilizers (Fig.5). Differences among treatments were significant (P < 0.01).

TABLE-3. Range and mean (±SE) of major water quality parameters in different dose treatments with (+W) and without (-W) waterhyacinth introduction.

	FERTILIZERS (P+N) ADDED							
Parameters	Dose-1 (7.5 mg/l)		Dose-2 (12.5 mg/l)		Dose-3 (17.5 mg/l)		Dose-4 (22.5 mg/l)	
	+W	-W	+W	-W	+W	-W	+W	-W
pH	8.1-8.7	8.7-9.1	8.0-8.8	8.7-9.3	7.9-8.6	8.6-9.2	7.8-8.5	8.5-9.2
Free CO_2 (mg/l)	1.31± 0.40	0.0	3.29± 0.43	0.0	5.18± 0.39	0.0	4.74± 0.50	0.0
$CO_3^=$ (mg/l)	11.65± 1.68	27.7± 1.25	7.04± 1.73	29.98± 1.12	5.41± 2.02	28.89± 1.74	4.92± 2.43	27.0± 1.37
HCO_3^- (mg/l)	227.6± 10.4	208.4± 9.44	242.5± 7.64	216.8± 10.14	239.7± 8.84	214.5± 7.69	234.3± 5.58	233.9± 7.32
Hardness (mg/l)	211.7± 9.06	191.3± 9.09	218.6± 11.13	192.4± 11.91	208.5± 9.48	199.2± 9.67	199.6± 7.22	205.5± 9.18
D.O. (mg/l)	7.70± 0.10	8.32± 0.17	8.08± 0.13	8.84± 0.07	7.37± 0.12	8.69± 0.14	8.10± 0.13	8.85± 0.09

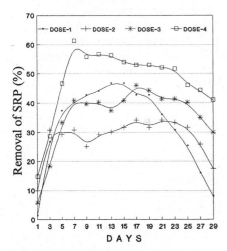

Fig.3: Fertilizer dose induced removal per cent of SRP

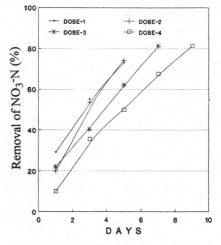

Fig.4: Fertilizer dose induced removal per cent of NO_3-N

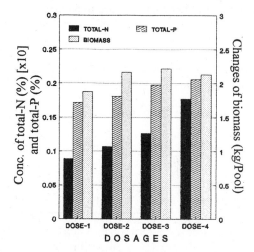

Fig.5: Fertilizer dose induced changes of nutrient contents and biomass of water-hyacinth

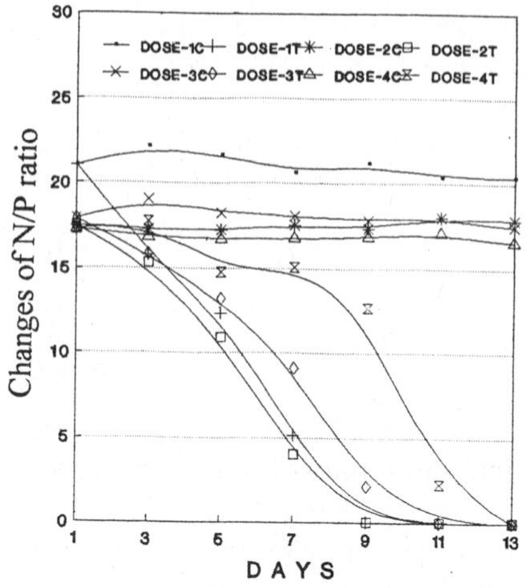

Fig.6: Fertilizer dose induced changes of N/P ratio in treated and control groups

DISCUSSION

Despite the highest removal rate of both P and N in the mixed enriched treatment, the maximal growth of waterhyacinth in N-enriched treatment was related to the highest tissue N/P ratio of waterhyancinth. In the mixed enriched treatment, waterhyacinth utilized nitrate at much faster rate than that of phosphate, resulting in complete removal of nitrate within seven days after their introduction. Because of fast rate of nitrate uptake than that of phosphate uptake, there was a sharp decline in N/P ratio in treated water, whereas, no marked change was observed in the control (Fig.6). Since removal efficiency of OP and SRP by waterhyacinth was directly related to the fertilizer dose tested, it is difficult to ascertain the optimal dose required for maximal efficiency of waterhyacinth in the present study. It is concluded that the waterhyacinth might be used to control the nutrient load of the eutrophic waters for profitable fish farming provided the optimal ratio between the waterhyacinth density and water volume is ascertained from different trophic water bodies. Further research is in progress.

REFERENCES

[1] J. Cairns Jr., USEPA Washington DC, 220-239 (1992).
[2] J. Mann, Oxford Univ.Press, 2nd edn. (1983).
[3] K.R. Aneja and K. Singh, Proc. Indian Nat. Sci. Acad. B **58**(6), 357-364 (1992).
[4] APHA (American Public Health Association), 17th edn. (1989).
[5] R.G. Wetzel and G.E. likens, Springer-Verlag, 2nd edn. (1991).
[6] M.L. Jackson, Prentice Hall of India Pvt. Ltd., New Delhi (1967).

Environmental Research Forum Vols. 5-6 (1996) pp. 337-344
© *1996 Transtec Publications, Switzerland*

Flax Pond Restoration Project

B. Josephson and J. Todd

Ocean Arks International, 1 Locust Street, Falmouth, MA 02540, USA

Keywords: Bioremediation, Pond, Eutrophication, Lake Restorer, Living Machine

ABSTRACT

Flax Pond, a 6.8 ha pond on Cape Cod has high levels of ammonia, high levels of iron which bind phosphorus, and low oxygen regime at the bottom. As a result, it is biologically impoverished. Using a combination of upwelling windmills, bacterial and mineral amendment, and a "Lake Restorer" floating ecological raft for biological filtration, bottom oxygen has been increased, deposition of organic sediment has been reversed, and species diversity has increased.

INTRODUCTION

Flax Pond is a 6.8 ha pond located in Harwich, Massachusetts on Cape Cod. The pond is situated immediately downgradient from the town landfill and, until recently, unlined lagoons for septic waste disposal. Cape Cod is made up of sandy glacial till, with pockets of clay. Hydrogeological studies [1] have shown that groundwater contamination from the landfill and septage lagoons is funneled into Flax pond by a clay layer. Although the septage lagoons were closed in 1992, the above study concludes that the pond will continue to be impacted by approximately 90 m^3/day polluted groundwater for another 30 years.

Flax pond water is used for irrigation of cranberry bogs - an important industry on Cape Cod. Poor water quality was adversely affecting the harvests of two growers. The pond was closed to swimming in the late 1980's due to high fecal coliform levels.

Ocean Arks International is a non-profit research and education organization located on Cape Cod. The mission of Ocean Arks is water purification through the use of ecological engineering. Ocean Arks was contracted by the town of Harwich in 1990 to study the pond and to restore water quality and biological balance within the pond.

THE PROJECT

In 1990 at the beginning of this project Flax Pond was highly polluted, being contaminated with fecal coliform bacteria, organic pollutants and extremely high levels of phosphorus, iron, and nitrogen. In the sediments, total phosphorus was 17-300 times greater than comparable Cape Cod ponds [2]. Iron levels were approximately 8-80 times higher. Ammonia levels averaged over 500 mg/kg and peaked at over 1000 mg/kg. These levels of ammonia are toxic to most bottom dwelling or benthic forms of life. Our first benthic survey in 1991 found

that bottom dwelling animals were completely absent from three of our seven
sampling stations.   The pond bottom routinely went anoxic for periods of the
year, making it additionally difficult for benthic animals to survive.

Flax Pond, based upon water chemistry parameters, is an enriched
eutrophic pond.   Based upon chlorophyll a readings, the pond is mesotrophic [3].
This represents a more biologically impoverished state, a consequence of the
high iron condition in the pond.

The iron in the sediments is so rich in areas it can be considered ore grade
(it has been measured up to 184,000 mg/kg).   This iron effectively binds up all
available phosphorus in the pond.   Orthophosphate in the water column is rarely
above detection limits.   Thus algae growth and eutrophication is not the
predominant problem in Flax Pond.   Rather, the nutrient imbalance has inhibited
all food chains, permitting ammonia levels to rise unchecked.

The primary focus of Ocean Arks' remediation efforts on Flax Pond have
been to reestablish food chains and natural processes of self regulation.
Maintaining positive oxygen levels at the bottom of the pond was one important
prerequisite.   Ammonia reduction through bacterial processes was the other
important step toward promoting biological diversity and balance.   Although we
regularly monitor a wide range of parameters in the water column as well as in
the sediments, this paper will focus on the oxygen and ammonia chemistry, as
well as a discussion of the biological diversity.

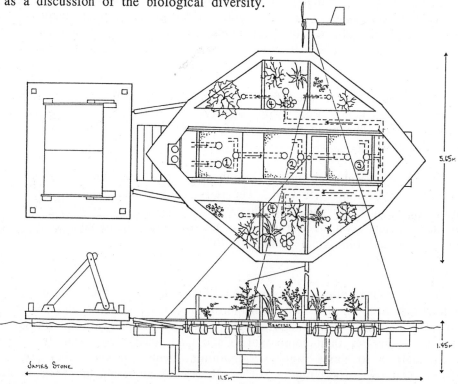

Figure 1: Plan and Elevation of Restorer 1 - Flax Pond, Harwich

In phase one of the project, Ocean Arks installed upwelling windmills on the pond and started regular inoculation of commercially prepared bacteria mixes into the sediments and into the water column. Then, in October of 1992, we installed the Lake Restorer on Flax Pond. The Lake Restorer was a development of our sewage treatment research. It is basically a high-rate ecological filter housed on a floating platform. Powered by wind and solar energy, large volumes of water are circulated through media and vegetation beds. High surface area within the beds permit bacterial breakdown of the ammonia and other organic pollutants in the water. Water is air lifted from the bottom of the pond up through the Restorer, creating upwellings and water turnover. Since air lift pumps do all the water circulation the water is directly aerated.

Flows through the Restorer are designed to be up to 75.7m^3/day. In fact, power generation systems were probably not functioning at full capacity for large portions of time and less than half that flow rate, or 30 - 38 m^3/day would be a more accurate estimate.

The system is designed with nine separate cells, the first three of which are 7.25 m^3 each filled with pumice stone media. Pumice was chosen as it is very porous, allowing for much surface area for microbial attachment, and is close to neutrally buoyant, which makes it easy to stir up or backflush as it becomes impacted with solids. These ecological fluidized beds have become key features in all of our subsequent designs for waste purification. The beds are inoculated on a weekly basis with commercially prepared mixes of bacteria. Freshwater clams have been introduced to the beds and snails have colonized the area as well. Kelp meal, dolomitic limestone, and rock phosphorus have been added to the cells to supply needed nutrients, micronutrients and minerals.

After water flows through the three fluidized beds in sequence it is divided and spread out into six vegetation cells. These cells have diverse species of higher plants suspended by a racking system into the water. The last four cells are bottomless, permitting the water to return to the pond.

Operation and maintenance of the Restorer, aside from the regular addition of bacteria, is minimal. Backflushing the pumice beds several times yearly and keeping the solar panels free of bird feces is the extent of maintenance. Backflushing is accomplished by pumping pond water backwards through the system with a portable gasoline pump.

The primary mechanism for ammonia removal by the Lake Restorer is the nitrification process carried on by aerobic and facultative bacteria in the filter beds. Although residence times within the Restorer itself are only several hours, significant removals of ammonia have been measured. The Table below shows ammonia concentrations measured this summer coming into the Restorer at an average of 1.90 mg/L and leaving at an average of 0.47, for a removal rate of 75%.

FIGURE 2: FLAX POND LAKE RESTORER
INFLUENT/EFFLUENT AMMONIA CONCENTRATIONS

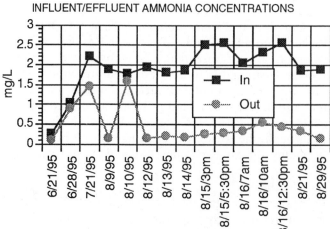

Nitrate levels during this same period rise from an influent average of 0.24 mg/L to an effluent average of 0.75 mg/L. The fact that the change in nitrate concentrations is less than the change in ammonia is evidence that some denitrification is also taking place in the Restorer. Anoxic zones within the beds probably support the communities necessary for the denitrification process.

Extrapolating the above ammonia removal rates, using a conservative figure of 30 m^3/day for flow rates, 43 grams of ammonia per day can be removed by the Restorer. During the three month summer period, approximately 4 Kg of ammonia can be removed. The Lake Restorer in Flax Pond operates year round, although more research remains to be done to determine reaction rates at different seasons and water temperatures.

PROGRESS

In analyzing the pond response to remediation efforts we are using sediment and water chemistry as well as biodiversity indices. We have also made rough calculations on sediment reduction. Sediment reduction calculations have been made by comparing measured depths of the water column to water level measurements to a fixed standard. Decreases in sediment depths are inferred from increases in water column depth. Our methodology may need some additional refining to increase its accuracy, nevertheless, the results are indicative of real changes in sediment depths. Sandy areas have begun to appear along the eastern edge of the pond, where five years ago the bottom was covered with undigested organic materials and mucks. Between 1990 and 1994 the depth of the sediment throughout the pond has decreased on the average 33 cm (the range is 23cm to 46cm). Approximately 19,000 cubic meters of sediments have been digested. This reversal in sediment deposition has been facilitated by the improved oxygen regime at the pond bottom.

Evidence shows that oxygen levels at the pond bottom have increased during the course of the study. There is still seasonal variation, but extended anoxic conditions at the sediment/water interface have not occurred since 1992.

We believe this is a result of the aeration and mixing provided by the Lake Restorer. Figure 3 shows oxygen levels at two of the sampling stations in the pond.

FIGURE 3: DISSOLVED OXYGEN PROFILES

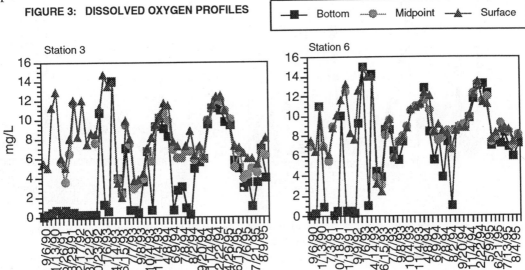

Sediment samples are taken quarterly from the seven sampling stations in the pond and analyzed regularly for Ammonia, Alkalinity, Nitrate Total Kjedahl Nitrogen, pH, Total Solids, Volatile Solids, Total Phosphorus, Orthophosphate and Iron. Annually, samples are also tested for Lead, Mercury, Copper and Volatile Organics. As shown in Figure 4, ammonia concentration in the .sediments exhibited dramatic reduction in the first year of the project. Subsequent years, however, show little change and, in fact, concentrations have generally increased since 1993. This may reflect a balance between the infusion of pollutants and internal digestion.

FIGURE 4: SUMMER SAMPLING SEDIMENT CHEMISTRY
AMMONIA

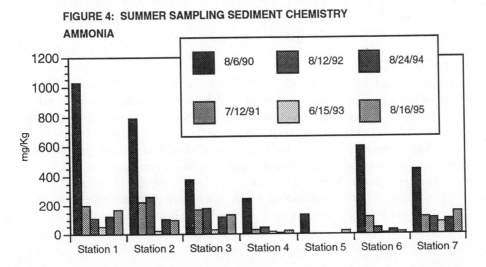

FIGURE 5: Ammonia Water Chemistry

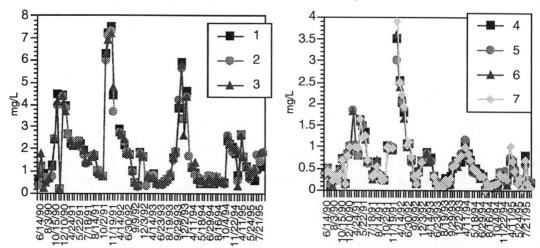

As shown in Figure 5, ammonia levels in the water vary seasonally, with peaks occurring in the late fall and early winter. The cold season peaks are probably related to a decrease in the rates at which the nitrifying bacteria, *Nitrosomonas* and *Nitrobacter*, metabolize. The magnitude of the peaks has been decreasing since 1992, definitely a trend toward stability and health. The pH of Flax Pond water ranges primarily between 7 and 8 standard units. At a pH of 8 in water of 20° C ammonia concentrations over 2 mg/L are toxic to fish [4].

Perhaps the best expression of the restoration of Flax Pond has been the dramatic increase in biodiversity as expressed by the macro-benthos or bottom dwelling organisms which are crucial to lake and pond metabolism. In 1991 sampling stations 2,3,5 and 7 were sampled. No live organisms were found at 2,5 or 7. Station 3 supported seven genera from three taxonomic groups. A year later one species from a new taxa (Mollusc) was added to the list. In 1994 benthic life was found at all stations. Number of species had more than doubled to include 15 genera from 6 taxa. 1995 sampling came up with two new genera, but slight decreases in diversity overall.

SUMMARY AND CONCLUSIONS

In five years of restoration efforts on Flax Pond using the techniques of ecological engineering, some improvements have been dramatic, while others remain elusive, in part because of continuing intrusion of polluted groundwater. Sediment chemistry has been less responsive, although calculations show that sediment is being digested. Indications are clear that bottom oxygen regimes have improved and that ammonia fluctuations in the water column are dampened. Species diversity has increased markedly since the project's inception. The Lake Restorer technology has been tested to be effective at ammonia removal and improving mixing and oxygenation in the pond.

REFERENCES

[1] Horsely, Witten, Hegeman, Inc., *Hydrogeologic Investigation for Groundwater Discharge Permit Application. Report to the Town of Harwich* (1991)

[2] K.V. Associates, Inc. and IEP, Inc., *Shallow Pond Diagnostic and Feasibility Study. Report to the Town of Barnstable* (1991).

[3] Olem, H and Flock, G. eds., *Lake and Resevoir Restoration Guidance Manual*, 2nd edition.　EPA 440/4-90-006.　Prepared by the North American Lake MAnagement Society (1990)

[4] Spotte, Stephen H., *Fish and Invertebrate Culture, Water Management in Closed Systems*, pp 103-104, (1970)

Posters

Environmental Research Forum Vols. 5-6 (1996) pp. 347-350
© 1996 Transtec Publications, Switzerland

The Subsurface Flow Constructed Wetland in Sobiechy in North Eastern Poland

R. Ciupa

Technical University of Bialystok, Civil and Environmental Engineering Faculty, Institute of Environmental Engineering, ul. Wiejska 45A, PL-15-351 Bialystok, Poland

Keywords: Constructed Wetland, Small Wastewater Treatment Plant, Ammonia Nitrogen Removal, Construction and Operation Costs

Abstract

The most adequate solution such a constructed wetland is shown in the paper. The efficiency of low technology according to Polish requirements cost evaluation of construction, operation cost have also been been considered. The constructed wetland is recomended for scattered areas and for small communities.

Introduction

North-Eastern Poland often called "Green Lungs of Poland" is characterized as unpolluted, agricultural and tourist area with high average degree of afforestation, a very low level of urbanization, industrialization and terrain investment. This area is differentiated from the rest of the country by its higher average of afforestation 30 % comparing to 28% for the whole country, considerable lake area amounting to 30% of total lake area within Poland, very high coefficient of bog grounds and large portion of terrain's under professional nature and landscape protection, (3 national parks, 5 landscape parks and 8 areas of protected landscape) amounting to 15.5% of area. Nowadays, more than 80% of municipal wastewater from scattered, villages, small towns areas are untreated and discharging to underground and surface water of the region. The constructing of the small wastewater treatment plants is the most important task for this unique, tourist, preserved region of Poland and for their small communities.

The small wastewater treatment plants should be characterized by significant resistance for fluctuation of wastewater flows and loads It is caused by the big non-uniform scattered area. It creates maintenance and operation problems and it increases the unit cost of treatment. The all solutions of small wastewater treatment plant, such trickling filter, a multiphase biological reactor operated on activated sludge, constructed wetlands and other should meet quality requirements since the end of 1991. The new outlet criteria has been established by Decree of Ministry of Environmental Protection, Natural Resources and Forestry dated 5th of Nov. 1991. The new requirements put huge attention for nutrients removal and they are presented as follows:

Parameters	Maximum Effluent Concentration
BOD_5	< 30 mg O_2/ l
COD-Cr	<150 mg O_2/ l
SS	< 50 mg / l
$N-NH_4$	< 6 mg N/ l
N-total	< 30 mg N/ l
P-total	< 5 mg P/ l

Moreover, the treated wastewater should reduce BOD_5, nutrients to 50% of requirements mentioned above and total phosphorus up to 1,5 mg P/l when the treated wastewater are discharged to close surface water reservoirs. Only the concentrations of pollution are considered by Ministry of Environmental Protection Decree. Load, year, seasons, the flow of recipient and other aspects are not taken into consideration. It cause that even wastewater treatment plant for single village houses have to meet requirements of Ministry Decree.

The constructed wetland in Sobiechy

Recently, alternative treatment methods for small wastewater producers have been considered even when environmental authorities are rather suspicious and distrustful about law technology wastewater treatment in Poland. The constructed wetlands are planted by wetland plants such as reeds, reed mace, rush etc. The wastewater is led through the root zone of the plants with a depth of 0.3-1.0 m dependent of the species. In the rhizosphere micro-organisms degrade organic matter which requires the presence of oxygen through their hollow roots. The most widely concept of constructed wetland in Poland is so-called constructed reed beds or root zone treatment plants with subsurface horizontal and vertical flow. The typical plant consist of a rectangular bed planted with common reed (*Phragmites australis*) and lined with an impermeable membrane.

The presented system in village Sobiechy (North-Eastern Poland) has operated at village school for 200 pupils and 10 permanent living persons since September 1994. It consists of primary clarifier and two stage biological parts, a single horizontal flow bed (448 m^2) and double (parallel) vertical flow bed (44 m^2) with average flow 7.6 m^3/d with 48 PE.

Table 1. Average indicators of technical efficiency
in constructed wetland in Sobiechy

	inlet	outlet	inlet	outlet
Period	March	1995	June	1995
pH	7.2	7.3	7.8	7.6
wastewater temp.° C	8.5	6.0	12.0	13.0
The air temp.° C	2.0	2.0	18.0	18.0
SS mg/l	63.8	21.0	132.0	20.0
BOD$_5$ mgO$_2$/l	110.2	17.2	370.0	20.5
COD-Cr mgO$_2$/l	318.5	50.7	420.0	70.0
N-NH$_4$ mgN/l	48.5	6.6*	44.3	4.7
N-total mgN/l	56.5	24.2	81.4	27.5
P- total mgP/l	11.2	0.9	16.2	4.2

* exceeded value of ammonia nitrogen
comparing to Ministry Decree

It consists of two chamber primary clarifier and two stage biological parts, a single horizontal flow bed H-CW (448 m^2) and double (parallel) vertical flow bed V-CW (44 m^2) which are planted with (*Phragmites australis*) with average flow 7.6 m^3/d with 48 PE. Wastewater are dosed onto vertical flow bed in portions for better aeration and nitrification processes by mean of dosing siphon device. The system is operated as gravitational because of natural steep slope (Fig.1). The example results of pollution removal, the construction and operation costs in the constructed wetland in Sobiechy are shown in Table 1,2.

Table 2. Construction and opration evaluation costs in constructed
wetland in Sobiechy

	CW
Average flow Q $[m^3/d]$	7.6
Person equivalent [PE]	48
Total construction cost in USD (1994)	8700.0
Unit construction cost in USD/ PE	181.25
Unit construction cost in USD/m^3/d	1145.0
Installed power in [kWh]	0.0*
Total year operation cost in USD/year	0.0*
Unit operation cost in USD/m^3/d	0.0*

* no power consumption

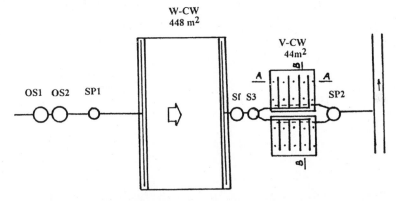

Figure 1. Scheme of constructed wetland in Sobiechy. Explanations: OS1,OS2-primary
clarifier, H-CW-horizontal flow bed,V-CW-vertical flow bed, Sf-syphon device,SP1,SP2,
S3-control wels.

Conclusions

The presented solution of constructed wetland in Sobiechy can afford the requirements according to
Ministry of Environmental Protection Decree. The outlet concentration of ammonia nitrogen has
exceeded the outlet criteria during winter period. Probably it is caused by low temperatures which
are characteristic for winter period in North-Eastern Poland. Other systems of small wastewater
treatment plants are better isolated and aerated and it permits to reach appropriate standards in
ammonia nitrogen removal. The presented constructed wetland has very low unit construction cost.
Practically it does not need continuous maintenance because of simple construction and it is not
effected by the cost of electricity consumption. This kind of solution has also not significant
operation cost what is very important for operation in scattered areas.

References

[1]H. Brix, K. Birkedal, N. H. Johansen, Wastewater treatment in construced wetlans-designers
manual, Transfer Low Technology Course,Gdansk, Poland (1993)

Environmental Research Forum Vols. 5-6 (1996) pp. 351-354
© *1996 Transtec Publications, Switzerland*

Phosphorus Removal by Horizontal Reed Beds Using Shale as a Substrate

A. Drizo[1], C.A. Frost[2], K.A. Smith[1] and J. Grace[1]

[1] Institute of Ecology and Resource Management, University of Edinburgh, Mayfield Road, Edinburgh EH9 3JU, UK

[2] Scottish Agricultural College, West Mains Road, Edinburgh EH9 3JG, UK

Keywords: Horizontal Reed Bed, Phosphate Adsorption, Phosphate Removal

ABSTRACT

The main objective of this research was to choose the most effective substrate for removal of P (as orthophosphate) from sewage water in a subsurface horizontal flow reed bed. Four substrates (bauxite, shale, light expanded clay aggregates and coarse fly ash) were compared for their physico-chemical properties. Shale (an argillaceous rock derived from the lower limestone group of the carboniferous system) was chosen based on its high phosphate adsorption capacity. Portable reed bed systems were set up using shale as the medium. Results from these experiments show that the maximum physico-chemical adsorption capacity of shale is 650 mg P kg^{-1} and that P reduction from sewage water by shale-based reed beds planted with *Phragmites australis* is as high as 99%.

INTRODUCTION

There is considerable uncertainty as to the effectiveness of reed beds for P removal from waste water [1,2]. Most of the studies agree that basic mechanisms for P removal are the adsorptive capacity of the substrate, and macrophyte, microbial and algal uptake [3,4]. However, in choosing the best potential conditions for removal, there has been little emphasis on rigorous comparison between substrates and their physico-chemical properties prior to the placement of material in a reed bed system. An earlier study tested these properties in different substrates [5]. This paper now focusses on the saturation point of P adsorption in the four substrates which showed the most suitable properties in Frost *et al*'s study, namely bauxite, shale, light expanded clay aggregates and coarse fly ash (Experiment 1). The capacity of shale, chosen for its high saturation point for P removal from sewage was then investigated, by setting up horizontal reed beds with shale as the substrate and monitoring phosphate removal over a period of 25 days (Experiment 2).

METHODOLOGY

For experiment 1, each of twelve Perspex columns (inner diameter: 100mm; height: 900 mm) was packed with substrate: three for each of the four substrates listed above. The columns were fed from below with a synthetic sewage solution containing 35-45 mg l^{-1} P (as KH_2PO_4) from a common reservoir, using a separate peristaltic pump for each column. The upward flow hydraulic regime was chosen to ensure good contact between effluent and substrate. The volume of material in each tube was approximately 4000 cm³ (50 cm long x 10 cm diameter) with a pore space of approximately 1600 cm³. The pumps were set to operate six times per day for 10 minutes each time, at a rate of 3 l hr⁻¹. The average contact time between the effluent and the substrate was therefore 12 hours. Samples were taken every second day and measured.

For experiment 2, *Phragmites australis* seedlings were obtained from the Yarningsdale Nursery Gardens, Warwick, and their growth followed in the glass house until they reached an average height of 0.51 m (max 0.61 m). Six rectangular pots were filled with 9 kg of shale. Three were planted with seedlings and three left unplanted as controls. Each of the planted pots contained between 26 and 30 new green stems, 5 to 8 young shoots and 70-80 dead culms. The volume of the material in each pot was approximately 0.0145 m³ (0.61 m long, 0.17 m deep and 0.17 m wide) with a pore space of approximately 0.0035 m³. Each pot was connected to a peristaltic pump, set to operate 4 times each day for 3.5 minutes, at a rate of 0.003 m³ h⁻¹. Therefore, 0.0007 m³ day⁻¹ of sewage passed through the system, which gives a retention time of 5 days, i.e. the system operated at a loading rate of 0.021 m³ m⁻² h⁻¹. Real sewage (settled sewage from Dunfermline STW) was used to introduce a typical microbial population. Orthophosphate ($H_2PO_4^-$) levels of the inlet and outlet sewage samples were monitored daily at the beginning of the experiment, followed by investigation on a five day basis.

RESULTS

As can be seen in Figure 1a), saturation was not reached after 40 days of running the experiment, although the results indicate that shale has the best phosphate adsorption capacity, followed by coarse fly ash, then by bauxite. It was therefore decided to continue the experiment for a further 40 days, doubling the phosphate concentration of the feeding solution as well as doubling the daily flow through the system. For this second phase only two substrates were tested, shale and bauxite (coarse fly ash has already been used for its high phosphate adsorption capacities in investigations carried out by other authors [6,3]). Saturation was reached in both after approximately 70 days, with shale absorbing 650-700 mg $H_2PO_4^-$ - P kg⁻¹ and bauxite 300 mg P kg⁻¹ (Figure 1 b).

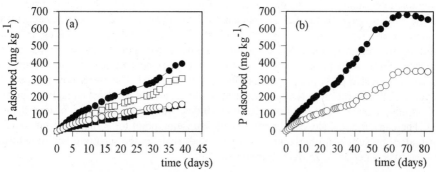

Figure 1: Cumulative P adsorption capacity for —●—Shale; —○—Bauxite; —■—LECA and —□—Ash over 40 days (a) and 80 days (b). The columns were fed from below with the synthetic sewage solution containing 35 - 42 mg P l⁻¹ (as KH_2PO_4) during the first 40 days of experimentation, followed by 70 - 90 mg P l⁻¹ (40 days) from a common reservoir, using a separate peristaltic pump for each column.

In Experiment 2, the planted horizontal flow reed beds with shale as the substrate showed a 99% reduction in phosphate after 25 days. The unplanted control beds showed a 98% reduction. Levels of reduction began to decrease in both towards the end of the experiment (see Figure 2).

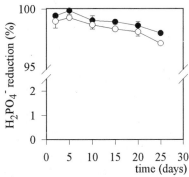

Figure 2: $H_2PO_4^-$ reduction (%) with time
—●— Plants; —○—Control.

DISCUSSION

Overall, the results from Experiment 1 show that saturation of shale and bauxite was reached at the same time (approximately 70 days). The experiments indicate that shale has a $H_2PO_4^-$ - P adsorption capacity in excess of 650-700 mg kg^{-1}. Applied at a field scale, one tonne of shale could absorb in excess of 650 g of phosphates. Septic tank effluent contains about 2 g of phosphorus per person per day. It is therefore likely that one tonne of shale would be adequate to treat the effluent from one person for 12 months, and possibly for much longer. The P adsorption capacity of 650 mg $H_2PO_4^-$ - P kg^{-1} was higher than for any other industrial substrata found in the literature. The closest reported values are of 240 mg P kg^{-1} for fly ash and 430 mg P kg^{-1} for blast furnace slag [3].

These results indicate the fundamental importance of the choice of substrate in constructed wetlands. The results from experiment 2 confirm this. The unplanted horizontal flow reed beds show a very high reduction in P of 98% after 5 weeks, with the addition of plants contributing an improvement of just 1% to 99%. The effects of shale alone on P removal is impressive. The observed decrease of phosphate levels towards the end of the experiment may have occurred due to the inevitable reduction in the efficiency of the substrate with time. However, the experiment will have to be pursued further to understand any long term trend. In particular, it will be interesting to note whether the planted beds maintain a high level of reduction through increased microbial activity associated with the reeds, and whether this is coupled with an unchecked decrease in removal in unplanted beds.

REFERENCES

[1] D.A. Hammer, *Constructed wetlands for waste water treatment. Municipal, industrial and agricultural.* (ed.) Lewis Publishers, Inc., Chelsea, Michigan (1989).
[2] G.A. Moshiri, *Constructed wetlands for water quality improvement.* (ed.) Lewis Publishers Inc., Baco Raton, Florida (1993).
[3] R.A. Mann and H.J. Bavor, *Water Science and Technology* **27** 107 - 113 (1993).
[4] C.E. Swindell and J.A. Jackson, in *Constructed wetlands in water pollution control: advances in water pollution control* (Eds: P.F. Cooper and B.C. Findlater). Pergammon Press, Oxford (1990).
[5] C.A. Frost, A. Drizo, R. Hood and T. Copland, *Simple passive systems for reducing phosphorus content of septic tank effluent.* Soils Department, Scottish Agricultural College, Edinburgh (1995).
[6] T.L.Theis and J.P. McCabe, *Journal of Water Purification Control Federation* **12** 2666-2676 (1978).

Environmental Research Forum Vols. 5-6 (1996) pp. 355-358
© *1996 Transtec Publications, Switzerland*

Aerobic Thermophilic Treatment of Blackwater, Mixed with Organic Waste and Liquid Manure: Persistence of Selected Pathogens and Indicator Organisms

G. Eller[1], E. Norin[2] and T.A. Stenström[1]*

[1] Swedish Institute for Infectious Disease Control, S-10521 Stockholm, Sweden
[2] Swedish Institute of Agricultural Engineering, Box 7033, S-75007 Uppsala, Sweden

Keywords: Aerobic Thermophilic Treatment, Hygiene, Organic Waste, Sewage, Liquid Manure, Pathogens, Survivial, Thermotolerant Indicator Organisms, Faecal Indicators, *Salmonella*, Ascaris, Vital-Stain, CTC, DAPI

Abstract: The survival of faecal indicators, *Ascaris suum* ova, and *Salmonella typhimurium* phage 28B (phage 28B) was tested in an aerobic thermophilic treatment process in pilot scale. Process material was blackwater from a vacuum toilet system mixed in different ways with organic kitchen waste and liquid manure. The sanitation effects of the treatment process were examined at 55°C and 60°C. The tests carried out in the bioreactor were supported by laboratory tests with *S. typhimurium* phage 28B and *Salmonella senftenberg 775w*. *Ascaris suum* eggs were tested enclosed in polyamide bags at 50°C, 55°C and 60°C in the reactor. All organisms examined were either thermotolerant species of a group of pathogens or indicator organisms for the respective group. Thermotolerant coliforms and faecal streptococci were reduced by approximately 4log-units (to less than 10 cfu ml⁻¹) after 12 h In the treatment process at 55°C. *Ascaris suum* eggs developed not to larvae after 5 min exposure in the treatment process at 55°C. Also at 55°C in the reactor, phage 28B was reduced by 1 to 2log-units after 24 h. Laboratory experiments showed, that the die-off rate of phage 28B did not only depend on the temperature, but also on the suspension medium and the conditions tested. Phage 28B showed in all tests at the laboratory longer survival than detected in the reactor.
S. senftenberg 775w's die-off was tested in nutrient broth at 55°C. The number of viable cells was estimated in parallel with plate counts on selective and nonselective agar and with vital-staining using 5-cyano-2,3-ditolyl tetrazolium chloride (CTC). As counterstain 4´,6-diamidine-2´- phenylindole-dihydrochloride (DAPI) was used. The results have to be stated by repeating experiments, but showed a trend to longer detection of viable bacteria using vital-staining. After 2h in 55°C the number of viable bacteria was lower than the detection limits of all used methods.

1. Introduction

The aim of the project was to examine the possibility for a decentral treatment of all kinds of organic wastes of a village. These wastes - organic kitchen waste, blackwater and liquid manure - should, after sufficient sanitation, be reused as fertiliser on farmland in the neighbourhood. Sanitation should be done by a composting process, which uses the self-heating potential of the material. Different mixtures of blackwater from a vacuum toilet system, organic kitchen waste and liquid cattle manure were used as substrates in the tests. The evaluations made were based on different groups of microbial indicators selected according partly to international standards or guidelines, partly to their thermotolerance. They should represent both endogenic organisms and exogenic tracers for the process.

2. Materials and Methods

The experiments were done in a well-isolated bioreactor, equipped with a heat-exchange system. The sludge volume in the tests was 1m³. The treatment was run under aerobic thermophilic conditions at 55°C and 60°C. Draw and fill periods (minimal retention times) of 24 h were tested at 55°C and 48 h at 60°C. The pH in the treatment process laid around 7.8. Material 1 (93% blackwater, 7% organic kitchen waste) had a dry matter content of 2.2% before and 1.5% after treatment and a COD (chemical oxygen demand) content of 32.5 gl⁻¹ before and 16.6 gl⁻¹ after treatment. Material 2 (60% blackwater, 4% organic kitchen waste, 36% liquid cattle manure) had a dry matter content of 4.4% before and 3.3% after treatment. The COD was 44.5 gl⁻¹ before and 35.4 gl⁻¹ after treatment.

Faecal indicators (thermotolerant coliforms, faecal streptococci and *Clostridium perfringens* spores) were quantified with spread-plate method (pour-plate for *Clostridium*) on the respective selective agar given in standard methods for the examination of water [1]. Phage 28B was enumerated with a plaque-forming unit method with *Salmonella typhimurium* 5 as host strain. All samples were examined on three parallel plates. The maturation of *Ascaris suum* eggs was examined by light microscopy. At least 300 eggs per sample were evaluated. The die-off of *S. senftenberg 775w* in nutrient broth at 55°C was followed by plate counts on selective (brilliant green) and nonselective (bloodagarbase) agar and in parallel with vital-stain, using 5-cyano-2,3-ditolyl tetrazolium chloride (CTC). As counterstain 4´,6-diamidine-2´- phenylindole-dihydrochloride (DAPI) was used. The arithmetical mean and sample standard deviation were calculated for the parallel plates. Exponential regression analysis was made for the die-off data detected for *S. senftenberg 775w* and phage 28B. Times for decimation of plaque-forming units respective viable cells by 1log₁₀ (90%, T₉₀-value) and 4log₁₀ were calculated with the regression curves.

---

*Author for correspondence

3. Survival of indicator organisms in the treatment process

Faecal streptococci and thermotolerant coliforms have been used or stated as indicators of treatment effi-ciency and for the die-off of pathogenic bacteria, especially Salmonella [3, 4, 5, 6]. Both indicator organisms were reduced by approximately $4\log_{10}$ after 12 h treatment at 55°C, which included the period of upheating the material from approximately 50°C to processtemperature after the draw and fill process. The die-off re-sults for these faecal indicators were supported by laboratory tests with S. senftenberg 775w. This Salmo-nella strain was isolated as the most thermotolerant strain among three other serovars and has also earlier been described to have a high thermotolerance [7, 8]. Its die-off was tested at the laboratory in nutrient broth at 55°C. After 2-3 h no viable bacteria were detected, equivalent to a $4\log_{10}$-reduction. The calculated decimation times for the different detection methods are given in table 2 (see part 4). The number of Clos-tridium perfringens spores was not reduced even after 48 h treatment at 60°C. It was stable at around 10^3 to 10^4 cfu ml^{-1}. For the die-off of enteric viruses, parvoviruses have been used as indicator organisms [9, 10, 11, 12]. They showed high thermotolerance compared with human enteric viruses. In an aerobic sewage sludge treatment at 60°C, bovine parvovirus (strain Haden) was reduced by $0.4\log_{10}$ per hour, whereas cox-sackievirus B5 was reduced by $1.67\log_{10}$ per min and rotavirus Wa by $0.75\log_{10}$ per min [2]. The bacterio-phage used in the experiments described here, showed an even higher thermotolerance. A titer-reduction by $0.4\log_{10}$ for phage 28B was reached after 13.6 h in material 1 and after 11.4 h in material 2 at 60°C at the laboratory. The corresponding T_{90}-values were 30.4 h and 25.8 h. The survival of phage 28B in the aerobic treatment process was in all tests shorter than detected at the laboratory. In material 1 a $0.4\log_{10}$-reduction was reached after 0.55 h at 60°C (T_{90}=3.5 h). This was the only experiment in which phage 28B showed a lower stability than described for parvovirus. In material 2 the corresponding values were 7.9 h ($0.4\log_{10}$-reduction) and 13.7 h (T_{90}). The results show a dependence of the die-off rate on the suspension material. The different decimation times detected for phage 28B in the treatment process and at the labora-tory are summarised in table 1.

Table 1: Die-off of S. typhimurium phage 28B in different materials and temperatures

Temperature	Suspension material	cor. coefficient	T_{90}-values [h]	$4\log_{10}$-reduction time [h]
50°C	Nutrient Broth	0.9796	198.6	697.4
55°C	treatment process, material 1	0.9999	23.4	93.6
	treatment process, material 1	0.9934	14.7	62.9
56°C	Material 1	0.9798	52.3	189
	Material 2	0.9757	36.7	125.6
60°C	Nutrient Broth	0.9824	43.6	164.5
	Material 1	0.9504	30.4	114.6
	Material 2	0.9662	25.8	86.0
	treatment process, material 1	0.9923	3.5	18.5
	treatment process, material 2	0.9563	13.7	42.9
65°C	Material 1	0.9797	13.8	48.8
	Material 2	0.9926	12.1	44.3
70°C	Nutrient Broth	0.9972	8.4	35.2

Material 1 ... 93% Blackwater, 7% organic kitchen waste; pH 7.8
Material 2 ... 60% Blackwater, 4% organic kitchen waste, 36% liquid cattle manure; pH 7.8

Due to the influence of the suspension material on the stability of phage 28B and viruses [2,3,4,5], direct comparing tests with parvovirus and phage 28B are desirable. It seems however to be possible to state a higher thermotolerance for phage 28B. Therefore it can beneficially be used as an indicator for the die-off of thermotolerant viruses in sanitation processes. Ascaris suum eggs were described to have comparable or even higher thermotolerance than the eggs of other pathogenic parasites [2]. After 5 min exposure in 55°C, Ascaris suum eggs developed not to larvae, whereas 2% of eggs developed to larvae after 4 h exposure in 50°C.

The goals for sanitation of sewage and slurry, thought to be used in agriculture, have been reported as less than 100 cfu ml^{-1} (respectively g wet-weight) thermotolerant faecal coliforms [4], enterococci [6] or entero-bacteriaceae [13] and no infective parasite egg in 1 litre (or kg wet-weight) material. The treatment process for 24 hours, of which at least 20 hours were at 55°C, led to satisfying sanitation in terms of thermotolerant coliforms, faecal streptococci, Ascaris ova and Salmonella, but spores of Clostridium perfringens were not reduced and phage 28B only by 1-2log pfu ml^{-1}. That indicated that more thermostable pathogens could sur-vive the treatment and would be spread if the material was used in agriculture.

4. Laboratory tests to the die-off of *S. senftenberg 775w*
During the initial incubation at 55°C, all three detection methods used gave approximately the same concentration of viable bacteria. With ongoing incubation deviations in the concentrations detected with the different methods became visible. The mean T_{90}-values and $4\log_{10}$-reduction times calculated are given in table 2. They illustrate the tendency to the longest detection of viable bacteria by using CTC-vital stain. This might be an indication for a progressively shift to sublethally damaged bacteria or to a viable but not cultureable state, which leads to an underestimation of viable bacteria with plate counts [14,15].

Table 2: Die-off rates for *S. senftenberg 775w* in nutrient broth at 55°C detected with different methods

Detection method	Mean T_{90} ± sample standard deviation	Mean $4\log_{10}$-reduction time ± sample standard deviation
CTC vital stain	(34.3 ± 4.7) min	(142.3 ± 12.7) min
plate count on nonselective agar	(38.0 ± 9.8) min	(88.3 ± 77.4) min
plate count on selective agar	(24.0 ± 6.6) min	(79.3 ± 10.4) min

However, the detection limit of vital stain was 903 cells ml^{-1} but 10 cfu ml^{-1} with plate count, so that in two of three experiments viability was longest detected with plate count on nonselective agar. After 2h in 55°C the number of viable bacteria was below the detection limits of all used methods. The results have to be stated by repeating experiments, but show a trend to longer detection of viable *S. senftenberg 775w* under this conditions using vital-staining. Different authors got varying results comparing CTC-vital staining with plate counts on nonselective agar for other bacteria. Some have found higher numbers of viable cells using vital stain [16, 17, 18] others lower, depending on the bacterial strain and the origin of the examined sample [14, 19].

5. Conclusions
The experiments with phage 28B showed, that it could be used as indicator for the die-off of thermotolerant pathogen viruses in treatment processes for slurry, sewage and similar materials. In the treatment process described here, thermotolerant coliforms and faecal streptococci were reduced by $4\log_{10}$-orders at 55°C, and *Ascaris suum* eggs developed not to larvae after 5 min exposure in 55°C. Also for *S. senftenberg 775w* can a $4\log_{10}$-reduction in the process be expected, due to its die-off detected in nutrient broth and the die-off detected for faecal coliforms and streptococci. However, the results for phage 28B and spores of *Clostridium perfringens* showed, that more thermotolerant organisms can survive under the tested treatment conditions.

References
[1] SS 028167 *E. coli*, SS 028195 faecal strepto., SS-EN 26 461-2:1992 spores of sulphite reducing anaerobes
[2] C. Birbaum, J. Eckert; Schweiz. Arch. Tierheilk. **127**, 25-44 (1985)
[3] H. J. Bendixen, S. Ammendrup; Safeguards against pathogens in biogas plants, Danish Ministry of Agriculture, Danish Veterinary Service (1992)
[4] D. D. Mara, S. Cairncross; Guidelines for the safe use of excreta in agriculture and aquaculture, World Health Organisation in collaboration with United Nations Environment Programme (1989)
[5] H. E. Larsen, B. Munch; Ugeskrift for Jordbrug, Selected Research Reviews, 57-66 (1986)
[6] H. E. Larsen, B. Munch, J. Schlundt; Zbl. Hyg. **195**, 544-555 (1994)
[7] W. Soldier, D. Strauch; J. Vet. Med. B **38**, 561-571 (1991)
[8] E. El-Gazzar, H. M. Elmer; J. Dairy Sci. (US), **75** (9), 2327-2343 (1992)
[9] F. Traub, S.K. Spillmann, M. Schwyzer, R. Wyler; Schr.-Reihe WaBoLu **78**, 107-119 (1988)
[10] S. Bräuniger, I. Fischer, J. Peters; Zbl. Hyg.**196**, 270-278 (1994)
[11] R. G. Feacham, D. J. Bradly, H. Garelik, D. D. Mara; Sanitation and Disease, The Intern. Bank for Reconstruction and Development (1983)
[12] K. Seidel; Zbl. Bakt. Hyg. I. Abt. Orig. B **178**, 98-110 (1983)
[13] C. Birbaum, J. Eckert; Schweiz. Arch. Tierheilk. **127**, 25-44 (1985)
[14] J. J. Smith, J. P. Howington, G. A. McFeters; Appl. Env. Microbiol. 2977-2984, Aug. (1994)
[15] J. P. Diaper, C. Edwards; Microbiology **156**, 307-311 (1994)
[16] P. S. Stewart, T. Griebe, R. Srinivasan, C.-I. Chen, F. P. Yu, D. deBeer, G. A. McFeters; Appl. Env. Microbiol. 1690-1692 May (1994)
[17] F. P. Yu, G. A. McFeters; Appl. Env. Microbiol. 2462-2466, July (1994)
[18] J. Coallier, M. Prévost, A. Rompré, D. Duchesne; Can. J. Microbiol. **40** (10), 830-836 (1994)
[19] G. G. Rodriguez, D. Phipps, K. Ishiguro, H. F. Ridgway; Appl. Env. Microbiol. 1801-1808, June (1992)

Environmental Research Forum Vols. 5-6 (1996) pp. 359-362
© *1996 Transtec Publications, Switzerland*

Water and the Environment:
Solutions with Decentral Wastewater Treatment

D. Glücklich

Technische Universität Hamburg-Harburg, Arbeitsbereich Angewandte Bautechnik,
D-21071 Hamburg, Germany

Keywords: Decentralized Wastewater Treatment, Small Water Circuit, Water Recycling, Septic Tank, Sand Filter, Percolation

Abstract: Answers to the increasing problems of water and waste water are only to be found through a variety of techniques within an ecologically and economically sound framework. Most important thereby is the preservation or reactivation of the Small Water Circuits, in order to protect the local and thus the global biosphere. The reform of the relevant laws and administrative orders is a prerequisite for optimized solutions.

1.   Basic Considerations

A situation where water is lacking both in quality and quantity is being brought to a head by the pollution of resources, the destruction of natural reservoirs and purifying systems, the way of treatment systems, and the excessive usage. The way we handle water will become an essential cultural and social factor in our society. There are not only a few solutions. We have to react with flexible systems accordance the local requirements. The most important methods of water conservation, treatment of waste water, water recycling, and sludge disposal are based on a network of biological processes and share one common factor: the Small Water Circuit in nature, that means the local water circuit, which has to be in optimal function without defects. This calls for a number of measures:

- Pollutants must only be introduced in a quantity and a quality that they do not harm the Small Water Circuit. All `heavy poisons` have to be kept out of the water.
- Biotopes as rain water treatment systems improve the water circuit and often avoid rain water pipes. Biotopes as natural storage avoid floods and often expensive rain water pipes.
- The treatment of waste water via passing the soil ground by percolation is effective and is generally speaking superior the treatment in an aquatic ambience.
- The anaerobic treatment in bioreactors produces less sludge than the aerobic.
- The biotope and natural systems of supplementary treatment for waste water can form a symbiosis.

2.   Water supply and waste water treatment

The weak points in the central sewage system remain:

- Fresh water and waste water transportation in long pipes is expensive and destroys the local water circuit. However in the centre of cities we cannot avoid them.
- It is an uncontrollable system for all kinds of effluents.
- It often causes unintentional drainage of the land and prevents the replenishment of ground-water.
- In heavy rain large amounts of untreated sewage end up in the rivers.
- The uncontrolled infiltration of the ground with waste water from the sewers cannot be prevented even by extensive technology.
- Valuable waste water and sludge become a mixture of refuse that is difficult to dispose of.
- The cost are too high and is a burden to the citizens as to the social system of the community.

On the other side:

- The central sewers represent an infrastructure which from the technical, organisational, and administrative point of view. In the short term only partial corrections can be made.
- Different qualities and greatly varying quantities of waste water can be disposed off rapidly an comfortably.
- Large parts of our building technology, standards and laws are made to fit the system.
- It appears futile to make a radical replacement of the system for economic reasons on account of the capital which has been invested.

Users are accustomed to easy waste disposal without being informed of its disadvantages. Responsible engineers for planning and administration are not educated to work in alternatives.

3.   Legal framework, tolerance levels and permissible release

The framework prescribed by the state of technics for water supply and waste water disposal are mostly fixed and hardly open to significant ecological and economic developments. So the engineers do not enter in competition to solve problems by using optimized technical solutions.

The percolation of domestic and other comparable effluents into the ground is often forbidden because of the principle risk of contaminating the ground-water. Usually the tolerance levels for contents of effluents of central plants with aquatic systems are applied bureaucratically to other purification systems, although scientific research shows that the controlled release of waste water represents a very minor threat to both ground-water and surface water compared with other sources of pollution.

4.   Technical Solutions

The overall concept: With the right administrative framework, given the boundary parameters, the engineer can employ a whole range of methods to reduce water consumption, to re-use water and to treat the waste water in a way that is ecologically sound for the planning area. The foremost criterion is the sustainable preservation or restoration of the local water Circuit in nature (Small Water Circuit). At first he has to develop an overall concept.

Decisions on detailed solutions can only be made within an optimized overall concept. However, some individual features have crystallized out of the general considerations and experience.

Rain water treatment: Water shortage is caused by the pollution and excessive use of reserves. Therefore, in the first place, those reserves must be increased and protected. This can be done economically and ecologically through percolation of rainwater on private and public ground (fig. 1).

1   Regenwasser

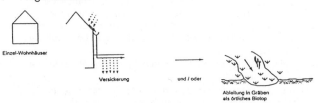

Einzel-Wohnhäuser

Versickerung   und / oder

Ableitung in Gräben
als örtliches Biotop

Waste water treatment with septic tanks and sand filters: „Benign" effluent -such as domestic waste water - should be treated locally whenever possible in order to reduce the amount of sludge from central sewage plants. In villages and on housing estates, an initial anaerobic purification in a three-chambered septic tank followed by aerobic treatment via a sand filter (fig. 2) has proved successful. The quantity of sludge as useful materials normally accounts for about 10% of the total volume of central sewage plants.

2   Unproblematisches Abwasser

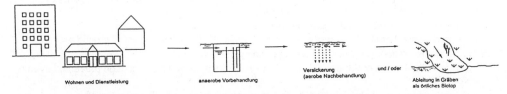

Wohnen und Dienstleistung   anaerobe Vorbehandlung   Versickerung   und / oder
                                                                          (aerobe Nachbehandlung)
                                                                                                    Ableitung in Gräben
                                                                                                    als örtliches Biotop

Waste water Treatment with reed bed and similar: Constructed wetlands preceded by septic tanks correspond to the treatment of waste water with sand filters as described above, provided that they are set up in a similar way.

Waste water Treatment with ponds: Ponds are simple constructed natural aquatic plants with similar problems as technical ones. Thats why the effluent should passage large sand filter (fig. 3). They can be bedded in the landscape close to nature.

3   Unproblematisches Abwasser

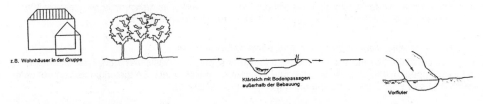

z.B. Wohnhäuser in der Gruppe

Klärteich mit Bodenpassagen
außerhalb der Bebauung

Vorfluter

<u>Improvement of existing sewage plants:</u> As many sewage plants no longer meet the necessary standards for the release of effluent, they are being refurbished at great cost. Thus, money is going into repairing a system rather than improving it. Therefore, before the development is implemented, the possibility of relieving the plant by means of local rainwater percolation and partial local treatment of the waste water should be explored.

<u>The combination of various methods of treating waste water:</u> The domestic waste water can be treated partially with local systems, while water with problematical inholds were dealt with centrally.

4   Problematisches Abwasser

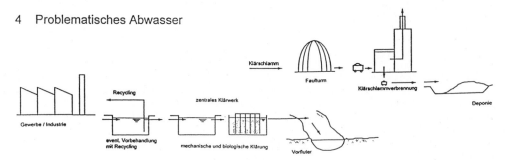

<u>The use of rainwater and grey water:</u>
The use of rainwater and grey water (fig. 5) are repair methods which do not lead to a lasting improvement of the whole system. It makes more sense to feed all the rainwater into the natural circuit: a natural reservoir is usually more effective than an artificial one.

5   Abwasserrecycling mit Regenwassernutzung

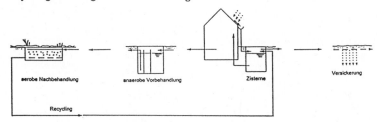

<u>The free of waste water building:</u> An environmentally sound development is called for: the reduction of consumption, the restoration of <u>Small Water Circuit</u>, and the avoidance of long term pollutants, taking the economy into account. Nowadays, houses which already have water-saving devices, decentralized treatment of waste water, followed by percolation, largely fulfill these conditions and are thus practically houses without waste water. As follows it is not allowed to bring back the purified water into the local water circuit by percolation or using wetland. The circuit cannot be closed. It is the pressure of this legal situation that gives rise to the special idea of a house that produces no waste water: If it is forbidden to release purified waste water into the ground or into public drains, most of it, after further treatment is necessary, will be used again. This often makes sense in spite of the ridiculous effort involved where the economy is concerned. The long pipes are very expensive. It would be more advantageous, however, to change the legal basics.

6.   **Prognosis**
Many firms and citizens want nature to be intact. At the same time they wish the supply of water and the disposal of waste water to be cheap. As the large systems can only be financed if there is a high turnover, which is in conflict with the demands of the <u>Small Water Circuit</u> with particular local systems and the sparing use of resources, there is no doubt that economic pressure will lead to a turning point. In the long term the economy cannot run contrary to nature.

Environmental Research Forum Vols. 5-6 (1996) pp. 363-366
© *1996 Transtec Publications, Switzerland*

From Wastes to Resources – Turning over a New Leaf:
Melaleuca Trees for Wastewater Treatment

M. Greenway and K.G.E. Bolton

Faculty of Environmental Sciences, Griffith University, Nathan, Queensland 4111, Australia

Keywords: *Melaleuca* Trees, Nutrients, Phosphorus, Sewage Effluent, Translocation, Wetlands

Abstract
Three species of Melaleuca trees are being investigated for their suitability in constructed wetlands receiving nutrient enriched wastewater. Highest growth rates occurred in pot trials receiving secondary treated sewage effluent and with added nitrogen; growth was slowest with added phosphorus (12mgL-1)suggesting toxicity. Prior to leaf fall nutrients are translocated back into plant biomass. Leaves of *M.alternifolia* accumulate the highest P concentrations. Harvesting the leafy foliage for tea-tree oil also removes nutrient biomass, thus a lucrative industry can be provided from a waste product.

Introduction
Constructed wetlands technology is now widely accepted throughout the world as an economical and effective means of wastewater renovation. Most studies have concentrated on the use of non-woody species (ie reeds and rushes) however, woody species may have additional advantages such as a higher biomass-sink volume, high productivity rates and the production of useful primary resources. Harvesting biomass to remove resources also removes nutrients.

In Australia, Melaleuca wetlands are important wildlife habitats that provide food and shelter for koalas, possums, bats, birds and bees. Melaleucas also provide resources for primary industries such as tea tree oil, honey and bark pictures. In sub-tropical/tropical Australia these woody swamps are seasonally flooded. They are highly productive, having annual litter falls of 700 gm^{-2} but very slow rates of litter decay; and appear to function as nutrient sinks [1]. Finlayson et al. [2] estimated the above-ground fresh weight of a Melaleuca wetlands in tropical northern Australia to be 263 t ha-[1]. Researchers at Griffith University, in Brisbane, Queensland are investigating the use of *Melaleuca* species as candidates for constructed wetlands for wastewater treatment. Pot trials consisting of 324 self-watering 40L pots receiving secondary treated effluent, have been established to examine the responses of three species *M. quinquenervia*, *M. alternifolia* and *M. leucadendra* to various effluent concentrations and waterlogging cycles. A plot trial consisting of 7 densely planted terraced beds examines the wastewater polishing potential of a mini-Melaleuca constructed wetland system.

M. quinquenervia forms highly productive and biologically diverse wetlands in subtropical coastal lowlands. *M. alternifolia* forms the base of a lucrative tea-tree oil industry; *M. leucadendra* is a fast-growing tropical species used ornamentally in urban settings. Thus constructed Melaleuca wetlands for wastewater treatment have potential for ancillary benefits such as habitat creation, primary industry production and aesthetics. This paper presents preliminary results on growth responses and leaf phosphorus content of Melaleucas in the pot trials experiments..

Materials and Methods
Pot trial setup
The site is located at the Loganholme sewage treatment works,SE Queensland. Secondary effluent is used for the pot trial experiments. Typical effluent contains $5mgL^{-1}$ N, $8mgL^{-1}$P. The pot trial is self watering from 3 x 3000L storage tanks containing: half effluent, half townwater - **50% treatment** ($2.5mgL^{-1}$ N, $4mgL^{-1}$ P); effluent enriched with $5mgPL^{-1}$ added as phosphoric acid - **100%+P** treatment ($5mgL^{-1}$ N, $12mgL^{-1}$ P), and effluent enriched with $5mgNL^{-1}$ added as Nitram fertiliser

(99% NH$_4$NO$_3$) - **100%+N** treatment (10mgL^{-1} N, 8mgL^{-1} P). A fourth system is connected directly to the effluent mains - **100%** treatment (5mgL^{-1} N, 8mgL^{-1} P). The trees are planted in 40L pots containing sand medium. In order to minimise genetic variation *M. quinquenervia* and *M. leucadendra* seedlings from single locations were selected from a nursery. The *M. alternifolia* seedlings were purchased from a nursery which is running a breeding programme to improve tea-tree oil production. The pot trial has a number of waterlogging regimes in the design, ie continuously waterlogged and cycles of 3 months dry; 6 months waterlogged; 6 months dry. However only the continuously waterlogged plants are considered in this paper.

Experimental design
The pot trial is comprised of 2 separate experiments, one being the edge row of the other. The results presented in this paper are from a portion of a larger split-plot trial examining effluent quality and waterlogging cycles. The design has thus been analysed as a Randomised Complete Block Design for the purposes of this paper where 6 replicates of each species are continuously subjected to 4 effluent qualities - a total of 72 trees. SAS was used to calculate ANOVAs and LSD values.

Field measurements
From May 1994, three-monthly measurements have been made of height, diameter at 15cm, branch number, new leaf number and adventitious root density.

P analysis of leaves
In May 1995, samples of young, mature and senescent leaves were collected, oven dried, ground, and subsamples acid digested. Total phosphorus was determined colormetrically.

Results
Growth parameters
After 12 months all 3 species of *Melaleuca* are responding similarly in growth to the different effluent treatments. In general, the 100%+N and 100% treatments resulted in the highest growth and the 100%+P and 50% treatments resulted in the lowest growth.

Adventitious root formation
Adventitious roots are produced when the plants are waterlogged for at least 2 weeks. Two forms occur: aerial adventitious roots which grow directly down from the stem into the water column and subterrainian adventitious roots which originate from within the sand-gravel medium and grow up into the water column. *M. alternifolia* only produced subterranean adventitious roots. The 100% and +N treatments resulted in significantly greater adventitious root formation than the +P and 50% treatments for *M. quinquenervia* and *M. leucadendra*. *M. alternifolia* had the lowest root densities for all treatments while and *M. quinquenervia* and *M. leucadendra* had similar densities. Some plants developed a thick algal/root mat association, particularly in the +N and 100% treatments.

Leaf phosphorus content

Table 1. Phosphorus concentrations (mgP/g dry wt.) in young, mature and old leaves of 3 species of *Melaleuca* growing in different secondary sewage effluent treatments.

	M.quinquenervia			M.leucadendra			M.alternifolia		
Effluent	young	mature	old	young	mature	old	young	mature	old
100%	2.221	0.867	0.733	2.860	1.003	0.859	3.887	2.100	1.372
100%+N	2.716	1.113	1.634	-	1.416	-	3.876	2.652	1.133
50%	2.550	1.670	-	3.250	2.090	-	5.641	5.773	5.042
100%+P	2.637	2.678	3.682	3.620	2.621	2.043	6.926	5.996	9.102

Table1 provides a comparison between P content (mg P/g dry wt.)of young, mature and senescent (old) leaves in the 3 species. *M. alternifolia* had the highest leaf P concentrations. The +P treatment generally resulted in significantly higher leaf P concentrations, and the young leaves were the best indicator of this. The younger leaves generally had the highest P concentrations and senescent leaves the lowest, however senescent leaves in the +P treatment had higher concentrations for *M. quinquenervia* and *M. alternifolia*. There is evidence of an N/P interaction in young leaves of *M. leucadendra* as the +N treatment resulted in the significantly lower P concentrations.

Discussion

All 3 species of *Melaleuca* are growing better in the 100% treatment (ie 5mgN and 8mgP L^{-1}) than in the 50% treatment indicating their suitability for growth in secondary treated sewage effluent. The 50% treatment may have been nutrient limiting. High growth in the 100%+N treatment ie $10mgL^{-1}$ N indicates the ability of the melaleucas to grow in N enriched wastewater. However the low growth and retarded adventitious root development in the 100%+P treatment ie $12mgL^{-1}$ P indicates phosphorus toxicity at such high P concentrations. As P is generally limiting in Australian soils and water this response is not surprising.

M. alternifolia demonstrated the greatest ability to accumulate P in leaves, with around twice the leaf P concentrations than the other two species. Since the process of tea-tree oil production requires foliar harvesting ,removal of plant parts also removes nutrients. This makes *M alternifolia* a particularly lucrative candidate for P stripping of wastewaters. Whilst the logistics of incorporating a harvesting schedule into constructed wetlands will require further agronomic and engineering research, effluent-fed *M. alternifolia* plantations have considerable potential for ecologically and economically-sound wastewater treatment.

In all but the +P treatment, leaf age and P concentration were inversely related. This bears evidence of P translocation out of leaves as they approach senescence, meaning less P is lost from the tree biomass-sinks when leaves abscise. Translocation of P and N from senescing leaves back into the tree also occurs in natural Melaleuca forests [1]. The higher leaf P concentrations in the +P treatment is evidence of a P exclusion mechanism in response to toxic concentrations.

Another major role of plants in a constructed wetland is to provide roots which support a community of microorganisms and physically filter the waste particles. The high rates of adventitious root development by all 3 Melaleuca species may increase the removal efficiency of both nutrients and suspended solids

This study suggests Melaleuca trees may be exceptional candidates for constructed wetlands for wastewater treatment. Of fundamental importance, they thrive in effluent-fed waterlogged conditions. Their high biomass potential, growth rates and ability to translocate nutrients from senescent leaves back into the plant , are promising indicators that Melaleuca trees function as biomass nutrient sinks. Their ability to produce adventitious roots mats provides an environment conducive for biofilm growth further enhancing the removal potential of nutrients.Furthermore, substantial fringe benefits including primary resource production ,aesthetics and wildlife habitat would be gained from constructed Melaleuca wetland systems.

References
[1] M. Greenway, Aust. J. Mar & Fresw.Res.**45**, 1509 (1994).
[2] C.M. Finlayson, I.D. Cowie and B.J. Bailey, Wet.Ecol. & Mgt. **2**, 177 (1993).

Environmental Research Forum Vols. 5-6 (1996) pp. 367-370
© *1996 Transtec Publications, Switzerland*

Artificial Wetlands for Municipal Wastewater Treatment and Water Re-Use in Tropical and Arid Australia

M. Greenway[1] and J.S. Simpson[2]

[1] Faculty of Environmental Sciences, Griffith University, Nathan, Queensland 4111, Australia

[2] Queensland Department of Primary Industries - Water Resources,
GPO Box 2454, Brisbane, Queensland 4001, Australia

Keywords: Artificial Wetlands, Australia, Effluent Polishing, Irrigation, Macrophytes, Municipal Wastewater

Abstract

Ten artificial Free Water Surface Flow wetlands have recently been constructed in Queensland, Australia to polish secondary municipal effluent. A variety of native aquatic macrophytes are being trialled for their ability to remove and store nutrients. All species growing in effluent had a higher phosphorus content than controls. In drier areas the polished effluent is being used to irrigate golf courses, parks, commercial treelots and pastures.

Introduction

Queensland, Australia has a subtropical-tropical climate with dry winters and wet monsoonal summers. Arid climatic conditions prevail inland with hot dry days and cold nights. The climatic conditions are conducive to high plant growth rates and hence offer great potential for constructed wetlands for water pollution control.

The water, a scarce resource during the dry season and in arid regions, can also be used to irrigate crops, playing fields, parks and gardens or golf courses. The water discharged from the wetlands is also of an acceptable quality to flow into estuarine and riverine environments. Many natural wetlands are only seasonally inundated and during the dry season wildlife has to seek alternative refuges. Artificial wetlands receiving sewage effluent provide permanent wildlife habitats and improve the landscape amenity. Effluent discharging into the waters of the Great Barrier Reef Marine Park, which extends for 2000km along the Queensland Coast, is of particular concern because of the sensity of the marine ecosystems to nutrient enrichment and increased turbidity. Three of the pilot wetlands now treat the secondary effluent before it is discharged into mangrove fringed creeks and estuaries.

The Queensland Government has initiated an Artificial Wetlands for Water Pollution Research Program. This was largely due to the fact that most of the wetland treatment systems, both in Australia and overseas have been installed in cold or temperate climates. It was felt that the current state of knowledge was insufficient for the design and management of artificial wetlands in Queensland conditions. Ten experimental pilot artificial wetlands have been established to treat secondary sewage effluent and monitor nutrient removal and bioavailability in native macrophytes.

Artifical Wetlands in Queensland

Most of the pilot constructed wetlands are Free Water Surface Flow i.e. open water channels; length, width, detention time and number of channels varies between wetlands. Each wetland contains a variety of macrophytes which are largely native species collected from natural wetlands in the locality. Exotic species such as water hyacinth which have been used overseas for wastewater treatment are declared noxious aquatic weeds in Queensland. Their use is prohibited to avoid spread of such weeds into natural waterways.The use of water lettuce and *Salvinia* is also discouraged though the latter is spread by birds Sections of the constructed wetlands have been planted out with different species in an attempt to investigate which is the best configuration for performance efficiency and to monitor natural changes in species composition over time. A variety of macrophytes are being used: free floating duckweeds mainly *Spirodela*;, submerged pond weeds eg *Ceratophyllum,* plants with submerged stems and floating leaves e.g. *Nymphoides*, *Marsilea* (a fern); emergent reeds, rushes and sedges including *Phragmites, Typha*, *Schoenoplectus, Eleocharis, Baumea, Cyperus*;, aquatic grasses including fodder species eg *Hymenachne*, and aquatic creepers with surface leaves, submerged stems and adventitious roots eg *Persicaria, Ludwigia, Ipomoea, Imperia*.

Plant species are being monitored for their ability to accumulate phosphorus.in plant biomass.

Case studies in Queensland

This paper will briefly describe 3 wetlands that have been in operation for 2 years.

Ingham Wetland (wet tropics)

The Ingham Wetland situated in tropical North Queensland, consist of 3 U-shaped channels 110m x 12 m wide x 500 mm depth. Fifteen species of aquatic macrophytes were successfully transplanted and another 5 species colonised naturally. The design detention time in the wetland is 7-12 days.

The primary objective of the Ingham wetland was to polish wastewater effluent to an acceptable standard for creek discharge. The wetland has achieved a reduction efficiency of 48% BOD, 52% Total N and 8% Total P. The local authority is now considering a joint treatment project with a nearby sugar mill. The final wetland effluent will be used to scrub flue gases in the sugar mill and to irrigate farms in the area.

Townsville Wetland (dry tropics)

The Townsville wetland is situated in the dry tropics, and consist of 4 straight channels and1 U-shaped channel 60 m x 4 m x 400 mm depth. Initially only 6 species of macrophyte (2 floating, 2 submerged, 3 emergent) were transplanted and all channels had an open water section. An additional 8 species have colonised and the channels are now completely vegetated. The design detention time was 5 days in the straight channels and 7 days in the U-shaped channel.

The objective of the Townsville pilot wetland was to polish the effluent from the wastewater treatment plant and then discharge to the adjacent Town Common, following the satisfactory performance of the pilot wetland. Drought conditions seriously deplete the natural wetland environment and wildlife populations in the Town Common - a popular tourist spot for bird watching, therefore additional sources of water are needed.

The pilot wetland with a 5-7 day detention, it is producing a high quality effluent with reduction efficiencies of 67% BOD, 44% SS, 74% Total N, 65% NH_4-N, 91% NO_xN, 35% Organic N, 6% Total P. Discharge onto the Town Common has been highly effective in achieving its re-use objective. It has promoted lush growth of aquatic and wetland vegetation which has attracted flocks of ducks, magpie geese and waders even during the drought and dry seasons.

Blackall Wetland (arid)
The Blackall Wetland situated in the arid Western Region consists of 4 linear channels 120 m x7 m x 600 mm. Initially only 3 species were planted out, one in each channel, with the fourth channel remaining as open water. However, this soon became colonised. Now the wetland supports 8 aquatic species. Detention time in the wetlands is 3-4 days.

Due to the lack of water, sewage effluent was considered a resource which was being wasted. The objective of the Blackall Wetland was to re-use the wastewater effluent for the betterment of the town community. Nutrient retention was considered more important than removal. A benefit of the removal of BOD and nutrients, was to reduce the effect of pollution on rivers and creeks.

A good standard of treatment is being achieved in spite of only 3 to 4 days detention time. The average reduction is 46% BOD and 68% suspended solids. On a concentration basis however, there is effectively no reduction (3%) in reactive PO_4 - P.

The Council is currently investigating the following effluent re-use schemes: irrigating plant nursery seedlings; irrigating commercial treelots, that is, different species of flowering gums for the export florist market, irrigating the local golf course, developing a riverbank eco-tourism wetland complex, using native plants and trees typical of Western Queensland irigated with effluent. As a longer term goal for this arid region of Queensland, it is proposed to irrigate all community parks and gardens with polished effluent from constructed wetlands in order to conserve the dwindling resources of the Great Artesian Basin

Phosphorus Removal and Macrophyte Bioaccumulation.

Although the wetlands show only a small percentage of phosphorus removal, phosphorus concentrations in plant tissue indicate an accumulation of phosphorus in the macrophytes receiving effluent compared with controls collected from nearby natural wetlands (Table 1.).

Table 1. Phosphorus content (mgP/g dry wt.) in plant components of macrophytes growing in artificial wetlands(treatment) and in natural wetlands(control). Pooled values for all species.

Plant component	Ingham			Townsville		
	Control	Treatment		Control	Treatment	
	1994	1994	1995	1994	1994	1995
leaves	1.70	4.12	3.89	0.64	2.25	3.67
stem	1.25	3.00	5.25	0.63	2.10	3.86
root	2.31	4.32	7.20	0.32	3.40	4.42
rhizome	1.80	5.00	5.55	0.74	2.36	2.15

The highest concentrations of phosphorus were found in duckweed (16mg/g), *Ceratophyllum* (14mg/g), *Monochoria cyanea,* a native relative of the water hyacinth, (13mg/g), *Nymphoides* (11mg/g) and the fern *Marsilea* (9mg/g).

In Australia most natural ecosystems are adapted to low levels of P and elevated levels of P can be toxic to many native plant species, thus it is encouraging that so many aquatic macrophytes can tolerate nutrient enriched artificial wetlands. When reusing effluent from these wetlands however consideration must be given to irrigation uses with respect to P tolerance. In most cases the irrigation of golf courses, recreational grounds, agricultural crops benefit from additional P but irrigation of native species should be carefully monitored.

Environmental Research Forum Vols. 5-6 (1996) pp. 371-374
© *1996 Transtec Publications, Switzerland*

Common Mussels for Oil Polluted Coastal Waters Treatment

A.V. Gudimov and T.L. Shchecaturina

Murmansk Marine Biological Institute, 17 Vladimirskaya Street, Murmansk, 183010, Russia

Keywords: Mussels, Chronic Pollution, Oil Residuals, Filtration, Accumulating, Removal and Treatment, Substrates, Faeces, Pseudofaeces, Oil-Degrading Bacteria

Abstract

The great contribution to the total input of petroleum hydrocarbons to the marine environment is from chronic inputs from rivers, sewage discharges, industrial effluents, oil refineries and terminals. Accidental spillages provide the largest transient inputs of oil to coastal environment.

Some bivalves, blue mussels for instance, are able to accumulate petroleum hydrocarbons in their tissues and to concentrate them to a marked degree over sea water levels. Besides, in the process of filtration mussels accumulate oil extracted from ambient water in pseudofaeces and faeces discharged after digestion. These facts, combined with an almost ubiquitous distribution and tolerance to oil pollution, render mussels as pollution indicators and efficient biofilters.

In this study we present some results of the investigations on the growth rate and hydrocarbon concentration in tissues of *Mytilus edulis* exposed to crude oil and oil spill residuals in the Black and the Barents seas conditions.

Introduction

City sewage commonly contains oil hydrocarbons, inflow of that in the estuarine and coastal waters may be significant. The greatest contribution to the total input of anthropogenic hydrocarbons to the marine environment is provided from chronic pollution.

These untreated mixed wastewaters are not suitable for raising fish, molluscs and seaweeds within coastal aquaculture. The wastewater aquaculture, producing valuable protein from waste, demands detoxification as a first step of treatment.

Since small water bodies like ponds have very limited purification capabilities and is very difficult to monitor the harmful components, the freshwater bivalves can be used on pre-treatment stage of integrated fish pond culture to accomplish reduction and control of toxic substances.

Some bivalve molluscs, blue mussels for instance, are able to concentrate and deposit oil hydrocarbons extracted from ambient water, accumulating its also in their bodies. It allows to use "the molluscs filter" at first stage of oil polluted wastewater treatment.

Accumulation

Mussels of the Black sea do not feel depressed at the oil concentrations hundred times as higher than allowable limited ones and continue to filtrate water as well as control specimens [1,2].

In the process of filtration mussels accumulate oil hydrocarbons in their tissues to a marked degree over sea water level. It was obtained, that accumulating coefficient of petroleum hydrocarbons in mussel tissues came to 700-1200 of the hydrocarbons concentration in water [2]. The total concentration of petroleum hydrocarbons decreased in two times at exit of running-water aquaria only due to filtration activity of mussels. The highest filtration rates were observed up to 0.5 ml of oil hydrocarbons per litre.

The pattern of accumulation of petroleum hydrocarbons in the body tissues was affected by the presence of algal food cells, the period of exposure, the physiological state of mussel, the hydrocarbon concentration in sea water, and the nature of the hydrocarbon [3,4].

The total amount of hydrocarbons in mussels from chronic oil polluted area of the Black sea coast was altered from 79-90 to 228 mg/100 g wet weight, and in mussels from relatively clean waters - from 19 to 60 mg/100g [2]. According to results of our recent field experiment conducted in April-May 1994 in the Kola bay, the concentrations of oil hydrocarbons in mussel tissues were reached 20-100 mg/100g after two weeks exposure at oil spill residuals. In the control point 5 (Fig. 1 C) with natural background of contamination, the amount of hydrocarbons in mussels tissues increased by 5 times, but in the basins with experimental oil spills treated biosorbents (see 1, 2, 4) and with crude oil (see 3) this amount increased by 1.5-2 times owing to oil toxic influence on filtration activity of mussels.

Growth rate and toxic effects

The distinct evidences to oil residuals influence on the coastal environment have been obtained according to results of the mussels linear growth rate. Mussels (*Mytilus edulis* L.) of 3 dimension classes were kept into the kapron net cages for 60 specimins each. In each basin and in the control there were 2 cages situated on the depth approximately 1.5 m (Fig. 2). The growth rate of mussels in cages was measured for 2 and 4 weeks of experimental exposure. The differences in mean increment of shell lengths have been estimated the biological effects of oil spill combatting. The growth rate were significantly altered by short-term (2 and 4 wk) exposure to oil residuals.

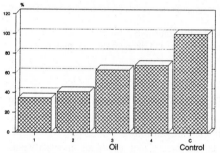 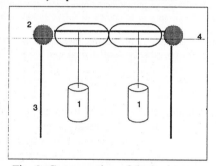

Fig. 1. The linear growth rate of mussels   Fig. 2. Cross-section of the basin:
(in % to Control) in two weeks: 1, 2, 4 -   1- mussel cages, 2 - float-basin,
basins with oil and biosorbents, 3- only   3- 'wall' of basin, 4- water surface.
crude oil, C -control.

Increments in mussel shells in each basin during two weeks were statistically valuable (p < 0.05) and smaller that in the control (Fig. 1). It is suggested, that there was existed some toxic effect of the not full bacterial degradation of the oil [5]. Stress indices such as length increments were negatively correlated with tissue aromatic hydrocarbons.

Removal and treatment of oil residuals

The main part of oil hydrocarbons concentrated due to filtration activity of mussels is discharged as faeces and thin mucous brown "cords" of pseudofaeces [1]. Faeces and pseudofaeces of mussels are becoming very suitable substrates for bacteria development.

Oil in tissues and excretions of mussels had a significant disintegration of aliphatic hydrocarbons. Such oil treatment makes improvement of oil as substrate for micro-organisms utilizing. This gives the oil-degrading bacteria the best chance of getting to both the hydrocarbon substrate and organic-mineral nutrients.

Intensity of faeces and pseudofaeces formation depends on concentration of oil hydrocarbons. When oil concentration was 10 ml per litre Milovidova [6] noted weak attachment of mussels by byssal threads, closing of shells and lack of pseudofaeces. When oil concentration was 1 ml per litre mussels lived for a long time, fed and excreted pseudofaeces. A few days before death adductor muscle was relaxed and mussels had wide gape. Restoring of adductor functioning occurred 1-2 days after placing them in pure water and byssus was secreted. Comparable results were found in our experiments on physiological responses of mussels exposed at 1% concentration of drilling fluids [7].

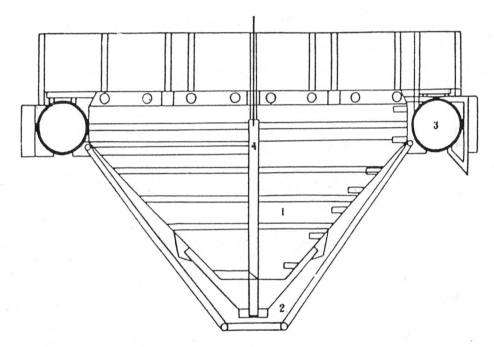

First attempts

Mussel tolerance to oil pollution allows to consider a possibility of using them in an integrated biological purification of sea water from oil hydrocarbons. This approach to oil polluted coastal water treatment have been used in pilot plant (Fig. 3) of hydrobiological purification constructed and successfully tested by us at the Black sea. The model of the plant was presented at the All Russian exhibition Centre and was awarded with a medal in 1994.

References

1. Alakrinskaya, I. O., Zool. jurn. V 45, 7, 998-1003 (1966).
2. Shchecaturina, T. L., Thesis of diss. Doctor chem.sc., 34 (1992).
3. Bayne, B. L., R. J.Thompson, J. Widdows. In: Mar.mussels: their ecol. and physiol., 494 (1976).
4. Shchecaturina, T. L., Mironov, O. G., Hydrob. jurn. V 27, 2, 71-76 (1987).
5. Platpira, V. P., 1 Vses. conf. ryboch. toxic. Riga. Ch 2, 73 (1988).
6. Milovidova, N. Ju. In: Biol. samooch. formir.kach. vody. 141-143 (1975).
7. Gudimov, A. V. In: Arctic seas: bioind. envir... Apatity, 36-44 (1993).

Environmental Research Forum Vols. 5-6 (1996) pp. 375-378

Wastewater Purification Efficiency and Growth of Aquatic Macrophytes – Possible Use in Aquaculture for Nutrient Recycling

A. Heeb[1], J. Staudenmann[2] and B. Schmid[1]

[1] Institut für Umweltwissenschaften, Universität Zürich, CH-8057 Zürich, Switzerland

[2] School of Engineering Wädenswil, CH-8820 Wädenswil, Switzerland

Keywords: Wastewater Treatment, Nutrient Uptake, Aquatic Macrophytes, Aquaculture/Horticulture System, Macrophytes as Food, Plant Production

Abstract
In three separate experiments, the possible use of several aquatic macrophytes in aquaculture was tested. Wastewater purification was measured by nutrient uptake, biomass production was examined under different harvesting methods, and was then utilized for food and to feed fish. Emersed and floating plant species were efficient at high nutrient levels, whereas submersed species had to be combined with zooplankton to control algae and were more efficient at lower nutrient concentrations of later steps in the purification process.

1.   Introduction

The idea of building artificial aquatic ecosystems to clean nutrient-enriched wastewater is based on the fact that plants take up nutrients from water to grow. By producing biomass, the plants "clean" the water.
The aquaculture project at the Engineering School in Wädenswil includes plants, phyto- and zooplankton, fish and sandfilters [1].The "aquatic macrophytes for wastewater treatment and production of biomass" section of the project is curried out as a diploma thesis at the University of Zurich.
The purpose of this work is to test indigenous aquatic macrophytes for their possible use in aquatic treatment systems. Furthermore, the possibility of plant biomass utilization as "food, feed, fiber, fuel, fertilizer" [2] is considered. According to "recycling the resource", nutrients from wastewater can be re-used for food production in an aquaculture, e.g. to feed fish, as it is done in this project [5].

2.   Experiment 1 (March 1995-April 1995): Testing Plant Species

2.1   Materials & Methods
In a first experiment the purification efficiency and growth of several aquatic macrophytes was examined. According to a literature study and the availability of plant material in February 1995, the following plants, grouped by growth habit, were chosen (nomenclature follows [6]):

"S-group", needing a substrate for rooting:
emerse plants:
- Alisma plantago
- Butomus umbellatus
- Sagittaria sagittifolia
submerse plants:
- Callitriche palustris

"W-group", without substrate:
emerse plants:
- Nasturtium officinale
floating plants:
- Azolla sp.
- Lemna minor
- Salvinia natans
submerse plants:
- Ceratophyllum submersum
- Elodea canadensis
- Myriophyllum spicatum
- Stratiotes aloides

Plants "S" and "W" were grown separately in small pots containing 5l and 1.2l water, respectively. There were 5 pots (replicates) for each species and nutrient level ("low" or "high"). All plants were initially grown in a 10% Hoagland solution [3], after t two weeks half of them were put into concentrated nutrient solution of 56 mg N_{tot} /l and 12 mg P_{tot} /l . (Two further tests with elevated nutrient levels of 112 mg and 210 mg N_{tot} /l and 24 mg and 46 mg P_{tot} /l respectively were made with one pot for each species and nutrient level.) During 21 days fresh weight of the plants and the electronic conductivity (EC), pH and O_2 concentration of the water were measured every 3 to 4 days.

2.2   Results
There were big differences in purification efficiency and growth among the species groups and between nutrient levels (p<0.01 for both effects in ANOVA): floating species like *Azolla*, *Lemna* and *Salvinia* as well as *Nasturtium* removed considerable amounts of nutrients when growing, whereas growth and "water cleaning" of submersed plants was suppressed, probably by the growth of algae that competed for light and nutrients (figures 1-4).

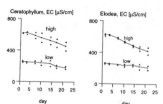

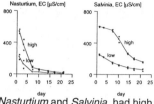

Fig. 1 *Ceratophyllum* and *Elodea* had low purification efficiencies.

Fig. 2 *Nasturtium* and *Salvinia* had high purification efficiencies.

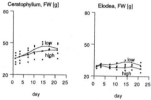

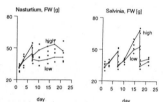

Fig. 3 *Ceratophyllum* and *Elodea* had low growth rates.

Fig. 4 *Nasturtium* (1 harvest) and *Salvinia* (2 harvests) had high growth rates.

The experiment with plants needing a substrate for rooting showed similar results for emersed and submersed plants, respectively: *Alisma*, *Butomus* and *Sagittaria* (all emersed species) showed good purification efficiency, whereas with *Callitriche* (submersed species) the EC-values did not decrease significantly (figure 5).

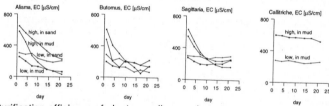

Fig. 5 Purification efficiency of plants needing a substrate

3.   Experiment 2 (May 1995-July 1995): Testing Harvesting Quantity

3.1   Materials & Methods
Of those species that had high purification efficiency in Experiment 1, the following were put together in a polyculture: *Alisma*, *Butomus*, *Sagittaria* (needing a substrate) *Azolla*, *Lemna* and *Salvinia*. *Nasturtium* was set aside for the later main experiment. To increase the growth and purification efficiency of the submersed species they were put together with zooplankton [4]. Experiment 2 consisted of 10 basins of approximatively 40 l of water and a 10 cm layer of gravel for the rooting plants. Growth and purification efficiency of the polycultures were examined under two different harvesting methods (5 basins each as replicates):
"large" harvests meant removing all aboveground biomass down to 5 leaves for rooting plants and reducing th cover of floating plants to 25% of the total watersurface,
"small" harvests" meant removing all aboveground biomass down to 10 leaves for rooting plants and reducing th cover of floating plants to 50% of the total watersurface.
During four 10-day periods the development of biomass under these two harvesting methods was observed. The purification efficiency was assessed by measuring EC, pH and O_2 values as well as additional analysis of NO_3^-, NH_4^+ and PO_4^{3-} concentrations in periods 1 and 3. Initial concentrations were about 56 mg NO_3^- /l and 12,5 mg PO_4^{3-} /l, slightly increased every 10 days by the addition of new nutrient solution.

3.2.   Results
In polyculture the waterbody was kept free from algae through the shading of the floating plants. pH values were slightly above 7, with a tendency to be higher in "large harvest" basins. O_2-values suddenly increased after harvesting the floating plants and decreased as they covered the watersurface again. With large harvests O_2 values were higher (p<0.05 in ANOVA) than with small harvests. A clear effect of different harvesting has been observed for biomass production: Measured in relative increase of water surface covered by floating plants (%), *Azolla* was predominant with small harvests, whereas with large harvests *Lemna* was able to cover free spaces fastest (p<0.01 for interaction term harvesting method x species in

ANOVA; figure 6). In terms of purification efficiency, no significant effect of different harvesting methods were observed. Both methods resulted in a clear decrease of NO_3^- and PO_4^{3-} concentrations (figure 7).

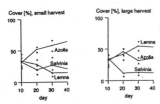

Fig. 6 Differences in watersurface covered by floating plants with small and large harvests.

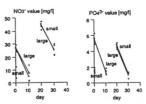

Fig. 7 Water purification shown with decreasing NO_3^- and PO_4^{3-} concentrations.with small and large harvests.

4. Main Experiment (July 1995-October 1995): Pilot-test of an artificial aquatic ecosystem for water purification

4.1 Materials & Methods

The pilot test of an artificial aquatic ecosystem for water purification [1] consists of several purification- and biomass-production steps: aquatic macrophytes occur at steps N, P, Z, phyto- and zooplankton at steps A, Z [4] and fish at step F [5].

Step N: Two different types of *Nasturtium* cultures were tested as first step of the purification and biomass production process. Harvest quantity and frequency depended upon growth rates. *Nasturtium* is an exellent vegetable, used as salad or cooked (We recommend Ravioli à la *Nasturtium*-Ricotta!).

Step P: All species used in the polyculture of Experiment 2 were taken over for the pilot test as second purification and biomass production step.

Step Z: Submersed macrophytes from Experiment 1, presumed to prove efficient when put together with zooplankton, were tested in the third step [4].

Measurements started in July and will presumably last until October 1995. All polyculture basins so far were harvested weekly. Harvested plant biomass was used for feeding fish [5]. Purification efficiency of *Nasturtium* and the polycultures was controlled every week by water analysis. A nutrient solution comparable to that in greenhouse cultures was used to simulate waste water.

4.2 Conclusions

Since the measurements are still running, no definitive statement can be given as yet. Due to technical problems at the beginning, it was difficult to reach and maintain the desired nutrient levels in the inlet. Furthermore, unusual climatic conditions in August slowed plants growth. This resulted in a decreasing amount of biomass - likewise feed shortage for the fish! Nevertheless, an obvious decrease of nutrients in the water by increasing biomass production can be assumed based on the results obtained from experiments 1 & 2.

5. Open Questions for further Research

The following questions will be addressed in the continuation of this work:
- how do plants develop and react in a long-term experiment (especially during the winter)?
- which other species could be used in aquacultures and have a high
    - purification efficiency?
    - production of biomass for further use?
- are polycultures more efficient or more stable than monocultures?
- how does adsorption in the substrate contribute to water "purification"?

6. References

[1] B. Züst, Workshop 1: "Recycling the Resource from Greenhouse Cultures", 2nd Int.Conf. on Ecol.Eng. f. Wastewater Treatment (Sept.1995).

[2] C.W. Hall, in K.R. Reddy & W.H. Smith (eds.), "Aquatic Plants for Water Treatment and Resource Recovery", 1013 (1987).

[3] K.R. Reddy et al., "Oxygen Transport through Aquatic Macrophytes: The Role in Wastewater Treatment", J. Environ. Qual. 19, 261 (1989).

[4] P. Stauffer, Poster: "Artificial Aquatic Ecosystems for Recycling of Nutrient Enriched Water and for the Production of Phyto- and Zooplancton as Food for Fish Stocking", 2nd Int.Conf. on Ecol.Eng. f. Wastewater Treatment (Sept.1995).

[5] K. Scheurer, Poster: „Macrophytes as Food for European Freshwater Fish (Rudd, *Scardinius Erythrophthalsmus* and Tench, *Tinca tinca*), 2nd Int. Conf. on Ecol. Eng. f. Wastewater Treatm. (1995).

[6] A.Binz & C.Heitz, "Schul und Exkursionsflora für die Schweiz", Schwabe & Co.AG, Basel (1990).

Environmental Research Forum Vols. 5-6 (1996) pp. 379-382
© *1996 Transtec Publications, Switzerland*

Nutrient Variability in Six Sewage-Fed Tropical Fish Ponds

B.B. Jana and S. Datta

Fisheries & Limnology Research Unit, Department of Zoology, University of Kalyani,
Kalyani-741 235, West Bengal, India

Keywords: Reclamation, Nutrient Variability, Sewage-Fed Ponds, Carp Culture

Abstract

The effectiveness of pond ecosystem has been assessed in terms of its nutrient removal capacity from the city sewage effluents before being discharged into the river system. A distinct nutrient gradient from high to low values were observed in a series of ponds starting from facultative to the last fish growing pond. The per cent of reduction was proportional to the distance covered from the anaerobic ponds.

INTRODUCTION

Sewage is well known for its vast array of high concentrations of nutrients necessary for biological systems. The so called 'waste' is an important source of fertilizers for aquaculture production through augmentation of natural food organisms. In India, the use of sewage effluents for fish production has gained momentum in recent years because of stringent measures of the Government to enforce the environmental laws to control pollution. As early as 1944, Hora [1] as well as Nair [2] separately investigated the potentialities of sewage-fed ponds for fish culture. In view of demographic scenario, sewage-fed fish farming is an age-old practice in West Bengal, and has advanced further in recent years. Calcutta and its adjoining areas cover an area of 3036 ha commercially viable sewage-fed fisheries and contribute about 4000 tonnes of fish per year. Since the direct use of untreated sewage in pisciculture involves the risks of disease transmission in the food chain, health hazards and possible damage of soil through salinity and alkalinity development, the scientific opinion does not favour the use of untreated sewage for fish culture. The sewage-fed fish culture is an unique system whereby the nutrients and COD values of water are abruptly reduced to safe limit before being discharged into the river canal as per requirements of the environmental laws. In the Kalyani township, a sewage-fed fish farm has been set up with a view to reducing the nutrient load of domestic sewage and to improve the water quality so as to cater the demand of environmental laws. While passing the sewage effluents through a series of wetlands used for fish farming, a distinct chemical gradient is known to develop between the inflow and outflow of the system. The purpose of this investigation was to monitor the nutrient variability of six sewage-fed fish ponds while the sewage effluents were on their way to river canal through a series of ponds used for carp farming.

MATERIALS AND METHODS

Six ponds in a newly constructed sewage-fed fish farm in Kalyani township were selected for the present investigation. The domestic sewage of Kalyani township is collected through underground sewerage system and is treated first by mechanically and then by biological means. Two facultative ponds were larger in area (1.2 ha) than their four fish growing counterparts

(1.0 ha). The ponds are located in two rows in a series in such a manner that the treated water from the facultative ponds passes through their outlets into the fish growing ponds placed in a series. The facultative pond is provided with an island at the centre so that the passing water from the inlet may cover maximum distance before reaching the outlet of the same. Similar to the facultative ponds, the fish growing ponds are provided with an inlet at the top and an outlet at the bottom. Finally, the water from the last pond (pond-6) is discharged through outlet into the river Ganges. The water flow is regulated through a sluice gate installed at the junction of anaerobic and facultative ponds. The time required for complete discharge of sewage water through the series of ponds is about ten days, and the mean depth of 1.5 m is always maintained. The water undergoes a series of chemical changes and mud-water interactions and improve its quality substantially before going into the river canal. A characteristic feature of the sediment of these sewage-fed ponds is extremely thin layer and hard due to rare accumulation of organic matter in the bottom sediment. Samples of water and sediment were collected twice a month from December,1994 until June,1995 for exanination of physico-chemical parameters following the standard method [3,4]. The biological materials were also collected and analysed; the results will be dealt separately.

RESULTS AND DISCUSSION

Water transparency was relatively low in two facultative ponds (18.8-24.0 cm) than the four fish growing ponds (21.6-27.0 cm).There was gradual improvement of transparency while sewage effluents passed from facultative ponds into fish growing ponds. Further improvement of transparency was noticeable from the first fish growing pond to the last one. The water pH (7.72-8.84) was less variable among six ponds investigated. The concentration of dissolved oxygen of water was fairly high in facultative ponds (10.9-18.2 mg/l) showing the dominance of algal bloom in response to nutrient enrichment. However, a distinct gradient from high to low values was observed from first fish pond (pond-3)to the last one (pond-6) on each day of sampling (Fig.1). The values of dissolved oxygen observed in pond-6 (7.4-11.7 mg/l) remained fairly well within the limit for carp culture in tropical conditions. The responses of COD were similar to that of DO of water. The COD values of 296-382 mg/l observed in facultative ponds underwent a reduction of 20-30% while passing from facultative ponds to fish growing ponds. This clearly indicates that the control of secondary sewage effluents by the carp culture in the pond wetlands has a profound influence in reducing the nutrient load before being discharged into the river canal.

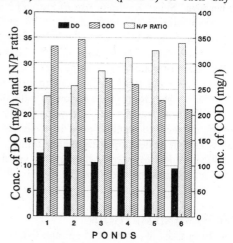

Fig.1: Variabilities of DO, COD and N/P ratio in six ponds investigated

The responses of ammonium, nitrite, nitrate and total nitrogen of water were similar to that of COD (Fig.2), showing the maximal and minimavalues in the facultative pond and the last fish growing pond, respectively. The per cent retduction of ammonium, nitrite and nitrate was 11-20%, 21-31%, and 2-18%, respectively (Fig.3). The amount of soluble reactive phosphate was about 23-39% higher in facultative ponds as compared to fish growing ponds. The concentrations declined as a function of distance from the facultative pond so that the lowest values were encountered in the last fish pond. The N/P ratio, on the other hand, was significantly higher in fish growing ponds than that of facultative ponds.

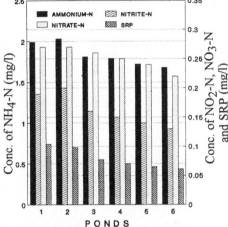

Fig.2: Variations of different nutrient parameters

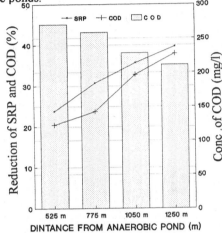

Fig.3: Variations of COD and SRP as functions of distance

The results of this study suggest that pond wetlands used for fish culture was highly effective system to reduce the nutrient load from the sewage effluents. The per cent of reduction was proportional to the distance covered from the anaerobic ponds. Among the chemical parameters, the reduction was maximal for soluble reactive phosphate and minimal for nitrate. The N/P ratio, a measure of biological productivity, showed the trend opposite to that of other chemical parameters. Available information suggests that the primary productivity and fish production were related to the nutrient load of sewage effluents. Since the microbial load specially the pathogenic forms and fish health have not been determined at present to assess the real state of reclamation before being passed into the river canal, at present, the environmental risks assessment in this sewage-fed fish farm was limited to the values of COD and BOD. Further research is in progress to assess the real state of reclamation using pathogenic microbial flora, fish disease and heavy metal contamination as parameters of environmental risk assessment.

REFERENCES

[1] S.L.Hora, Fish Farm. Cal. Beng. Fish. Dept. Cir. **7** (1944).
[2] K.K.Nair, Proc. Nat. Inst. Sci. India **10**, 459-462 (1944).
[3] APHA (American Public Health Association), 17th edn. (1989).
[4] M.L.Jackson, Prentice Hall of India Pvt. Ltd., New Delhi (1967).

Environmental Research Forum Vols. 5-6 (1996) pp. 383-386
© *1996 Transtec Publications, Switzerland*

Proposed Aquatic Treatment System for Polishing Domestic Wastewater of Sangli City

V.R. Joglekar

Shivsadan Renewable Energy Research Institute,
192/3, Industrial Estate, Sangli 416 416, Maharashtra, India

Keywords: Water Hyacinth, Biogas, Organic Manure, Aquatic Treatment System (ATS), Advanced Wastewater Treatment (AWT) System Operated Mechanically

ABSTRACT

Based on the experience gained from an experimental project entitled 'Application of water hyacinths for domestic wastewater treatment, generation of biogas and organic manure' which was erected close to the existing Sangli Municipal Oxidation Ponds, SRERI has now designed a low capital, low energy and minimal O&M input aquatic treatment system to treat 6 MLD wastewater of Sangli city. The system envisages to upgrade the existing 4 ha ponds and to provide biomass harvesting and composting system, sprinkler system for mosquito control and final chlorinating tank at an estimated cost of Rs.7.5 million (U.S. $ 0.25 million). It is estimated that the sale of water hyacinth composted manure alone would fetch Rs.1.4 million per annum covering almost the yearly O & M cost.

INTRODUCTION

It is an important task of any city Municipal administration to treat the city wastewater to bring down the pollution load within the regulatory standards specified for its safe disposal on land or into river. To establish a proper technology to achieve the above objectives, an experimental project was carried out by SRERI under the sponsorship of the Govt. of India [1] which has brought forward the following conclusions :

1. One ha of hyacinth pond can effectively treat 1.25 MLD of wastewater bringing down the pollution parameters within the regulatory limits (Table 1).

Table 1
Effect of Growing Water Hyacinth on Domestic Wastewater

Parameter	Before Treatment	After Treatment	Percent Change
pH	7.5	7.1	-
Turbidity	352.0	5.0	98.6
Dissolved O_2	< 0.1	8.2	-
C O D	280.0	80.0	71.4
B O D	187.5	20.0	89.3
Nitrogen	22.5	2.5	88.9
Phosphorus	5.0	2.5	50.0
MPN 100 mL^{-1}	>2400	>2400	Nil

Except for pH and MPN all the values are in mg L^{-1}.

2. Daily harvested hyacinths from the above pond could either be composted to produce 1.45 MT of quality manure or anaerobically fermented to generate biogas equivalent to 1 million K Cal of heat energy.

These conclusions were the basis of designing a 6 MLD wastewater treatment system for Sangli Municipal Council.

PROPOSED SCHEME

The primary treatment plant preceding the existing Municipal oxidation ponds meant for the removal of floating matter and sludge will be revamped. The ponds having 1.5 M depth would be properly lined and upgraded to allow uniform flow of wastewater throughout its width.

Two sets of harvesters (Fig.1) equipped with primary belt conveyors and chaff cutters mounted on chassis plying on rail tracks will be provided along the two sides of ponds. They would harvest nearly 30 MT of wet biomass every day. A device consisting of a buoy with spikes and wire ropes with winches would maneuver the hyacinths from the ponds onto the primary belt conveyor. At the end of the secondary belt conveyor from where the chaffed biomass would drop down to form a compost stock pile, a mechanized device would be fitted to add nutrients and bacterial cultures to the chaffed biomass to accelerate the composting process. The average daily compost yield would be 6 MT.

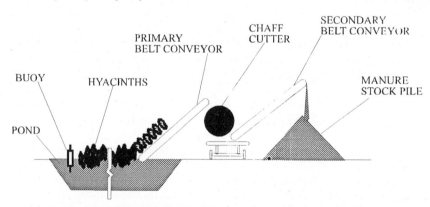

BIOMASS HARVESTING AND COMPOSTING EQUIPMENT

Fig. 1

The entire pond area will be brought under an elaborate sprinkler system for spraying insecticide emulsions, besides two larvivorous fish species - Guppi and Gambusia would be introduced in the ponds to control mosquito breeding.

The treated water would then be chlorinated to eliminate the pathogens before being released either for irrigation purpose or into the river.

Besides construction of the envisaged works, Shivsadan would operate and maintain the project for the initial period of two years and train the Municipal staff in the meantime before handing over the project to the Municipal Council.

COST- BENEFIT COMPARISON FOR ATS AND AWT SYSTEM

1. A typical AWT system treating one MLD of wastewater presently costs Rs 2.25 million whereas the proposed ATS would cost only Rs. 1.25 million pei MLD as the capital cost.

2. The prevailing O & M charges for an AWT system are very high i.e. Rs.0.43 million per MLD per annum out of which electrical energy cost alone amounts to Rs.0.35 million. On the contrary, O & M charges for ATS would be as low as Rs.0.24 million out of which electrical charges would be remarkably low i.e. Rs.0.03 million (1/12 th of the AWT system).

3. In case of ATS , the resource recovery in the form of composted manure sold at Rs.640 per MT will generate an yearly revenue of Rs.1.4 million offsetting the entire amount of yearly O & M expenditure. In case of AWT system , however, there is no recovery of any resource of significant value.

ENVIRONMENTAL IMPACT

In short, unlike high capital and recurring cost of AWT system plagued with frequent power failures and rising cost of electrical energy, ATS requires low capital and low energy inputs and depends merely on nature's unfailing scavenging ability and photosynthesis mechanism. Furthermore, it produces very valuable resources in the form of salable treated water for agricultural use and composted biomass as priceless organic fertilizer. The perennial coverage of lush-green water hyacinths on the ponds tremendously contributes in maintaining the natural balance in the environment.

REFERENCE

1. Joglekar, V.R. and V.G. Sonar. 1987. Application of water hyacinths for treatment of domestic wastewater, generation of biogas and organic manure. p. 747-753. In K.R. Reddy and W.H. Smith (ed.) Aquatic Plants For Water Treatment and Resource Recovery. Magnolia Publishing Inc., Orlando, Florida, U.S.A.

Environmental Research Forum Vols. 5-6 (1996) pp. 387-392
© 1996 Transtec Publications, Switzerland

Nitrogen and Phosphorus Removal in Old Created Biopond

A. Kontautas, K. Matiukas and R. Liepinaitis

University of Klaipeda, Centre for System Analysis, Sportininku 13, LT-5813 Klaipeda, Lithuania

Keywords: Biopond, Restoration, Nitrogen, Phosphorus, Nutrients Removal

ABSTRACT

A small ponds are able to reduce the load of Nitrogen and Phosphorus, especially in a long time perspective. This study shows, that in a small catchment areas such old ponds can be used as a simple way to reduce nutrients transport and as one of measures for restoration of small rivers basins. The removal rate depends on climatic conditions and discharge. In low temperatures and high water removal of N and P was lower.

INTRODUCTION

The eutrophication of internal waterbodies and coastal waters is among the most important environmental problems in Baltic Sea basin. As it was noted in the Baltic Sea Environmental programme the load of nutrients has been accredited as well as to point sources and non-point sources. The total estimated discharges of N_{tot} T/year and P_{tot} T/year is 7780 and 870 respectfully. Total river transport based on calculated monitoring data is 20 000 T/year of N_{tot} and 3300 T/year P_{tot} [1]. That means the importance of reduction N and P from non-point sources of pollution.

Since many of small streams were canalised in Lithuania during this century, the self-cleaning capacities are reduced.

This study was based on the idea that lakes, ponds and wetlands have the ability to reduce the load of nitrogen and phosphorus during runoff especially in long-term perspective. Importance of it was noted by other authors [2,3,5].

THE MAIN GOAL OF THIS STUDY

- to determine the influence of previously created pond for the transport of nutrients in the Smeltaitė river during the year,
- to determine the forms of nitrogen and phosphorus being removed more efficiently,
- to determine the importance of created ponds to the restoration of canalised stream ecosystem.

THE MAIN CHARACTERISTICS OF STUDY AREA

- average of temperature in January is -3 to -5° C, in July +17° C;
- average annual level of precipitation 700-750 mm;
- average area runoff 8-10 $l/s/m^2$ [4];
- the natural stream ~ 3 km;
- the main sources of pollution - non-point agriculture and household wastewater both;

- drainage area (above the pond) - 22.9 km^2 [4];
- the surface area of biopond ~ 5300 m^2;
- the max. depth - 0.9 m, average depth - 0.5 m;
- the depth of the silt 0.2-0.3 m;
- the date of construction 1975;
- the main species of vegetation: Potomogeton lanceolatum, P. crispus, P. natans, Lemna minor, Typha latifolia, Sparganum sp., Carex sp., Elodea canadensis, Phragmites australis.

THE RESULTS

Normally, discharge varies more than nitrogen and phosphorus concentrations. From different studies we know that the losses from the catchment are very high during the rest of the year. (Except the strong increase of discharge during the rainfall when huge amounts of water flow through the canalised stream during very short time).

Annual inflow and outflow concentrations are presented in Fig.1 - Fig.5.

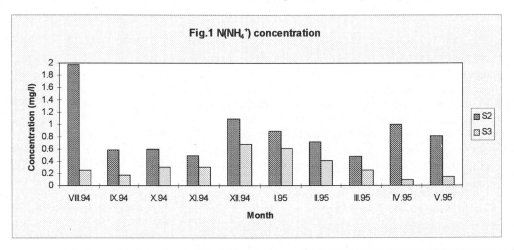

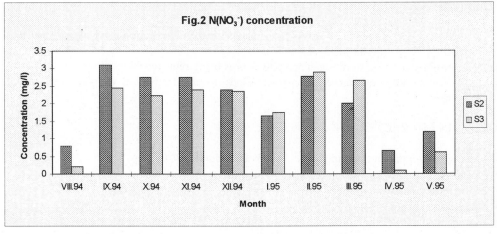

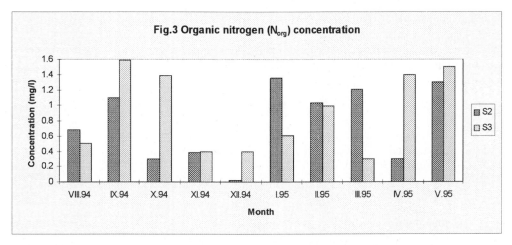

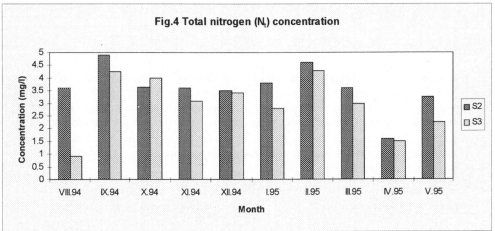

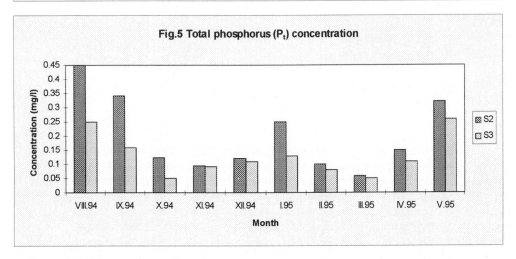

The comparison of concentrations above and below the pond (S_2 and S_3) show that for the $N(NH_4)$ and P_{tot} during time of the studies the pond "catch" some amount of nutrients. But for the $N(NO_3)$ during winter months and for the N_{org} during late autumn and spring months the pond have influence as a "producer" of that form of nutrients.

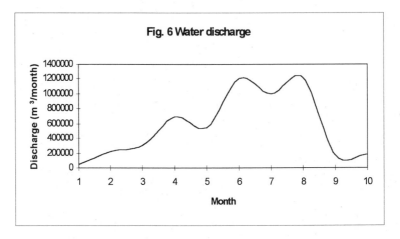

The increasing of $N(NO_3)$ concentrations in the pond after the winter time (January - March) could be compensated by decreasing the concentrations of $N(NH_4)$ and N_{org} at the same time. This is the reason for N_{tot} concentration decrease during all study time. However, in October 94 we found the increase of the N_{tot} concentration after the end of vegetation season in the pond and increase of the water discharge. It is due to the big amount of N_{org} which has been removed from the pond by water flow. That means the importance N_{org} in the total transport of nitrogen during different seasons of year.

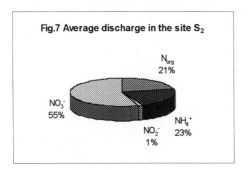
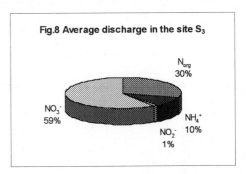

The analyses of discharge (Fig.6 - Fig.8) explain the low possibilities of nutrients removal processes by denitrification and retention processes by uptake of nitrogen into plant material during the time of high discharge (autumn, winter and early spring months). Only 6-10% of N_{tot} and P_{tot} could be removed in the pond by sedimentation and biological processes during period November - March. Such situation was observed in other water bodies [3]. The other reason is that retention time decrease during this period from 10-12 to 2-3 days.

CONCLUSIONS

- the removal $N(NH_4)$ and N_{org} is significant in the total removal of nitrogen in created biopond;
- the lowest rate of the removal of N and P in pond is during the period of low temperatures and high discharge (November - March);
- the creation of ponds in the canalised small catchments could be important measure to reduce the nutrient transport in streams;
- this kind of renaturalisation is also important for the increasing of the biodiversity in agricultural landscape.

ACKNOWLEDGEMENTS

This research was supported financially by CCB (Coalition Clean Baltic) and the authors want to express their great thanks for this.

REFERENCES

1. Baltic sea Environmental Programme. Pre-Feasibility Study of the Lithuanian Coast and the Nemunas River basin. Technical Report. May 1992.
2. Fleisher S., Gustafson A., Joelsson A., Pansar J. and Stibe L. Nitrogen Removal in Created Ponds. Ambio, No.6, 1994.
3. Jansson M., Andersson R., Berggren H. and Leonardson L. Wetland and Lakes as Nitrogen Traps. Ambio, No.6, 1994.
4. Lasinskas M., Macevičius J., Jablonskis J. Lietuvos TSR upių kadastras. Vilnius. Valstybinės politinės ir mokslinės literatūros leidykla. 1959.
5. Weisner S.E.B., Eriksson P.G., Graneli W. and Leonardson L. Influence of Macrophytes on Nitrate Removal in Wetlands. Ambio, No.6 1994.

Environmental Research Forum Vols. 5-6 (1996) pp. 393-396
© 1996 Transtec Publications, Switzerland

Solar Drying of Sewage Sludge – System IST

U. Luboschik

IST ANLAGENBAU GmbH, Ritterweg 1, D-79400 Kandern-Wollbach, Germany

Keywords: Water Sewage Plants, Sludge Drying, Solar Drying, Solar Energy

Abstract: A solar sludge dryer for 450 t with 20% solid parts is (SP) run since Dec.94 in Hammerstein, Germany, near Basel. The annual drying capacity is monitored and found to be about 700 kg water per m². The end product is a granulate with 90% SP. The system requires about 6000 kWh per year electricity for the machine and small ventilators.

General Situation

The sludge from a water treatment plant in Germany is mostly mechanically dehydrated. Solid parts are typically between 20 and 40 %. This sludge is mostly dumped on land filling sites. However the legal situation forces the municipalities that from the year 2005 onwards the residue of the water treatment plants may only deposited on land filling areas if the biological parts are below 5 % of the solid mass. This excludes any deposit of waste water sludge on land filling sites. Two kinds of final treatments will be then possible:

1st   incineration with or without heat recovery

2nd   recycling in agriculture

Why Dry Sewage Sludge?

The mechanical dehydration of sewage sludge with different pressing methods has its limits at about 40 % of solid mass. Most of the processes stop around 30 %. A further dehydration is only possible with thermal processes, at least for the time being.

Solar sludge dryer Hammerstein, 56 x 8 m, 450 t/a wet sludge 300 t water evaporation

Drying must be seen under different aspects:

> 1st   as a reduction of volume and weight
>
> 2nd   handling aspects, especially storage problems when used in agriculture
>
> 3rd   improving the gross calorific value, which is important in the case of incine-
> rating the sludge

Any thermal drying process requires large quantities of energy, about 2.450 kJ/kg (0,68 kWh/m^2), plus the efficiency of the drying process. This energy can be provided by various energy sources:

- fossil fuel
- ambient air e.g. high rate ventilation
- renewable energy, solar radiation, bio mass

Concerning a CO_2 neutral drying - the best solution is obviously the third method - the use of renewable energy.

Solar Drying - System IST

In the frame of a research programme - supported by the German Federal Ministry for Research and Technology - several large scale convection dryers were monitored and analyzed. The first pilot application for sludge drying was built in 1991 in Köndringen/ Emmendingen, South Germany.

Since December 1994 a fully automatic system is run in Kandern-Hammerstein, see picture. (56 m x 8,0 m chimney 7,80 m 2,0 m diameter) The system is commercially available.

The solar sludge dryer is constructed with the following elements:

- water resistant ground, built like a conventional road
- greenhouse construction, with high transparent plastic foil
- chimney, light constructed
- turning and transporting machine with programmable control unit
- charging and decharging equipment

The sludge from the sewage pit is mechanically dehydrated by a press and then spread out at the end of the dryer. The machine turns and transports the sludge to the other end of the dryer. The aeration of the sludge starts an additional composting effect, which neutralizes very quickly the smell of fresh sludge and which has a self-heating effect accentuating the drying process. The radiation of the sun falls on the sludge and heats the surface up, just like any absorber in a solar collector. The warm surface heats up the air, which is always cooler than the sludge surface. The driving force for the drying process is the pressure difference between the water vapour in the air and in the sludge. A certain drying potential exists whenever this difference is positive, also during dry nights and in winter. The air is only used to evacuate the moistness rising out of the sludge. In winter it happens frequently that fog is formed due to the above mentioned compost effect. The evacuation of the air is augmented with ventilators having a total electrical power of 1,5 - 2 kW. They are switched on/off according to the internal enthalpie conditions.

Capacity of the IST Solar Dryer

The monitoring of our system shows, that we have an annual efficiency of approx. 30 % calculated against the total radiation. This enables us to evaporate in middle Europe 500 - 700 kg water per m^2 dryer surface.

The size of the dryer is based on this drying capacity and the final % of the solid parts. A good rule is 1 m^2 per annual ton of sludge. The result shows that the drying is relatively constant up to 70 % solid parts and after that requires more time. Values over 90 % are possible and easily achieved in summer. During winter we store the sludge over 4 - 5 months. If turned regularely and well aerated, the sludge does not turn moldy even kept for months.

Due to the turning process, the sludge is formed into pellets of different sizes. This facilitates the handling of the dried product. In principal it would be possible to mix these pellets with wood chips or other materials to obtain a product, which may be used for recultivating processes.

Cost of the Solar Drying Process

In Germany we achieve a specific drying cost per m^3 water of about 150,-- DM/t. The basis of this calculation is 10 years write-off period without cost of land. So whenever the disposing cost are higher than 150,-- DM/t, the solar dryer is cost effective.

Environmental Research Forum Vols. 5-6 (1996) pp. 397-400
© *1996 Transtec Publications, Switzerland*

Overview of Constructed Wetland Applications and Research in Australia

C. Mitchell

Department of Chemical Engineering, The University of Queensland, Brisbane 4072, Australia

Keywords: Constructed Wetlands, Artificial Wetlands, Municipal Wastewater Treatment, Domestic Wastewater Treatment, Stormwater Runoff, Industrial Waste Treatment

Abstract

The technology of constructed or artificial wetlands has long been recognised as one which holds great promise. It has the benefits of lower construction and operation costs, lower energy costs and lower maintenance than many other mechanical wastewater treatment options (particularly in tropical and sub-tropical climates). In addition, constructed wetlands offer the not-so-tangible benefits associated with systems bases on natural principles, relying on naturally mediated mechanisms, and which rely on and thrive from a diverse ecology. In Australia, artificial wetlands are now being used to treat:

- domestic waste water of varying influent qualities to varying degrees, including nutrient removal;
- stormwater runoff from rural and agricultural land;
- industrial waste water such as those associated with the timber products industries, food industries, petroleum refineries, and mining activities.

However, in Australia, as in the rest of the world, the development of wetlands is being constrained by a lack of understanding of the fundamental mechanisms involved in wetland pollutant removal processes, which inevitably leads to conservative design processes, and ultimately, over-design. This paper highlights some Australian examples of constructed wetlands technology and research.

Facts and Figures

Australia is the world's largest island, and with an area of 7 682 300 km², the sixth largest country. Geographically, it spans 113 - 154°W, 10 - 43°S. As such, it has an extraordinarily diverse range of climates and ecosystems, from desert to alpine to tropical rainforest, and is consequently recognised as one of the few megadiverse countries. After Antarctica, Australia is also the world's driest continent. Its population of roughly 18 million live mostly where it rains: on the east coast of the mainland. The remainder of the country is sparsely populated. These facts and figures are important when considering constructed wetlands, since climate and rainfall do have a significant impact on the design and performance of wetlands.

When political surveys have been conducted in Australia recently, environmental issues have consistently been a major concern for the public. The environmental benefits of constructed wetlands for wastewater treatment fit very well with such heightened concern.

Municipal Waste Water Treatment

The majority of constructed wetlands in Australia involve municipal wastewater treatment.

Early work at the University of Western Sydney focussed on secondary effluent polishing in a series of 100m long open water, gravel, and mixed trenches, both planted and unplanted. This research provided a strong basis for more detailed investigations now being carried out under the auspices of the Cooperative Research Centre for Waste Management and Pollution Control, including elucidation of the mechanisms associated with phosphorus and nitrogen removal and storage; disinfection capability; potential of vertical flow multi media beds; and development of SWAMP, a decision support system for wetland evaluation and design [6].

In 1990, large sewage treatment wetland trial began at Byron Bay, on the north coast of New South Wales. This system includes a variety of wetlands, ranging from upflow gravel beds, through horizontal flow gravel beds, to open water trenches, and rejuvenated natural wetlands. The wetlands are fed with effluent from an intermittent aerated reactor which is chemically dosed for phosphorus removal. Byron Bay is a holiday town whose population is 4000 during most of the year but swells to 15000 during school holidays. The broad objectives of the research at Byron Bay are to develop parameters for wetland design and appropriate operation and management strategies. To date, Byron Bay wetlands have successfully mitigated the impacts of step changes in load to the sewage treatment plant, and greatly improved the final outlet quality [1].

The University of Queensland has just completed construction of a set of pilot scale gravel bed wetlands in conjunction with the Brisbane City Council. These beds have been designed to facilitate extensive sampling for hydraulic and water quality parameters [5]. This project presumes that the current design methods for constructed wetlands are based on limited understanding of pollutant removal mechanisms, and therefore inappropriate assumptions. The aim is to develop hydraulic models with additional layers of complexity incorporating pollutant removal mechanisms. We believe that one of the major opportunities for wetlands in Australia and Asia is in improving primary effluent, so this system treats primary settled sewage. This project will develop a rational basis for selection of gravel size and shape, for plants, and for design and operation.

In Queensland, in contrast to other Australian states, much of the work involving constructed wetlands and municipal wastewater treatment has been undertaken as part of an ongoing program coordinated and administered by the Queensland Department of Primary Industries. Under this program, 10 separate systems have been designed and constructed in a wide range of climatic conditions, ranging from the tropical monsoonal regions, through subtropical coastal areas, to semi-arid inland areas. Nine of these systems are open surface wetlands receiving secondary effluent, with the objectives of nutrient polishing and disinfection. Overall, these wetlands are providing valuable performance data and experience and are performing very well. The tenth is a small scale gravel bed system treating combined black water and grey water from a single household. A new septic tank handles primary treatment, then two sets of gravel beds are loaded alternately on a one-week-on/one-week-off roster. An open water wetland completes the treatment prior to disposal through a soakage pit. The system is now consistently producing better than secondary effluent.

Researchers from Griffith University are investigating the potential of native *Melaleuca* species in natural and constructed wetlands under various operating regimes, including polishing biological nutrient removal effluent, nutrient uptake, various wetting and drying cycles, in stream stormwater management, and potential for tea tree oil production [2].

In northern New South Wales, there is strong community support for holistic environmental solutions. As a result, there is considerable interest in individual household water treatment and reuse. In these situations, composting toilets are often used, and grey water alone is treated in some form of wetland or reed bed. Plant growth in these beds is often far from luxurious. We suspect nutrient (probably nitrogen) limitation may be occurring. Research at Southern Cross University is showing very promising results for combining the liquid effluent from composting toilets with grey water for treatment in very small scale constructed wetlands (Marshall pers. comm.)

Stormwater Management

Much of the work involving stormwater and wetlands has been done in drier areas of Australia. In Canberra, the approach of treating stormwater as a valuable resource rather than a waste, and incorporating a range of community values, including access corridors, recreation and aesthetics, has led to a system of lakes and green space which attracts some 3 000 000 visitors each year - quite an achievement in a city of about 300 000 people and an annual rainfall of about 500mm [3].

In Adelaide in the early 1970's, far-sighted landscape architects designed The Paddocks for community recreation and stormwater management. The wetlands here incorporate meandering channels, grass swales and open ponds. Their pollutant removal performance is impressive - more than 80% of suspended solids, 30% of phosphorus, 50% of total nitrogen [7].

An excellent example of a stormwater wetland is the Plumpton Park system in Sydney's western suburbs. In 1994, a 180m x 25m linear wetland was built into a creek line upstream of a major flood mitigation pond (100 year Average Return Interval event), incorporating bird refuge islands and other aesthetic features. Preliminary results are very encouraging, indicating more than 40% removal of suspended solids and almost 30% of phosphorus [4].

In 1994, the largest constructed wetland to date in Australia was installed at Rouse Hill in Sydney's west. This is a series of in stream open surface wetlands covering some 3.6 ha. Rouse Hill is the first major development in Australia to adopt dual reticulation for water reuse in all toilet flushing and outdoor uses. The wetlands receive effluent in excess of reuse requirement from a biological nutrient removal plant with additional tertiary treatment. In addition, because the wetlands were constructed in a degraded stream bed, they play a significant stormwater management role in terms of both pollutant control and flood mitigation.

Industrial Waste Treatment

Constructed wetlands are increasingly being used for industrial waste water treatment in Australia. Examples include a waste water treatment and reuse scheme at a medium density fibreboard plant, treating sugar mill effluent, oil terminal wastewater, dairy shed waste, piggery waste, various metalliferous mine wastes. Since the climatic conditions at most Australian coal mines are quite different to those in North America, the application of constructed wetlands for acid mine drainage treatment in Australia has great potential but plenty of challenges.

References

[1] H J Bavor, E F Andel, T Schulz, and M Waters (1992) 'Byron Bay Constructed Wetlands Research Program, Progress Review Report'
[2] K Bolton and M Greenway (1995) 'An investigation of *Melaleuca* as candidates for constructed wetlands' Wetlands for Water Quality Control, Townsville, September
[3] P Cullen (1993) 'Canberra - How drainage problems become community assets' 15th Federal AWWA Convention, Gold Coast, pp79-83
[4] G Hunter (1995) 'Plumpton Park Wetland - Eco-Engineering in Action' Blacktown City Council
[5] A C King and C A Mitchell (1995) 'Subsurface Flow Artificial Wetlands: Design and construction oat pilot scale to enable rigorous performance analysis and modelling' Wetlands for Water Quality Control, Townsville, September
[6] P Maslen and T Schulz (1995) "The CRC advanced constructed wetland project: an Australian Cooperative Research Program' 16th Federal AWWA Convention, Sydney, pp 421-428
[7] G W Tomlinson, A G Fisher and R D S Clark (1993) 'The Paddocks: Stormwater - A Community Resource' EWS Report 93/11

Environmental Research Forum Vols. 5-6 (1996) pp. 401-404

Nutrient Considerations for Commercial Greenhouse Plants in Hydroponic Cultivation with Sewage Water

H. Morin

Östergötlands Natural Resource Use Programme,
Utställningsvägen 28, S-602 36 Norrköping, Sweden

Keywords: Greenhouse, Crops, Nutrients, Sewage, Tomatoes

Abstract

Different strategies can be used to make cultivation of commercial greenhouse crops in sewage water possible. From a greenhouse-growers point of view, levels of accessible nutrients are often low and it is necessary to optimise the nutrient uptake conditions. In this matter control of climate conditions are of the same importance as control of pH- and oxygen in the solution.
Except for low nutrient levels, the relations between NO_3-N and NH_4-N and antagonism between different ions will complicate growers work.

Requirements

When designing systems for ecological treatment of sewage water, utilisation of valuable nutrients from the water is a main point. The nutrients can be used to produce biomass for direct consumption or in processed form as compost, biogas, etc. Thereby there is a possibility to make a profit out of the sewage treatment plant

Now, there are wastewater treatment plants built with greenhouse technology for year-round operation in temperate climates, therefor it is interesting to see if there are any possibilities to produce conventional greenhouse crops of good quality with sewage water as growing media and main nutrient resource.
To be successful in this field there are certain conditions to be considered.

1) From beginning it is clear that the plant nutrient considerations is the most difficult thing to deal with. The grower must have knowledge of nutrient demands for each type of plant.

2) Climatic factors as temperature, light, RH, CO_2, and air movements, has to be controlled. Practically, a good way is to divert sewage water from the treatment plant in to a separate greenhouse for commercial crops. The drainage from the crops can be returned to the treatment system or used in for other purposes.

3) It is difficult to reach top yields in this kind of production. The investment in greenhouses must stay at a reasonable level. Use reliable, conventional greenhouse equipment.

4) Biological control of pests is used with few exceptions. If nutrient solution is returned in to the treatment system pesticides may cause damage to micro-organisms etc.

5) Are there any possibilities to get a better price for the product compared with the market for conventional products?

6) Products for human consumption must be good and healthy. High quality is always desirable.

The nutrient resource

When the sewage water is evaluated as growing media and nutrient resource it is necessary to look at the conditions for the plant to take up and assimilate the nutrients.

The growth is limited by the particular plant nutrient witch appears with the largest deficiency. It can never be compensated whit increased levels of other nutrients.

Except for being toxic, high concentration of one nutrient can cause problems for the plant to get the right uptake of another. This antagonism is known to occur in many ways, e.g. NH_4 prevent the plants uptake of K and Mg. High concentrations of specific nutrients is also known to disturb the transport of others, e.g. Mg in high concentrations reduce the production of auxin and the transport of Ca is reduced.

The plant sometimes compensate the lack of a nutrient by uptake of "wrong" nutrient. When the levels of S is too low, the plant will take up Cl instead, and then the uptake of NO_3 is disturbed.

The pH-level is of great importance. Many greenhouse crops work best when pH-level is between 5.5 and 6.5. High pH-levels cause micro-nutrient deficiency.

The content of oxygen in the water is of great importance for root function. Lots of organic materials and slow movements in the water will cause lack of oxygen

Except for these characteristics of the nutrient solution, the nutrient uptake can depend on climatic factors. Ca is taken up in a passive way, depending on the transpiration. Because of that the grower always try to avoid RH over 90%.

Tomatoes as an example

In 1993 a study was made at the Stensunds Wastewater Aquaculture, Trosa, Sweden. The idea was to look at the sewage water with a growers eyes.

Tomato was chosen as reference crop, depending on the fact that tomatoes already was grown in the hydroponics at Stensund. The tomato is of interest in many ways. The crop is one of the biggest in Sweden and is produced in different ways; hydroponics, active and inactive(most common in Sweden) media is used. The production starts at the beginning of the year and runs to October. The tomato plant is grown with a kind of "lay-down" system in witch the plant will reach a length of approximately 10 meters at the end of the season.

In conventional production the tomato plant is supplied by nutrients in high levels to make sure that the roots always can reach the different ions. The drainage from the greenhouse is recycled, used for other purposes or just let out.

At Stensund the plants is grown in hydroponics, integrated with the biomass in the wastewater treatment system. They are grown year-round, not as a optimised crop, but fruits are harvested continuously. The concentration of different nutrients depends on the levels in the income sewage water and by the biological processes in the different treatment steps.

Visual inspection of the plants indicates nutrient deficiency. Samples was taken from the water in the hydroponics. The samples were analysed as a common water test for conventional greenhouse growing. The results was evaluated according to recommended nutrient levels for conventional tomato growing in inactive media

1)A general low nitrogen status with to much NH_4-N compared with NO_3-N. NH_4-N causes complications when it competing with K, Mg and Ca.

2)Low levels was measured for several nutrients and antagonism problems seems to appear.

3)pH-level at 7.5 indicates Fe uptake problems.

4)Na and Cl occurred in high levels and there is always a risk for toxic effects. The tomato plant is particularly sensible for Na(low K-levels accelerate Na uptake). Tomato is more tolerant with Cl but Cl disturb the uptake of NO_3-N.

Discussion

As we see there are lots of complications to deal with. How can the grower compensate the low levels of some nutrients and limit the consequences of high levels of others?

Of course there is possible to add missing nutrients to the water, organics or fertilizers. A quite simple way to manage the problem, the difficulties are more of principle art. Adding nutrients is maybe in opposition to ideas witch are fundamental for the managing of the ecological wastewater treatment. Comparing with conventional greenhouse production, the consumption of fertilizers will be very low.

Another way is to try different ways of producing greenhouse crops. There are lots of techniques; hydroponics, active growing media(with neutralisation effect on variations in the nutrient solution, inert growing media. thin-film techniques, aeroponics(root-spray nutrition, gives roots maximum of oxygen).

The crop can be produced in compost soils and the sewage water is used in a diluted form.

Look for the greenhouse crop that suites the water best. Ornamentals, vegetables, pot plants, cut flower and much more.

Environmental Research Forum Vols. 5-6 (1996) pp. 405-408

Prospects and Challenges of Wastewater Recycling in Kampala - Uganda

M. Nalubega[1], F. Kansiime[2] and H. van Bruggen[3]

[1] Makerere University, Civil Engineering Department, P.O. Box, 7062 Kampala, Uganda

[2] Makerere University, Institute of Environment and Natural Resources,
P.O. Box, 7062 Kampala, Uganda

[3] IHE., Delft, P.O. Box 3015, NL-2601 DA Delft, The Netherlands

Keywords: *Cyperus papyrus*, Wastewater, Nutrients, Wetlands

Abstract
Although direct wastewater recycling in Uganda is under-exploited, indirect reuse is common in urban areas. In Kampala, the capital city, wastewater is reused for subsistence agriculture and wetland macrophyte enrichment. This city discharges 60,000 to 400,000 m^3/d of wastewater into the Inner Murchison Bay of L.Victoria, via a predominantly *Cyperus papyrus* Nakivubo swamp. About 4 km south-west in the same bay is the city's water supply intake. *Cyperus papyrus*, has a growth cycle of 4 to 8 months and when harvested, removes nutrients from the system (about 8 mg P/m^2.yr and 103 mg N/m^2.yr). Wastewater channelization within the swamp has led to lower pollutant removals (20-60% P, 20-55% N and 55% faecal coliform). Increased canalization along major wastewater streams for crop cultivation improves distribution and nutrient recycling. However, possible accumulation of deleterious levels of organic pollutants and/or heavy metals and occupational health hazards to the farmers are unknown.

Wastewater reuse in Uganda
Uganda's sewered towns (eg., Kampala, Entebbe and Jinja) discharge semi-treated effluent into rivers and drains leading into bigger water bodies like L.Victoria and R.Nile. Wastewater resource utilization is very limited.

Forms of wastewater reuse include:
- Wastewater diversions into agricultural plots with crops like sugarcane, yams, sweet potatoes, etc., that are mostly sold for domestic consumption.
- Sludge from conventional wastewater treatment plants (eg., in Kampala) is bought at a small fee (about US$ 5 per pick-up load) for soil conditioning.
- Fish, caught from bays that receive wastewater is also sold.
- Biogas production is localised and from wastes other than human excreta.
- Water supply is sometimes subject to consequential wastewater recycling (E.g., for Kampala, due to its location 4 km downstream in the same bay as the sewer "outfall". Dilution and the lake's self-purification improve raw water quality.

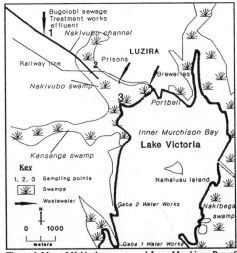

Figure 1 Map of Nakivubo swamp and Inner Murchison Bay of Lake Victoria, showing field sampling sites 1 to 3. These are respectively: Upper & lower swamp inlets and swamp- lake interface.

The under-utilization of wastewater resources in Uganda may be attributed to:
- Negative social attitudes (due to origin of water).
- Relative abundance of resources (water, food, etc.).
- Inadequate information on recycling technologies.

With changing quality of surface water bodies [4, 6], increased coverage of lakes and rivers (eg., L.Victoria and R.Nile) by the noxious water hyacinth (*Eichhornia crassipes*) and increasing number of fish processing and exporting firms, it imperative to identify and practice means to effectively improve wastewater treatment and resources recycling. Aquaculture, based on the abundant wastewater nutrient-recycling and aqua-forestry to improve nutrient removal from wastewater will be important options.

Kampala's wastewater purification by Nakivubo swamp

Kampala, Uganda's capital city is located on the northern shores of L.Victoria. The city discharges 60,000 to 400,000 m³/d of combined wastewater southward into the Inner Murchison Bay of L.Victoria, via the predominantly papyrus Nakivubo swamp. Wastewater flows into the lower swamp via the upper swamp (see Fig. 1). The city's water abstraction point is located about 4 km away in the same bay, at Gaba. The pollutant levels in the water entering, within and leaving Nakivubo swamp measured between 1994 and 1995 are shown in Table 1.

There is significant pollutant removal in the upper swamp. In the lower swamp however, contrary to common speculations that the swamp effectively depletes nutrients in waste-water filtrating through [3, 8], on-going research shows that channelization has resulted into lower treatment efficiency. This could be one cause for the reported changes in the lake [4, 6, 7].

Nutrient removal due to growth of *cyperus papyrus*

Cyperus papyrus is a tropical sedge with high biomass productivity rates. It has a standing crop of 48-143 t/ha.yr and P and N uptake rates of about 7.8 g/m².yr and 103 g/m².yr respectively [1, 3] . However, growth and nutrient storage values are dependant on the nutrient availability. In the ongoing study, papyrus' growth cycle was seen to be 4-8 months with a lower skew.

TABLE 1. WASTEWATER PURIFICATION IN NAKIVUBO SWAMP

Location	TRP mg P/L	TP mg P/L	NH$_4$-N mg N/L	BOD mg/L	FC no/100mL	EC µS/cm	DO mg/L
Upper swamp inlet	0-15	2-22-30	2-25	120-180	2-7.2E6	410-650	0-1.5
Lower swamp inlet	0.5-2.3	0.6-2.4	9-18	42-80	2.4-9.5E5	351-570	0-2.0
Lake-swamp interface	0.41-1.8	0.45-2.0	6-11	28-40	0.3-8.2E5	181-435	0-0.1
'Unpolluted' bay	0.01	0-0.11	0.51-4.0	5-20	0-20	80-100	6.0-7.2

Effective wastewater purification by papyrus is however only possible if nutrients are removed through harvesting. The harvesting cycle in Nakivubo swamp was observed to be about 6 months, intensifying in the drier periods of December to February and July to September. Papyrus is harvested for fuel, building material and for crafts (Figs. 2 and 3).

Agricultural wastewater reuse in Nakivubo swamp

Nakivubo swamp is increasingly under cultivation. This directly provides the livelihood of at least 40 low-income families of mostly urban migrants, and supports some wildlife especially monkeys. The crops grown include yams (most abundant), sugarcane,and rice. Nutrient-rich wastewater is directed into the plots.The effect of possible accumulation of deleterious levels of organic pollutants and/or heavy metals and occupational health hazards to the farmers are unknown. Many of the consumers of the crops are unaware of their source and would possibly reject it if they were aware.

Future of wastewater recycling in Kampala

Changing water quality in Lake Victoria as well as increasing industrialization and population growth necessitates:
- reduction of nutrient and pollutant loads into the lake

Fig. 2 Papyrus harvesting in Ugandan swamps , used for fuel, roofing an crafts

- increase in fish stocks
- increased food production
- cheaper and environmentally friendly energy sources

Managed wastewater resource recycling will ensure achievement of this. To improve nutrient removal from wastewater, there are proposals to introduce aqua forestry for tertiary treatment. Some challenges however include:
- Identification and control of occupational health hazards to farmers
- Research into ways for better farming methods including optimum harvesting for maximum nutrient removal
- Selection of alternative crops for optimum nutrient removal and economic benefits
- Identification of health effects due of wastewater reuse products consumption.
- Overcoming the social attitudes of many people leading to objection.

Fig.3 Papyrus stems used for thatching

References
[1]F.M.M. Chale. Plant biomass and nutrient levels of a tropical macrophyte (*Cyperus papyrus*) receiving domestic wastewater. *Hydrobiol. Bull. 21:* 167-170 (1987).

[2]P. Denny Wetlands of Africa: Introduction. *In:* Wihigham et al.(eds.).*Wetlands of the World, I.* pp 1- 46 (1993).

[3]J.J. Gaudet. Uptake, accumulation and loss of nutrients in tropical swamps. *Ecology 50:* 415 - 422 (1977).

[4]R.E. Hecky. The Eutrophication of Lake Victoria. *Int. Ver. Theor. Angew.Limnol Verh. 25:* 39-48 (1993).

[5]C. Howard-Williams and J.J Gaudet. The structure and functioning of African swamps. *In:* Patrick Denny, (ed.). *Ecology and management of African wetland vegetation,* Dr. Junk Publishers, Dodrecht. (1985).

[6]R. Mugidde. The Increase in Phytoplankton productivity and Biomass in Lake Victoria (Uganda). *Ver. Intern. Verein. Limnol. 25:* 846-849 (1993).

[7]P.B.O Ochumba and D.I. Kibaara. Observations on blue green algal blooms in the open waters of lake Victoria, Kenya. *Afr. J. Ecol. 27:* 23-34 (1989).

[8]A.R.D. Taylor. Status of Nakivubo and Luzira swamps in Kampala. Report prepared for the Ministry of Environment Protection, Natural Wetlands conservation and management programme, Kampala- Uganda (1991).

Environmental Research Forum Vols. 5-6 (1996) pp. 409-412
© 1996 Transtec Publications, Switzerland

Occurrence and Persistence of Faecal Microorganisms in Human Urine from Urine-Separating Toilets

A. Olsson[1], T.A. Stenström[1]* and H. Jönsson[2]

[1] Swedish Institute for Infectious Disease Control, S-105 21 Stockholm, Sweden

[2] The Swedish University of Agricultural Sciences, Department of Agricultural Engineering, Box 7033, S-750 07 Uppsala, Sweden

* Corresponding Author

Keywords: Faecal Indicators, Human Urine, Urine-Separating Constructions, Persistence, Die-Off, Microorganisms, *Salmonella typhimurium*, Bacteriophage

Abstract

Urine from seven urine-separating constructions was collected in October -94 and May -95 and the content of faecal indicators quantified. No coliphages could be detected. The content of *Escherichia coli* was low in all the samples, while faecal streptococci was present in 85% of the samples in amounts between 7 - 13000 cfu/ml. The highest amounts were obtained from the bottom portion of the storage tanks. The pH in the stored urine ranged from 7.7 - 9.2. A trend towards higher counts of faecal streptococci with lower pH values existed.

The die-off of *E coli*, faecal streptococci, *Salmonella typhimurium* SH 4809, *Campylobacter jejuni* S201/94 and bacteriophage *S. typhimurium* 28B was tested by adding these strains to urine in amounts of about 10^6 organisms (cfu) per ml. The samples were incubated at 4 and at 20 °C. The die-off of *E coli, S typhimurium* and *C jejuni* in the urine was rapid. *C. jejuni* could not be recovered at all. After 2 days *S typhimurium* could not be detected, neither in the urine incubated at 4 °C nor at 20 °C. *E coli* could not be detected after 3 days in urine stored at 20 °C and after 7 days in 4 °C. The persistence of faecal streptococci was considerably better. After 35 days the amounts in urine incubated at 20 °C were reduced to less than 1 cfu/ml. At 4 °C the logarithmic reduction time (T90-value) was 14-17 days due to the prevailing conditions. A subfraction may show extended persistence in the urine, as shown for urine stored in barrels at one of the test units. The T90-value here was 65 days for low persisting amounts. No significant reduction was obtained for bacteriophage *S. typhimurium* 28 B in urine incubated at 4 or at 20 °C during 49 days.

Background.

Human urine is the organic waste fraction from densely populated areas containing the largest amount of plant nutrients. Of the plant nutrients that agriculture delivers to consumers in urban areas 60-70% is later refound in toilet waste, just urine accounts for 40-60%. Thus, by source separation and recirculation of the urine, a high degree of recirculation of plant nutrients can be achieved. Urine-separating constructions, with their specially designed toilet bowls, started to appear in Sweden during the 1980's. These toilets are a prerequisite for urine separating solutions of wastewater recirculation.

When recircualting the human urine for agricultural production the risk of disease transmission has to be taken into account. The design of the toilet bowl is not optimal. A small amount of faecal material may be introduced into the urine fraction. In addition the dilution is minimal compared to traditional wastewater collection and treatment. Therefore the presence and survival characteristics of faecal microorganisms in the stored urine are essential to validate for the future use of human urine as fertiliser and in relation to regulations related to storage as well as use in agricultural production.

Hygienic quality of stored human urine and die-off of added organisms.

The investigations were performed on field samples and as laboratory investigations during the autumn -94 and spring -95. Samples were collected from 7 different human urine storage tanks, connected to urine-separating toilet bowls. The content of faecal indicator organisms in the stored urine from the bottom and the surface of the tanks, were analysed for *E coli*, faecal streptococci and coliphages. No coliphages were detected in the samples. Generally the content of *E coli* in the stored urine was below the detection limit (1cfu/ml). Positive samples with amounts ≥10 cfu/ml were only found in some bottom samples (20% of the total samples) and always coincided with elevated amounts of faecal streptococci (2000-5000 cfu/ml). Of the samples, 85% were positive for faecal streptococci and 35% had counts above 1000 cfu/ml. The arithmetical mean pH of the samples with high numbers of faecal streptococci was lower (pH 8.4) compared to the samples with low numbers (pH 8.9), and two of the lowest conductivity values coincided with elevated faecal streptococcal counts. The numbers of faecal streptococci were always higher in the bottom portion from the urine storage tanks than in the corresponding surface portion samples (Fig 1).

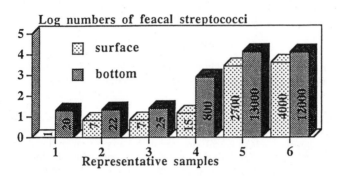

Fig 1. Amounts of faecal streptococci in the bottom and surface portion of urine storage tanks

Supplementary laboratory investigations on the persistence and die-off of the indicator bacteria *E coli* and faecal streptococci as well as with *Salmonella typhimurium* SH 4809, *Campylobacter jejuni* S201/94 1Band the *Salmonella typhimurium* bacteriophage 28B were done. The respective organism was added separately to a compiled sample of urine from the 7 different urine storage tanks tested, in an amount of 10^6 organisms per ml. Two batches of samples were incubated at 4 °C and 20 °C respectively. The strains of *E coli* and faecal streptococci added to the urine were prior isolated from urine storage tanks. *Salmonella typhimurium* SH4809 and the bacteriophage *S. typhimurium* 28B were from the strain collection at the Swedish Institute for Infectious Disease Control. *Campylobacter jejuni* was isolated in connection with a waterborne outbreak in Sweden.

The survival times of *C.jejuni* , *E coli*, and *S. typhimurium* and respectively added to the urine were very short. *C. jejuni* added to the urine could not be recovered at all. No *E coli* could be detected after 3 days in the urine incubated at 20 °C and after 7 days at 4 °C. After 2 days no *S. typhimurium* could be detected neither in the urine incubated at 4 °C nor at 20 °C. No significant reduction was noticed of bacteriophage *S. typhimurium* 28B in urine incubated at 4 °C or 20 °C (Fig 2a).

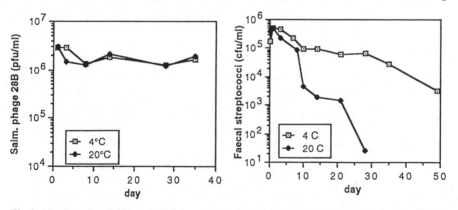

Fig 2. Persistence of phage 28B (a) and faecal streptococci (b) in stored urine at 4 °C and 20 °C.

The survival of feacal streptococci was considerably better, than for the other tested bacteria (Fig 2b). After 35 days at 20 °C the selected strain was reduced to less than 1 cfu/ml. In 4 °C the amount was reduced from initially 3 x 10^6 cfu/ml to 3 x 10^3 cfu/ml at day 49. This represent a T90-value (one

log_{10}-reduction) of 16.3 days. When faecal material was added to the compiled urine sample, the die-off of faecal streptococci gave similair values. T_{90} was 14 days. Samples were also taken from separatly stored urine, after the urine tank had been emptied. The urine was stored outdoors in sealed plastic barrels. The results from this storage investigation is summarised in Tab 1. The die-off is slower than in the laboratory experiments. The T_{90}-value during the first 3 months of storage is around 65 days assuming the same initial contamination level. The lower die-off rate here is anticipated, since this experiment represent a small fraction of the initial population, adapted to the prevailing conditions.

Table 1. Survival of faecal indicators in separatly stored urine at Stensund.

storage time (months)	E coli (cfu/ml)	Faecal streptococci (cfu/ml)	pH
1	<1	234	8.7
2	<1	90	8.8
3	<1	30	8.7
12	<1	<1	8.7

Further extended investigations are in progress at the Swedish Institute for Infectious Disease Control in collaboration with Swedish University of Agricultural Sciences. In these investigations the persistence of 10 selected microorganisms, representing different microbial groups, is investigated in relation to different pH-values, dilution rates and temperatures. Preliminary results support the earlier hypothesis that dilution enhance the survival time, as well as pH-values closer to neutral.

Conclusions
The aim of the urine-separating toilets, to get urine and faecal material as two totally independent fractions, is at present not ultimatly fulfiled due to the design of the toilet bowl. Since the dilution in the system is low, a small fraction of faeces introduced in the urine may constitute a risk for secondary transmission of potential human and zoonotic pathogens, when used as fertilizers for agricultural production.
In the present investigation, no growth of pathogenic bacteria has been noticed in the urine storage tanks. This means that the prevailing conditions in relation to composition , dilution, temperature, pH and the sediment fraction will govern the die-off kinetics of different groups of potentially pathogenic microorganisms. Some Gramnegative bacteria including *E coli* , *Salmonella typhimurium* and *Campylobacter jejuni* seem to die-off rapidly in the stored urine. *E coli* does not fulfil the criteria of an indicator of the sanitary effect, since the persistence of model viruses and faecal streptococci is extensively longer. To judge the sanitary quality, samples should include the bottom sediment phase. This often contain higher numbers than the top liqued phase. Feacal streptococci seem to be appropriate to use as an indicator of the sanitary effect, however, viruses may be more resistant as well as parasitic cysts and oocysts (not yet investigated). Based on the faecal streptococci a storage time between 3 to 6 months seem to be appropriate for sanitation even at low temperatures. Initial one log_{10} reduction (T_{90}-value) is 14-17 days at a temperature of 4 $^{\circ}$C. For a resistant subfraction the T_{90}-value was around 65 days in urine stored outdoors. Factors that may enhance a rapid die-off are undiluted urine and elevated pH-values. Optimization of these factors and related storage times need, however, to be judged in relation to the potential loss of nitrogen from the urine solution, as well as the choise of crops, and ways of application to agricultural lands.

References
Kirchman, H., Jönsson, H., Stenström, T.A., Olsson, A & Pettersson, S. Separated human urine - plant nutrients and hygiene. (In Swedish). Fakta Mark/Växt No 1. Swedish University of Agricultural Sciences, Uppsala, 1995.

Olsson, A., Human urine - a resource or a risk? (Report in Swedish; English abstract). Swedish University of Agricultural Sciences, Department of Agricultural Engineering and Swedish Institute for Infectious Disease Control (in press) 1995.

Environmental Research Forum Vols. 5-6 (1996) pp. 413-416
© *1996 Transtec Publications, Switzerland*

New Vermitechnology Approach for Sewage Sludge Utilization in Northern and Temperate Climates All Year Round

N.F. Protopopov

Department Ecology & Safety, Tomsk Polytechnical University,
P.O. Box 2845, 634045 Tomsk-45, Russia

Keywords: Vermitechnology, Conventional Systems, Sewage Sludge, Coniferous Woodwaste, *Eisenia foetida spp.*, Western Siberia

Abstract

New vermitechnology approach for sewage sludge utilization is presented. Vermitechnological bloc (Vermibloc) is added to the conventional system of wastewater treatment. Vermibloc includes some steps which are easily incorporated into the system. The field experiment was carried out in Waste Water Treatment Plant, Kyslovka, Tomsk. The positive results of this experiment are presented.

Introduction

Vermistabilization of sewage sludge has been considered as an attractive option of sludge treatment [1, 2, 3]. Earthworms accelerate the conversion of sewage sludge into a rich fertilizer, conditioning soil properties. The known methods of sludge vermistabilization suggest mixing sludge with bulking materials [4] such as solid urban wastes, peat, straw, crushed old woodbark [5], and leafy wood sawdust. Most vermitechnologies are developed and applied in regions with a warm climate. In Russia the surplus activated sludge from many wastewater treatment plants is brought as a rule to dumps, where it pollutes the environment and represents the real risk for human ecology. Fresh coniferous sawdust is known to be toxic for earthworms, and in Western Siberia it is also dumped. The approach is developed to the creation of utilizing conventional system sewage sludge under extreme conditions of Siberia all year round.

New vermitechnology approach

The all-year-round method designed for northern and temperate climatic conditions is characterized by the inclusion of Vermibloc into conventional systems of wastewater treatment. The Vermibloc consists of the following steps:

(a) dehydrating sewage sludge from secondary settlers by filtration through fresh coniferous sawdust beds;

(b) rejecting filtrate into airtanks for recycling;

(c) vermistabilizing of sludge+sawdust;

(d) producing and marketing vermicompost.
The introduction of Vermibloc into the active conventional system is easily accomplished, requiring no essential investments. Fig.1 illustrates the scheme of conventional system + Vermibloc.

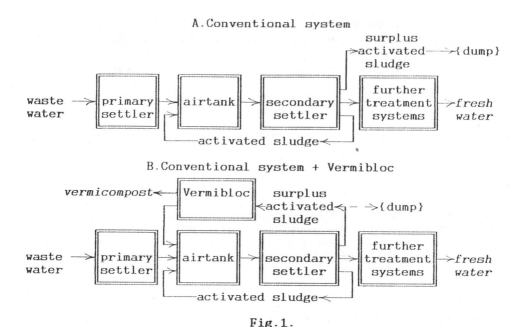

Fig.1.

(a) Field and laboratory experiments which were conducted in Kyslovka Waste Water Treatment Plant (WWTP) during 1989-1992, proved fresh coniferous woodwastes to be applicable for vermicomposting sewage sludge. Fresh sawdust and small size woodchips appeared to be rather good sludge filter and bulk material. In Kyslovka WWTP (treatment facility: 700 m^3 per 24 hours; domestic wastewater; activated sludge and further biological treatment) the surplus activated sludge was taken by airlifts from secondary settlers and was spilled through woodwaste bed (10 m^3).

(b) Filtrate with undetained sludge and extracted wormtoxic substances is drained into airtanks.

(c) There are two ways for vermistabilization of sewage sludge: vermicomposting beds in an open area or indoor containers. High efficiency of vermicomposting sludge+woodwaste mix all year round in the open area under extreme Siberian conditions has been found. Pilot experiments in Kyslovka WWTP (average environmental air temperature of the coldest month is -19.2°C and of the wamest month is +23.7°C) carried out in 1989-1990 have demonstrate unexpected results. It is self-heating of sludge+woodwaste mix in the winter period that is the essential solution for vermicomposting in northern and temperate climates. Even when the difference between bed and environmental air temperatures approaches 60°C, the earthworms remain in good physical condition. Earthworms *Eisenia foetida spp.* were used in our experiments. The earthworms were in reproductive and feeding activity and they selected bed layers of preferable temperature and feeding conditions. Vermistabilization indoors is suitable and preferable if free area is available and/or it is for nothern regions in extra continental climate. The period of vermistabilization in summer beds in open areas and containers is 20-30 days, and in winter beds, a few months.

(d) After separating the vermicompost remainds and worms were used again. Laboratory experiments show that the vermicompost made from activated sludge and coniferous woodwaste has good physical and chemical properties. It had a light-olive-grey colour, porous structure, the smell of wet soil, pH 6.0-6.5, and containd available nutrients: N-(NO_3+NH_4) 1.5 g per kg dry matter (d.m.), P_2O_5 5-6 g per kg d.m., K_2O 0.20-0.25 g per kg d.m.; exchange Ca 5-6 g per kg d.m. Vermicompost is recommended as the fertilizer for ornamental and horticulture plants, and as a conditioner for soil restoration and soils of loamy and clay texture.

Conclusion

Positive experiments resulted in The Russian Patent Application for 'All-year-round method of sewage   sludge and woodwaste utilization by earthworms, with Vermicompost produced from sewage sludge and coniferous woodwastes'. The investigations of sewage sludge utilization under extreme Siberian climatic conditions not finished.

Acknowledgements

I am grateful to V.G.Avzalov, S.G.Tuchak, V.N.Tuchak, V.V.Lukyantzev, L.Y.Chikina, Dr.B.R.Stryganova and N.Windmann who helped me to present the paper.

References

[1] M.J.Mitchell, R.M.Mulligan, R.Hartenstein,   E.F.Neuhauser, Compost Science. 18, 4 (1977).
[2] M.J.Mitchell, S.G.Hornor, B.I.Abrams, J.Environ.Qual. 9, 3 (1980).
[3] J.E.Hall, R.Bland, E.Neale, Compost: Production, quality and use (1987).
[4] N.F.Protopopov,   Bioconversion   of   organic wastes. III Int.Congr.Org.Wastes Bioconversion, Moscow (1994).
[5] J.Haimi, V.Huhta, Pedobiologia. 30, 2 (1987).
☐

Environmental Research Forum Vols. 5-6 (1996) pp. 417-420
© *1996 Transtec Publications, Switzerland*

High Quality Compost from Composting Toilets

T. Rohrer

Centre for Applied Ecology Schattweid, CH-6114 Steinhuserberg, Switzerland

Keywords: Composting Toilets, Second Composting Step, High Quality Compost, C:N Ratio, Coarse Organic Material, Faeces, Urine Separation, Hygienic Safety

Abstract

From an agronomists point of view, only high quality compost proves beyond doubt the long term reliabilty of composting toilets. However, the buyer's decision is often guided by the simplicity to install the system and other short term factors (like toilet design, odors, price...). For the operation, mainly the capacity has been important.
In order to get high quality compost which can be used for gardening, the biological principles of composting must be obeyed (C:N ratio, humidity, structural stability of raw materials). From these principles, three requirements are derived: Composting toilets must be used in combination with urine separation (not urine drainage). After every bowel movement, coarse structured organic material must be added to the composting toilet. The composting toilet must be emptied after 3-12 months. The material must be mixed, loosend and composted in a second step.

Composting of feces: Biological principles

The composting of feces is subject to the same biological principles as the composting of organic wastes from kitchen or garden. Essentially, the attempts have to focus on providing the microorganisms, which are responsible for the composting process, with suitable living conditions. The most important factors are:

a) Balanced nutrition

Microorganisms need a certain ratio between Carbon (C) and Nitrogen (N) in the organic raw material to build up their body substance. The C:N ratio must be 20-30 [1,2]. For the composting process this means: Material rich in nitrogen (such as human feces) must be mixed with carbon rich material.

b) Adequate humidity

In order to achieve an optimized composting process, the water content in fecal compost should lie between 40 and 50%. Below 25% the composting process stops. Above 70% anaerobic fermentation starts [2]. For the composting process this means: Material with high water content (such as human feces) must be mixed with dry material.

c) Adequate air (oxygen) supply

The oxygen content in the compost pile depends on the pore volume, and thus on the structural stability of the raw material ("structural stability" stands for the resistance of the material against collapsing). If the structural stability is low (e.g. feces), the material tends to stick together, thus preventing air from entering the compost pile. In this case, even mechanically breaking up and turning the compost pile will only lead to a short term improval of the oxygen supply. The oxygen in the pores will be rapidly used up by the microorganisms [1], and the material will collapse again . For the composting process this means: Material with low structural stability (such as human feces) must be mixed with materials with high structural stability.

Characteristics of additives

The term "additive" is used for materials which are added to the fecal compost. In order to create adequate conditions for the composting of feces we first must examine a number of potential additives on C:N ratio, humidity and structural stability. In Table 1 a number of organic materials are listed, which are commonly added to fecal compost.

Material	C:N ratio	Humidity	Structural stability
Human feces	< 10	too humid	low
Human urine	0.8	liquid	low (liquid)
Lawn cut (fresh)	12-25	too humid	low
Organic wastes (kitchen)	12-20	humid to medium	low to medium
Organic wastes (garden)	20-60	medium	medium
Leaves	30-60	medium to dry	medium to high
Ash	(no C)	dry	low
Straw (wheat, barley)	100	dry	high
Woodchips	100-150	dry	high
Bark	100-130	rather dry	high

Table 1: Characteristics of selected organic raw materials for composting [from [3] and [4]]

Human feces have a C:N ration below 10, are too humid and have a low structural stability. Therefore, only materials with a high C:N ratio, low humidity and a high structural stability should be used as additives. These requirements are only fulfilled by straw, bark and woodchips. Leaves and organic garden wastes can be applied within limits, due to C:N ratio and structural stabilty which are sometimes low. The other additives are not recommended. They need an addition of straw, bark or woodchips themselves, to be properly composted.

An example - addition of straw to fecal compost

In order to give an idea, how much additive must be added for the proper composting of feces, an example has been calculated in Table 2.

Origin / material	Total weight [g]	Water content [g]	C-content [gC]	N-content [gN]	C:N ratio
1 kg feces	1000	750 (75%)	123	12.3	10
1 kg straw	1000	120 (12%)	350	3.5	100
0.33 kg urine	330	314 (95%)	2.5	3	0.83
Mixture of feces, straw and urine	2330	1184 (50.8%)	475.5	18.8	25.3

Table 2: Calculation of total weight, water, carbon and nitrogen content and C:N ratio in a mixture of 1 kg feces (weekly production of 1 person), 1 kg straw (wheat or barley) and 0.33 kg urine (1/25 of weekly production of 1 person) (from [3] and [4])

The example shows that for the composting of feces about an equal weight of straw as additive is needed. Only traces of urine can be tolerated. As far as structural stability is concerned, 25-30 % of additive has been considered necessary. In the example we have 40 % which should be sufficient.

The excretion of 1 kg feces needs 12 bowel movements at average [4]. A handful of straw weighs about 30 g. Thus after every bowel movement 2-3 handfuls of straw-additive must be given into the composting toilet.

Requirements for the operation of composting toilets
In order to achieve a high quality compost from composting toilets, a number of requirements must be fulfilled:

a) Urine
Urine, if introduced into the compost pile, will lead to water logging and a surplus of nitrogen. The limit for anaerobic fermentation (70% water content) is easily passed, and the C:N ratio drops below 20. Requirement: Every composting toilet should be combined with urine separation. A urine drainage in the composting chamber is of little use.

b) Additives
A sufficient amount of additive is required for the composting process. In order to guarantee the entry of oxygen into the compost pile, the additive has to be distributed as evenly as possible. Requirement: After every bowel movement the additive has to be added to the composting chamber.

c) Second composting step
Even if the additive is added regularly by the users, an absolutely even distribution of feces and additive cannot be expected. In analogy to the composting of organic waste from kitchens, there will always be a surplus of water (combined with oxygen deficiency) in the lower layers of the pile, and a tendency to dry out in the outer parts of the pile. High quality compost can only be produced if the content is mixed, loosened and composted a second time outside of the composting toilet. This second composting step is also necessary to achieve high temperatures in the pile, which are necessary for hygienic reasons. Requirement: After 3 to 12 months composting toilets must be emptied. The contents must be mixed, loosened and set up in a pile for a second composting step.

Open questions to the second composting step
Temperatures of 50-60 °C for 2-3 weeks in the second composting step are needed, for the compost to be hygienically safe. With material from composting toilets of only 1 or 2 households this has not yet been reached, probably due to the small amounts (< 1m3) that were examined. The use of young material from composting toiletts (2-3 months old), an addition of easily degradable substances (proteins, sugars, carbohydrates), which are available in lawn cut or a urine-sugar-solution (e.g. 2 l urine, 1 kg sugar) might be a possibility. Also, the use of closed wooden boxes, with a cover (to keep off precipitation) and holes in the bottom (to allow the entry of air) will minimize heat losses. The second composting step should not begin in the cold season.

Literature
[1] Gottschall, G., 1992. Kompostierung. Optimale Aufbereitung und Verwendung organischer Materialien im ökologischen Landbau. Alternative Konzepte, 45. C.F. Müller Verlag, Karlsruhe (in German).
[2] Schriefer, T. 1988. Kommunale Kompostierung und Qualitätssicherung, Bodenökologische Arbeitsgemeinschaft Bremen e.V. (Hrsg.) (in German)
[3] Pfierter,A., Hierschheydt, A. et al.,1982. Kompostieren - Anleitung für eine sinnvolle Verwertung von organischen Abfällen, Verlag Genossenschaft Migros, Aargau/Solothurn (in German).
[4] Documenta Geigy, Wissenschaftliche Tabellen, 7. Auflage (1968) S. 653-673, J.R.Geigy A.G., Basel.(in German)

Environmental Research Forum Vols. 5-6 (1996) pp. 421-424
© 1996 Transtec Publications, Switzerland

Macrophytes as Food for European Freshwater Fish
(Rudd, *Scardinius erythrophthalmus* and Tench, *Tinca tinca*)

K. Scheurer[1] and J. Staudenmann[2]

[1] Zoologisches Museum der Universität Zürich, CH-8057 Zürich, Switzerland

[2] School of Engineering Wädenswil, CH-8820 Wädenswil, Switzerland

Keywords: Aquaculture, Fish Production, Fish Growth, Macrophytes as Food, Rudd (*Scardinius erythrophthalmus*), Tench (*Tinca tinca*), Polyculture of Fish, Protein Production

Abstract
In an aquaculture plant for wastewater treatment at Ingenieurschule Waedenswil the growth of two indigenous fish species (rudd, tench) feeding on macrophytes is studied.

1. Introduction
1.1 General
Wherever you look at aquaculture systems for wastewater treatment, you will find most likely some fish ponds as a part of it. One of the aim of ecological engineering approaches for wastewater treatment is to convert nutrients in the wastewater into utilizable products. Fish is the main product and therefore the last organism in the multi-layered system. To achieve maximal transformation of resources and energy, food chains should be short.

Up to now, polyculture systems with different Asian or African fishes have been used. The leading fishes in such polyculture systems are mainly herbivores (e.g. grass carp). As a result of the climate of their original habitat, herbivory could develop because of vegetation growing throughout the year. Since the idea behind the project at Ingenieurschule Waedenswil [1] is to use indigenous organisms for the aquaculture system, the main question for my diploma thesis at University of Zurich is: Do indigenous fishes exist which can feed and grow predominantly on plants?

With this condition, the wide range of European fish species reduces to the family Cyprinidae.

1.2 Biology of cyprinids [2]
The family Cyprinidae is the one with most numerous species in the world. Cyprinids are primary freshwater fishes, sometimes found in brackish water as well. They are widespread and show high tolerance to water temperature and oxygen concentrations.

Cyprinids generally tend to omnivory. Even when the principal food consists of invertebrates, they can supplement their diet with plant material. All of them have a specialized feeding mechanism which involves a protrusible and toothless jaw, a branchiostegal pumping system and a pharyngeal mill. This pharyngeal mill is one key to the plant feeding of cyprinids. Cyprinids feeding on macrophytes can only digest the contents of broken cells, because the cell walls are resistant to the digestive enzymes of fish and because cellulolytic activity of symbionts is negligible.

2. Preliminary experiment
The aim of the preliminary experiment was to find indigenous fish species which feed predominantly on plants in their natural biotopes and can be kept in experimental basins with relatively low water quality. Further, the fish should be usable as food or animal feed.

For the preliminary experiment, four species were chosen (see Tab.1):

Tab. 1: Chosen fish species for the preliminary experiment. Prefered biotope and food are
indicated as well.

fish	species	biotop	natural feed
carp	*Cyprinus carpio*	lakes, ponds and slow running rivers	**invertebrates**, fishes, plants
roach	*Rutilus rutilus*	lakes, ponds and slow running rivers	zooplankton, invertebrates, plants
rudd	*Scardinius erythrophthalmus*	lakes, ponds and slow running rivers	**plants**, invertebrates
tench	*Tinca tinca*	lakes, ponds and slow running rivers	**invertebrates**, plants

Five individuals of each species were kept in four compartments of 500 l each. They were fed both
animal feed and plant feed. Macrophyte species were chosen out of the variety of plants tested in
Heebs [3] preliminary experiments (*Callitriche, Ceratophyllum, Elodea, Hippuris, Lemna*). The
feeding behaviour of all fish species was noted qualitativly by persistence or disappearance of the
different plants.
Rudd and tench could be stocked without any problems. In the basins of rudd *Elodea, Hippuris,
Callitriche* and *Lemna* disappeared, whereas in the basins of the tench only *Lemna* was eaten.
I received carp in very bad condition and they fed on neither plants nor animals. Roach seemed to be
very sensitive to changes of habitat and water quality. Soon they were infected by fungus disease and
died within two weeks. Plants which were fed by roach were *Hippuris. Elodea, Callitriche and
Lemna*. Nothing can be said about the preference of any of the four investigated fish species for any
plant species.
While stocked together for one week rudd and tench seemed to complement one another with respect
to food and habitat choice.

3. Main experiment
Since the stocking of rudd and tench was not problematic in the preliminary experiment, the
experiments has been carried out with the two species held together in the same basins. The aim of
this main experiment was to find out whether these fish can grow on plant food alone or if they need
some animal food in addition.

3.1 Questions
- Are there any differences in the growth of the fish (rudd, tench) when fed:
        a.) only with plants
        b.) with plants as well as animal food?
- Which plants do rudd and tench prefer?
- Does the addition of animal food influence the amount of plants beeing eaten?

3.2 Methods
Rudd and tench has been stocked in a polyculture of 10 and 5 individuals, respectively, in each of
eight 1100 l fish basins. The fish basins were placed in a greenhouse at Ingenieurschule Wädenswil.
There has been two treatments with four replicates each (see Fig. 1).

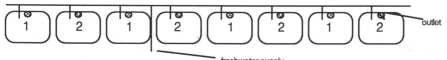

Fig. 1: Scheme of the eight fish basins. Treatment 1: Feeding with plants only; Treatment 2: Feeding
with a mixed diet of plant and animal food supply.

Fish were acclimatized for one week. Two days before analyzing weight and length, they went
without food to empty the intestines (food evacuation time). Fish were anesthetized before measuring
to lower stress. Besides weight both maximum total length (TL) and maximum standard length (SL)
were measured. Length and weight are parameters to record possible growth, so every three weeks
weighing and measuring was repeated.

Plants were taken from the aquaculture plant. Therefore the selection of plants for food was made
according to the experimental design of Anuschka Heeb [3]. Two groups of plants were offered as
food: - floating plants: *Lemna* sp., *Azolla* sp., *Salvinia natans*
                          - submersed plants: *Elodea canadensis, Ceratophyllum demersum*

Floating plants has been offered ad libitum in a styrofoam ring floating on the water surface of each
basin. As a control, a similar ring, with its plants protected by a screen from the fish, was placed on
the water in the same basin. Both rings obtained the same controlled amount of floating plants.
The submersed plants, planted in pots, were placed into every basin. The length and number of
shoots was noted.

In the basins with additional animal food supply, a defined amount of frozen mosquito larvae was fed
daily. For a period of three weeks fish recieved a daily ration of mosquito larvae of 2.5%, 5%, 10%
and 15% of fish weight, respectively.

Besides the measurement of fish and the controlling of the amount of plant food eaten, the following
parameters of water quality were measured every week: O_2, pH, T^0C, EC, NH_4, NO_2, NO_3.

3.3 Results
When publishing this poster measurements and observations still were running. A slight increase in
weight could be noted though for the fishes with animal food supply, while the fishes with only plant
food lost weight. Nevertheless a significant change in weight could only be noted for tench in
treatment 1.

For the amount of plant food eaten a difference between the two treatments was perceived for whole
trial. While fed with > 10% mosquito larvae of fish weight per day significant less plant food was
eaten and when fed with 15% hardly any plants were taken anymore.

Preferences for the two floating plants *Azolla* sp. and *Lemna* sp. could be observed.

3.4 Discussion
Plant food does not seem to have sufficient available nutritive value to the examined fishes. Only with
a specific amount of animal food supply (> 10% of fish weight) growth could be noted. At the same
time plant food hardly was eaten anymore.

Other ways of recycling the nutrients into utilizable products has to be found for indigenous species.

4. References
[1] Züst, Brigitta: Recycling the resource from greenhouse culture.
    Workshop 1, this volume.
[2] Winfield, I. J. & J. S. Nelson (1991): Cyprinid Fishes: Systematics, biology and exploitation.
    Chapman & Hall, London.
[3] Heeb, Anuschka: Wastewater Purification Efficiency and Growth of European
    Water Plants - Possible Use in Aquaculture for Nutrient Recycling.
    Poster, this volume.

Environmental Research Forum Vols. 5-6 (1996) pp. 425-428
© *1996 Transtec Publications, Switzerland*

Liquid Fertilization of Nightsoil Sewage by Photosynthetic Bacteria

J.L. Shi and H.G. Zhu

Department of Environmental Science, East China Normal University, Shanghai 200062, China

Keywords: Nightsoil Sewage, Liquid Fertilization, Photosynthetic Bacteria

Abstract: Liquid fertilization of high density nightsoil sewage by photosynthetic bacteria makes special organic fertilizer rich with photosynthetic bacteria. Therefore, zero discharge is accomplished as well as the rise in crop output and improvement of crop quality. Liquid fertilization of nightsoil sewage through photosynthetic bacteria is simple in technology with less investment, easier management, lower consumption of energy and higher availability, which will satisfy social, economical and environmental requirements.

1. INTRODUCTION

Nightsoil has been the main organic fertilizer in China and is considered as a farmer's treasure. In recent years as chemical fertilizer developed greatly and inconvenience existed in use of nightsoil, much less urban nightsoil is used in rural areas. Thus problems arose in nightsoil disposal and lack of organic fertilizer in countryside. Photosynthetic bacteria (PSB) treatment of high density nightsoil sewage is a direct treatment without dilution with high removal rate of BOD. The sludge of PSB created in the process can be recycled in agriculture. But COD and NH_3-N are still high in the outlet water and need further treatment with great dose of reagent. More reagent makes the rise in cost and it is hard to reutilize the sludge with so much chemicals. So the research on PSB approach of liquid fertilization of nightsoil sewage started, which transforms high density nightsoil sewage directly into PSB liquid fertilizer.

2. METHODS AND MATERIALS

2.1 Bacteria Species

The species used in the experiment are mixed bacteria of purple non-sulphur bacteria and purple sulphur bacteria. Growing condition of PSB is in weak illumination, low aeration, with sodium acetate and inorganic salt as culture medium.

2.2 The Flow of Liquid Fertilization of Nightsoil sewage by PSB

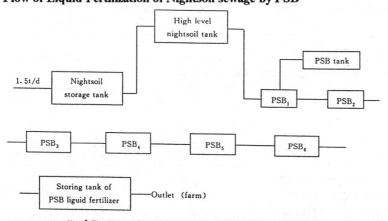

Fig. *1* The flow of Liquid Fertilization of Nightsoil Sewage by PSB

Intermittent aeration was adopted to constrain anaerobic bacteria. The PSB can grow under either the aerobic or anaerobic condition. So its advantage was manifested.

2.3 Analysis

Samples were taken from each tank in the flow and the concentrations of CODcr, BOD$_5$, TVA, NH$_3$-N, NO$_3$-N, T-PO$_4$, SS, MLSS, MLVSS, pH were analyzed. The composition of PSB liquid fertilizer, hygiene quotas-the number of *Escherichia coli* and ascaris eggs were also tested.

2.4 Test on the effect of PSB liquid fertilizer in the farm

The test was carried out in the vegetable fields of No.5 Brigade of Tang Zijing and No.3 Brigade of Sanjiaodi, Yangsi Town, Chuansha County. The vegetable at the experiment were greens (pakchoi), Jimaocai, autumn lettuce, spring lettuce, cucumber and sword bean.

3. RESULTS AND DISCUSSION

3.1 The Characteristics of Nightsoil Sewage

Table 1 Summary of the characteristics of nightsoil sewage

testing items	average content (mg/l)	Range (mg/l)
CODcr	2190	11424-39200
NH3-N	5787	4026-7500
BOD5	3775	1064-5292
T-PO43-	620	1096-4288
T-N	4666	3868-5820
TVA	8479	2070-5250
pH	8.5	8.0-9.0

The average quotient among 11 groups of BOD$_5$/CODcr is 0.27. It shows that the value of un-biodegradable CODcr in nightsoil sewage is high. The C:N expressed by BOD$_5$/T-N is 1.23, It is very low comparing to 2-2.5 as published in Japanese literature. These is great amount of impurity in nightsoil sewage. The listed features of high-density nightsoil tells the difficulty in its treatment.

3.2 The Effect of The Liquid Fertilization Treatment and The Properties of The Fertilizer

Table 2 Effect of PSB liquid fertilization treatment of high density nightsoil sewage

	CODcr	BOD$_5$	NH$_3$-N	T-PO$_4^{3-}$
high density nightsoil (mg/l)	21910	5737	3775	620
PSB liquid fertilizer (mg/l)	4030	944	1503	287
removal rate (%)	81.6	83.5	60.2	53.7

Most of the bio-degradable organic materials has been decomposed to stable state(table 2), and as a liquid fertilizer, its state is also stable. The content of MLVSS in liquid fertilizer is around 4.3 g/l and the number of PSB is over 1.5×10^8 per ml. The N, P, K content in PSB fertilizer is shown in Table 3. It illustrates that the fertilizer contains the nutrition needed by crops.

Table 3 N, P, K content in PSB liquid fertilizer

samples	Nitrogen	Phosphorus (P$_2$O$_3$)	Potassium (K$_2$O)
PSB liquid fertilizer (mg/l)	0.237	0.038	0.017
nightsoil of human (mg/l)	0.237	0.068	0.030
feces of pigs (mg/l)	0.111	0.023	0.010

The number of Esc. co. in the PSB fertilizer is approximately 9.2×10^3 MPN/ml and the removal rate is 99.7%. The number of ascaris eggs is 2 per liter with the rate of removal at 99.6%, much better than 95% as regulated by the domestic sanitary standard on the allowed number of asccaris eggs in compost. Besides, there is no refuse in PSB liquid fertilizer. So the fertilizer is applicable to spraying, sprinking, watering and irrigating and is far safer and cleaner than raw nightsoil.

3.3 Experiment on The Effect of PSB Liquid Fertilizer in The Field

Effect on Autumn Lettuce

The harvest in experimental field prolonged for 3 days comparing to the comparison field. Experimental lettuce had bigger leaves and thicker stems. The output in 0.9 mu field was 1981.4 kg while only 1701.9 kg of output was harvested in the comparison field. The increase rate of 14.1% was obtained.

Effect on Tomatoes

The outputs of the sections with 75ford, 50ford and 25ford dilution of PSB liquid fertilizer were increased by 13.7%, 17.7% and 27.0% respectively, which means the output increased with the increase of fertilizer concentration. The result of quality examination is shown in Table 4.

Table 4 Improvement of quality of tomato

	total acidity	Vit.C (mg/100g)
experiment	0.48	15.32
comparison	0.56	15.52
increase rate (%)	-14.3	+0.46

4. SUMMARY AND DISCUSSION

In summary PSB fertilizer is a clean, effective and good-quality organic fertilizer and is comparable to pig feces. PSB may help with the increase in output and improvement of quality of crops if used correctly. The facilitates is simple in technology, low in energy consumption and high in availability. If medium and small sized PSB fertilizer plants are chosen to located correctly in the joint zone of urban and rural area, the high density nightsoil sewage can be treated conveniently and the PSB liquid fertilizer produced can be used by near farms.

The established PSB liquid fertilization process is one way to disposal the nightsoil sewage. It can not only solve the problems of sanitary and environmental protection but also support the agricultural production.

5. REFERENCES

1. H. Kitamura, etc., "Photosynthetic Bacteria", Academic Press Center, Japan, (1984).
2. A. Hirayashi, J. L. Shi, H. Kitamura, Purple nonsulfur bacteria and other microorganisms in photosynthetic sludge reactors for the purification of soybean curd wastewater. Nippon Mizushori Seibutsugakkaishi, 19 (2), 1-8, (1983).
3. A. Hirayashi, J. L. Shi, H. Kitamura, Effects of organic nutrient strength on the purple nonsulfur bacterial content and metabolic activity of photosynthetic sludge for waste water treatment, 68 (4), 269-267, (1989).
4. J. L. Shi, Y. T. Xu, etc., A study in treatment of waste water from soybean product-making by using photosynthetic bacteria, Vol. 4, 27-36, (1989).
5. J. L. Shi, A. Hirayashi, H. Kitamura, Acetic acid metabolism of *Rhodobacter sphaeroides*, J. East China Normal University, Vol. 3, (1989).
6. M. Kuboyashi, Ferment. & Industri., 36 (9), (1978).

Environmental Research Forum Vols. 5-6 (1996) pp. 429-430
© *1996 Transtec Publications, Switzerland*

Productive Artificial Aquatic Ecosystems for Recycling of Anorganic Nutrients in Wastewater and Production of Zooplankton and Plants as Food for Fish Stocking

P. Stauffer, J. Staudenmann and J.-B. Bächtiger

School of Engineering Wädenswil, CH-8820 Wädenswil, Switzerland

Keywords: Aquaculture, Wastewater Treatment, Plankton Production, Productive Ecosystems, Aquatic Macrophyts, Nutrient Recycling, Zooplankton

Abstract

The phytoplankton-zooplankton system is an important part of the aquatic ecosystem and a basic source of food for fish in natural lakes and rivers. It is also the main transformer of anorganic nutrients to organic material. Those organisms have a high productivity and are able to adapt to changing environment conditions by varying the size of their population.
Zooplancton is well suited to serve as animal protein for fish stocking.
In this experiments the possibilities of practical application for elimination of anorganic nutrients out of greenhouse culture wastewater was testet.
In a preceeding experiment we tested the possibilities of a low technology continuous culture of Algae and daphnids was tested. The main experiment is one step of a aquaculture pilot plant [1,2,3]in wich plants and fish for food purpose is produced. In this presently running main experiment we test the operation of smal aquatic ecosystems (300 l each) with different planctonic crustaceas and submersed vascular plantsare cultivated for use as fish feed.

Preceding experiment

The operation of a continuous culture of planctonic Algae (*Chlorophytae*, mixed culture of 6 species) and daphnids in a low technology experimental plant was tested. Thereby the nutrient uptake of algae (NO_3 and PO_4) and the growth of daphnids was analysed. Also the whole development of the organisms was observed and information on the possibilities and problems aswell as experience in this field was gained.
This experiment gave us hints to find a well working and reliable method for the pilot plant.

Results

The reduction of the analysed nutrients varied strongly. The reduction of NO_3 was neglectible, whereas the reduction of PO_4 was up to 70%. At certain times the algaes grew strongly, but over the whole time the growth varied much. Problems with settlement of cells and too strong dilution made the plant unreliable and made a continuous culture with the used low tecnology impossible.
Due to difficult procurement of daphnids during the winter season the inoculated population was too small. The growth of algae in the daphnid tank was too strong, so the pH raised up to 10 and more and the daphnids didn't develop. Different attempts to improve the conditions failed. The population stayed stable for a long period and suddenly died of at one time. The reason for the death was presumed in toxic concentrations of NO_2 and NH_4.

Presently running main trial

Based on the results of the preceding trial we chose for the cultivation of zooplankton no monoculture, but small aquatic biotops including submersed vascular plants (*Potamogeton lucens, Elodea canadensis, Ceratophyllum demersum*), different planctonic crustaceas

(*Daphnia magna, D. pulex, Cyclops sp. Ostracoda sp., Bosmina sp.*) and different water snails (*Lymnaea stagnalis, Planorbius corneus, Anisus sp., Radix spc.*).
Volunteer organisms also occure like different larvae of insects, water mites, rotaria and others. Those cultures are much more stable to changing environmental conditions then monocultures. Many of the organisms have positive influences on others, which makes a good growth possible. So are submersed plants dependent on zooplankters who feed on planctonic algae and keep the water transparent, so enough light for growing gets through. Macrophites provide O_2 for the animals and prevent algae from bloom through nutrient competition.
In this experiment we test the influence of different densities of submersed plants on elimination of nutrients and growth of zooplancton aswell as influences of retention time on water cleaning.

Construction

In the pilot plant the water is divided into 5 parts to provide replicas for the plant cultures. The pretreated water is led into the 5 tanks *Z1 a-e*. From the tanks *Z1a, Z1b* and *Z1c* the water runs in second tanks (*Z2 a-c*) (*fig 1.*). This system provides two tracks with a retention time of 8.7 d and 17.4 d respectively.

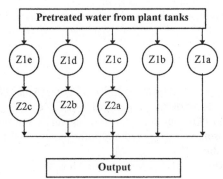

fig 1.: Scheme of Zooplancton-Submersed Macrophytes Experiment

First results

At this stage it is too early to give any statements on efficency and productiviy of our pilot plant. First observations showed good development of plancton, but in different tanks it varied quite much. While in tanks with a high starting population of daphnids the water staid clean and the daphnids developt well untill now, in other tanks a bloom of algae was observed. First harvests of small amounts of plankton were made.
Now it is important to get some results of the development over a longer term.

References
[1] B. ZÜST ET AL., Workshop 1: "Recycling the Resource from Greenhouse Cultures", 2nd. int. Conf. on Ecol. Eng. f. Wastewater Treatment (sept. 1995)

[2]A. HEEB, Poster: "Wastewater Purification Efficency and Growth ~f Aquatic Macrophytes, Possible Use in Aquaculture for Nutrient Recycling", 2nd. int. Conf. on Ecol. Eng. f. Wastewater Treatment (sept. 1995)

[3] K.SCHEURER, Poster: "Effects of Macrophytes from Wastewater -fed Aquaculture as Food on the Growth of European Omnivorous Fish", 2nd. int. Conf. on Ecol. Eng. f. Wastewater Treatment (sept. 1995)

Environmental Research Forum Vols. 5-6 (1996) pp. 431-434

Procedure and Pilot Plant for the Disposal of Inorganic Loads from Circulating Water in Aquacultures by Microalgal Cultivation

R. Storandt[1], I. Färber[1], O. Pulz[1] and R. Hahlweg[2]

[1] IGV Institut für Getreideverarbeitung GmbH,
Arthur-Scheunert-Allee 40-41, D-14558 Bergholz-Rehbrücke, Germany

[2] Ingenieurbüro Hahlweg, Erlenhof 28, D-14478 Potsdam, Germany

Keywords: Aquacultures, Closed Circulating Plants, Elimination of Nitrates and Phosphates, Assimilation Ability of Microalgae, Microalgal Biomass, Valuable Biomaterial for Aquacultures, Algae

Introduction

Despite bacterial water treatment, in the water of closed aquaculture circulating plants especially nitrates and phosphates are accumulating. The development of the procedure was aimed at the utilization of the assimilation ability of microalgae for the elimination of the nitrates and phosphates, harmful to fish and environment, as well as at upgrading to valuable biomaterial, applicable in aquacultures, and at the minimization of fresh water supply.

Materials and methods

For the analyses and laboratory tests, fish water of two plants with closed circulation (15 m^3 and 20 m^3) was used at an initial fish stock of 22.3 $kg \cdot m^{-3}$ and a feed ration of 1.5-3 $kg \cdot d^{-1}$.

Microalgal strains from international collections (*Spirulina platensis* 85.79, *Spirulina platensis* IPPASB-433, *Chlorella vulgaris* IPPASC-1) and the strain IR1 selected on fish water, were cultivated in laboratory and open-air photobioreactors (0.3-1800 l) at temperatures ranging from 18 °C to 33 °C and gassing by CO_2-air mixture, in a batch (3-22 days) or semi-continuous process (harvests 3 per day, 1 per day, and every 2nd day, respectively).

For the control of the growth and condition of the algae, the pH-value, extinction (750 nm), dry mass, nitrates, phosphates (DIN 38405-D 19/20) were determined and monitored in the light and electron microscope.

Results

- In the circulating fish water, a medium accumulation of 9.5 $mg \cdot l^{-1} \cdot d^{-1}$ nitrate and 4.7 $mg \cdot l^{-1} \cdot d^{-1}$ phosphate as well as pH-values of 4.0 to 6.8 were determined.
- In the analyzed fish water, the ratio of nitrate and phosphate significantly deviates from that of synthetic nutritive suspensions for *Chlorella vulgaris* to the disadvantage of the phosphate.
- The low pH-values of the fish water neither inhibit the growth nor prevent the complete nitrate and phosphate elimination by microalgae.

- With beginning growth of the cells the pH is increased to
  7.0-7.6 at constant CO_2-supply.
- A stable cultivation of the *Spirulina* strains was impossible
  without the addition of sodium hydrogen carbonate to the fish
  water.
- The adaptation of the strains C1 and IR1 to fish water without
  nutritive salt additions took several months, at continuous
  increase of the growth and elimination rates during adaptation.
- The selected microalgal strains eliminated 80-208 $mg \cdot l^{-1} \cdot d^{-1}$
  nitrate at optimal conditions.
- *Chlorella vulgaris* strain C1 and the selected strain IR1 had
  a better nitrate and phosphate elimination efficiency than
  the *Spirulina* strains, phosphate having been eliminated first.
- The phosphate deficiency, observed after a few cultivation
  days, did not prevent the complete elimination of the nitrates
- Strain IR1 had the highest resistance to bacterial effects.
- The following maximum growth parameters and elimination and
  growth rates, respectively, were achieved:

Table 1

	Strain C1	Strain IR1
Extinction (750 nm)	3.568	3.716
d.m. $(mg \cdot l^{-1})$	3.6	4.9
$NO_3{}^{2-}$ $(mg \cdot l^{-1} \cdot d^{-1})$	208	185.4
$PO_4{}^{3-}$ $(mg \cdot l^{-1} \cdot d^{-1})$	71	76
Extinction (750 nm) $(\cdot d^{-1})$	0.777	0.63
d.m. $(mg \cdot l^{-1} \cdot d^{-1})$	0.463	0.611

Discussion

Of special interest are the facts that the elimination rates
increase continuously, after several months adaptation, and that
the strains C1 and IR1 exceeded the results of *Spirulina*, after
six months, although no phosphate was available. The strains C1
and IR1 are evaluated as suitable approach, the activities being
concentrated to IR1 because of the better stability. Any
additions to the fish water will not be necessary, with the
exception of the continuous CO_2-air mixture.

It is estimated that the selected strains also are able to
eliminate completely inorganic loads from waters of higher
contamination, with higher phosphate contents, but also at
lowest concentrations.

Further investigations should find out how to adapt to a
possibly low and stable level the nitrate and phosphate contents
for the transition to semi-continuous processes.

Conclusions

The results achieved allow for the development, establishment and testing of a continuous 3000 l photobioreactor, consisting of a sun-oriented, photosynthetically active surface in form of parallel 32 mm plexiglass plates with meandrian flow, a gas exchange, measuring and controlling module, a harvesting module, and a separate cleaning cycle. This is the change-over to the production scale.

Acknowledgement

The project was supported by the Federal Foundation for Environment of the Federal Republic of Germany.

Environmental Research Forum Vols. 5-6 (1996) pp. 435-438
© *1996 Transtec Publications, Switzerland*

A Northern Constructed Wetland for Treatment of Wastewater

L. Thofelt and T. Samuelsson

Mid Sweden University, S-832 25 Östersund, Sweden

Keywords: Constructed Wetland, Northern Climate, Cold Environment, Winter Performance, Surface Flow System, Biological Dams, Månsåsen

Abstract.

In the Swedish mountain area, at about 63 degr. north, a small constructed wetland, consisting of three biological dams (25x25 m) was constructed in 1991. The plant is situated in an agricultural area near the lake Storsjön, Jämtland county. The plant serves about 100 persons.

The first dam in the constructed wetland is S-formed, about 3 m deep with a combined anaerobic and aerobic treatment. The second stage is a vegetated (Carex sp), rather shallow dam (0,4 m) and the third dam is about 2 m. deep and equipped with a few islets. In the third dam, waterfowls are breeding each summer. Around the constructed wetland, trees has been planted and in the wetland the hardy species crucian carp (Carassius carassius) has been introduced in summer 1994 in order to extend the food-web.

The constructed wetland is not yet fully established. Consequently, especially in winter, the microbial community and the plants are not able to take up mineralized nutrients. However, in summer, the constructed wetland meets the requirements.

This experiment is during the summer of 1995 being extended with an surface flow system. The effluent is meandering through a small wooded (Salix sp.) area in which the nutrients are to be absorbed. The constructed wetland plant at Månsåsen, Åre municipality, Jämtland county, Sweden is an experiment in which a wetland system is combined with a terrestrial system.

The Månsåsen plant at Hallen, Sweden.

In 1965, the Åre municipality decided to treat the domestic wastewater from the village of Månsåsen by using a system of biological dams, each about 25 metres in square. In the rather harsh climate of the Swedish mountain area at about 63 degrees north, however, an agricultural district, this simple system for treating domestic wastewater did not meet the the swedish water authorities criteria for treated wastewater. Thus, it was in 1990 decided to improve the purifying capacity of the Månsåsen wastewater treatment by trying a rather radical approach, which was not heartily approved by the Swedish environmental protection agency. However, in Europe and U.S.A. similar approaches are being attempted (1,2,3,4).

The three biological dams were in 1991 totally changed into the shape as is shown in figure 1. Dam (1) and (2) and between (2) and (3) is separated by a bank in order to aerate the flowing water. The first dam (1) is about 3 metres deep and does a S-turn in order to combine anaerobic and aerobic treatment of the in all respects untreated wastewater from the about 100 persons in the village of Månsåsen. In the next shallow dam (2) with a depth of 0,4 - 1,0 metres, emergent plants as the dominating sedge (Carex rostrata), the common reed (Pharagmites australis) and bulrush (Typha latifolia) are planted. The islets in dam (3) increases the litoral, which may enhance the metabolism of the plant and also provide breeding places for waterfowls. The wastewater passes the plant in about 20 days.

During 1995, the terrestrial area marked (4) in figure 1 is being transformed into a channel system, surrounded by small trees, herbs and grasses, which are to suck up the mineralized nutrients. In the farther end of the windling channel, the rectangular area is a gravel ditch, which is to aerate the water and also have a suitable growing bed for the roots of the surrounding terrestrial plants. The treated water is let into a about 4 kilometres ditch, which at the far end empties its by that over and over again treated wastewater.

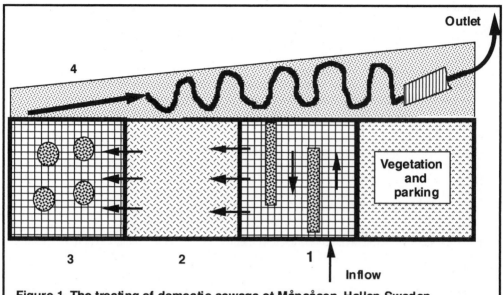

Figure 1. The treating of domestic sewage at Månsåsen, Hallen Sweden.
Anaerobic and aerobic treatment(1), vegetated dam (2), biodam with islets (3)
and absorption of nutrients by terrestrial vegetation (4). (1) - (3) 25 by 25 metres.
(4) at start bredth 10 metres; at other end 30 metres; length about 100 metres.

The wastewater plant is to be surrounded by a windshield of trees. An extensive planting of Norway spruce (Picea abies), birches (Betula verrucosa) and rowan trees (Sorbus aucuparia) and bird-cherry (Prunus padus) has been done during 1994. However, the summer of 1994 was extremely hot and dry, which gave a rather high mortality of the small trees. During 1995, new trees have been planted, which may brave the climate in order to survive.

In the Månsåsen wastewater plant, the animal life is teeming with zooplankton, especially Daphnia, and insects of various kinds. In 1994, some 250 crucian carps were introduced in an attempt to by biological means have the bottom thoroughly sought and ploughed. The carps seem to have enjoyed their nutritious new home.

As for the performance of the Månsåsen plant, chemical sampling has revealed that performance in winter still is rather poor, but during summer meeting the criteria. The temperature of the water may partly explain this winter depression. At about 20 degrees Celsius in July, the temperature drops to about zero degrees Celsius in winter (December to April).

As for phosphorus and nitrogen, during winter Månsåsen leaks phosphorus but reduces nitrogen. During the months July to October 10, 1994, as a mean, phosphorus was reduced from 4,1 mg per litre to 1,6
(- 40%) and nitrogen reduced from 29,6 to 17,6 mg per litre (-40,5 %). During winter October 10 1994 to April 1995, phosphorus was increased from 4,6 mg per litre to 4,8, but nitrogen reduced from 30,1 mg per litre to 24,5 (-19,6 %). It is to be mentioned that the performance is measured without a functional terrestrial part (cf 4 in figure 1), which still is under construction.

The Månsåsen plant is not yet at full capacity, which depends on a much longer establishment time for the emergent and surrounding plants than expected. Also, the meandering terrestrial part is far from being completed. The wastewater plant at Månsåsen, Hallen, Sweden is to have a aquatic, anaerobic and aerobic part with partly stagnant and partly flowing water and a terrestrial part, absorbing the mineralized nutrient. Due to our experiences, the Månsåsen plant will be fully established in about two to three years time.

References.

1.   Hammer, D.(ed.) Constructed Wetlands for Wastewater Treatment. Municipal, Industrial and Agricultural. Lewis Publ. Inc. Chelsea. Mich. USA.(1989).

2.   Reed,S.C. , Middlebrooks, E.J. and Crites, R.W. Natural Systems for Waste Management & Treatment. MacGraw-Hill Book Co. New York. (1988).

3.   Schierup, H.H. and Brix,H. 1990. Plantebaserade rensemetoder i Europa. Stads- och havneingeniören 1:18-22.

4.   Seidel,K. Macrophytes and Water Purification. Bio.Contr. Water Poll. Conf. Biol. Water Quality Improv. Alt. Phil. PA. U:S.A.March 3-5, 1975. Un. Phil.Press.Phil.USA. (1975).

Environmental Research Forum Vols. 5-6 (1996) pp. 439-442

Perspectives in the Laminarian Aquaculture: Biomass Production, Restoration of Marine Ecosystems and Usage as Biofilters

G. Voskoboinikov and V. Makarov

Murmansk Marine Biological Institute, Kola Science Centre of the Russian Academy od Sciences, Murmansk, 183010, Russia

Keywords: Algologically Degraded Marine Ecosystems, Aquaculture, Biofilters, Laminarian Plantation, Heavy Metals, Restoration, Algae

Abstract

Possibility to use seaweeds aquaculture on the Barents Sea coast for biomass production, for seeding algologically degraded marine ecosystems, as biofilters is discussed on the basis of authors and literature data.

Algae of the genera *Laminaria* are a raw material for alginat and mannit and also for medical food stuffs and medical preparation. Requirement in products of that algae is very high.

Until recently a trade of the algae has been carried out with a drag causing a large ecological damage. A grow of Laminaria in the plantation can not to change the natural trade in a full volume but in some extent can be an alternative to trade.

At the Murmansk shore of the Barents sea the most perspective species for cultivation is *Laminaria saccharina* which is well adapted to sircumstences of that region. This species in the rate of growth, productivity, contents of alginal acid and mannit, taste is similar to *Laminaria japonica*; at two-years cycle of grow it is possible to obtain 50-100 t of row *Laminaria saccharina*.

Laminaria saccharina is a species well adapted to the severe conditions of the north. It posseses high rate of growth and short life cycle. In the life cycle of *Laminaria* there are two strict alternating forms: mucroscopical sporophyte and microscopical gametophyte. Trade under brush (thicket) in general is plants of length 2-3 m; the length of some specimen can reach 7 m. Most of the plants do not live more 3 years. Thalli of *Laminaria* is differentiated in rizoids which are root-like organs of fastening; trank wnich is long (lenthing) usually roundish or ovalish pivot carring out in general mechanical function; and a plate which is leaf-like one-year-old formation carring out a function of assimilation and reproduction.

Process of cultivation of *Laminaria* begins from the selection of mother plants with natural sporangia. For the preparation of first plantation mother plants are selected from natural thicket, then the air-temperature stimulation of outlet of spores is carried out which consists in drying of mother plants during 12-32 hours. The time of the end of stimulation is determined by the periodical control of dinamics of outlet of spores. Stimulated mother plants are put in tanks with marine water. Obtained suspension is filtered and concentration of spores is brought to the calculating earlier. Rope substrates are put in suspension and after sedimentation and fastening of spores on them are hang on installations in the sea where futher development of *Laminaria* proceeds.

Next stages of the grow consist in the rewoval of overgrowers, rarefy and transfer of *Laminaria*, technical service of installations and harvest.

Preparatory works. Before the cultivation of *Laminaria* it is necessary to prepare the sea installations for the grow, sowing tanks, sheds or rooms for the stimulation of mother plants and to carry out the processing of the substrates.

Installations for the grow of *Laminaria* should be put at the sites with the intensive dinamics of water owe to flood-tidinal currents and relatively protected from the influence of ones are the end and closed parts of gulfs. The depth at the sites should be enough to exclude the touching of the substrates with *Laminaria* to the ground. In other case the substrates will always be entangled that can embarras a care after plantation and also cause diminushion of biomass for too large tightness of *Laminaria*.

For sowing it is possible to use some tanks from un spoiled material. For preventation of intoxication of spores by soluble things, which can be in a material of a tank the spores should be thoroughly washed out and soaked during 1-2 weeks with obligatory change of water not less than once a day.

For the stimulation of mother plants rums with hangers for *Laminaria* with well natural or artifical ventilation must prepared at the shore. The optimal temperature for the stimulation is 10 C. The preparation of substrates to sowing concludes in their soaking and sterilization for preventation of intoxication of spores by soluble things remaining on synthetic fiber after its production. The substrates are soaked in water during 3-4 days and after that sterilizited during 3-7 days.

The most large plants with mature spore spots are selected. Selected mother plants or their parts with spore spots are washed out in filtered marine water and hanged for 30-40 min until the dissappearance of drops of water. After that the plates of *Laminaria* are rolled in a tube interlaying by a paper and put in cold room at the temperature close to 0 C for 12-20 hours. After that the plants or their parts put into marine water.

Our investigation show that after true made stimulation a nature thallus of *Laminaria* saccharina brings about 130000 zoospores from 1 mm of spore spot.

After the end finish of sowing the substrates are fastened on marine installations. To transport the substrates to the installations tanks with marine waters should be used.

During winter a control for technical condition of plantation after strong storms is need.

A harresting from plantation can be begun from second part of July. To that time *Laminaria* reachs conditional sizes and mass. After harvesting the substrates and horizontal ropes are cleaned. At random the durability of all the elements of construction is controlled.

The leading abiotic factor is distribution of granulometric structure of grounds along shore line and over depths such as in the absence of suitable hard grounds under the other fardrable conditions the laminarian algae do not form trade thicket. It is enough for an optimal development of *L.saccharina* that bottom is covered by large block material on 3-10%. At high covering of the bottom by hard substrate the average sizes of plants and biomass as a rule diminish.

Other important factor is the intensity of flood-tidinal movements of water which designates the correlation of species of *Laminaria* and their size-age characteristics.

The profile of depth and shape of shore line cause the hydrodinamics of region influencing the biomass of algae and area of their thicket and also determine the technical conditions of exploitation of plantations. With enlargement of depth the force of wave movements of water quickly decreases. If in the upper part of sublitoral the algae are damaged by surf in deeper they can have well development. Besides a profile of depth the area orientation of the slopes has large importance because the force of wave influence is related to rose of winds of the region. On the vertical and very sharp slopes the breadth of the band of algae and their biomass decrease due to the sharp strengthening of hydrodinamic activity and change for the worse of light conditions. It is possible to divide the bays of Murman on the base of the distribution of the depth thus: bays with equal increase of the depth; bays with the threshold at the entrance and bays of open type. End parts of first two types of bays are always poor with algae.

The algae, bred on the plantation till the formation of the sporogenic tissues might be used for seeding algologically degraded marine ecosystems. The areas of moderately silted seafloor (70 x 70 m), chosen for in-situ experimentation were cleared of macroalgae. Boulders, being the substrates for expected algal populations were installed on one of the experimental sites. After 5 to 7 days the seeding of substrates was carried out as following: the thalli, containing the mature sporogenic tissue were stimulated and their stipes were attached to the boulders right on the bottom. To check the quality of seeding and rates of sporogenesis the holders with glass slides were placed on the controlled boulders. After one year, numerous young plants were observed on the experimental substrates. At the control sites, where the artificial seeding did not conducted, the young algae were very scarce. All stages of the experiments were snapshotted.

Simultaneously to the described above experiments the observations of the reconstruction of thichet of *Laminaria* were held in the places of trade harvest of algae during 6 years. On the sites the square until 0.8 h where the meliorative works and artificial sowing of spores on boulders have been held in three years the full reconstruction of algae has been observed. In sites where that works have not been held in three years only some plants and on some sites silted seafloor have been noted. Gradual reconstruction of degraded sites were observed after 5 years only.

The possibility of the use of algal plantations as biofilters is widely discussed in the periodicals. One of the structural materials of cellular membranes of the brown algaes including *Laminaria* is algin acid and its salts-alginates. Alginates contain as in cellular membranes as in intercellular space. The content of algin acid in *Laminaria* from plantation is 22-24%. Exeptionally large period of half-concluding of biologically activ metalls is explained by the interaction with algin acids. Taking into account the huge ability of phaeophytes, including the laminarian algae, to accumulate the heavy metals from the ambient aquatic environment, these plants are considered to be a perspective object to deal with by studies in this direction, in particular in relation to extraction of shelf ores. Recently in phaeophyte species the relationship between the metal content and some ecophysiological parameters such as age, location on thallus, time of the day etc. was shown. The algae are able to accumulate and expel aback to the seawater in course of a day different groups of metals. This seems to be related to the diurnal rhythms of the metabolism. The latter property make the usage of the laminarian plantations as biofilters for heavy metals very problematic.

Environmental Research Forum Vols. 5-6 (1996) pp. 443-446
© 1996 Transtec Publications, Switzerland

Hydrogen Production from Tofu Wastewater by *Rhodobacter sphaeroides* and *Clostridium butyricum* in Rest Culture

H.G. Zhu[1], J. Miyake[2] and Y. Asada[2]

[1] Department of Environmental Science, East China Normal University, Shanghai 200062, China

[2] National Institute of Bioscience and Human Technology, 1-1-3 Higashi, Tsukuba, 305 Japan

Keywords: Hydrogen Production, Tofu Wastewater, Photosynthetic Bacteria, Anaerobic Bacteria, Co-Culture, Immobilized Cell

Abstract

Tofu wastewater with no any dilution was used for hydrogen production by immobilized photosynthetic bacteria *Rhodobacter sphaeroides*, immobilized anaerobic bacteria *Clostridium butyricum* and immobilized co-culture of these two bacteria respectively. The maximum hydrogen production rate from the wastewater by these three systems was 2.07, 1.98, 2.11 l/hr/m^2 respectively. These rates were even higher than the hydrogen production from glucose (1.85 l/hr/m^2) by the *Rb. sphaeroides* system reported as a high effective hydrogen evolution system. The hydrogen evolution was repressed only in initial 5 hours. The ammonium was consumed from 2.3 mM to below than 0.1 mM within 10 hours. Nevertheless, the time maintaining in high rate from wastewater was shorter than from glucose. The co-culture system enhanced the total hydrogen produced than the single culture systems. The maximum TOC removal from the wastewater was 40.5% in the case of *Rb. sphaeroides* as biological phase. After hydrogen production for 85 hours, there were large quantity of organic acids accumulated. The major organic acids were butyric acid and acetic acid. The level of organic acids in the systems in which the *C. butyricum* played a role were more than in the system in which only *Rb. sphaeroides* played a role.

1. INTRODUCTION

The research on hydrogen production from wastewater is a very important aspect in the fields of biological hydrogen production and waste recycling. The great quantity of the organic material in waste water like the wastewater from beans product factory (tofu wastewater) can be served as carbon sources and electronic donors for hydrogen production. In the traditional treatment of the waste water like this, it is usually needed to add additional nitrogen source to increase the efficiency. Using this kind of waste water as raw material to produce hydrogen can overcome this shortcoming. Some photosynthetic bacteria can production hydrogen even when their growth is restricted in the surrounding of nitrogen starvation[1,2,3]. Especially the photosynthetic bacteria can directly use the light energy for hydrogen metabolism[4]. Hydrogen production from anaerobic bacteria depends on NAD-depending hydrogenase[5,6]. Ammonia is not a critical inhibitory factor to them.

Many researcher already referred their studies to hydrogen production from waste water. For example. Zurer and Bachofen etc., A. Mitsui etc., Vrati and Verma, Liu and Yang etc.[7,8,9,10]. The immobilized cells can effectively resistant the inhibitory factors, its quantity can be easily controlled and can easily separated from the liquid. There are a lot of beans product factories in east Asia. A great deal of such kind of waste water is discharged to natural environment from these factories every day. In this paper, we illustrate some immobilization hydrogen production systems for tofu wastewater and hope to find an effective way to reuse wastes and release environmental stress.

2. MATERIAL & METHODS

Tofu wastewater was sampled from Putuo Beans Product Factory of Shanghai.

A photosynthetic bacterium Rb. sphaeroides RV strain and an anaerobic bacterium C. butyricum were employed in the experiment.

Growth medium for Rb. sphaeroides contained ammonium sulfate, sodium succinate, yeast extract. The growth condition was 28 C, 2 klux, anaerobic. The growth medium for C. butyricum contained glucose, meat extract, peptone, yeast extract. The growth condition was 32 C, dark, anaerobic. Inoculum 20 ml were quickly mixed with 20 ml melted agar (4%) in 45 C in a reciprocal shaker, then transferred to a 200 ml tissue culture flask to develop an agar immobilization membrane. The inoculum was changed according to hydrogen production system we intend to check. In the case of co-culture, 10 ml of photosynthetic bacteria and 10 ml of anaerobic bacteria (the total inoculum volume kept to 20 ml) were used. Then the flask was filled with pre-culture medium to initiate the nitrogenase for hydrogen production. The pre-culture medium for photosynthetic bacteria contained L-glutamic acid monosodium salt, sodium DL-lactate and sodium hydrogen carbonate . The pre-culture medium for anaerobic bacteria contained L-glutamic acid monosodium, glucose, meat extract, yeast extract, sodium phosphate buffer (pH=7.9). After that, the flask was sealed by a silicon stopper. The hydrogen produced was collected in 100 ml syringe. The purification of the hydrogen gas was analyzed by gas chromatography. All flasks were kept in a glass water bath, 30 C, illuminated with 8 klux.

About 48 hours later when the hydrogen was stably evolved from pre-culture medium the used media were exchanged by sterilized tofu wastewater to check the hydrogen production from the wastewater. In the waste water phosphate buffer (pH=7.9) was added.

The change of ammonium concentration was estimated by ammonium electrode (AOC). TOC changement was analyzed by TOC autoanaliser. Organic acids were assayed by gas chromatography.

3.RESULTS & DISCUSSION
Table 1 shows the result of content examination of tofu wastewater.

Table 1 The contents of tofu wastewater

contents	mg/l	contents	mg/l
sugar reductive	250	alcoholity(v/v%)	nd*
sucrose	800	total alkalinity(w/w%)	0.749
starch	6750	PO_4-P	6.8
protein	630	NH_3-N	52(3.71 mM)
volatile acids	47	NaCl	127
Vit.B1	0.119	CODcr	27440
ethanol	38	BOD_5	9800

nd means "not detectable".

Figure 1 shows the time courses of hydrogen production from wastewater in the different hydrogen production systems (immobilized *Rb. sphaeroides*, immobilized *C. butyricum* and immobilized co-culture of these two bacteria). Hydrogen production is depressed only in the initial 5 hours (around). It is because the main inhibitory factor-- ammonium is always consumed within 10 hours (see figure 2), After that, hydrogen is effectively evolved. In this period, the maximum hydrogen production rate from waste water is 2.07, 1.98, 2.11 $l/hr/m^2$ in these three systems respectively. These rates are even higher than the maximum rate from glucose by Rb. sphaeroides RV strain (1.85 $l/hr/m^2$). These high hydrogen production rates maintain for about 30 hours from tofu wastewater.

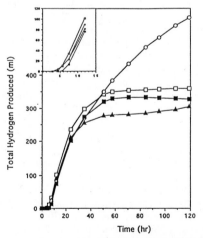

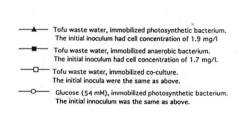

—▲— Tofu waste water, immobilized photosynthetic bacterium. The initial inoculum had cell concentration of 1.9 mg/l

—■— Tofu waste water, immobilized anaerobic bacterium. The initial inoculum had cell concentration of 1.7 mg/l.

—□— Tofu waste water, immobilized co-culture. The initial inocula were the same as above.

—○— Glucose (54 mM), immobilized photosynthetic bacterium. The initial innoculum was the same as above.

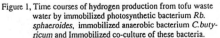

Figure 1, Time courses of hydrogen production from tofu waste water by immobilized photosynthetic bacterium *Rb. sphaeroides*, immobilized anaerobic bacterium *C.butyricum* and Immobilized co-culture of these bacteria.

The total hydrogen produced from tofu wastewater by co-culture was more than from tofu wastewater by immobilized *Rb. sphaeroides* and by immobilized *C. butyricum*. It can be attributed to the mutual benefit of the carbon utilization and corporation of resistance of inhibitory factors in the multi-enzyme system.

In the bottle of *C. butyricum*, the total hydrogen produced almost immediately decreases after the maximum is arrived. But in other bottles participated by photosynthetic bacteria, the total hydrogen keep slowly increasing for a longer time. It implies that hydrogen production system of *C. butyricum* is not so stable comparing to the system of *Rb. sphaeroides*.

Figure 3 shows the TOC removal within 85 hours. The majority of the TOC is removed in initial 10 hours. Considering the ammonia is also nearly consumed out in this period, we can be convinced that bacteria mainly use organic materials for growth and accumulate the reduction power in this period. Only when the ammonia concentration is very low and the growth of the bacteria is repressed the nitrogenase takes the function of hydrogen production. The TOC removal ratio after 85 hours arrived to 40.5% in the case of *Rb. sphaeroides* as biological phase. The TOC removal ratio in the bottle of co-culture (32.2%) was higher than in the bottle of *C. butyricum*.

Table 2 shows that, accompanying with the hydrogen production, large quantity of organic acids and ethanol are accumulated.

The major organic acids are acetic acid and butyric acid. The concentration level of organic acids in the bottles in which the *C. butyricum* plays a role are much higher than in the bottles in which only *Rb. sphaeroides* plays a role. This phenominum illustrates that the paths of the transformation of carbon sources in the hydrogen production systems of photosynthetic bacteria and anaerobic bacteria are quite different. There are not obvious differences of organic acid level between immobilized *C. butyricum* and immobilized co-culture. The organic acids consumed by *Rb. sphaeroides* can be replenished rapidly by *C. butyricum*.

Table 2 The accumulation of organic acids after H$_2$ production for 85 hours (mg/l)

items	tofu waste water by *Rb.sphaeroides*	tofu wastewater by *C.butyricum*	tofu wastewater by co-culture	tofu wastewater by *Rb.sphaeroides*
ethanol	170	255	64	272
acetic acid	711	2140	2220	345
propionate acid	nd*	nd*	nd*	nd*
butyric acid	2310	5380	5470	45
valeric acid	nd*	nd*	nd*	nd*
caproic acid	nd*	nd*	nd*	nd*

* nd means "not detectable".

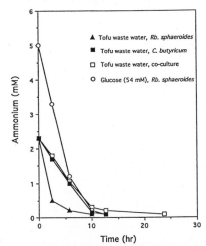

Figure 2, The time courses of ammonium consumption during the processes of hydrogen production from tofu waste water by immobilized photosynthetic bacterium *Rb. sphaeroides* and anaerobic bacterium *C. butyricum*

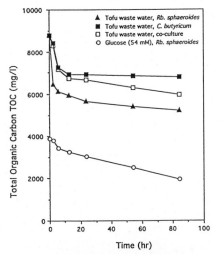

Figure 3, The time ciurses of TOC removal during the processes of hydrogen production from tofu waste water by immobilized photosynthetic bacterium *Rb. sphaeroides* and anaerobic bacterium *C. butyricum*.

REFERENCE:

1. **P.H. Hillmer** and **H. Gest**, H_2 metabolism in the photosynthetic bacterium *Rhodopsudomonas capsulata*: H_2 production by growing cultures, J. Bacteriol. 129, No. 2, 724-731, (Feb. 1977).

2. **J. G. Ormerod, K. S. Ormerod, H. Gest**, Light-dependent utilization of organic compounds and photoproduction of molecular hydrogen by photosynthetic bacteria; relationships with nitrogen metabolism, Archiv. Biochemistry and Biophysics, 94, 449-463, (1961).

3. **D. C. Yoch** and **J. W. Gotto**, Effect of light intensity and inhibitors of nitrogen assimilation on NH_4^+ inhibition of nitrogenase activity in *Rhodospirillum rubrum* and *Anabaena sp.*, J. Bacteriol., 151, No. 2, 800-806, (Aug. 1982).

4. **P. E. Weaver, S. Lien, M. Seibert**, Photobiological production of hydrogen, Int. J. Hydrogen Energy, 24, 3-45, (1980).

5. **M. Kuhn, A. Steinbuchel** and **H. G Schlegel**, Hydrogen evolution by strictly aerobic bacteria under anaerobic condition, J. Bacteriol., 159 (2), 633-639, (Aug. 1984).

6. **J. E. Zajic, A. Margaritis** and **J. D. Brosseau**, Microbial hydrogen production from replendishable resources, Int. J. Hydrogen Energy, 4, 385-402, (1979).

7. **A. Misui, T. Matsunaga, et al.**, Organic and inorganic waste treatment and simultaneous photoproduction of hydrogen by immobilized photosynthetic bacteria, Developments in Industrial Microbiology, 26, 209-222, (1985).

8. **R. Bolliger, H. Zurrer** and **R. Bachofen**, Photoproduction of molecular hydrogen from waste water of a sugar refinery by photosynthetic bacteria, Appl. Microbiol. Biotechnol., 23, 147-151, (1985).

9. **S. Vrati** and **J. Verma**, Production molecular hydrogrn and single cell protein by *Rhodopsudomonas capsulata* from cow dung, J. Ferment. Technol., 61 (2), 157-162, (1983).

10. **Liu Shuangjiang, Yang Huifang, et al.**, Beans product wastewater treatment and simultaneous hydrogen production by immobilized photosynthetic bacteria, Environ. Science, 16 (1), 42-44, (1995).

Case Studies

Environmental Research Forum Vols. 5-6 (1996) pp. 449-452
© *1996 Transtec Publications, Switzerland*

Case Study: Emmi Cheese Factory
– Summary and Results by Chairperson & Secretary –

A. Marland[1] and B. Josephson[2]

[1] Watershed Systems Ltd., ETTC, Kings Buildings,
Mayfield Road, EH9 3GL Edinburgh, Scotland, UK

[2] Ocean Arks International, 1 Locust Street, Falmouth, MA 02540, USA

Keywords: Dairy Products, Wastewater Treatment, Resource Recovery, BOD Reduction, Nitric Acid, Protein Recycling, Nitrification, Anaerobic Treatment, Aerated Fish Tanks, Aquaculture/Horticulture System, Industrial Sewage

1. Introduction

1.1. Situation and site description

The Emmi Milch AG factory at Dagmersellen processes 150 - 400 tons of milk daily into Mozzarrella cheese and dried milk products. The wastewater flow from the plant averages 412 m^3/day, and ranges between 93 m^3/d and 710m^3/d. The load is high in BOD and fat (Tab.1).

BOD$_5$	1720 mg/l	NO$_3$-N	122 mg/l
COD(soluble)	2360 mg/l	P(total)	37 mg/l
COD(total)	2491 mg/l	TOC	853 mg/l
Fat	100 mg/l	TSS	75 mg/l
N(total)	213 mg/l		

Tab. 1 Average concentrations of various parameters in the wastewater

The wastewater from the plant consists primarily of wash water. When "clean in place"(CIP) takes place, this contains caustic soda, nitric acid, hydrogen peroxide and peracetic acid. Other process waters generated are fed to pigs. There is an 80 m^3 equalization tank currently used for wastewater neutralisation by mixing. After this rough equalization, the wastewater is discharged to the town sewer.

1.2. Requirements and improvements desired

The plant is currently under considerable pressure to treat its wastewater. The local EPA has ordered them to reduce their load output by 80% due to overloading at the municipal wastewater treatment plant (WWTP). In addition, and despite these reductions, the cost of discharge to sewer is high and will increase severely in the near future. In 1994, Emmi paid the WWTP CHF 0.79 per m^3. By 2005, this is expected to have reached CHF 4.80 per m^3, even though they release an 80% cleaner water, with respect to BOD.

2. Proposals and scenarios

2.1. Possible reduction of wastewater and loads

We first looked into ways in which the loading or volumes of wastewater could be reduced within the plant. Nitric acid is currently used in the cleaning process, and it contributes around 50% of the total nitrogen load in the plant's wastewater. According to the plant manager, there is no alternative to nitric acid for the cleaning process. However, perhaps 30% of the nitric acid used could be replaced with hydrochloric acid if that was added downstream in the equalization tank. This measure would reduce the nitrogen loading in the wastewater. All other measures to reduce water or chemical uses appear to have been fully investigated by the company.

2.2. Wastewater Treatment Options Considered

We first discussed the limitations of land at the factory site. The only ground space available is a greenfield site 30m x 45m. This determined that an intensive rather than extensive system was required. In addition, it is worth noting that there is also considerable available roof area.

We discussed two design options
1. the norm of 80% removal and sewer disposal
2. full treatment for discharge to the river.

We decided that the full purification was most desirable, and would be the most cost-effective option, but that a design option for sewer discharge should be included at the point where an 80% reduction was achieved to provide protection for the river in the event of a system failure.

Option A: Anaerobic pretreatment followed by aerobic treatment

Stage One: Anaerobic Pretreatment:

We agreed that anaerobic pretreatment was a viable option, but the retention time was difficult to predict because of the lack of detailed information on wastewater characteristics.

Stage Two: Aerobic cell options:

1. Oxidation after anaerobic treatment with clarification and mical precipitation of phosphorus, followed by ecological tertiary treatment.
2. Oxidation after anaerobic treatment with aquaculture/horticulture system and then clarification and chemical precipitation of P.
3. Step feed/high recirculation aquaculture cells in series followed by clarification, ecological fluidized beds and lastly chemical precipitation of P.
4. Same as above but with addition of rooftop horticulture in recycle loop.

Option B: Direct protein concentration and removal followed by aerobic treatment to remove remaining nutrients and low molecular weight organics

1. flocculation and clarification
2. ultrafiltration

Option C: No pretreatment, recycling the resource

1. feed directly to an aquaculture/horticulture system, with a recycle to reduce influent concentrations
2. direct protein removal in accordance with recent research by Professor Yan.

3. Evaluation and discussion

The evaluation procedure considers cost, resource recovery, land use, energy use, aesthetics, reliability of effluent quality and impact on the wider environment, for example making user of waste heat.

Option A

A1 incorporates minimal resource recycling and lacks broad spectrum ecosystems for efficiency enhancement

A2 lacks a means of removing nitrate

A3 can remove nitrogen, but is weak in resource recovery

A4 could require excessive maintenance.

Option B

B1 would require extensive chemical use and its efficiency is questionable without further wastewater quality information and bench testing

B2 could require high energy inputs.

Option C

C1 would have a larger footprint

C2 Prof. Yan will carry out some experiments with protein removal from a 5 litre sample of the waste water. If protein can be extracted efficiently, it would provide resource recovery and would be a good treatment step.

4. Conclusions

The majority of the group felt that anaerobic treatment followed by aerated tanks with whole system ecologies and media beds for denitrification would be the most satisfactory option.

Using this system, we believe that the effluent from the ABR could be as follows in Tab. 2:

COD	750 mg/l	BOD	500 mg/l	NH_3	50 mg/l
NO_3	5 mg/l	Tot P	35 mg/l	Fats	0 mg/l
pH	5 - 6	N_{org}	50 mg/l		

Tab. 2 possible parameters after treatment with proposed ABR

The first series of aeration tanks would not have fish due to high ammonia levels. Rapid recycle would dilute both BOD and ammonia, allowing nitrification to begin. Denitrification would take place in the media-filled tanks. Varying amounts of nitrate could be taken up by crops grown in roof top hydroponica.

We believe that there is adequate land available at the factory to construct an intensive natural treatment system.

The benefits of a natural system at Emmi would be:

1. With a longer retention time, the system would provide consistent, high quality treatment
2. There will be a lower sludge yield (0.2kg per kg BOD removed)
3. It will offer positive solutions to a wider range of environmental concerns
4. It will look very good
5. It could be more cost effective both for capital and operating costs.

5. Group members

R.Cheung (HongKong), Aleksandra Drizo (Scotland), Maciej Dzikiewicz (Poland), Carl Etnier (Sweden), John Furze (Danmark), Yan Jingsong (PR China), Eric Leuenberger (Switzerland), Y.Liang (HongKong), D. Lückens (Germany), Cynthia Mitchell (Australia), Hanna Obarska-Pempkowiak (Poland), Dan Porath (Israel), Ming H. Wong (HongKong)

Environmental Research Forum Vols. 5-6 (1996) pp. 453-456
© *1996 Transtec Publications, Switzerland*

Case Study: Lustenberger Butchery
– Summary and Results by Chairperson & Secretary –

Ü. Mander[1] and P. Wyss[2]

[1] Institute of Geography, University of Tartu, 46 Vanemuise St., EE-2400 Tartu, Estonia

[2] Centre for Applied Ecology Schattweid, CH-6114 Steinhuserberg, Switzerland

Keywords: Slaughtery / Butchery Wastewater, Water Reuse and Recycling, Separate Treatment, Fat and Blood, Storm Water, Cooling Water, High Temperature, Storage Tank

1. Introduction

1.1. Situation and site description

Metzgerei Lustenberger is a family-owned slaughterhouse and butcher shop (35 employees) which celebrated its 50th anniversary in May 1995. In 1966, the present owner, A. Lustenberger, took over the business from his father, the founder.

The slaughterhouse operates three or four days a week. On a normal slaughter day, 15-30 cows, 150-200 pigs are slaughtered. An average of 22 m³ highly polluted wastewater (BOD_5 = 2350 - 5510 mg/l, TKN = 150 mg/l) is generated on every slaughter day, with a normal flow of 3 - 4 m³/h and a peak flow of 12 - 15 m³/h (Fig. 1). The wastewater is from cleaning and contains some blood (up to 5% of the total possible amount) and others slaughter wastes (Tab. 1).

Type of water	cleaning water from slaughterhouse and butcher shop (35 employees)		
Characteristics	high in BOD, fat and temperature (up to 60° C)		
Amount (Q)	01. - 25 m³/day	pH	6.4 - 7.3
BOD_5	500 mg/l	TKN	2 - 153 mg/l

Tab. 1 Average wastewater parameters

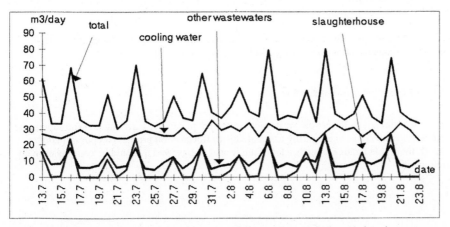

Fig. 1 Typical amounts of wastewater from the different sources during 40 days in summer

About 2 m³ is from the hot (60°) water bath used to clean the pig carcasses. A small additional amount comes from the dish washing, toilettes etc. in the butcher shop.

Wastewater from cleaning the forecourt of the slaughterhouse (up to 12 m³ a day) and from the barn will be collected and treated separately. The cooling water in the refrigerators (25 - 35 m³ per day) is drained to stormwater collecting systems into the combined sewer system (Fig. 1).

1.2. Requirements and improvements desired

The cleaning water from the butchery and slaughters is sent to the village wastewater treatment plant, where its high fat content and temperature causes problems in the biological processes. A new sewer system has been approved, uniting the region including Menznau and leading the wastewater to Dagmersellen (22 km from Menznau).

The municipality proposed Lustenberger to build a holding tank large enough to handle the wastewater generated on the slaughter day and released it slowly over two days into the sewer system, in order to solve the acute problem that the Menznau wastewater treatment plant has with the peak flow. It would also lead to lower charges in the new system, where charges are apparently affected by peak flows.

The disadvantages would be the cost of about CHF 137'000.- for construction, and Mr. Lustenberger is afraid that such a long detention time would lead to foul smells, a sanitary nuisance. Also, the cost for running the new wastewater treatment plant will be distributed among the producer of the wastewater. For the Metzgerei Lustenberger the following costs have been calculated depending on the release time of the holding tank and the load of the wastewater :

release time	2 days	4 days	4 days	4 days	7 days
load reduction	0 %	0 %	45 %	70 %	0 %
costs [CHF]	68'500.-	36'000.-	21'500.-	12'500.-	11'600.-

Tab. 2 Cost calculation for holding tank, depending on release time and load reduction

2. Proposals and evaluation

2.1. Main objectives

The main objectives of the case study were:

- to decrease the BOD value in the outlet from the slaughterhouse as much as possible
- to decrease the proposed construction costs
- to find alternatives to the officially proposed solution

During the discussion between the team members and Mr. A.Lustenberger and Mr. U. Ackermann (the technician of the slaughterhouse), various alternative solutions were worked out

2.2. Solutions without wastewater holding tank

Proposal (1)
sewage water will be sent directly to the central wastewater treatment plant after purifying it in the supersonic or ozone treatment unit, which are technologically associated with flocculation process (proposal made by Heguang Zhu; Fig.2); the whole treatment needs a lot of additional energy; and it should be comprehensively tested in local conditions.

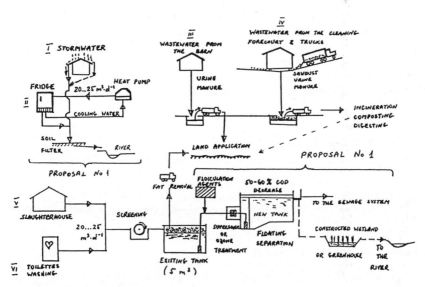

Fig. 2   *Proposal (1)* (by Heguang Zhu): sheme of purification and treatment measurements
(I - VI: different wastewater sources)

2.3. Solutions with wastewater holding tank

Proposal (2)
wastewater treatment in the holding tank (30 m³) with chemicals (e.g., iron-salts; method used in Lithuania
for treatment of wastewater with high BOD value; proposed by E.Davidavicius). This is to be tested in the
specific conditions of the Lustenberger slaughterhouse.
main problem: how to manage with the sludge with certain content of added chemicals (composting,
incineration, biogas production ?)

Proposal (3)
Aeration of wastewater in the holding tank by pumping it over a double-bed cascade of small biological filter
units with macrophytes, combined with flow-forms (proposed by T.Mauring and Ü.Mander, Fig. 3).

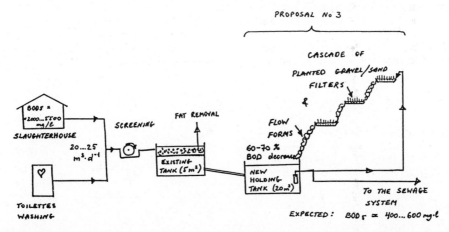

Fig.3   *Proposal (3)* (by T.Mauring and Ü.Mander): sheme planted gravel/sand filters with Flow-Forms

positive aspects: pumping is foreseen to empty the holding tank. Two 20 m^3/h pumps will be used intermittently, they could periodically (e.g. up to 5 times a day) be used for circulating the water between the tank and cascade. There is enough space in the neighbourhood. A double-bed system enables to regenerate filter units.

negative aspects: despite aeration there could still be a deficit of oxygen in the holding tank; sludge needs to be removed from the tank. Periodical change of filter material in the cascade filters may be required.

Proposal (4)
The combination of the proposals (2) and (3).

Proposal (5)
Greenhouse as "living machine" to recycle wastewater (*recycling the resource*): in this case, the cooling water could be used for regulating concentration in the wastewater (proposed by A.Lustenberger).

3. Conclusions and general remarks

- cooling water and storm water should be separately treated, recycled and used several times
- due to deficit of suitable area in the village of Menznau, there is impossible to use ecological engineering measures for wastewater treatment with a great demand of space (e.g., constructed wetlands)
- all proposed solutions need a more detailed feasibility study of the technical and economical aspects.

4. Group members

Eduardas Davidavicius (Lithuania), Warren Poole (USA), Antanas Kontautas (Lithuania), Tommy Samuelsson (Sweden), Ülo Mander (Estonia), Jüri Tenson (Estonia), Tõnu Mauring (Estonia), Philippe Wyss (Switzerland), Gunnar Noren (Sweden), Heguang Zhu (China), Albertas Peciukenas (Lithuania)

Environmental Research Forum Vols. 5-6 (1996) pp. 457-460
© *1996 Transtec Publications, Switzerland*

Case Study: Bischofszell Cannery
– Summary and Results by Chairperson & Secretary –

P. Adler[1] and B. Züst[2]

[1] USDA-ARS, 45 Wiltshire Road, Kearneysville, WV 25430, U.S.A.

[2] Centre for Applied Ecology Schattweid, CH-6114 Steinhuserberg, Switzerland

Keywords: Food Processing, Sludge Pasteurization, Filtration, Floculation/Flotation, Heat Recovery, Counter Current Recycling, Water Conservation, Energy Forest, Nutrient Recycling, Industrial Waste Treatment

1. Introduction

1.1. Situation and site description

The *Bischofszell Canning Factory* processes a number of foods by pasteurizing, sterilizing, concentrating, freezing, drying, and of course canning. The products range from iced tea, fruit juices, canned and frozen vegetables to jams and jellies and many other things.

The processes produce 3'000-6'000 m^3 wastewater per day. The current daily water consumption averages about 5,000 m^3. Most of the water used in the Bischofszell Cannery is for cleaning the produce, from removing soil initially to processing residue later on in the operation. The wastewater temperature is about 25-30 C.

All the raw wastewater is treated in the factory through:

- filtration down to 0.25 mm
- a sequenced batch reactor, and
- a floculation / flotation treatment

After this pretreatment the effluent is discharged into the sewer system and lead to the municipal wastewater treatment plant. More than 60% of the wastewater treated there comes from the cannery.

1.2. Requirements and improvements desired

The cannery pretreatment system has operated since the beginning of 1994. The effluent discharged to the sewer system still is relatively high in BOD content (500-1'000 mg/l) and causes high discharge costs for the Bischofszell Cannery, since new Swiss laws stipulate that wastewater treatment fees must be measured by the amount of load.

Therefore the two main objectives are:

1) Reduce the quantity of water used
2) Reduce the quantity and improve the quality of wastewater going to the municipal wastewater treatment plant.

2. Proposals and discussion

The cost charged to the Bischofszell Cannery by the municipal wastewater treatment plant (MWTP) for the quantity of wastewater, BOD, and sludge, are equally weighted. Therefore, a given reduction in any of the 3 will result in an equal savings in water treatment costs. However, a reduction in water use will also result in savings in the cost of water. Therefore the greatest savings result from a reduction in water use.

2.1. Water Conservation and employee instruction

Currently the produce is washed with clean water at each step. At the initial cleaning step, the produce is the dirtiest. Since clean water does not need to be used for each cleaning step, this water can be recycled. We suggest that counter-current flow technology be used to recycle water. In this procedure clean water is only used to clean the cleanest or final product. Then this water is recycled and used to clean the next dirtiest product and so on. This procedure typically results in savings of 50% of the water used.

Employee instruction also needs to be included in this program. People in the food industry are accustomed to using lots of water and typically do not think in terms of conservation. Employees need to be educated on how to conserve water from simple steps of turning the water off after it has been used to clean a piece of equipment, to how to properly operate counter-current flow technology.

Rainwater may be able to be used to supply a portion of the water needs. However, it needs to be determined whether rainwater meets cleanliness standards and how the volume available can be economically recovered. Use of the first flush separator may permit rainwater to meet cleanliness standards.

2.2. Optimizing Sludge Removal

Currently the screens for removing solids are overloaded. The 50% reduction of water to be treated due to the counter current recycling method will reduce the hydraulic loading on the screens by 50 %, thereby permiting the use of a recycle step back to the screens. By including the recycle step, the efficiency of the screens can be increased. Flocculation can be optimized by selecting a coagulant based on the particular food being processed, with few additional energies. Currently the same coagulant is used year round. By optimizing the filtration hydraulics and coagulant selection to flocculate and float off solids, the volume of sludge can be reduced by about 50%.

Solids generated at the Bischofszell Cannery are currently fed to pigs, a good ecological engineering approach. When feed prices are high, farmers will pay for the sludge. Currently, feed prices are low, and although farmers will not pay for the sludge, they will take it, thereby eliminating the cost of disposal. However, the sludge must be pasteurized (at a cost of about CHF 900.-/day or 225'000.-/year, respectively) before it is given to the farmers. If this sludge was sent to the municipal wastewater treatment plant (which currently, by the way, needs more sludge to operate its biogas recovery system), the pasteurization step and the associated costs could be eliminated. Since the cannery would be charged more for water treatment if solids were not removed as currently practised, it is cost effective to continue to remove the solids at the cannery and then send them down to the wastewater treatment plant for biogass recovery.

2.3. Heat Recovery and Economics

Heat recovery from wastewater is a readily available technology. About 45,000 KWh/day (average temperature: 25-30° C) can be recovered from a flow of 2,500 m3/day, which is the current wastewater volume reduced by 50% (due to counter current recycling). This will return about CHF 6'000.-/day or about CHF 500'000.-/year if the energy costs are based on electrical costs.

2.4. Nutrient Removal Technologies

More than 60% of the wastewater treated by the local wastewater treatment plant comes from the cannery. Another 10-15% comes from similar food processing industries and 20-25% from the domestic load of the town (about 6,500 person equivalents(p.e.)). For this reason, the cannery may consider purchasing and operating the wastewater treatment plant. The quality of this mixed wastewater from food processing industries and domestic sources are well suited for recycling in an agricultural or aquacultural system. Although there are currently no regulations on nutrient discharge, future regulations are inevitable.

One solution would be to reclaim the wastewater and its plant nutrients, nitrogen (N) and phosphorus (P), by irrigation of an energy forest. There could be a combined solution of conventional operation at the wastewater treatment plant during the winter season and irrigation during the vegetation period (approx. April - October, i.e. 7 months). The details on mass balances for N and P could be calculated. These types of recycling solutions would give the company much credit for being environmentally concerned and contributing to sustainable water and wastewater schemes.

3. Summary and conclusion

Currently the water treatment costs for the Bischofszell Cannery Plant are about CHF 500'000.-/year. By using counter-current flow technology and optimizing solids removal, water treatment costs can be reduced by 75 %, i.e. about CHF 375'000.-. There will also be a 50 % savings in water use costs since water consumption will be reduced by 50% or about CHF 375'000.-, respectively.

Heat recovery from wastewater could generate about CHF 500'000.-/year.

Pasteurizing solids for pig feed costs about CHF 225'000.-/year. This cost could potentially be eliminated by using waste heat recovered from the wastewater to pasteurize the solids. The solids sent to the wastewater treatment plant for biogas recovery would not need to be pasteurized.

4. Group members

Urs Schori (Switzerland), Simon Gerber (Switzerland), Erik Kärrman (Sweden), Kenth Hasselgren (Sweden), Mark Prein (Philippines), Björn Guterstam (Sweden), Maimuna Nalubega (Uganda), D. Chatterjee (India), J. Zeisel (Germany), Urs Baier (Switzerland)

Environmental Research Forum Vols. 5-6 (1996) pp. 461-462
© *1996 Transtec Publications, Switzerland*

Case Study: Wädenswil Aquaculture Research Project
– Summary and Results by Chairperson & Secretary –

T. Ozimek[1] and G. Rasmussen[2]

[1] University of Warsaw, Institute of Zoology, Department of Hydrobiology,
ul. Stefana Manacha 2, PL-02-097 Warszawa, Poland

[2] JORDFORSK - Centre for Soil and Environmental Research, N-1432 Aas, Norway

Keywords: Indoor Aquaculture, Productive Wastewater Treatment, Algae and Zooplankton, Polyculture, Fish Culture, Indigenous Species, Soil-Independent Cultures, Sand Filter, Greenhouse, Research Activity, Biomass Production

1. Introduction

1.1. Situation and site description

In the Waedenswil Aquaculture Plant nutrient solution from indoor-soil-independent tomatoe cultures is treated using aquaculture. Only inorganic nutrients are used. This also simulates the effluent from a wastewater treatment plant.

The Waedenswil Aquaculture Research Project has three main goals:
- Remove nutrients and/or pathogens from water of the soil-independent tomatoe cultures (low in organic compounds, high content of nitrogen and phosphorus)
- Education
- Research in aquaculture

How it works (see also Züst et al., Heeb et al., and Scheurer et al.; this volume):
The aquaculture system in Waedenswil consists of several steps. The water first flows through containers with *Nasturtium*-cultures, then to a community of floating plants (*Azolla*, *Lemna* and *Salvinia*), followed by polycultures of floating and rooted plants. The water then continues through polycultures with phytoplankton, macrophytes, zooplankton and snails. Fish tanks are also used, but they are yet not connected. Water will be further treated in a sandfilter followed by algae tanks. This effluent will be returned to the zooplankton tanks. Some fish are fed produced plants, while some are in addition fed zooplankton.

1.2. Questions and tasks

Jürg Staudenmann, who works on the aquaculture project, had prepared questions to be discussed by the group.

His major concerns were:
- goals for further research
- wether polycultures or monocultures should be used, and
- wether indigenous or well studied (tropical) species should be used and tested in the system.

He also questioned if the industry should be involved at this point.

2. Proposals and discussion

2.1. Data available

There are few existing results on nutrient removal. First priority should be to perform more analysis to find out how it is working. Documentation is also important if the system will be used for wastewater treatment.

The group felt that in order to document that the system is working for treatment (polishing) of wastewater, real wastewater should be used and tested. It might contain compounds that are not accounted for in the nutrient mixture used now. These compounds could for instance be harmful to the organisms.

2.2. Performance of the aquaculture plant

Existing results show rather poor removal of nutrients. The reason for this should be found, and improvements should also be made. The group suggested that the existing polyculture (*Azolla, Lemna* and *Salvinia*) should be changed to a monoculture with *Azolla*. Present results show that *Azolla* is abundant in the culture, and that it inhibits growth of *Lemna* and *Salvinia*. Nutrient uptake and harvest rate should thus be tested for *Azolla*. We propose to add *Glyceria maxima* to the tanks with communities of emergent and floating plants because of its high accumulation rate of nitrogen.

Gravel (limestone) is currently being used as a sediment in several tanks. If higher phosphorus removal is a goal, these sediments could be substituted with a phosphorus adsorbing substrate. Norwegian produced Leca (Light expanded clay aggregates) show promising results for P-removal. Normal Leca has a high pH, but the dolomite causing this can be washed out. The disadvantage with Leca is that energy is used for production and is thus less ecological. Other media should also be tested.

In the containers with submerged macrophytes, phytoplankton and zooplankton, the density of zooplankton should be higher to limit the growth of phytoplankton.

2.3. Further research and development activity

At present it is too early to offer the aquaculture system tested a Waedenswil as a solution to the industry or horticulture. Further basic research is needed, and applied projects should be a long term goal.

From an educational- and scientific point of view the system is very interesting. Monocultures can be used to see how every species function, and how efficient they are for nutrient uptake. By putting them together in various combinations one can find out how the different species interact. However, for the purpose of water treatment there should be as few steps as possible to make it affordable.

The group agreed that if the goal is to remove as much nutrients as possible, well known, efficient tropical species (plants and fish) should be used. However, to achieve high removal rates, more energy might be needeed to warm up the greenhouse. This will make the system less ecological and more costly. From a scientific point of view it will be more interesting to test out native species.

To create and develope an effecient acuaculture system, an interdisciplinary team is needed. It should consist of ecologists, engineers and economists. It is also important to include people with knowledge within horticulture, food production, chemistry, hydrobiology and social science.

3. Closing remark

It is of major importance that institutes working with this kind of aquaculture cooperate and exchange information. There are aquaculture systems in warmer climates that are working. Visiting them could give new knowledge. People working in colder climates should exchange results and experience so the same mistakes are not conducted twice. The group agreed that this was one of the most urgent needs for the aquaculture centre in Waedenswil at this point.

4. Group Members

Göran Dave (Sweden), Erik Tybirk (Denmark), Henrik Morin (Sweden), Werner Stirnimann (Switzerland), Thorsten Reckerzügl (Germany), Dianne Hughes (USA), Alexander Gudimov (Russia), Grigorii Voskoboinikov (Russia), Nikolai Protopopov (Russia), R. Nagendran (India), B. B. Jana (India), V. S. Govindan (India), K. Forster (Switzerland), Guido Maspoli (Switzerland), Andreas Schönborn (Switzerland)

Environmental Research Forum Vols. 5-6 (1996) pp. 463-466
© *1996 Transtec Publications, Switzerland*

Case Study: Treatment of Organic Waste on a Farm
– Summary and Results by Chairperson & Secretary –

P.D. Jenssen[1] and A. Finnson[2]

[1] Department of Engineering, The Agricultural University of Norway, N-1432 Aas, Norway

[2] Swedish Environmental Protection Agency, S-106 48 Stockholm, Sweden

Keywords: Organic Waste, Nutrient Recycling, Liquid/Solid Fermentation, Liquid/Windrow Composting

1. Situation and tasks

1.1. Motivation

In order to utilize the resources in organic waste in an ecologically acceptable way treatment and use of organic waste on farms adjacent to population concentrations is explored. The need for organic fertilizer products will also increase as organic farming becomes more widespread. The possibility of treating organic waste on a farm in Switzerland was considered.

1.2. Constraints

The farm has 28,5 ha arable land and 75 cattle equivalents (CE). This equals 2,6 CE/ha. The maximum permissible number is 3,0 CE/ha. The evaluation is made in a restricted time and with limited possibility to obtain exact figures especially regarding economy.

2. Proposals and Scenarios

2.1. Capacity to import waste

According to the constraints the farmer an excess capacity to use nutrients equalling 0,4 CE/ha (0,4CE/ha x 28,5ha = 11,4CE) 11,4 CE. We have assumed 1 CE equal to 50 kg of N and 20 kg of P. The nutrient content in organic household waste is based on preliminary Norwegian figures (SFT 1993) given as 0,72 kg of nitrogen (N) per person per year and 0,12 kg of phosphorus (P) per person per year. Based on these figures the farmer can import organic household waste from 1800 persons in terms of P-load or 800 persons as N. This equals 180 respectively 80 tonnes of waste. Since P is the main limiting factor 180 tonnes is used in our assessment.

Waste resource:	Tonnes
- Vegetable processing factory	150
- Municipality of Abtwil (500 persons)	50
- Municipality of Sins (3000 persons)	300
Sum	*500*

Tab. 1 waste resource options

If we assume that the slurry from a vegetable processing factory in Hünenberg has the same content of nutrients per tonne as the organic household waste the farm can process waste from this factory and the village Abtwil (Tab. 1)

464   Recycling the Resource

2.2. Waste processing methods

In Table 2 waste sources and possible waste processing methods are shown.

	Liquid fermenta- tion	Solid fermenta- tion	Liquid compost- ing	Windrow compost- ing	Edge of farm com- posting	Direct into manure pit
Vegetable factory liquid	x	(x) in mix	x			x
Garden waste		(x)	(X)	x	x	
Org. househ. waste	x	(x) in mix	x	x	x	
Manure dry		x		x	x	
Manure slurry	x		x			
Ashes		?	?	x	x	
Positive energy output	x	x	x			
Investment cost	- -	- -	-	-	+	
Nutrient losses					-	- (?)
Export potential of				++		
Hygiene					- -	
Possible methane loss						- (?)

Tab. 2   Waste sources and possible waste processing methods
        (x= suitable (x) = suitable under certain conditions)

3. Evaluation of methods

3.1. Direct discharge to the manure slurry pit

Direct discharge of organic waste to the manure slurry pit was considered. To our knowledge this method has not been in use. There might be problems connected to methane formation and subsequent losses, also loss of ammonia is possible. This is the cheapest method, but we do not recommend it because of the the little practical experience and large ucertainties.

3.2. Edge of the field composting

Composting on the edge of the fields is a possible method, but we outrule this method because of hygienic problems that may occur when household kitchen waste is spread over the perimeter of the farm. This method may also produce runoff and subsequent losses of nutrients to surface- or ground water.

3.3. Windrow composting

This is a well known method. Use of this method will require investment in facilities to avoid runoff or for safe disposal of possible runoff. These investments are most likely defended by the economic income from the waste handling (180 tonnes at 100SFr/tonne). It is not possible to utilize energy produced by this method. This method opens the possibility for export of compost.

3.4. Liquid composting

Liquid composting in small reactors are developed in Scandinavia for use in waste handlng on farms. The method operates with a positive energy budget in the order of 60 - 130 KWh/tonne of processed waste. The energy is in terms of heat and can be used for heating of adjacent buildings. The investments for this method is of the same magnitude as for windrow composting.

3.5. Dry fermentation and liquid fermentation

These methods yield biogas thus a higher quality energy than the liquid composting. Methane fermentation is sensitive to temperature and the process needs a stable temperature around 35¡C. In the Swiss climate this method will require temperature control and thus the method can become costly. The dry fermentation is more dependent on preparation and mixing of the waste than the liquid fermentation.

3.6. Combination of methods

One of the group members suggested to use the supernatant from the biogas reactor for production of biomass in ponds. This biomass can be fed to the biogas reactor thus increasing the total biogas yield.

4. Concluding discussion

The farm can handle organic household waste from 1800 persons. This yields 180 tonnes of waste at 100 SFr/tonne.

Composting on the edge of the field and direct discharge of household waste to the slurry pit are not recommended due to hygienic risks and environmental concerns. It is likely that the fermentation methods are too expensive in this small scale. The most suitable methods are liquid and windrow composting.

In order to defend the investment in a fermentation plant, the farmer will have to reduce the number of CE. If he reduces the number to 40 CE he can process nutrients from waste equalling 45 CE. This equals organic household waste from 7500 people. With this amount the farmer can probably defend investment in a fermentation plant. However, we think that this requires that he can be paid 100SFr/tonne of waste delivered at the farm.

The present limit is 3 CE/ha. This yields a high nutrient load (in the order of 150 kg N/ha). This high load is environmentally questionable and in Germany the number is now decreased to 1,5

5. Reference

SFT 1993. Gjenvinning av matavfall til dyrefor. State Pollution Control Authority, report TA - 954, Oslo.

6. Group members

Kenji Furukawa (Japan), Kazuhiro Terai (Japan), Tutaka Morii (Japan), Takeshi Yamasaki (Japan), Ziquing Ou (China), Tieheng Sun (China), Uwe Klaus (Germany), Kaiser Jamil (India), Ulrika Geber (Sweden)

Environmental Research Forum Vols. 5-6 (1996) pp. 467-470
© *1996 Transtec Publications, Switzerland*

Case Study: Analyzing the Planning Process
– Summary and Results by Chairperson & Secretary –

R. Fismer and L. Wendeler

Zentrum Prinzhöfte, Simmerhauser Str. 1, D-27243 Harpstedt, Germany

Keywords: Planning and Working Process, Social Aspects, Ecological Engineering, Holistic, Brain Storm

1. Introduction

1.1. Motivation

Our work group had a double subject: first a case study (the FOCUS housing community project in Uster, canton Zürich, Switzerland [1]), and on the other hand, and primarily, an analysis of the process of planning itself.

If we claim, as ecological engineers, to achieve solutions that meet certain standarts of ecology and holism, what demands are needed for our working process? What characteristics are necessary for it to lead towards success ? Are there systematic deficiencies in our methods, which periodically affect our success?

1.2. Thesis and working procedure

Our *1st thesis* was formulated: *Every planning process is at the same time a social process.*

Planning goes on in several social relations: users, deciders, neighbours etc., diverse planners and specialists are involved. The process is a *material working process* and at the same time an *interpersonal process of communication and learning.*

This means that the process is sensitive to perturbations by communication errors, antipathy, negative emotions and personal constitution of the people involved. But simultaneously, it gains creativity and liveliness from a positive group process, from a joyful cooperation. Planning is a game!

So it is necessary to gain a self-consciousness of this social process. A wide range of methods are at our disposition here, which should be incorporated into our work as Ecological Engineers. This subject was not pursued more detailed in this workshop.

But even the material side of the working process does not only consist of combining data and formula correctly, as it appeared to be in classical Engineering. Every working process possesses characteristic structures of its organization and its sequence. We tried to specify these characteristics more precisely, based on our common work on the Uster housing project and on our personal experiences.

We encountered a number of typical problems that affect the planner's work, make it unproductive, let its quality often be inferior to the formulated standards. They all result from a lack of consciousness about the course of our process, about the point we have just reached in it, about the level of abstraction we are working on. This is working without self- consciousness.

2. Characterization of an adequate working procedure for Ecological Engineering

2.1. Theses

From the phenomena we found, we formulated *theses* to characterize a working process that should be suitable for Ecological Engineering:

1. Every planning process is at the same time a social process.

2. Favourite ideas block our view.
3. No idea is forbidden.
4. Think in possibilities and alternatives.
5. Think in qualities instead of products.
6. The role of the planner is that of a servant.

These are to be illustrated further now.

2.2. Discussion and proposals

Favourite ideas (thesis 2) in our subject are, for example, the compost toilet, biogas reactor, graywater reuse, urine separation, greenhouses, aquaculture. They all may be wonderful solutions in some cases; but in other applications they will surely be dislocated. It often happens to the planner (and, even more often, to the houseowner) that he falls in love with one solution right from the beginning, instead of starting from the problem. For example, a biogas system would certainly make no sense for the FOCUS project, regarding its minor size and the swiss climate. In this sense, for the planner's flexibility of mind, "nothing is more dangerous than a successfull solution". People have the tendency to solve the problems they are able to solve, and not those they are needed to solve.

Instead, it must be a principle in the first phases of a planning process that *no idea is to be excluded in advance* (thesis 3) only because it seems to be stupid at first glance. Prohibitions of thought are always unproductive. A small example from our group: In discussing the installation of compost toilets, someone remarked, 'How about imitating the noise of the water flush electronically?' How unecological and artificial! But in the course of the debate, this idea led us to the perception that a major barrier for the acceptance of compost toilets lies in the users' customs in this intimate and emotionally very sensitive area, and not only in technics.

This open collection of ideas is needed to take into regard all the possibilities (thesis 4) that are inherent in the system. We should first inspect all the possibilities of the system not yet looking for solutions, but exploring all the qualities and structures of the system for their own sake. In this phase, understanding the dynamics of the system is the main topic. In our example, this meant studying the water cycle in all its elements: precipitation, potentials for runoff and drainage, possible use for wastewaters, the different qualities of water envolved, the locally typical aquatic flora and fauna,...

This means to think in *qualities* and not yet in *products* (thesis 5). There is a fine example for this by Frederik Vester: People say they need washing machines. That is not true; what they need is clean clothing. This advance may open up new solutions.

Starting from these *possibilities* of the system, *alternative* solutions can be developed (thesis 4). Now we look for solutions of our problem that are consistent and realistic. This is classical engineering work: A solution is constructed according to the conditions and the technical and scientific laws. But it is still too early to make a choice; all feasible solutions should be pursued (see thesis 3).

Then, the consequences of every alternative have to be outlined: what costs, what benefits, which side effects are implied, what is excluded. (Because every decision for one path is always a decision against others, too.)

In the preceding phases of work we were to *understand* the problem and the structure of the system. Only now follows the *evaluation*. The evaluation does not result from the qualities of the system itself, but only from the criteria of the observer (planner, customer, user). (To understand this differenciation fully, the ideas of Maturana, Varela et al. are very helpful.)

Now it is important to remember the role of the planner. He is not the one to decide. But he is employed to *serve the deciders* (thesis 6). With his competence, he has to enable the customer to make a profound and reasonable valuation of the alternatives, on which to base his decision. Only too easily the planner becomes the real decider. This is undemocratic, deprives people of their responsibility, it is expertocracy. As soon as a planner presents only a single solution to his customer, it must be suspected that he wants to sell a product and not to solve a problem and give advice.

Furthermore, the planner has to enter into communication with other partners affected by the project: possible users, inhabitants, neighbours, authorities, financers,...

3. Conclusions

The real process does not linearly follow the sequence lined out here, but jumps and reprises will again and again be necessary. For example, working out one alternative may induce a still more detailed study of the possibilities; or the dialogue with the users may show up totally new ideas. The movement shows the form of a vortex: the work is repeated several times on a continuously higher level.

This makes it all the more important to have a consciousness of the state of the process in every moment of the planning. It has to become one of the qualities distinguishing an Ecological Engineer that he is not only an expert for ecosystems and for engineering, but also for the social planning process. Else the work will continue to often produce unsatisfactory results.

4. Final remarks

We intend to drive on the analysis of the characteristics of the planning process systematically at Zentrum Prinzhöfte (which we have been doing the last years, in the cadre of Permaculture). We invite all persons who feel inspired by these preliminary ideas to comment and participate.

[1] The FOCUS project is a cooperative effort of a group of architects and lay people who wish to provide affordable, environmentally friendly housing for themselves and their families. A four-story appartment building has been designed with attention given to the latest technology in building materials, energy and water conservation (water-saving fixtures, composting and low-flush toilets, rainwater use), as well as ecological treatment possibilities for wastewater and organic waste treatment.

5. Group members

Anna Olsson (Sweden), Gundula Eller (Germany), Carola Petrikowski (Germany), Boris Zhukov (Russia), Erwin Nolde (Germany), Robert Ciupa (Poland)

Environmental Research Forum Vols. 5-6 (1996) pp. 471-472
© *1996 Transtec Publications, Switzerland*

Case Study: Wädenswil Wastewater Treatment Plant
– Summary and Results by Chairperson & Secretary –

J. Heeb[1] and A. Peters[2]

[1] Centre for Applied Ecology Schattweid, CH-6114 Steinhuserberg, Switzerland

[2] Swedish Environmental Protection Agency, S-106 48 Stockholm, Sweden

Keywords: Conventional Treatment Plant, Legal Requirements, BOD Overload, Household and Industrial Sewage, Nutrient Recovery, Awareness of People, Brain Storm, Ecological Engineering Solution, Social Aspects

1. Introduction

1.1. Situation and site description

The Waedenswil wastewater treatment plant is designed to treat the community waste water (77'000 pe, measured in terms of BOD, coming from household and industrial sources. The plant is today overloaded due to industrial sewage, which constitutes the larger part of the total load.

The number of connected inhabitants in the catchment is very high, about 97 percent. The sewage network is relatively new. The amount of water leaking into the system is thus very small. There are no problems with the hydraulic load in the plant. Quality of effluent: Suspended solids 14 mg/l (limit 5), Total Phosphorus 0,36 mg/l (limit 0,2), Dissolved Organics (DOC) 8,5 mg/l (limit 10) and Biological Oxygen Demand (BOD_5) 9 mg/l (limit 10).

Sludge quality is good. Most of the sludge is being used in agriculture, but in the winter time some sludge is incinerated. About 1/3 of the energy consumption at the plant is covered by the biogas production. Heat is extracted from the water to heat some 150 apartments nearby.

1.2. Requirements and improvements desired

Due to the overload the local government requirements on suspended solids and total phosphorus cannot be met. There is also a problem to reach the required discharge level of BOD.

Different strategies to deal with these problems are presently being discussed in the community. One possibility would be that one of the three industrial sources (food industry) disconnected its wastewater, whereas the second one would contribute far less to the BOD load, due to other measures. This would on one hand lead to a substantial reduction of the total BOD load, but on the other hand it would also mean a decrease in income to the wastewater treatment plant.

Another possibility would be to enlarge the treatment plant at an estimated cost of 12 million Swiss francs. This idea has already been worked out to a certain extent.

2. Definition of problems

The first step of the group was a brain storming session with the goal to define the present problem areas of the plant. Some examples for the problems areas:

- present regulations somewhat questionable
- social and political awareness for changes lacking
- reuse of nutrients, especially nitrogen, insufficient
- present system does not favor load reductions
- time perspective (What would be the best solution in the short term and the long term perspective, respectively?)

The various problem areas were then grouped thematically and for each group one objective was formulated (Fig. 1).

3. Scheming solutions and evaluation

In a similar way various possible measures were suggested and then formed into three groups:

Zero solution: No changes to the present state

Technical (conventional) solution: Technical improvements on site in order to meet the legal requirements. This solution corresponds to the present plan of the community to enlarge the plant.

Ecological engineering solution: Here a number of measures were grouped together, comprising local treatment (e.g. with local small wastewater treatment plants or aquacultures, local infiltration of rainwater) and source reduction (e.g. local treatment of industrial wastewater, use of urine separation and composting toilets in the households). The present plant would be used for the treatment of the remaining wastewater.

Because of the limited time the group decided not to go into detail on the different measures but rather evaluate the three types of solutions against the important objectives. Both objectives and the three solutions were put into a matrix (see Fig. 1). As a third component of major importance the time aspect was added. The short term perspective was chosen to be less than five years, thus approximately corresponding to the time available to fulfill the present requirements- The long term perspective was chosen to be more than five years.

Evaluation of proposed solutions + : positive effects estimated - : negative effects estimated	zero solution		conventional solution		ecological engineering solution	
	short term	long term	short term	long term	short term	long term
Meeting legal requirements	-	-	+	+	-	+
Recycling nitrogen	+	+	+	+	+	+
Recycling phosphorus	+	+	+	+	+	+
Diminishing wastewater	-	-	-	-	-	+
Diminishing primary pollution	-	-	-	-	-	+
Improving awareness of people	-	-	-	-	+	+

Fig. 1 Matrix for the evaluation of the proposed solution

4. Conclusions

In the opinion of the group it would in the long term be possible to achieve improvements on most of the objectives by realizing the solution based on principles of ecological engineering. Some important objectives can also be achieved with the other two solutions. The big advantage of an ecological engineering approach is that it increases the awareness of the people, thereby influencing both water use and the extent of water pollution.

In reality, source reduction could also be part of a conventional solution. Also, costs for the different measures were not considered.

5. Group members

C. Bunge (Germany), V. Schultz-Jochens (Germany), R. Hübner (Germany), J. Laber (Austria), L. Hoesch (Switzerland), L. Jarlöv (Sweden), D. Wilcke (Germany), T. Herrmann (Germany)

AUTHOR INDEX

KEYWORD INDEX

Visit us on the Internet

World Wide Web

Please visit us on the World Wide Web at

http://www.transtec.ch

where detailed information on all
published titles is provided as well as:

- Tables of Contents
- New and Forthcoming Titles
- Instructions for Authors and Editors
- Information on all Periodicals
- Online Order
- Various Files for Downloading

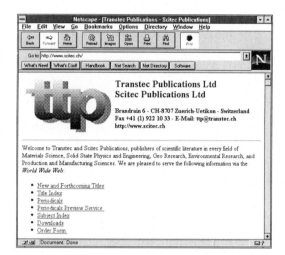

Periodicals Preview Service

Transtec Publications' preview service offers automatic delivery of tables of contents - several
weeks before the actual release of the respective publication.

Included are the following periodicals:

- Advanced Manufacturing Forum
- Defect and Diffusion Forum
- Environmental Research Forum
- GeoResearch Forum
- Key Engineering Materials
- Materials Science Forum
- Molten Salt Forum
- Production and Logistics Forum
- Solid State Phenomena

This service is free of charge. For details please send an e-mail message containing the line
help
to **preview@transtec.scitec.ch**
or use the online registration at http://www.transtec.ch

For regular e-mail please use the address **ttp@transtec.scitec.ch**

 Transtec Publications Ltd
Materials Science • Solid State Physics • Engineering

Brandrain 6
CH-8707 Zuerich-Uetikon, Switzerland
Fax +41 (1) 922 10 33
e-mail: ttp@transtec.ch

Environmental Research – New Title

Integrated Water Management in Urban Areas
Searching for New, Realistic Approaches with Respect to the Developing World

Ed. Janusz Niemczynowicz

Proceedings of the International Symposium held in Lund, Sweden, September 1995

Environmental Research Forum Vols. 3-4

ISBN 0-87849-736-6
1996, 492 pp, CHF 180.00 (ca. US$ 144.00)

Continuing growth of urban population in the world and, especially in developing countries, is one of the most frightening problems of today. Megacities, i.e. cities with more than 10 millions inhabitants, are growing fastest in developing countries and their problems are to be put on top of all well known environmental problems of the world.

Two basically different philosophies with different ways of approaching increasing environmental problems of the world are distinguished. Both of them claim an ability to solve the problems. The first approach, a traditional one, is a continuation of present line of development and a wider implementation of technology known and used in developed countries. The other approach states that basic changes on all levels of the societies, as well as different novel technologies are needed in order to bring progress leading to conservation of resources and sustainability.

The books concludes that there is no real contradictions between these two approaches. Simply, both of them are simultaneously valid, and both of them will be applied in the future. The real challenge is to find the way how to gain environmental and social benefits of using well balanced mix of them that is adequate at a given location with a given culture, climate and socio-economical conditions.

Some Highlights:

Challenges and Interactions in Water Future, *J. Niemczynowicz*

Forty Years of Urban Water Management in Calabria (Italy), *F. Calomino, P. Veltri and G. Santoro*

Strategic River Management with Environmental Considerations in Brunei Darussalam, *C.T. Yong*

Problems of Water Supply in Estonia, *L. Vallner*

Transtec Publications Ltd
Brandrain 6 • CH-8707 Zuerich-Uetikon • Switzerland
Fax +41 (1) 922 10 33 • e-mail: ttp@transtec.ch
http://www.transtec.ch